# Methods in Molecular Biology



*Series Editor*
**John M. Walker**
**School of Life Sciences**
**University of Hertfordshire**
**Hatfield, Hertfordshire, AL10 9AB, UK**

For further volumes:
http://www.springer.com/series/7651

# Ras Signaling

## Methods and Protocols

Edited by

**Lorenza Trabalzini**

*Department of Biotechnology, Chemistry, and Pharmacy, University of Siena, Siena, Italy*

**Saverio Francesco Retta**

*Department of Clinical and Biological Sciences, University of Torino, Torino, Italy*

Humana Press

*Editors*
Lorenza Trabalzini
Department of Biotechnology, Chemistry, and Pharmacy, University of Siena
Siena, Italy

Saverio Francesco Retta
Department of Clinical and Biological Sciences, University of Torino
Torino, Italy

ISSN 1064-3745 ISSN 1940-6029 (electronic)
ISBN 978-1-4939-6054-5 ISBN 978-1-62703-791-4 (eBook)
DOI 10.1007/978-1-62703-791-4
Springer New York Heidelberg Dordrecht London


Softcover reprint of the hardcover 1st edition 2014


*Cover Illustration Caption*: Ribbon diagram of H-ras, PDB ID 121p, generated with UCSF Chimera. Strands are purple, helices aqua, loops gray. Also shown are the bound ligand (GTP analog) and magnesium ion. UCSF Chimera development is funded by the NIH (grant P41-RR01081).

Humana Press is a brand of Springer
Springer is part of Springer Science+Business Media (www.springer.com)

# Preface

The Ras GTPase superfamily includes monomeric, low molecular weight GTP-binding and hydrolyzing proteins that act as molecular switches by coupling extracellular signals to different cellular responses, thus controlling cellular signaling pathways responsible for growth, migration, adhesion, cytoskeletal integrity, survival, and differentiation.

The activities of RasGTPases are controlled by a large number of regulatory molecules that affect either GTP loading (guanine nucleotide exchange factors or GEFs) or GTP hydrolysis (GTPase activating proteins or GAPs). In their active state, they interact with a continually increasing, functionally complex array of downstream effectors.

In addition to their prominent role in regulating virtually all fundamental cellular processes, Ras GTPases have acquired medical relevance because of their participation in several diseases, such as cancer, cardiovascular disorders, aging, neurodegeneration, and developmental syndromes. Thus many efforts have been made to understand molecular mechanisms of Ras signaling both in the physiological and in the pathological state and to develop novel strategies for the treatment of many pathological conditions where RasGTPases play a role.

This book focuses on experimental approaches aimed at shedding light on the complexity of the biological functions of RasGTPases. In particular, it contains general overviews and detailed applications of both well-established and recently developed research techniques, including biochemical, biophysical, molecular biology, genetic and behavioral approaches, advanced high resolution fluorescence and electron microscopy imaging, and "omics" technologies, providing information on expression, posttranslational modifications, subcellular localization and dynamics, regulatory mechanisms of upstream and downstream signaling pathways, and, ultimately, biological activities and functions of RasGTPases in different model systems, including high- and low eukaryotic organisms.

The individual chapters are organized in order to comprise an introductory overview, a list of the materials and reagents needed to complete the experiments, as well as a detailed procedure supported with a troubleshooting section.

We are extremely grateful to the investigators who have generously contributed their time and expertise to bring a wealth of technical knowledge into this volume, which is intended as an aid for investigators of different backgrounds and interests related to the multiple physiological and pathological functions of the large superfamily of RasGTPases.

*Siena, Italy* — *Lorenza Trabalzini*
*Torino, Italy* — *Saverio Francesco Retta*

# Contents

## Contributors

DANIEL ABANKWA • *Turku Centre for Biotechnology, Åbo Akademi University, Turku, Finland*
IAN M. AHEARN • *Department of Medicine, NYU Langone School of Medicine, New York University, New York, NY, USA; Department of Cell Biology, NYU Langone School of Medicine, New York University, New York, NY, USA; Department of Biochemistry and Molecular Pharmacology, NYU Langone School of Medicine, New York University, New York, NY, USA; Cancer Institute, NYU Langone School of Medicine, New York University, New York, NY, USA*
PONTUS ASPENSTRÖM • *Department of Microbiology, Tumour, and Cell Biology, Karolinska Institutet, Stockholm, Sweden*
EVA BALDINI • *Department of Biotechnology, Chemistry, and Pharmacy, University of Siena, Siena, Italy*
FIORELLA BELOTTI • *Department of Biotechnology and Biosciences, University of Milano-Bicocca, Milan, Italy*
CHRISTOPH BISKUP • *Biomolecular Photonics Research Group, University Hospital Jena, Jena, Germany*
ENRICO BRACCO • *Department of Oncology, University of Torino, Torino, Italy*
RICCARDO BRAMBILLA • *Division of Neuroscience, Institute of Experimental Neurology, San Raffaele Scientific Institute and University, Milan, Italy*
SHARON L. CAMPBELL • *Department of Biochemistry and Biophysics, University of North Carolina at Chapel Hill, Chapel Hill, NC, USA; Lineberger Comprehensive Cancer Center, University of North Carolina at Chapel Hill, Chapel Hill, NC, USA*
TANMAY S. CHAVAN • *Department of Medicinal Chemistry, College of Pharmacy, University of Illinois at Chicago, Chicago, IL, USA*
JEOU-YUAN CHEN • *Institute of Biomedical Sciences, Academia Sinica, Taipei, Taiwan*
LUKE CHISHOLM • *Department of Biology and Biochemistry, University of Bath, Bath, UK*
AUDREY COLOMBA • *Randall Division of Cell and Molecular Biophysics, King's College London, London, UK*
SONIA COLOMBO • *Department of Biotechnology and Biosciences, University of Milano-Bicocca, Milan, Italy*
ADRIENNE D. COX • *Department of Radiation Oncology, Lineberger Comprehensive Cancer Center, University of North Carolina at Chapel Hill, Chapel Hill, NC, USA; Department of Pharmacology, Lineberger Comprehensive Cancer Center, University of North Carolina at Chapel Hill, Chapel Hill, NC, USA*
VALENTINA CUTANO • *Department of Clinical and Biological Sciences, University of Torino, Torino, Italy*
RAFFAELE D'ISA • *Division of Neuroscience, Institute of Experimental Neurology, San Raffaele Scientific Institute and University, Milan, Italy*

Molly J. DeCristo • *Department of Radiation Oncology, Lineberger Comprehensive Cancer Center, University of North Carolina at Chapel Hill, Chapel Hill, NC, USA; Department of Pharmacology, Lineberger Comprehensive Cancer Center, University of North Carolina at Chapel Hill, Chapel Hill, NC, USA*
Valeria de-Turris • *Center for Life Nanoscience - IIT@Sapienza, Rome, Italy*
Malgorzata Dobosz-Bartoszek • *Department of Biochemistry and Molecular Genetics, College of Medicine, University of Illinois at Chicago, Chicago, IL, USA*
Stefania Fasano • *Division of Neuroscience, Institute of Experimental Neurology, San Raffaele Scientific Institute and University, Milan, Italy*
Elisa Ferro • *Department of Biotechnology, Chemistry, and Pharmacy, University of Siena, Siena, Italy*
Boris Fichtman • *Faculty of Medicine in the Galilee, Bar-Ilan University, Safed, Israel*
Beatrix Fürst • *Division of Cell Biology, Biocenter, Innsbruck Medical University, Innsbruck, Austria*
Annica K.B. Gad • *Department of Microbiology, Tumour, and Cell Biology, Karolinska Institutet, Stockholm, Sweden*
Vadim Gaponenko • *Department of Biochemistry and Molecular Genetics, College of Medicine, University of Illinois at Chicago, Chicago, IL, USA*
Luca Goitre • *Department of Clinical and Biological Sciences, University of Torino, Torino, Italy*
Giulia Guarguaglini • *Institute of Molecular Biology and Pathology, Consiglio Nazionale delle Ricerche (CNR), c/o Sapienza University of Rome, Rome, Italy*
Gianni Francesco Guidetti • *Department of Biology and Biotechnology, University of Pavia, Pavia, Italy*
Harsha P. Gunawardena • *Program in Molecular Biology and Biotechnology, University of North Carolina at Chapel Hill, Chapel Hill, NC, USA*
Camilo Guzmán • *Turku Centre for Biotechnology, Åbo Akademi University, Turku, Finland*
Amnon Harel • *Faculty of Medicine in the Galilee, Bar-Ilan University, Safed, Israel*
G. Aaron Hobbs • *Department of Biochemistry and Biophysics, University of North Carolina at Chapel Hill, Chapel Hill, NC, USA*
Lukas A. Huber • *Sektion für Zellbiologie, CCB—Centrum für Chemie und Biomedizin, Biozentrum der Medizinischen Universität Innsbruck, Innsbruck, Austria*
Lauren P. Huff • *Department of Radiation Oncology, Lineberger Comprehensive Cancer Center, University of North Carolina at Chapel Hill, Chapel Hill, NC, USA; Departments of Pharmacology, Lineberger Comprehensive Cancer Center, University of North Carolina at Chapel Hill, Chapel Hill, NC, USA*
Catherine N. Kaminski • *Department of Molecular and Cellular Biology, University of Kentucky College of Medicine, Lexington, KY, USA*
David F. Kashatus • *Department of Microbiology, Immunology, and Cancer Biology, University of Virginia, Charlottesville, VA, USA*
György Kéri • *Vichem Chemie Ltd., Budapest, Hungary*
Patrizia Lavia • *Institute of Molecular Biology and Pathology, Consiglio Nazionale delle Ricerche (CNR), c/o Sapienza University of Rome, Rome, Italy*
Enzo Martegani • *Department of Biotechnology and Biosciences, University of Milano-Bicocca, Milan, Italy*
James O. Meyer • *Department of Biochemistry and Molecular Genetics, College of Medicine, University of Illinois at Chicago, Chicago, IL, USA*

JOHANNES NEU • *ONCOTYROL, Center for Personalized Cancer Medicine, Innsbruck, Austria*
TRIPAT KAUR OBEROI-KHANUJA • *Cell Death Signaling Group, Institute for Biochemie II, Goethe University Medical School, Goethe University, Frankfurt, Germany*
LÁSZLÓ ŐRFI • *Vichem Chemie Ltd., Budapest, Hungary*
BARBARA PERGOLIZZI • *Department of Clinical and Biological Sciences, University of Torino, Torino, Italy*
MARK R. PHILIPS • *Department of Medicine, NYU School of Medicine, New York University, New York, NY, USA; Department of Cell Biology, NYU School of Medicine, New York University, New York, NY, USA; Department of Biochemistry and Molecular Pharmacology, NYU Langone School of Medicine, New York University, New York, NY, USA; Cancer Institute, NYU Langone School of Medicine, New York University, New York, NY, USA*
KRISHNARAJ RAJALINGAM • *Cell Death Signaling Group, Institute for Biochemie II, Goethe University Medical School, Goethe University, Frankfurt, Germany*
SAVERIO FRANCESCO RETTA • *Department of Clinical and Biological Sciences, University of Torino, Torino, Italy*
ANNE J. RIDLEY • *Randall Division of Cell and Molecular Biophysics, King's College London, London, UK*
DANIEL RÖNNLUND • *Department of Biomolecular Physics, Applied Physics, School of Engineering Sciences, KTH-Royal Institute of Technology, AlbaNova University Centre, Stockholm, Sweden*
IGNACIO RUBIO • *Institute for Molecular Cell Biology, University Hospital Jena, Jena, Germany*
SIMON SCHNAITER • *ONCOTYROL, Center for Personalized Cancer Medicine, Innsbruck, Austria*
LIHI SHAULOV • *Faculty of Medicine in the Galilee, Bar-Ilan University, Safed, Israel*
GENG-XIAN SHI • *Cancer Center, University of Hawaii, Honolulu, HI, USA*
MAJA ŠOLMAN • *Turku Centre for Biotechnology, Åbo Akademi University, Turku, Finland*
PUTTY-REDDY SUDHIR • *Institute of Biomedical Sciences, Academia Sinica, Taipei, Taiwan*
RENATA TISI • *Department of Biotechnology and Biosciences, University of Milano-Bicocca, Milan, Italy*
MAURO TORTI • *Department of Biology and Biotechnology, University of Pavia, Pavia, Italy*
LORENZA TRABALZINI • *Department of Biotechnology, Chemistry, and Pharmacy, University of Siena, Siena, Italy*
ELIANA TRAPANI • *Department of Clinical and Biological Sciences, University of Torino, Torino, Italy*
FREDERICK D. TSAI • *Department of Medicine, NYU Langone School of Medicine, New York University, New York, NY, USA; Department of Cell Biology, NYU Langone School of Medicine, New York University, New York, NY, USA; Department of Biochemistry and Molecular Pharmacology, NYU Langone School of Medicine, New York University, New York, NY, USA; Cancer Institute, NYU Langone School of Medicine, New York University, New York, NY, USA*
FRIGYES WÁCZEK • *Vichem Chemie Ltd., Budapest, Hungary*
JERKER WIDENGREN • *Department of Biomolecular Physics, Applied Physics, School of Engineering Sciences, KTH-Royal Institute of Technology, AlbaNova University Centre, Stockholm, Sweden*
WINFRIED WUNDERLICH • *ONCOTYROL—Center for Personalized Cancer Medicine GmbH and ADSI—Austrian Drug Screening Institute, Innsbruck, Austria*

Joseph P. Wynne • *Department of Medicine, NYU Langone School of Medicine, New York University, New York, NY, USA; Department of Cell Biology, NYU Langone School of Medicine, New York University, New York, NY, USA; Department of Biochemistry and Molecular Pharmacology, NYU Langone School of Medicine, New York University, New York, NY, USA; Cancer Institute, NYU Langone School of Medicine, New York University, New York, NY, USA*

# Chapter 1

# The Ras Superfamily of Small GTPases: The Unlocked Secrets

**Luca Goitre, Eliana Trapani, Lorenza Trabalzini, and Saverio Francesco Retta**

## Abstract

The Ras superfamily of small GTPases is composed of more than 150 members, which share a conserved structure and biochemical properties, acting as binary molecular switches turned on by binding GTP and off by hydrolyzing GTP to GDP. However, despite considerable structural and biochemical similarities, these proteins play multiple and divergent roles, being versatile and key regulators of virtually all fundamental cellular processes. Conversely, their dysfunction plays a crucial role in the pathogenesis of serious human diseases, including cancer and developmental syndromes.

Fuelled by the original identification in 1982 of mutationally activated and transforming human Ras genes in human cancer cell lines, a variety of powerful experimental techniques have been intensively focused on discovering and studying structure, biochemistry, and biology of Ras and Ras-related small GTPases, leading to fundamental research breakthroughs into identification and structural and functional characterization of a huge number of Ras superfamily members, as well as of their multiple regulators and effectors.

In this review we provide a general overview of the major milestones that eventually allowed to unlock the secret treasure chest of this large and important superfamily of proteins.

**Key words** Small GTPases, Ras superfamily, Posttranslational modifications of Ras GTPases, Subcellular dynamics of Ras GTPases, Function and regulation of Ras GTPases, Signaling of Ras GTPases

## 1 Introduction

The Ras GTPase superfamily includes low molecular weight (20–30 kDa), monomeric GTP-binding and hydrolyzing (GTPases) proteins that act as binary, GDP-/GTP-regulated, molecular switches in coupling extracellular signals to intracellular signaling networks that regulate a wide range of fundamental cellular processes, including proliferation, differentiation, morphology, polarity, adhesion, migration, survival, and apoptosis. Indeed, dysregulation of numerous Ras superfamily G-protein-dependent regulatory cascades underlies the development of human diseases [1]. Ras proteins localize to discrete membrane microdomains

Lorenza Trabalzini and Saverio Francesco Retta (eds.), *Ras Signaling: Methods and Protocols*, Methods in Molecular Biology, vol. 1120, DOI 10.1007/978-1-62703-791-4_1, where their cellular activity is cyclically regulated in response to either outside-in or inside-out signals by the opposing action of guanine nucleotide exchange factors (GEFs) and GTPase-activating proteins (GAPs), which promote the acquisition of the active GTP- and inactive GDP-bound conformation, respectively. Multiple GEFs and GAPs coexist in most cells, increasing the diversity of signals that regulate the activity of Ras proteins. In the active conformation, Ras proteins bind with differential affinities to a variety of effectors, which are defined as downstream signaling molecules that interact selectively with the GTP-bound form of Ras GTPases, and become activated as a consequence of this interaction. The existence of a large number of Ras effectors, as well as of extensive cross talk with multiple intracellular signaling pathways, further increases the signaling complexity of Ras GTPases.

The story of small GTPases started more than three decades ago with the discovery of the Ras oncogenes, which was soon followed by the discoveries of related proteins now forming the Ras superfamily [2, 3]. The three human Ras proteins, H-Ras, K-Ras, and N-Ras, are the founding members of this large superfamily of small GTPases, comprising over 150 human members with evolutionarily conserved orthologs found in *Drosophila, Caenorhabditis elegans, Saccharomyces cerevisiae, Saccharomyces pombe, Dictyostelium,* and plants. This superfamily is divided into families and subfamilies on the basis of sequence and functional similarities (Table 1). The five major families are Ras, Rho, Rab, Arf, and Ran [1, 4].

Ras family proteins are almost universal components of signaling pathways in eukaryotic organisms, including vertebrates, invertebrates, and yeasts, where they play critical roles in development, proliferation, differentiation, and survival. Besides the ubiquitously expressed major Ras isoforms, H-, K-, and N-Ras, which are highly conserved, yet exhibiting different biological outputs [5, 6], the Ras family includes Rap, R-Ras, Ral, and Rheb proteins, also functioning as pivotal signal transduction nodes in various signaling networks.

Rho GTPase family includes the well-characterized family members Rac1, RhoA, and Cdc42, which are involved in signaling networks that regulate cytoskeletal organization, cell polarity, cell-cycle progression, and gene expression [7].

Rab proteins comprise the largest branch of superfamily and regulate intracellular vesicular transport and trafficking of proteins [8, 9].

Like the Rab proteins, Arf family proteins are involved in regulation of vesicular transport [10].

The Ran protein is the most abundant small GTPase in the cell and is best known for its function in nucleocytoplasmic transport of RNA and proteins [11].

Sequence comparisons of Ras superfamily small GTPases from various species have revealed that they are conserved in primary structures at the level of 30–55 % similarity. Furthermore, various crystallographic analyses revealed the presence of a structurally

**Table 1**
**The Ras superfamily of small GTPases**

| Family | Subfamily | Function | Members |
|---|---|---|---|
| Ras | Ras | Cell proliferation, differentiation, survival, apoptosis, gene expression | E-Ras; N-Ras; H-Ras; K-Ras; M-Ras; R-Ras; R-Ras2; Di-Ras1; Di-Ras2; Di-Ras3; NKIRas1; NKIRas2; RasD1; RasD2; RasL10A; RasL10B; RasL11A; RasL11B; RasL12, Rerg |
| | Ral | GTP-dependent exocytosis | RalA, RalB |
| | Rap | Cell–cell and cell–matrix adhesion | Rap1A; Rap1B; Rap2A; Rap2B; Rap2C |
| | Rad | Cell shape remodelling, cell-cycle checkpoint | Rad, Gem, Kir, Rem1, Rem2 |
| | Rheb | mTOR pathway, cell growth and cell-cycle progression | Rheb; RhebL1 |
| | Rit | Neuronal differentiation and survival | Rit1; Rit2; Rin; Ric |
| Rho | | Cytoskeletal dynamics; cell shape, polarity, adhesion, and movement; cell-cycle progression; gene expression | RhoA; RhoB; RhoBTB1; RhoBTB2; RhoBTB3; RhoC; RhoD; RhoF; RhoG; RhoH; RhoJ; RhoQ; RhoU; RhoV; Rnd1; Rnd2; Rnd3; Rac1; Rac2; Rac3; Cdc42 |
| Rab | | Membrane and protein traffic in the endocytic and secretory pathways | Rab1A; Rab1B; Rab2; Rab3A; Rab3B; Rab3C; Rab3D; Rab4A; Rab4B; Rab5A; Rab5B; Rab5C; Rab6A; Rab6B; Rab6C; Rab7A; Rab7B; Rab7L1; Rab8A; Rab8B; Rab9; Rab9B; RabL2A; RabL2B; RabL4; Rab10; Rab11A; Rab11B; Rab12; Rab13; Rab14; Rab15; Rab17; Rab18; Rab19; Rab20; Rab21; Rab22A; others |
| Arf | | Vesicular trafficking, endocytosis, and exocytosis | Arf1; Arf3; Arf4; Arf5; Arf6; Arl1; Arl2; Arl3; Arl4; Arl5; Arl5C; Arl6; Arl7; Arl8; Arl9; Arl10A; Arl10B; Arl10C; Arl11; Arl13A; Arl13B; Arl14; Arl15; Arl16; Arl17; TRIM23, Arl4D; ArfRP1; Arl13B |
| Ran | | Nucleocytoplasmic transport; mitotic spindle organization | Ran |

preserved GTP-binding domain that is unique for the whole superfamily. Conversely, this superfamily contains proteins characterized by highly divergent functional specializations, which control a vast number of important molecular and cellular processes, such as sorting and amplification of transmembrane signals, synthesis and translocation of proteins, vesicular traffic through the cytoplasm, and proliferation and differentiation. The functional diversity of these proteins is based on minor modifications in sequence, structure, and/or regulatory posttranslational modifications, which dictate specific subcellular localizations and interactions with proteins that act as either regulators or effectors [1, 4, 12].

Identification and structural and functional characterization of members of the Ras superfamily, as well as their relationships with regulatory and effector molecules, have been made possible by a variety of powerful experimental techniques, which eventually allowed to unlock the secrets of this large and important superfamily of proteins. Indeed, following the use of gene transfer, DNA sequencing and DNA mapping techniques that allowed the identification of Ras genes as key players in experimental transformation as well as in human tumor pathogenesis [3, 6], advances in structural biology techniques, including X-ray crystallography, nuclear magnetic resonance (NMR) and electron paramagnetic resonance (EPR), coupled with modern biochemical, biophysical, molecular genetic, cell biology, system biology, and bioinformatic approaches, made possible key discoveries regarding the structure, biochemistry, and regulatory and signal transduction molecular mechanisms of Ras proteins, as well as the identification and characterization of novel Ras superfamily members and their multiple regulators and downstream effectors, each with a distinct pattern of tissue-specific expression and a distinct set of intracellular functions [3, 13].

In addition, upon the sequencing of human and other genomes, in silico analyses have identified the complete repertoire of genes encoding Ras-related small GTPases in distinct eukaryotic genomes. Notably, while the genomes of some plants were reported to contain Ras-related small GTPases of the Rab, Rho, Arf, and Ran families, no Ras GTPases have been found in any plants [3]. Furthermore, thanks to the availability of complete genome sequences of species representing important evolutionary branch points, phylogenetic analysis of gene families at the organism and sequence level revealed complex relationships between the evolution of this protein superfamily sequence and the acquisition of distinct cellular functions, helping to identify features important for the recognition of molecular partners and the functional specialization of different members of the Ras superfamily [4, 14].

## 2 Structure of Ras Superfamily GTPases

Small GTP-binding proteins of the Ras superfamily maintain a structurally and mechanistically preserved GTP-binding core and a universal conformational switch, despite considerable divergence in sequence and function. Indeed, they share a set of characteristic conserved sequence motifs at the N-terminus, named "G boxes" (G1 through G5), which form five polypeptide loops lined up along the nucleotide-binding site. Together with five alpha helices (A1–A5) and six beta-strands (B1–B6), these motifs make up a 20 kDa "G domain" (Ras residues 5–166) that carries out the basic function of guanine nucleotide binding and hydrolysis [1, 13, 14].

The G1 box (aaaaGxxxxGKS/T, where a = L, I, V, or M and x = any amino acid), also known as a P-loop (phosphate-binding loop), is a purine nucleotide-binding signature; the G2 box (T) is located in one of two segments that reorient as a function of GDP or GTP binding and provide major components of effector binding surface; the G3 box (blbbDxxGQ/H/T, where l = hydrophilic and b = hydrophobic) is involved in binding a nucleotide-associated $Mg^{2+}$ ion; G4 box (bbbbN/TK/QxD) makes hydrogen bond contact with the guanine ring and provide stabilizing interactions with G1 box residues; G5 box (bbEC/S/T/ASA/K/L) makes indirect associations with the guanine nucleotide [1, 14].

Ras superfamily signaling is dependent on the binding of specific regulators and effectors. Thus, minor modifications in sequence, structure, and/or cellular regulation of members of the superfamily will affect binding to regulators and effectors and consequently cell signaling. Accordingly, an important goal in studies of Ras superfamily signaling has been to identify the determinants of these specific associations [15].

Elucidation of the crystal structure of H-Ras contributed significant insight into how Ras functions as a GDP-/GTP-regulated switch. The first accurate structure of the "G domain" was reported in 1989 [16, 17], whereas subsequent structural studies identified two key regions of conformational differences between GDP- and GTP-bound H-Ras [18, 19]. Referred to as switch I and II, these regions coincide with sequences critical for Ras interaction with its regulators and effectors [13].

The structural details of GAP-mediated and GEF-mediated regulation of Ras activity, as well as a molecular basis for the biochemical consequences of the Ras-activating mutations mimicking the GTP-bound state and GAP insensitivity (e.g., mutations at G12 and Q61 in Ras isoforms), were ultimately laid out by Wittinghofer and Kuriyan groups in the late 1990s, when the structures of H-Ras in complex with the p120RasGAP GRD (GAP-related domain) and that of H-Ras in complex with the catalytic domain of the RasGEF SOS were reported [20, 21].

In addition, the combination of structural biology with biochemical and molecular genetic approaches has allowed to assign discrete functions of several GTPases to specific structural elements and even to individual amino acids [4, 12, 15].

## 3 Posttranslational Modifications of Ras Superfamily GTPases

Posttranslational modifications (PTMs) of Ras proteins direct them to various cellular membranes and, in some cases, modulate GTP–GDP exchange. These modifications include the constitutive and irreversible remodelling of its carboxy-terminal CAAX motif by farnesylation/geranylgeranylation, proteolysis and methylation, reversible palmitoylation, and conditional modifications, including phosphorylation, peptidyl–prolyl isomerization, monoubiquitylation, diubiquitylation, nitrosylation, ADP ribosylation, and glucosylation [22].

Ras represents the archetypal CAAX protein. This class of protein terminates in a CAAX sequence, where C is Cys, A is usually an aliphatic amino acid, and X is any amino acid. The CAAX sequence directs the PTM of the carboxyl terminus of the protein with a polyisoprenoid lipid, which in the case of Ras is a farnesyl moiety. This modification converts an otherwise globular, hydrophilic protein to one that associates with the cytoplasmic leaflet of cellular membranes, a process that is crucial for proper Ras activation and biological function.

The functional connection between this C-terminal processing and Ras function was made by Lowy's group in 1984, which showed that lipid binding and membrane association were actually required for the transforming activity of the viral H-Ras oncoprotein [23], whereas working with cellular H-Ras, Aaronson's group demonstrated that this C-terminal processing and membrane recruitment of Ras is a prerequisite to its biochemical activation [24].

The molecular mechanisms of Ras lipid processing were laid out over the subsequent 5 years through a series of observations using yeast genetics, protein biochemistry, and in vitro cellular systems [22].

## 4 Regulation of Ras Superfamily GTPases

Major milestones in understanding Ras regulatory mechanisms were as follows: the finding in 1984–1985 that mutant Ras proteins are impaired ~10-fold in their intrinsic GTP hydrolysis (GTPase) activity; the discovery in 1987 that the activity of a cytosolic GAP was responsible for a 300-fold acceleration of the hydrolysis of GTP bound to normal Ras but not to tumor-associated mutant Ras proteins; and the biochemical demonstration in 1987–1991 that yeast CDC25, *Drosophila* Son of Sevenless (SOS), and their mammalian homologous proteins may function as GEFs for Ras, enhancing GDP release and rebinding of GTP, thereby acting as upstream activators of Ras proteins.

All Ras superfamily members are binary on–off molecular switches that alternate between two conformational states: a GDP-bound inactive state and a GTP-bound active state. The active form typically binds effector molecules, while GTP hydrolysis disrupts this interaction. Because small GTPases exhibit high-affinity binding for GDP and GTP and low intrinsic GTP hydrolysis activity (which is why small GTPases are better described as "small GTP-binding proteins"), they require the helping of two types of regulatory proteins: GTPase-activating proteins (GAPs) and guanine nucleotide exchange factors (GEFs) [1, 25–28]. GAPs accelerate GTP hydrolysis, thus promoting the formation of the inactive, GDP-bound state, whereas GEFs facilitate GDP dissociation, promoting the formation of the active, GTP-bound state due to the high intracellular ratio of GTP:GDP [28]. The transition of Ras proteins between the GDP- and GTP-bound state is accompanied by a conformational change that greatly enhances their affinity for downstream effectors, leading to activation of downstream signal transduction pathways, and consequently generating cellular responses, until GTP hydrolysis returns the protein in the inactive GDP-bound state [15].

Other factors, called guanine nucleotide-releasing factor (GRF) [29], ras guanine nucleotide exchange factor (rGEF) [30], GDP dissociation stimulator (GDS) [31], and Ras exchange promoting (REP) [32], have been described that accelerate release of GDP and subsequent rebinding of GTP to some GTPases (Ras and Rapl) [33]. These factors have been proposed to serve as "activators" of small GTPases [33] and are generally called guanine nucleotide-releasing proteins (GNRPs) [2, 34].

Besides the two key classes of regulatory proteins, GEFs and GAPs, some small GTPases of the Ras superfamily are also regulated by a third class of proteins called guanine nucleotide dissociation inhibitors (GDIs), which bind to small GTPases in the cytosol and mask their C-terminal hydrophobic prenyl tail, thus preventing membrane association and interaction with regulators/effectors [1, 26, 35].

The factors that regulate Ras GTPases are themselves affected by diverse upstream signals, such as those elicited by growth factors and hormones, which serve to activate or inactivate Ras GTPase signaling pathways.

Interestingly, recent work on a pair of GTPases involved in the signal recognition particle (SRP) pathway has revealed a distinct mode of GTPase regulation. They do not need either to recruit GEFs to facilitate GDP → GTP exchange or to recruit GAPs to facilitate GTP hydrolysis. They are instead regulated by nucleotide-dependent dimerization [36].

## 5 Functional Specialization of Ras Superfamily GTPases

Collectively, the small GTPases of the Ras superfamily heavily influence cellular organization and signaling, playing critical roles in normal cellular growth, differentiation, and development in addition to their potential for malignant transformation. Indeed, the elucidation of the biochemical and molecular mechanisms by which Ras GTPases facilitate signal transduction established many of our fundamental concepts of how a normal cell orchestrates responses to extracellular cues.

However, although all Ras superfamily members can be identified through the presence of the conserved G box sequence motifs involved in guanine nucleotide binding and GTP hydrolysis and share a conserved alpha/beta core structure [1], a high functional specialization exist among different superfamily members, resulting in multiple and divergent roles. The functions of Ras GTPases are therefore best described by referring to the division of Ras GTPases into families and subfamilies based on both phylogenetic and functional relationships. In particular, the Ras superfamily has traditionally been subgrouped into five major protein families, including Ras, Rho, Rab, Arf, and Ran [1, 4], whose specific identities reside mainly in the presence of distinct sequence motifs that mediate interactions with family-specific regulators and downstream effectors, leading to distinct biological functions. Moreover, members of the same family differ from each other in their "variable" membrane targeting domains, which dictate subcellular localization and dynamic spatiotemporal regulation. Conversely, distinct members of the Ras superfamily transact and interconnect to each other through complex signaling networks [37].

### 5.1 The Ras Family

The identification of Ras (*Ra*t *s*arcoma) genes emerged in 1970s during the extensive study of acutely transforming retroviruses isolated from rats, mice, and other animals, which eventually led to the identification in 1982 of mutationally activated and potently transforming human Ras genes in human cancer cell lines [38]. Subsequently, oncogenic forms of Ras constitutively activated by missense mutations (K-Ras, H-Ras, and N-Ras) have been found in 15–20 % of human tumors and shown to be characterized by either disabled GAP responsiveness (e.g., H-Ras$^{G12V}$) or enhanced guanine nucleotide exchange (e.g., H-Ras$^{N116H}$) [14, 39]. In addition, in some tumors Ras overactivation is caused by either deletion of negative regulators (GAPs) or overexpression of positive regulators (GEFs) [14].

The Ras family is composed of over 36 members (Table 1), which are activated in response to multiple extracellular stimuli. Activated Ras interacts with effector proteins that regulate cytoplasmic signaling networks controlling gene expression and cellular proliferation, differentiation, and survival [1]. Most of the Ras family proteins localize predominantly to the plasma membrane as a result of the C-terminal prenylation [14, 40].

The best characterized Ras signaling pathway is the activation of Ras by the epidermal growth factor receptor (EGFR) tyrosine kinase, with activated Ras then stimulating the ERK mitogen-activated protein kinase (MAPK) cascade. Specifically, EGFR dimerization induces its autophosphorylation on tyrosine residues, thereby creating docking sites for the SH2 domain of the adaptor protein growth-factor-receptor-bound protein 2 (GRB2), which contains also SH3 domains that mediate its association with the RasGEF

SOS. Recruitment of GRB2 to the activated EGFR localizes SOS to the plasma membrane in close proximity to Ras, facilitating its activation. Led by genetic studies in *C. elegans* and *Drosophila*, the key findings demonstrating that GRB2 links EGFR and SOS for Ras activation were obtained in 1993 [3]. Many other receptor types, including the G-protein-coupled receptors, were subsequently demonstrated to activate Ras through stimulation of GEF proteins. In some cases, this has been shown to involve transactivation of growth factor receptor tyrosine kinases [41]. In the same 1993 milestone year, further seminal genetic and biochemical studies of Ras proteins in distinct model organisms identified the Raf serine/threonine kinase as a protein that bound preferentially to activated Ras-GTP, suggesting that it might play a role in selective transmission of signals from the active form of Ras [3]. Indeed, active Ras promotes Raf recruitment to the cell membrane and its activating phosphorylation by Ser/Thr and Tyr kinases. In turn, activated Raf phosphorylates and activates the MEK1/2 dual-specificity protein kinase, which phosphorylates and activates the ERK1/2 mitogen-activated protein (MAP) kinase. Finally, ERK phosphorylates and activates target transcription factors, resulting in the induced expression of genes that promote cell-cycle progression, including D-type cyclins [1, 14, 41]. In addition to the Raf–MEK–ERK cascade, Ras has also been found to activate several other effector pathways, including phosphatidylinositol 3-kinase (PI3K), RalGEF-Ral, and phospholipase C epsilon (PLCε) pathways [41, 42].

A structural feature common to the majority of Ras effectors is the presence of a Ras-binding domain (RBD) or Ras association (RA) domain, which is characterized by a conserved ubiquitin-fold structure and promotes association of the effector with GTP-bound Ras. This characteristic feature has been useful for the discovering of novel RBD/RA domain-containing proteins by in silico searches, leading to the identification of important effectors of Ras, including Tiam1, a RacGEF, and PLCε, as well as for the development of the now-classic pull-down assay to measure Ras activation in cell lysates [3]. Further RBD/RA domain-containing proteins were searched and identified by yeast two-hybrid screening [43].

Within the Ras family are branches with transforming capability, including those of Ras and R-Ras, and those that are nontransforming, including those of Rap, Ral, and Rheb. Members of these branches are covalently modified by either farnesyl or geranylgeranyl isoprenoid lipids and targeted to subcellular membrane compartments for proper biological function [44]. Remarkably, the use of FRET (fluorescent resonance energy transfer)-based sensors allowed the spatiotemporal imaging of growth-factor-induced activation of Ras and Rap1 in living cells, showing that growth factors activate Ras at the peripheral plasma membrane and Rap1 at the intracellular perinuclear region, thus helping to resolve long-standing questions as to how Ras and Rap1 induce different cellular responses and how the signals for differentiation and survival are distinguished by cells [45]. Furthermore, the combination of biochemical and molecular biology assays with time-lapse fluorescence microscopy experiments led to the identification of a pivotal role for Rap1 in the cross talk between cadherins and integrins [46, 47].

Recently, a novel branch of the Ras family has been described, whose members share highly conserved GTP-binding motifs but do not contain any known lipidation motifs. Members of this novel branch include Rit (Ras-like protein in tissues), Rin (Ras-like protein in neurons), and the *Drosophila* protein Ric (Ras-related protein which interacted with calmodulin), which form a distinct subfamily more highly related to the Ras and R-Ras branches than to the Rap, Ral, or Rheb branches [48]. The conservation of Rit and Rin from flies to humans suggests conservation of important physiological functions. Indeed, studies using transgenic and knockout animal models have begun to elucidate the physiological roles for the Rit subfamily, including emerging roles in the regulation of neuronal morphogenesis and cellular survival signaling, whereas genetic studies have implicated Rit and Rin signaling in human disorders such as cancer, Parkinson's disease, autism, and schizophrenia [49, 50].

### 5.2 The Rho Family

The Rho (*R*as *ho*mologous) family of small GTPases is closely related to the Ras family and is composed of 23 members, which are involved in signaling networks regulating actin cytoskeleton organization, cell adhesion, polarity and motility, cell-cycle progression, and gene expression [7]. The best characterized members of the family are RhoA, Rac1, and Cdc42, which act as molecular switches in either distinct or interconnected signal transduction pathways linking plasma membrane receptors to actin cytoskeleton rearrangements and consequent biological effects in a wide variety of mammalian cell types as well as in yeast, flies, worms, and some plants [27]. Indeed, these three proteins play crucial roles in regulation of cell polarity, gene transcription, cell-cycle progression, microtubules dynamics, vesicular transport, extracellular matrix (ECM) remodelling, and a variety of enzymatic activities, including NADPH oxidase activity in generating reactive oxygen species (ROS) [27, 51].

The first *Rho* gene was discovered by mistake, in 1985, by a group trying to clone a homologue of human chorionic gonadotropin from the sea snail *Aplysia* with a low stringency cDNA screening. Using the *Aplysia* Rho sequence, three closely related mammalian genes were cloned, now called RhoA, RhoB, and RhoC [52, 53]. Rac was next identified in 1989 as a substrate for the C3 transferase that modifies Rho [54], followed by Cdc42 in 1990 [55]. The Rho regulators, GEFs, GAPs, and GDIs, and downstream targets were found through a combination of biochemistry, genetics, and molecular biology studies and linked to Rho-regulated cellular responses. In parallel, a wide range of bacterial proteins were found to modify Rho proteins or alter their activity in cells, many of which turned out to be useful tools to study Rho functions [53]. Indeed, the combination of cell microinjection with constitutively active and dominant-negative Rho and Rac proteins, and treatment with the ADP-ribosyltransferase C3 from *Clostridium botulinum*, together with stimulation with different types of growth factors, lipids, and cytokines led to identify the roles of Rho and Rac in regulating the actin cytoskeleton dynamics, including the formation of actin stress fibers and focal adhesions, and lamellipodia and membrane ruffles, respectively [56–59]. Moreover, the use of constitutively active and dominant-negative forms of Cdc42 allowed to identify the role of this GTPase in the formation of highly dynamic fingerlike actin-rich protrusions known as filopodia [60].

Rho proteins are activated by multiple stimuli, including those transduced by growth factor receptors and integrins. Conversely, active Rho proteins interact with many effectors to modulate their localization and activity. Many Rho effector proteins are kinases, including the well-characterized p21-activated kinase (PAK), which binds active Cdc42 or Rac1, and Rho-associated coiled-coil-forming kinase (ROCK), which binds active RhoA. Other effectors are scaffold proteins, which control cellular functions through protein–protein interactions [51]. The activities of these effectors are mainly linked with actin cytoskeleton dynamics; however, many effectors are able to activate signaling pathways implicated in the regulation of a wide range of biological processes, including development, differentiation, transformation, apoptosis, and response to oxidative stress [61–63].

Rho GTPases are activated by two unrelated families of GEFs, carrying either a Dbl-homology (DH) domain (at least 70 members) or a Dock homology region (DHR) domain (11 members); moreover phospholipase D2 has been also proposed as a GEF for Rho proteins [26]. On the other hand, Rho GTPases are downregulated by GAPs with a unique RhoGAP catalytic domain (70 members in mammals) [26]. Moreover, Rho proteins carry either a farnesyl or a geranylgeranyl lipid at their C-terminus, which locates them to cellular membranes. Rho GDIs (three members in mammals) bind to geranylgeranyl lipids and maintain inactive pools of Rho proteins in the cytosol [26].

Recent investigations have unveiled important additional regulatory mechanisms: microRNA (miRNA) regulating post-transcriptional processing of Rho GTPase-encoding mRNAs; palmitoylation and nuclear targeting affecting intracellular distribution; posttranslational phosphorylation, transglutamination, and AMPylation impacting Rho GTPase signaling; and ubiquitination controlling Rho GTPase protein stability and turnover. These modes of regulation add to the complexity of the Rho GTPase signaling network and allow precise spatiotemporal control of individual Rho GTPases [64].

### 5.3 The Rab Family

First described as *Ra*s-like protein in *b*rain, the Rab family is the largest small GTPase family (over 60 members in human) [1, 26]. Rab GTPases regulate intracellular vesicular transport and the trafficking of proteins between different membrane compartments [1]. In particular, through their effectors, these proteins coordinate consecutive stages of transport, such as vesicle formation, vesicle and organelle motility, and tethering of vesicles to their target compartment [9].

Rab GTPases are activated by at least four distinct types of Rab GEFs (a subfamily that carry a Vps9 domain or a DENN domain, the TRAPP complex, and Sec2) and downregulated by Rab GAPs belonging to a large family of TBC (Tre2/Bub2/Cdc16) domain-containing proteins. Rab3 has an unrelated and still poorly characterized GAP, Rab3GAP, whose mutations cause Warburg micro syndrome [26].

Individual Rab proteins function at specific sites within the cell, for example the endoplasmic reticulum (ER), the Golgi apparatus, and the early, late, or recycling endosomes. Specificity in subcellular localization and function is dictated by C-terminal sequences carrying geranylgeranyl groups [1] and is often conserved from yeast to humans, underscoring the significant contributions that simple model organisms, such as *Caenorhabditis elegans*, can make towards a better understanding of human disease caused by Rab and vesicle trafficking malfunction [65].

### 5.4 The Arf Family

Like the Rab proteins, the *A*DP-*r*ibosylation *f*actor (Arf) family proteins are major regulators of vesicle biogenesis in intracellular traffic. The Arf family comprises also Arl (Arf-like), Arp (Arf-related proteins), and the remotely related Sar (secretion-associated and Ras-related) proteins [1, 26, 66]. These proteins share a much larger GDP/GTP conformational change than that occurring in the other Ras families [1, 26, 66]. Arf proteins are activated by ArfGEFs (16 members) containing a catalytic Sec7 domain, and inactivated by GAPs (31 members) containing a conserved ArfGAP domain [26].

Differently from Rab proteins, which function at single steps in membrane trafficking, Arf proteins can act at multiple steps. For example, Arf1 is able to control (1) the formation of coat protein I (COPI)-coated vesicles involved in retrograde transport between the Golgi and ER; (2) the formation of clathrin/adapter protein 1 (AP1) complex-associated vesicles at the trans-Golgi network (TGN) and on immature secretory vesicles; and (3) the formation of AP3-containing endosomes [1].

### 5.5 The Ran Family

The *Ra*s-like *n*uclear (Ran) protein is the most abundant small GTPase in the cell. It regulates nucleocytoplasmic transport of RNA and proteins through nuclear pore complexes [67, 68]. Moreover, Ran proteins are involved in the assembly of mitotic spindle and in the formation of nuclear envelope [26]. Recently, they have been endowed with novel cytoplasmic functions including trafficking of an ephrin receptor homolog in nematode oocytes, control of neurite outgrowth in *Drosophila* and mammalian neurons, and retrograde signaling in nerve axons after injury [69]. Ran functions in nucleocytoplasmic transport rely on the segregation of its GEF RCC1 (there are also alternative GEFs: importin beta and RanBP10) in the nucleus and of RanGAP in the cytosol, which establishes a Ran-GTP gradient. The high concentration of Ran-GTP in the nucleus facilitates the directionality of nuclear import and export. Nuclear Ran-GTP interacts with importin beta to promote importin-cargo dissociation into the nucleus. On the other hand, exportins load their substrates in response to RanGTP binding in the nucleus and traverse nuclear pore complexes as ternary RanGTP–exportin–cargo complexes to the cytoplasm, where GTP hydrolysis leads to export complex disassembly [70].

## 6 Breakthrough Techniques for Uncovering the Secrets of Ras GTPases

Many biological models and research techniques have contributed to the identification and structural and functional characterization of Ras GTPases. Indeed, the numerous potential biological functions for these GTPases and the organization of signal transduction pathways that they mediate have been established using several different types of cell grown in culture, whereas several animal model systems have helped to examine the function of the signaling pathways of Ras proteins in an in vivo context.

On the other hand, a number of methods have been used to identify Ras proteins and their multiple regulators and effectors, to characterize their specificity and activity and, ultimately, to gain insights into their biological functions. In particular, the major discoveries have been facilitated by the development of some powerful techniques that finally offered researchers easy ways to analyze activation of Ras GTPases by monitoring their switching between the "inactive" GDP-bound form and the "active" GTP-bound form under different cellular conditions. Indeed, activation of Ras GTPases can be investigated and quantified in cell extracts using so-called pull-down assays based on affinity-precipitation of the GTP-bound form from whole cell or tissue lysates. These assays take advantage of the fact that effector proteins interact only with GTP-bound Ras GTPases. Specifically, effector proteins or their RBD/RA domains that bind specifically to the activated form of a given Ras GTPase are used to capture it onto a bead support. Western blotting of the captured samples with specific antibodies then allows for quantification of the Ras GTPase activation levels in the sample. These assays are now considered standard and have been used extensively to identify the Ras GTPases that are activated in response to specific signals or cellular processes.

The first generation of Ras-GTP assays directly measured the ratio of GTP to GDP bound to Ras extracted from cells [71]. The assay was used to confirm that oncogenic mutants of Ras did indeed have much higher levels of bound GTP in cells than wild-type Ras and to demonstrate that growth factors thought to act via Ras actually increased Ras-GTP levels in cells. Although this assay has been useful in studies of Ras function, it suffers from some important drawbacks [72], which were overcome by the development of a pull-down assay exploiting the specificity with which Ras-GTP interacts with the Ras-binding domain (RBD) of one of the major effectors of Ras in cells, the Raf-1 protein kinase. Indeed, Ras-GTP binds to Raf RBD with an affinity approximately three orders of magnitude higher than Ras-GDP [73]. Specifically, in the pull-down assay for measuring Ras-GTP levels in cells, the Raf RBD is expressed as a glutathione S-transferase (GST) fusion protein, immobilized on glutathione-Sepharose beads and then used to affinity precipitate Ras-GTP from cell lysates. The affinity-precipitated Ras is then detected by immunoblotting with specific anti-Ras antibodies. The binding of the GST-RBD fusion protein to GTP-Ras inhibits its conversion to GDP; thus, the assay provides a reliable measure of the total amount of active GTP-loaded Ras present in the cell at any given time.

The same principle was later adapted to GTPases of the Rho family, such as RhoA, Rac, and Cdc42, by using effector domains specific for these proteins, including the RhoA-binding domain of rhotekin and the Rac1-/Cdc42-binding domain of PAK [74–76]. A biochemical assay was later developed to monitor the temporal activation and specificity of the direct regulators of Ras GTPases, the GEFs and the GAPs, in response to upstream signals [77].

Important insights into the roles of specific Ras proteins in the activation of downstream pathways have been also facilitated by the creation of dominant-negative (DN) and constitutively active (CA) forms, as well as by using siRNA-mediated knockdown approaches. Furthermore, the combination of yeast two-hybrid methods, including the dual-bait system, which simultaneously compares the interaction of two distinct baits with one interactive partner [78], with the generation of Ras mutants defective for different target interactions has allowed to discover novel distinct Ras-binding proteins and dissect their involvement in Ras function. In this way was discovered at first Ras interaction with Raf, Rin1, and RalGDS [79–81]. Moreover, the identification of mutations that selectively uncouple the association of Ras GTPases with specific binding proteins has led to information about which target interactions are required to generate the different downstream effects.

Finally, a breakthrough technical approach that allowed to visualize the spatiotemporal changes in the activity of Ras GTPases in living cells was the development of biosensors based on the FRET principle, a radiation-free transfer of excited-state energy from a donor fluorophore to a neighboring acceptor fluorophore [82–84]. In particular, either intramolecular or intermolecular FRET probes have been designed [85, 86]. In intramolecular FRET probes, also referred to as unimolecular biosensors, both the donor and acceptor are fluorescence proteins and incorporated into a fully genetically encoded single chimeric protein containing the Ras GTPase linked to the RBD of an effector protein. Such FRET biosensors have been successfully used to monitor growth-factor-induced spatiotemporal activation of Ras and Rap1 in living cells and collectively designed "Ras and interacting protein chimeric unit" (Raichu) probes [45, 87]. A similar unimolecular biosensor has been developed to monitor the spatiotemporal activation of RhoA [88]. Conversely, in intermolecular FRET probes, frequently referred to as a dual-chain or bimolecular biosensors, the donor and acceptor fluorescence proteins are incorporated into two distinct molecules containing either the Ras GTPase or the RBD of an effector protein. An example of such bimolecular FRET probes is the biosensor for Rac1 activation, named Rac1 FLAIR [89]. The intramolecular or intermolecular binding that occurs between the active GTP-bound GTPase and the effector RBD upon cell stimulation brings the donor and acceptor fluorescence proteins close to each other, thereby permitting FRET. Two fluorescent proteins, Clover and mRuby2, which are the brightest green and red fluorescent proteins to date and have the highest Förster radius of any ratiometric FRET pair yet described, have recently replaced the CFP-YFP pairs in reporters of small GTPase activity, significantly improving photostability, FRET dynamic range, and emission ratio changes and enhancing thereby detection of transient biochemical events [90].

## 7 Conclusions

Over the past 30 years since the discovery of the founding member of the Ras superfamily, the study of this large and important superfamily of proteins has seen the development and applications of many advanced cellular and animal models and a plethora of new powerful research techniques that allowed important discoveries, including the identification of novel members, as well as of new regulators and effectors, and the characterization of their structure and biological functions. Indeed, besides defining functional specificities and roles in complex signaling networks, these tools have provided growing evidence of biological and biochemical interplay and cross talk among members of seemingly divergent branches of the Ras superfamily, with GEFs such as Tiam1 and Rin leading the way and linking Ras with Rho and Rab GTPases, respectively. Moreover, new technological developments such as fluorescence imaging and interfering RNA, together with the continued emergence of "omics" applications and computational methods, have advanced our ability to study old questions in new ways, as well as to address many new questions, allowing further dissection of the function of Ras proteins and contributing to establish many of our current fundamental paradigms of signal transduction. However, despite great progress in our understanding of the molecular and cellular mechanisms of action of the Ras proteins, the expanding roster of their upstream regulators and downstream effectors and the complexity of the signaling cascades that they regulate indicate that much remains to be learnt. Hence, seeking a solution to the many remaining unanswered questions and identifying the as-yet unknown actions of many of the 150 Ras superfamily members will surely open new exciting research paths in the coming years.

## References


1. Wennerberg K, Rossman KL, Der CJ (2005) The Ras superfamily at a glance. J Cell Sci 118:843–846
2. Bourne HR, Sanders DA, McCormick F (1990) The GTPase superfamily: a conserved switch for diverse cell functions. Nature 348:125–132
3. Cox AD, Der CJ (2010) Ras history: the saga continues. Small GTPases 1:2–27
4. Rojas AM, Fuentes G, Rausell A et al (2012) The Ras protein superfamily: evolutionary tree and role of conserved amino acids. J Cell Biol 196:189–201
5. Haigis KM, Kendall KM, Wang Y et al (2008) Differential effects of oncogenic K-Ras and N-Ras on proliferation, differentiation and tumor progression in the colon. Nat Genet 40:600–608
6. Karnoub AE, Weinberg RA (2008) Ras oncogenes: split personalities. Nat Rev Mol Cell Biol 9:517–531
7. Heasman SJ, Ridley AJ (2008) Mammalian Rho GTPases: new insights into their functions from in vivo studies. Nat Rev Mol Cell Biol 9:690–701
8. Grosshans BL, Ortiz D, Novick P (2006) Rabs and their effectors: achieving specificity in membrane traffic. Proc Natl Acad Sci U S A 103:11821–11827
9. Zerial M, Mcbride H (2001) Rab proteins as membrane organizers. Nat Rev Mol Cell Biol 2:107–117
10. Donaldson JG, Jackson CL (2011) ARF family G proteins and their regulators: roles in membrane transport, development and disease. Nat Rev Mol Cell Biol 12:362–375
11. Moore MS (1998) Ran and nuclear transport. J Biol Chem 273:22857–22860
12. Biou V, Cherfils J (2004) Structural principles for the multispecificity of small GTP-binding proteins. Biochemistry 43:6833–6840
13. Vetter IR, Wittinghofer A (2001) The guanine nucleotide-binding switch in three dimensions. Science 294:1299–1304
14. Colicelli J (2004) Human RAS superfamily proteins and related GTPases. Sci STKE 2004:RE13
15. Herrmann C (2003) Ras-effector interactions: after one decade. Curr Opin Struct Biol 13: 122–129
16. Pai EF, Kabsch W, Krengel U et al (1989) Structure of the guanine-nucleotide-binding domain of the Ha-ras oncogene product p21 in the triphosphate conformation. Nature 341: 209–214
17. Tong L, Milburn MV, de Vos AM et al (1989) Structure of ras proteins. Science 245:244
18. Milburn MV, Tong L, deVos AM et al (1990) Molecular switch for signal transduction: structural differences between active and inactive forms of protooncogenic ras proteins. Science 247:939–945
19. Schlichting I, Almo SC, Rapp G et al (1990) Time-resolved X-ray crystallographic study of the conformational change in Ha-Ras p21 protein on GTP hydrolysis. Nature 345:309–315
20. Boriack-Sjodin PA, Margarit SM, Bar-Sagi D et al (1998) The structural basis of the activation of Ras by Sos. Nature 394:337–343
21. Scheffzek K, Ahmadian MR, Kabsch W et al (1997) The Ras-RasGAP complex: structural basis for GTPase activation and its loss in oncogenic Ras mutants. Science 277:333–338
22. Ahearn IM, Haigis K, Bar-Sagi D et al (2011) Regulating the regulator: post-translational modification of RAS. Nat Rev Mol Cell Biol 13:39–51
23. Willumsen BM, Christensen A, Hubbert NL et al (1984) The p21 ras C-terminus is required for transformation and membrane association. Nature 310:583–586

24. Srivastava SK, Lacal JC, Reynolds SH et al (1985) Antibody of predetermined specificity to a carboxy-terminal region of H-ras gene products inhibits their guanine nucleotide-binding function. Mol Cell Biol 5:3316–3319
25. Bernards A, Settleman J (2004) GAP control: regulating the regulators of small GTPases. Trends Cell Biol 14:377–385
26. Cherfils J, Zeghouf M (2013) Regulation of small GTPases by GEFs, GAPs, and GDIs. Physiol Rev 93:269–309
27. Etienne-Manneville S, Hall A (2002) Rho GTPases in cell biology. Nature 420:629–635
28. Schmidt A, Hall A (2002) Guanine nucleotide exchange factors for Rho GTPases: turning on the switch. Genes Dev 16:1587–1609
29. Wolfman A, Macara IG (1990) A cytosolic protein catalyzes the release of GDP from p21ras. Science 248:67–69
30. West M, Kung HF, Kamata T (1990) A novel membrane factor stimulates guanine nucleotide exchange reaction of ras proteins. FEBS Lett 259:245–248
31. Yamamoto T, Kaibuchi K, Mizuno T et al (1990) Purification and characterization from bovine brain cytosol of proteins that regulate the GDP/GTP exchange reaction of smg p21s, ras p21-like GTP-binding proteins. J Biol Chem 265:16626–16634
32. Downward J, Riehl R, Wu L et al (1990) Identification of a nucleotide exchange-promoting activity for p21ra. Proc Natl Acad Sci U S A 87:5998–6002
33. Burstein ES, Macara IG (1992) Characterization of a guanine nucleotide-releasing factor and a GTPase-activating protein that are specific for the ras-related protein p25rab3A. Proc Natl Acad Sci U S A 89: 1154–1158
34. Bourne HR, Sanders DA, McCormick F (1991) The GTPase superfamily: conserved structure and molecular mechanism. Nature 349:117–127
35. Seabra MC, Wasmeier C (2004) Controlling the location and activation of Rab GTPases. Curr Opin Cell Biol 16:451–457
36. Saraogi I, Akopian D, Shan SO (2011) A tale of two GTPases in cotranslational protein targeting. Protein Sci 20:1790–1795
37. Ferro E, Trabalzini L (2010) RalGDS family members couple Ras to Ral signalling and that's not all. Cell Signal 22:1804–1810
38. Malumbres M, Barbacid M (2003) RAS oncogenes: the first 30 years. Nat Rev Cancer 3: 459–465
39. Pylayeva-Gupta Y, Grabocka E, Bar-Sagi D (2011) RAS oncogenes: weaving a tumorigenic web. Nat Rev Cancer 11:761–774
40. Casey PJ (1992) Biochemistry of protein prenylation. J Lipid Res 33:1731–1740
41. Downward J (2003) Targeting RAS signalling pathways in cancer therapy. Nat Rev Cancer 3:11–22
42. Kelley GG, Reks SE, Ondrako JM et al (2001) Phospholipase C(epsilon): a novel Ras effector. EMBO J 20:743–754
43. Peterson SN, Trabalzini L, Brtva TR et al (1996) Identification of a novel RalGDS-related protein as a candidate effector for Ras and Rap1. J Biol Chem 271:29903–29908
44. Cox AD, Hisaka MM, Buss JE et al (1992) Specific Isoprenoid Modification Is Required for Function of Normal, but Not Oncogenic Ras Protein. Mol Cell Biol 12:2606–2615
45. Mochizuki N, Yamashita S, Kurokawa K et al (2001) Spatio-temporal images of growth-factor-induced activation of Ras and Rap1. Nature 411:1065–1068
46. Retta SF, Balzac F, Avolio M (2006) Rap1: a turnabout for the crosstalk between cadherins and integrins. Eur J Cell Biol 85:283–293
47. Balzac F, Avolio M, Degani S et al (2005) E-cadherin endocytosis regulates the activity of Rap1: a traffic light GTPase at the crossroads between cadherin and integrin function. J Cell Sci 118:4765–4783
48. Campbell SL, Khosravi-Far R, Rossman KL et al (1998) Increasing complexity of Ras signaling. Oncogene 17:1395–1413
49. Cai W, Shi GX, Andres DA (2013) Putting the Rit in cellular resistance: Rit, p38 MAPK and oxidative stress. Commun Integr Biol 6:e22297
50. Shi GX, Cai W, Andres DA (2013) Rit subfamily small GTPases: regulators in neuronal differentiation and survival. Cell Signal 25: 2060–2068
51. Sahai E, Marshall CJ (2002) Rho-GTPases and cancer. Nat Rev Cancer 2:133–142
52. Madaule P, Axel R (1985) A novel ras-related gene family. Cell 41:31–40
53. Ridley AJ (2012) Historical overview of Rho GTPases. Methods Mol Biol 827:3–12
54. Didsbury J, Weber RF, Bokoch GM et al (1989) Rac, a novel ras-related family of proteins that are botulinum toxin substrates. J Biol Chem 264:16378–16382
55. Munemitsu S, Innis MA, Clark R et al (1990) Molecular cloning and expression of a G25K cDNA, the human homolog of the yeast cell cycle gene CDC42. Mol Cell Biol 10: 5977–5982
56. Ridley A (2013) GTPase switch: Ras then Rho and Rac. Nat Cell Biol 15:337
57. Ridley AJ, Hall A (1992) The small GTP-binding protein rho regulates the assembly of focal adhesions and actin stress fibers in response to growth factors. Cell 70:389–399
58. Ridley AJ, Hall A (2004) Snails, Swiss, and serum: the solution for Rac 'n' Rho. Cell 116:S23–5, 2 p following S25

59. Ridley AJ, Hall A (1992) The small GTP-binding protein rac regulates growth factor-induced membrane ruffling. Cell 70:401–410
60. Nobes CD, Hall A (1995) Rho, rac, and cdc42 GTPases regulate the assembly of multimolecular focal complexes associated with actin stress fibers, lamellipodia, and filopodia. Cell 81:53–62
61. Gonzalez-Ramos M, Mora I, de Frutos S et al (2012) Intracellular redox equilibrium is essential for the constitutive expression of AP-1 dependent genes in resting cells: studies on TGF-beta1 regulation. Int J Biochem Cell Biol 44:963–971
62. Karin M, Gallagher E (2005) From JNK to pay dirt: jun kinases, their biochemistry, physiology and clinical importance. IUBMB Life 57:283–295
63. Pinkus R, Weiner LM, Daniel V (1996) Role of oxidants and antioxidants in the induction of AP-1, NF-kappaB, and glutathione S-transferase gene expression. J Biol Chem 271:13422–13429
64. Liu M, Bi F, Zhou X et al (2012) Rho GTPase regulation by miRNAs and covalent modifications. Trends Cell Biol 22:365–373
65. Gallegos ME, Balakrishnan S, Chandramouli P et al (2012) The *C. elegans* rab family: identification, classification and toolkit construction. PLoS One 7:e49387
66. Pasqualato S, Renault L, Cherfils J (2002) Arf, Arl, Arp and Sar proteins: a family of GTP-binding proteins with a structural device for 'front-back' communication. EMBO Rep 3:1035–1041
67. Stewart M (2007) Molecular mechanism of the nuclear protein import cycle. Nat Rev Mol Cell Biol 8:195–208
68. Weis K (2003) Regulating access to the genome: nucleocytoplasmic transport throughout the cell cycle. Cell 112:441–451
69. Yudin D, Fainzilber M (2009) Ran on tracks—cytoplasmic roles for a nuclear regulator. J Cell Sci 122:587–593
70. Guttler T, Gorlich D (2011) Ran-dependent nuclear export mediators: a structural perspective. EMBO J 30:3457–3474
71. Gibbs JB, Schaber MD, Marshall MS et al (1987) Identification of guanine nucleotides bound to ras-encoded proteins in growing yeast cells. J Biol Chem 262:10426–10429
72. Satoh T, Kaziro Y (1995) Measurement of Ras-bound guanine nucleotide in stimulated hematopoietic cells. Methods Enzymol 255:149–155
73. Herrmann C, Martin GA, Wittinghofer A (1995) Quantitative analysis of the complex between p21ras and the Ras-binding domain of the human Raf-1 protein kinase. J Biol Chem 270:2901–2905
74. Benard V, Bohl BP, Bokoch GM (1999) Characterization of rac and cdc42 activation in chemoattractant-stimulated human neutrophils using a novel assay for active GTPases. J Biol Chem 274:13198–13204
75. Ren XD, Kiosses WB, Schwartz MA (1999) Regulation of the small GTP-binding protein Rho by cell adhesion and the cytoskeleton. EMBO J 18:578–585
76. Suryavanshi N, Ridley AJ (2013) Determining Rho GTPase Activity by an Affinity-Precipitation Assay. Methods Mol Biol 1046:191–202
77. Garcia-Mata R, Wennerberg K, Arthur WT et al (2006) Analysis of activated GAPs and GEFs in cell lysates. Methods Enzymol 406:425–437
78. Serebriiskii I, Khazak V, Golemis EA (1999) A two-hybrid dual bait system to discriminate specificity of protein interactions. J Biol Chem 274:17080–17087
79. Han L, Colicelli J (1995) A human protein selected for interference with Ras function interacts directly with Ras and competes with Raf1. Mol Cell Biol 15:1318–1323
80. Van Aelst L, Barr M, Marcus S et al (1993) Complex formation between RAS and RAF and other protein kinases. Proc Natl Acad Sci U S A 90:6213–6217
81. White MA, Vale T, Camonis JH et al (1996) A role for the Ral guanine nucleotide dissociation stimulator in mediating Ras-induced transformation. J Biol Chem 271:16439–16442
82. Miyawaki A (2003) Visualization of the spatial and temporal dynamics of intracellular signaling. Dev Cell 4:295–305
83. Pollok BA, Heim R (1999) Using GFP in FRET-based applications. Trends Cell Biol 9:57–60
84. Tsien RY, Miyawaki A (1998) Seeing the machinery of live cells. Science 280:1954–1955
85. Hodgson L, Shen F, Hahn K (2010) Biosensors for characterizing the dynamics of rho family GTPases in living cells. Curr Protoc Cell Biol, Chapter 14, Unit 14.11: 1–26
86. Nakamura T, Matsuda M (2009) In vivo imaging of signal transduction cascades with probes based on Forster Resonance Energy Transfer (FRET). Curr Protoc Cell Biol, Chapter 14, Unit 14.10
87. Bos JL (2001) Glowing switches. Nature 411:1006–1007
88. Pertz O, Hodgson L, Klemke RL et al (2006) Spatiotemporal dynamics of RhoA activity in migrating cells. Nature 440:1069–1072
89. Machacek M, Hodgson L, Welch C et al (2009) Coordination of Rho GTPase activities during cell protrusion. Nature 461:99–103
90. Lam AJ, St-Pierre F, Gong Y et al (2012) Improving FRET dynamic range with bright green and red fluorescent proteins. Nat Methods 9:1005–1012

# Chapter 2

# A Novel Method for the Production of Fully Modified K-Ras 4B

**Tanmay S. Chavan*, James O. Meyer*, Luke Chisholm, Malgorzata Dobosz-Bartoszek, and Vadim Gaponenko**

## Abstract

Post-translational modifications in proteins play a major functional role. Post-translational modifications affect the way proteins interact with each other, bind nucleotides, and localize in cellular compartments. Given the importance of post-translational modifications in protein biology, development of methods to produce post-translationally modified proteins for biochemical and biophysical studies is timely and significant. At the same time, obtaining post-translationally modified proteins in bacterial expression systems is often problematic. Here, we describe a novel recombinant approach to prepare human K-Ras 4B, a protein that is post-translationally farnesylated, proteolytically cleaved, and methylated in its C-terminus. K-Ras 4B is a member of the Ras subfamily of small GTPases and is of interest because it is frequently mutated in human cancer.

The method relies on separate production of two structural domains—the N-terminal catalytic domain and the C-terminal peptide chemically modified with S-farnesyl-L-cysteine methyl ester. After the two domains are prepared, they are ligated together using the transpeptidase enzyme, sortase. Our procedure starts with the use of the plasmid of K-Ras 4B catalytic domain containing the sortase recognition sequence. After this, we describe the bacterial expression and purification steps used to purify K-Ras 4B and the preparation of the conjugated C-terminal peptide. The procedure ends with the sortase-mediated ligation technique. The produced post-translationally modified K-Ras 4B is active in a number of assays, including a GTP hydrolysis assay, Raf-1 binding assay, and surface plasmon resonance-based phospholipid binding assay.

**Key words** K-Ras 4B, Post-translational modification, Farnesylation, Sortase ligation, Expression, Purification, Cross-linking

## 1 Introduction

Approximately 0.5 % of all eukaryotic proteins undergo a form of post-translational modification known as prenylation [1]. Prenylation is an enzymatic reaction that results in addition of either a farnesyl or geranylgeranyl moiety by farnesyltransferase (FT) or geranylgeranyltransferase (GGT) enzymes, respectively. These prenyl groups

*Tanmay S. Chavan and James O. Meyer have equally contributed to this Chapter.

Lorenza Trabalzini and Saverio Francesco Retta (eds.), *Ras Signaling: Methods and Protocols*,
Methods in Molecular Biology, vol. 1120, DOI 10.1007/978-1-62703-791-4_2, © Springer Science+Business Media, LLC 2014

facilitate protein–protein and protein–lipid interactions and regulate signal-transduction cascades.

Proteins modified with a farnesyl moiety include members of the Ras family of proto-oncogenes [2]. Ras proteins are small GTPases that play an essential role in many signaling processes. Three isoforms of classical Ras proteins have been discovered: H-, N-, and K-Ras. Ras proteins share a basic structure, comprising a nucleotide binding catalytic domain and a C-terminal hypervariable region (HVR), which harbors the farnesyl modification. The farnesyl group directs Ras proteins to specific locations in the cellular membrane where signaling is initiated [3, 4].

To study the details of farnesyl-mediated anchoring in the membrane and protein–protein interactions, sufficient amounts of modified proteins must be produced for in vitro experiments. While it is possible to prepare sizeable quantities of unmodified K-Ras [5], synthesis of fully modified K-Ras is far more problematic. It is hard to purify naturally modified K-Ras from mammalian cells because it has many interaction partners.

K-Ras 4B is a ubiquitously expressed splice variant of K-Ras. Its 3-step process of post-translational modification begins with the addition of a farnesyl group to a C-terminal CAAX sequence in the HVR, after which the AAX peptide is removed by Rcel protease. Finally, the terminal cysteine is methylated by the isoprenylcysteine carboxyl methyltransferase (ICMT enzyme). The methylation step removes the C-terminal negative charge, renders the protein more hydrophobic, and aids in membrane binding [6, 7].

Recent progress in solid- and solution-phase procedures for preparation of synthetic lipidated peptides has allowed a more practical method to produce modified K-Ras 4B [8]. In this instance, the lipidated peptide corresponds to the K-Ras 4B HVR and can be attached to truncated K-Ras 4B via intein-mediated protein splicing or maleimidocaproyl-controlled ligation. This method requires custom production of a peptide with the post-translational modifications introduced during the peptide synthesis procedure.

Our method uses commercially available components for preparation of fully modified K-Ras 4B. To create the post-translationally modified HVR, we conjugate S-farnesyl-L-cysteine methyl ester to the isolated HVR peptide of K-Ras 4B. The bacterial transpeptidase, sortase, is subsequently used to ligate the modified HVR peptide to the truncated K-Ras 4B catalytic domain.

We demonstrate that the fully modified K-Ras 4B possesses GTPase activity, effector (Raf-1) binding capability and associates with membrane phospholipids. In conclusion, the approach that we describe to produce fully modified K-Ras 4B provides an opportunity to study the effect of post-translational modifications on protein–protein and protein–membrane interactions. The developed procedure can potentially be applied to other proteins that undergo similar post-translational modifications.

## 2 Materials

Prepare all solutions using ultrapure deionized water, with reagents stored at room temperature (21–25 °C) unless specified otherwise.

### *2.1 Bacterial Expression*

1. Cloned K-Ras 4B catalytic domain construct with a C-terminal sortase recognition sequence "LPXTG," where X is any amino acid (*see* **Note 1**), followed by a His-Tag, and two stop codons. We express this protein in a PET 42a vector (Novagen) stored at −20 °C.
2. One-shot BL 21AI cells (Invitrogen) stored at −80 °C.
3. Sterile SOC medium (Invitrogen, *see* **Note 2**) stored at 4 °C.
4. 50 mg/mL kanamycin sulfate stored at −20 °C.
5. LB cell medium: 25 g/L Luria Broth (LB)-Miller powder.
6. 10 % L-arabinose solution.
7. 100 % ethanol, molecular grade.
8. 1 M isopropyl β-D-1-thiogalactopyranoside (IPTG) stored at −20 °C.

### *2.2 Protein Extraction*

1. First Extraction Buffer—Make one solution containing the following: 10 mM magnesium chloride ($MgCl_2$), 2 mM phenylmethylsulfonyl fluoride (PMSF) stored at 4 °C (Pierce), 50 μg/mL DNaseI stored at −20 °C, 1 tablet of EDTA-free Complete (Roche Protease Inhibitor cocktail) per 50 mL of buffer volume, 10 mg lysozyme per 50 mL buffer volume, stored at −20 °C (*see* **Note 3**). Make the dilution in B-PER solution (Pierce), not water.
2. Second Extraction Buffer—Make one solution containing the following: 10 mM Tris–HCl pH 7.6, 50 mM KCl, 10 mM ethylenediaminetetraacetic acid (EDTA), 2 mM β-mercaptoethanol (βME), 100 μM guanosine diphosphate (GDP) (*see* **Note 3**). Make the dilution in water.

### *2.3 Protein Purification*

1. His-binding buffer containing glycerol: 20 mM Tris–HCl pH 7.6, 500 mM NaCl, 5 mM βME, 400 mL glycerol (*see* **Note 4**).
2. His-binding buffer: 0.5 M NaCl, 20 mM Tris–HCl pH 7.6.
3. His-wash buffer: 0.5 M NaCl, 10 mM imidazole (*see* **Note 5**), 20 mM Tris–HCl pH 7.6.
4. His-elution buffer: 1 M Imidazole, 20 mM Tris–HCl pH 7.6.
5. His-purification nickel-binding resin (Novagen). Store at 4 °C.
6. Disposable plastic columns (*see* **Note 6**).

#### 2.4 Farnesylation Reaction for HVR Peptide of K-Ras 4B (for 500 μL Reaction Volume)

1. 68 μL of 73.6 mM S-farnesyl-L-cysteine methyl ester in DMSO.
2. 50 μL of 50 % *N*-octyl glucopyranoside in $dH_2O$.
3. 381 μL of phosphate buffer, pH 7.4: 137 mM NaCl, 2.68 mM KCl, 10 mM $Na_2HPO_4$, 1.76 mM $KH_2$ $PO_4$.
4. 1 μL of 50 mM 4-dimethylaminopyridine (DTAP) stored at 4 °C.
5. 2 mg of Sulfo-SMCC crosslinker (Thermo Scientific).
6. Chemically synthesized HVR peptide (GKEKMSKDGKKK KKKSKC).

#### 2.5 Sortase-Mediated Ligation

1. 15 μM sortase A (*see* **Note 7**).
2. 15 μM K-Ras catalytic domain stored at 4 °C.
3. Tris–EDTA buffer pH 9: 50 mM Tris–HCl, 10 mM EDTA, 10 mM $MgCl_2$, 5 mM $CaCl_2$
4. 10 mM βME.
5. 30 μM HVR peptide of K-Ras conjugated with S-farnesyl-L-cysteine methyl ester using Sulfo-SMCC crosslinker.

#### 2.6 GTPase Activity Assay

1. 0.1 M GTP stored at −80 °C (*see* **Note 8**).
2. 10 μM fully modified K-Ras 4B stored at 4 °C.
3. 1 mM 2-amino-6-mercapto-7-methylpurine riboside (MESG) stored at −20 °C.
4. 0.5 M EDTA, pH 8.
5. Purine nucleoside phosphorylase.
6. 1 M $(NH_4)_2SO_4$.
7. UV spectrophotometer.

#### 2.7 Raf-1 Binding Assay

1. 0.1 M GTPγS or GDP stored at −80 °C.
2. 0.25 M EDTA solution.
3. Raf-1 Ras-binding domain (RBD) agarose beads.
4. 1 M $MgCl_2$ solution.
5. HEPES buffer: 25 mM HEPES, 150 mM sodium chloride, 5 mM $MgCl_2$, 10 mM βME, 5 % glycerol.
6. SDS loading buffer supplemented with 5 % βME (710 mM).

#### 2.8 SPR Membrane Binding Assay

1. Running buffer: 50 mM Tris–citrate pH 6.5, 50 mM NaCl, 5 mM $MgCl_2$, 10 mM β-mercaptoethanol.
2. CM-5 SPR chip immobilized with phospholipid nanodiscs (e.g., DOPC, DPPC, DPPE, or a mixture of any of these) on one flow cell and ethanolamine on another to act as reference.
3. 2 M sodium chloride.
4. Biacore T100 SPR instrument.

## 3 Methods

### 3.1 Bacterial Expression

1. Place a single vial of one-shot BL21AI cells on ice and allow thawing. Add 1 μL of K-Ras 4B PET 42a plasmid to vial. Keep vial on ice for 30 min.
2. Heat the vial to 42 °C for 45 s and return to ice for a further 2 min.
3. After this, add 350 μL of sterile SOC medium to the cell vial and incubate with agitation at 37 °C and 250 rpm for 1 h.
4. Make a 350 mL solution of LB Broth supplemented with 350 μL of 50 mg/mL kanamycin solution. Filter the solution through a 0.22 μm pore-size membrane into a 1 L autoclaved conical flask.
5. After the SOC-cell culture has been incubated for 1 h, inoculate the 1 L LB Broth medium with the full 350 μL of SOC-cell culture.
6. Incubate the cell culture at 37 °C and 250 rpm on the orbital shaker overnight.
7. Fill six 2,800 mL autoclaved conical flasks with 1 L of M9 medium in each [9]. Ensure that the medium has been filtered through a 0.22 μm pore-size membrane and autoclaved beforehand.
8. Add kanamycin to each 2,800-mL flask to a final concentration of 50 μg/mL.
9. Inoculate the 1 L of medium in each flask with 50 mL of the BL21AI/LB Broth culture and incubate at 37 °C, 250 rpm on an orbital shaker until the optical density (OD) measured at the wavelength of 600 nm reaches 0.6.
10. Induce cells by adding 200 μL of 1 M IPTG, 40 mL of 10 % L-arabinose, and 20 mL of ethanol to each 2,800 mL flask.
11. Incubate the flasks at 18 °C for 22–24 h at 250 rpm on an orbital shaker.
12. Centrifuge cell culture at 14,000 × *g* for 15 min to form cell pellet. Discard supernatant. The cell pellet can be stored overnight at −80 °C if not used immediately.

### 3.2 Protein Extraction

#### 3.2.1 First Extraction

1. Resuspend cell pellet in 25 mL of first extraction buffer per 1 L of cell culture. If pellet has been frozen, thaw it to room temperature first.
2. Incubate suspension at room temperature for 1 h with agitation.

3. Centrifuge suspension at 38,000 × *g* for 35 min at 4 °C, then transfer supernatant (first extraction) to a separate vessel. Keep the pellets for the second extraction.
4. Add EDTA, KCl, βME, and GDP to all of first extraction obtained, such that their final concentrations are 10 mM EDTA, 50 mM KCl, 5 mM βME, and 100 μM GDP.
5. Incubate the solution at room temperature for 2 h with agitation.
6. Dialyze first extraction in His-binding buffer containing glycerol overnight at 4 °C.

#### 3.2.2 Second Extraction

1. Resuspend pellet from first extraction in 25 mL of second extraction buffer per 1 L of cell culture.
2. Incubate the suspension with agitation for 1 h then centrifuge at 38,000 × *g* for 35 min.
3. Transfer supernatant (second extraction) to a separate vessel; the pellet can be discarded.
4. Dialyze the second extraction in His-binding buffer containing glycerol overnight at 4 °C.

### 3.3 Protein Purification (Use for Both First and Second Extractions Separately)

1. Prepare a disposable gravity column by washing with at least twice the total column volume of distilled water.
2. Add 4 mL of suspended nickel-binding resin to the column and allow storage ethanol to drain.
3. Wash the resin with at least 10 mL of His-binding buffer, then transfer resin to vessel containing the extraction.
4. Incubate the vessel at room temperature for 2 h under agitation.
5. Pour the extraction into the gravity column and wash with 500 mL of wash buffer in order to remove nonspecific binding to the nickel beads.
6. From the elution buffer stock, make 30, 60, 100, 200, and 1,000 mM imidazole elution buffers by diluting with His-binding buffer.
7. Elute the protein with elution buffer and collect the eluate in a falcon tube. Begin with 30 mL of 30 mM elution buffer followed by 20 mL of each remaining elution buffers in ascending order of imidazole concentration (*see* **Note 9**).
8. Run these purified fractions on an SDS-PAGE gel to confirm the presence of K-Ras catalytic domain.
9. Select the fractions with highest purity; these can then be pooled and dialyzed into Tris–EDTA buffer pH 9.

Figure 1 shows catalytic domain expressed and purified using this protocol on an SDS-PAGE gel developed with silver staining. The purity of the protein preparation is further confirmed using MALDI-TOF mass spectrometry (Fig. 2).

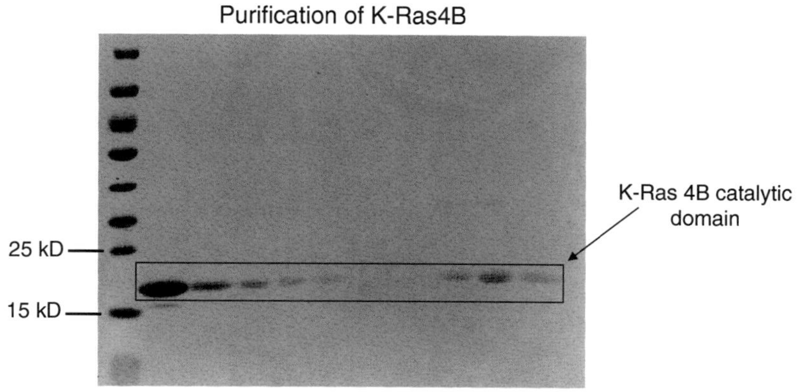


**Fig. 1** Purification of K-Ras 4b. K-Ras 4B catalytic domain with N-terminal LPETG sequence and 6(His) tag, purified on a nickel column from the first extraction of 6 L of cell culture

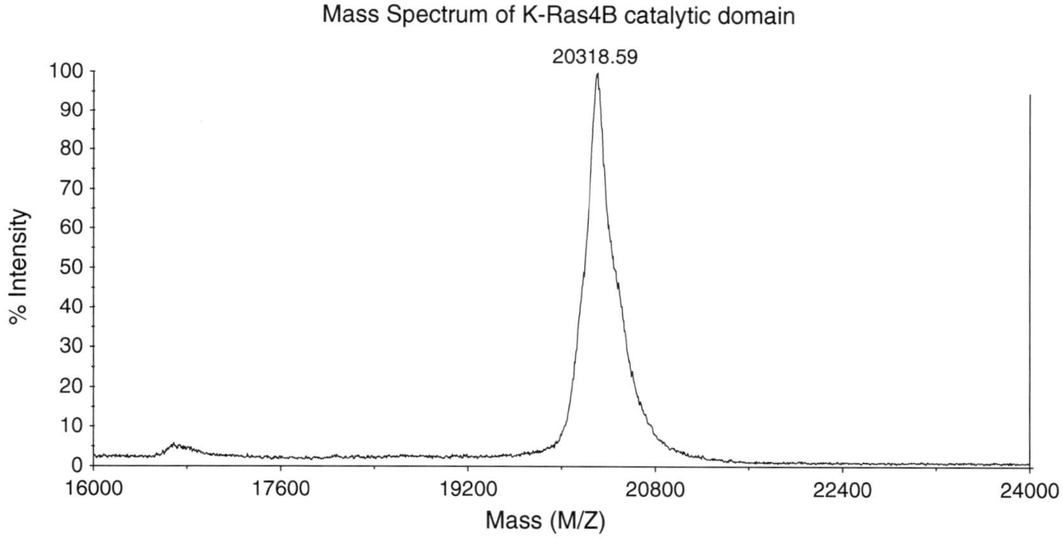


**Fig. 2** Mass spectrum of K-Ras 4B catalytic domain. Mass spectrum of K-Ras 4B catalytic domain appended with N-terminal LPTTG sequence and 6(His) tag in 50 mM Tris–HCl pH 9. Sample analyzed using MALDI-TOF mass spectrometry

### *3.4 Farnesylation of Hypervariable Region of K-Ras4B*

1. Take 68 μL of 73.6 mM S-farnesyl-L-cysteine methyl ester in 100 % dimethyl sulfoxide (DMSO) and add 2 mg of no-weigh sulfosuccinimidyl-4-(*N*-maleimidomethyl)cyclohexane-1-carboxylate (Sulfo-SMCC). Mix thoroughly till all of the Sulfo-SMCC dissolves.
2. Add 50 μL of 50 % (w/v) *N*-octyl-β-D-glucopyranoside to the above solution, followed by 381 μL of phosphate buffer, pH 7.4, and 1 μL of 50 mM di-tert-amyl peroxide (DTAP).
3. Incubate the reaction mix for 1.5 h and thereafter dilute to a final volume of 8 mL with phosphate buffer.

4. Centrifuge the solution at 3,000 × *g* for 30 min at 4 °C.
5. Discard the supernatant and dissolve the pellet in 150 μL of 100 % ethanol followed by the addition of 100 μL of 50 % (w/v) *N*-octyl-β-D-glucopyranoside.
6. Add the above solution to 2.009 mg of chemically synthesized HVR peptide dissolved in 750 μL of 1× phosphate buffer and incubate it at room temperature overnight.
7. Apply the reaction mix to a C-18 column and purify it by reversed phase high-performance liquid chromatography (RP-HPLC). Briefly, equilibrate the column with 90 % Buffer A (0.1 % trifluoroacetic acid) and 10 % Buffer B (90 % acetonitrile) before applying the sample. Thereafter, apply a linear gradient up to 80 % Buffer B. Elute the fractions at 66–68 % Buffer B and conduct a mass spectrometric analysis for purity.
8. Pool the pure fractions of farnesylated and methylated HVR and dry using Eppendorf Vacufuge.

### 3.5 Sortase-Mediated Ligation

1. Carry out reaction in 2 mL plastic microfuge tubes. To increase yield, carry out more individual reactions rather than directly scaling up the process.
2. Before proceeding with the reaction, dialyze K-Ras catalytic domain and sortase transpeptidase into pH 9 Tris–EDTA buffer (*see* **Note 10**).
3. Add 10 μM K-Ras catalytic domain, 15 μM sortase, 30 μM F-HVR, 10 mM βME, and 5 mM $CaCl_2$ to a 2 mL microfuge tube.
4. Place reaction mixture under agitation at room temperature for 48 h (*see* **Note 11**).
5. Halt the reaction by storing the reaction mixture at 4 °C. It is recommended that a sample of this is run on an SDS-PAGE gel to indicate the presence of reaction product (*see* **Note 12**).

   Figure 3 shows the sortase-mediated ligation after 48 h at room temperature.
6. Separate the modified full-length recombinant K-Ras from the reaction mixture by equilibrating once again with His-nickel-binding resin (200 μL/1 mL) for 1 h at room temperature and running through a gravity column. The product does not bind to the beads and can be immediately collected without need for elution.
7. To minimize the occurrence of nonspecific binding of the product to the nickel resin, it is recommended that NaCl is first added to the mixture to a concentration of 0.5 M before loading onto the column.

SDS PAGE gel of the reaction mixture

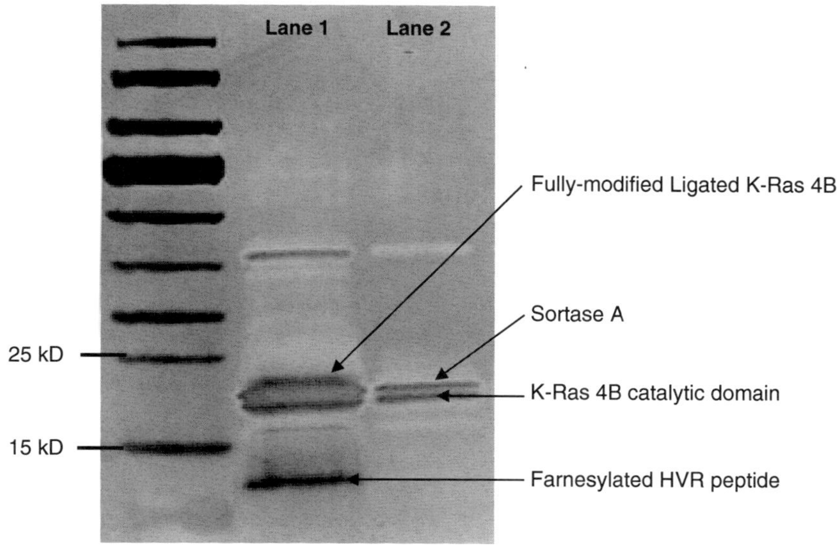


**Fig. 3** SDS-PAGE gel of the reaction mixture. *Lane 1* shows the sortase-mediated ligation of K-Ras 4B catalytic domain after 48 h at room temperature. This reaction mixture contains modified product and sortase as well as unreacted HVR and catalytic domain. *Lane 2* shows sortase and K-Ras 4B catalytic domain

SDS PAGE gel of purified fully modified K-Ras4B

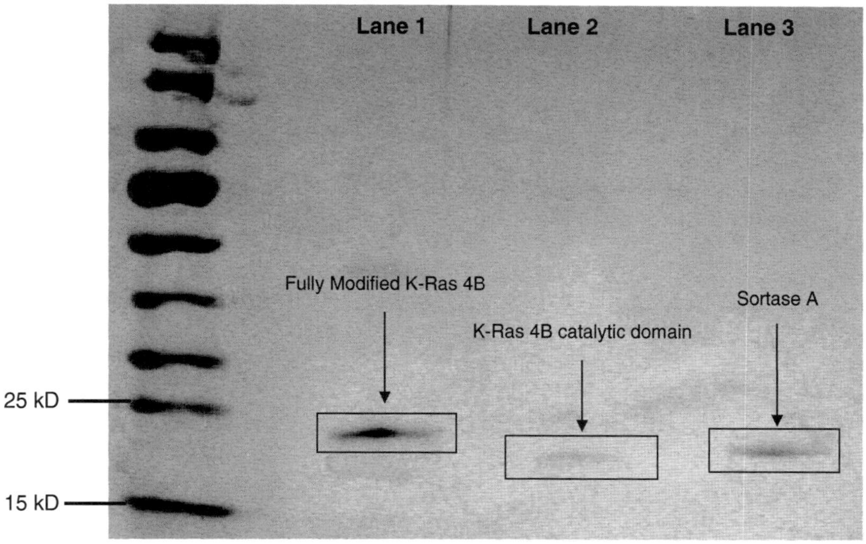


**Fig. 4** SDS-PAGE gel of purified fully modified K-Ras4B. The figure shows the fully modified product after components of the sortase reaction were removed using nickel affinity purification. *Lane 1* contains fully modified K-Ras 4B. *Lanes 2* and *3* are controls with unreacted K-Ras 4B catalytic domain and sortase A as labelled

Figure 4 shows the fully modified K-Ras 4B construct purified from the sortase-mediated ligation reaction using this protocol.

Figure 5 shows the mass spectrum of the fully modified K-Ras 4B present in Fig. 4 (lane 1).

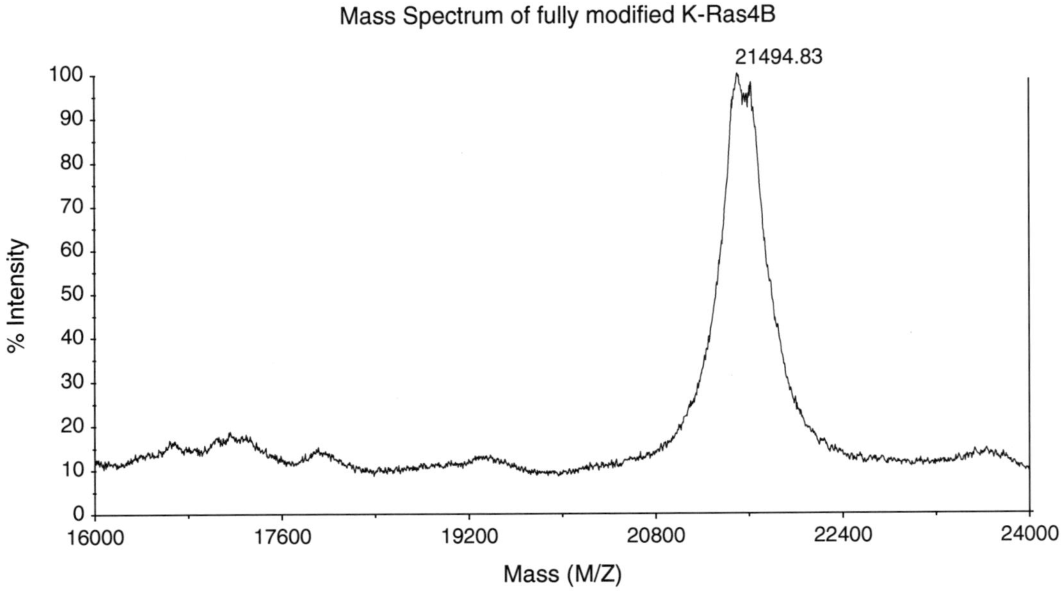


**Fig. 5** The mass spectrum of fully modified ligated K-Ras 4B in Tris–EDTA buffer

### 3.6 GTP Hydrolysis Assay

1. Prepare the reaction mixture for the sample: 10 μM fully modified K-Ras 4B, 0.2 mM MESG, 1 unit/mL purine nucleoside phosphorylase, 10 mM $(NH_4)_2SO_4$, 0.1 mM EDTA, and 0.5 mM GTP. Fresh GTP should be added to both sample and control mixtures at the same time.
2. Prepare the control mixture with the same components of the reaction mixture excluding modified K-Ras, with the remaining volume made up using the K-Ras storage buffer (Tris–EDTA buffer pH 9). It is beneficial to make two volumes of the control reaction to blank the UV spectrophotometer prior to following the reaction.
3. Once these reaction mixtures are made up, immediately transfer samples to UV cuvettes (use quartz if available) and measure absorbance of the sample mixture against the control at 360 nm for at least 600 min. GTP should naturally hydrolyze in both cuvettes, though a far greater rate should be observed in the presence of an active GTPase.

Figure 6 shows the spectrophotometric recording of K-Ras 4B modified using the sortase-mediated ligation reaction as well as a positive control using the catalytic domain of K-Ras 4B.

### 3.7 Raf-1 Binding Assay

1. Carry out the GTP or GDP loading of protein samples by incubating them with 10 mM EDTA and 0.1 mM GTPγS or 0.1 mM GDP, respectively. Incubate this mixture at room temperature for 1 h with agitation.

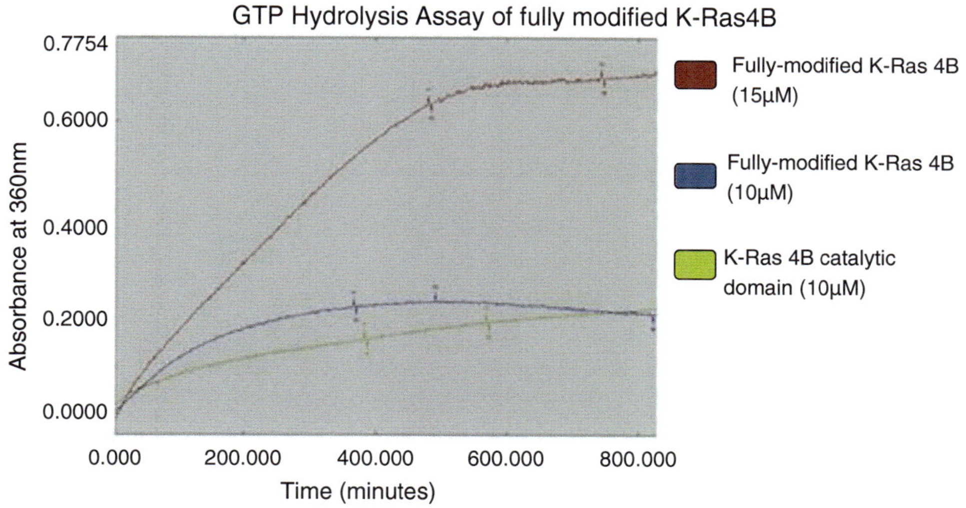


**Fig. 6** GTP hydrolysis assay of fully modified K-Ras4B. Absorbance at 360 nm reflecting the release of inorganic phosphate as GTP is hydrolyzed to GDP. Free inorganic phosphate is absorbed by MESG which increases absorbance at 360 nm. Shown in *yellow* is the degree of GTP hydrolysis with 10 μM K-Ras 4B catalytic domain in a 1 mL reaction. Shown in *blue* and *red* are GTP hydrolysis reactions with fully modified K-Ras 4B at 10 and 15 μM concentrations, respectively

2. In order to stop the reaction, add $MgCl_2$ to a final concentration of 50 mM and three times the volume of HEPES buffer. Also add Raf RBD agarose resin to the mixture at this step and incubate it for 45 min at room temperature.
3. Wash the beads with HEPES buffer at least two times (*see* **Note 13**).
4. Add 20 μL of HEPES buffer to the beads and 5 μL of SDS loading dye and heat the reaction mixture at 90 °C for 7–10 min.
5. Centrifuge this mixture at 2,500 × *g* for 3 min and then load the supernatants on an SDS-PAGE gel.
6. Perform a western blot on the gel and probe it with an anti K-Ras antibody.

Figure 7 shows a western blot performed on samples after carrying out the Ras activation assay.

### *3.8 SPR Membrane Binding Assay*

1. Apply the running buffer to the sensor chip containing phospholipids for equilibration (*see* **Note 14**).
2. To study the binding of modified and non-modified proteins, dissolve the protein samples to 2.4 μM in the running buffer.
3. Apply the protein samples on the sensor chip. Use a flow rate of 10 μL/min with a contact time of 240 s, a dissociation time of 240 s, and a regeneration time of 30 s.

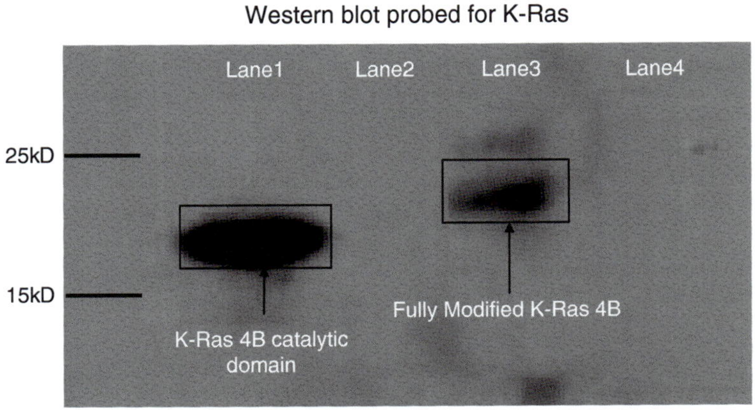


**Fig. 7** Western blot probed for K-Ras. The western blot was run on GTPγS-loaded K-Ras 4B catalytic domain (*Lane 1*), GDP-loaded K-Ras 4B catalytic domain (*Lane 2*), GTPγS-loaded fully modified K-Ras 4B (*Lane 3*), and GDP-loaded fully modified K-Ras 4B (*Lane 4*) after carrying out the Ras activation assay. Since the GDP-loaded forms of K-Ras 4B do not bind to Raf, no band is seen in *lanes 2* and *4*. Since fully modified K-Ras 4B has a higher molecular weight as compared to its catalytic domain, it runs higher than the catalytic domain of K-Ras 4B

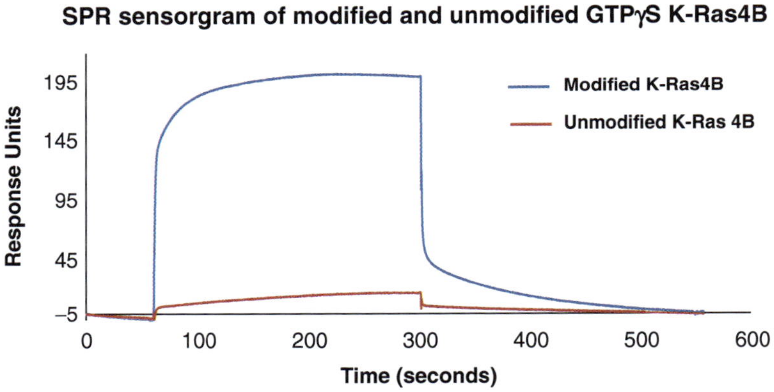


**Fig. 8** SPR sensorgram of modified and unmodified GTPgS K-Ras4B. Solutions of 2.4 μM fully modified GTPγS-loaded K-Ras 4B (*blue*) and 2.4 μM unmodified K-Ras4B (*red*) were applied to immobilized DPPC nanodiscs at a flow rate of 10 μL/min. Binding of the fully modified version of the protein leads to an increase in response units as compared to the unmodified protein

4. Use 2 M NaCl as the regeneration buffer to bring the response back to baseline before successive rounds of association and dissociation.
5. Use the affinity wizard to test the binding and dissociation of protein samples. Perform data analysis by plotting baseline corrected response units at equilibrium versus concentration of proteins for each cycle.

Figure 8 shows an SPR sensorgram comparing membrane binding abilities of the modified and unmodified K-Ras 4B.

## 4 Notes

1. The sortase recognition sequence is LPXTG where X is any amino acid with the exception of glycine or proline. Our constructs contain a threonine residue at this position.
2. Sterile SOC medium is packaged along with one-shot BL21AI cells from Invitrogen.
3. It is recommended to make the extraction buffers fresh every time.
4. 10 % glycerol is present in the His-nickel-binding buffer used to dialyze the K-Ras 4B catalytic domain but is not present in the His-nickel-binding buffer used to wash the nickel-binding resin. Glycerol serves as an osmolyte in the dialysis buffer and stabilizes the protein structure.
5. As imidazole is light sensitive, it is highly recommended that the elution and wash buffers in which it is present be stored in lightproof containers at 4 °C if they are to be kept for a prolonged period of time.
6. Purification can be done using an HPLC fitted with a His-nickel-binding resin cartridge and may yield more efficient purification. However, if this equipment is unavailable, good results can still be obtained using conventional gravity columns.
7. The sortase construct can be readily obtained from Addgene and transformed in standard competent *E. coli* cell lines [10].
8. GTP stocks must be fresh or stored at −80 °C and thawed to room temperature just prior to use. Once thawed, it must not be refrozen as GTP inherently hydrolyzes to GDP and inorganic phosphate making it unsuitable for hydrolysis assays.
9. Some K-Ras 4B may misfold during expression. Misfolded K-Ras 4B may have a higher nickel-binding affinity as compared to the correctly folded form, such that it will only elute at the highest concentrations of imidazole. If protein elution is observed only at low and high imidazole concentrations but not intermediates (100–200 mM), then the high imidazole fractions should be discarded as they are likely misfolded protein states.
10. The K-Ras 4B catalytic domain fractions purified for the sortase reactions should be dialyzed into Tris–EDTA buffer using dialysis cassettes and not buffer exchange columns. We have found that using these columns results in a significant loss of K-Ras 4B.

11. Some precipitation may be observed during the course of the reaction; if this becomes apparent, the precipitate should be separated using centrifugation with the reaction stalled by transferring to 4 °C. Then, proceed with the purification steps as described.
12. Protein yields can be relatively low for the reaction so it is recommended that silver staining be used to develop SDS-PAGE gels containing sortase reaction mixtures or purified product.
13. For washing the beads, centrifuge them at 2,500 ×*g* for 3 min and discard the supernatant. Then, add new buffer and repeat the process.
14. Run all experiments at 25 °C on a Biacore T100 SPR instrument. For making the phospholipid nanodiscs, follow the protocol by Nath et al. [11].

## Acknowledgement

This work was supported by the American Cancer Society grant RSG 09-057-01-GMC, the NIH grants R01CA135341 and R21HL118588. We thank Meredith Peters for her help in protein expression and purification.

### References

1. Cox AD, Der CJ (1992) Protein prenylation: more than just glue? Curr Opin Cell Biol 4:1008–1016
2. Casey PJ, Solski PA, Der CJ et al (1989) p21ras is modified by a farnesyl isoprenoid. Proc Natl Acad Sci U S A 86:8323–8327
3. Resh MD (2006) Trafficking and signaling by fatty-acylated and prenylated proteins. Nat Chem Biol 2:584–590
4. Hancock JF, Paterson H, Marshall CJ (1990) A polybasic domain or palmitoylation is required in addition to the CAAX motif to localize p21ras to the plasma membrane. Cell 63:133–139
5. Campbell-Burk SL, Carpenter JW (1995) Refolding and purification of Ras proteins. Methods Enzymol 255:3–13
6. Trueblood CE, Boyartchuk VL, Picologlou EA et al (2000) The CaaX proteases, Afc1p and Rce1p, have overlapping but distinct substrate specificities. Mol Cell Biol 20: 4381–4392
7. Clarke S, Vogel JP, Deschenes RJ et al (1988) Posttranslational modification of the Ha-ras oncogene protein: evidence for a third class of protein carboxyl methyltransferases. Proc Natl Acad Sci U S A 85:4643–4647
8. Chen YX, Koch S, Uhlenbrock K et al (2010) Synthesis of the Rheb and K-Ras4B GTPases. Angew Chem Int Ed Engl 49:6090–6095
9. Neidhardt FC, Bloch PL, Smith DF (1974) Culture medium for enterobacteria. J Bacteriol 119:736–747
10. Mao H (2004) A self-cleavable sortase fusion for one-step purification of free recombinant proteins. Protein Expr Purif 37:253–263
11. Nath A, Atkins WM, Sligar SG (2007) Applications of phospholipid bilayer nanodiscs in the study of membranes and membrane proteins. Biochemistry 46:2059–2069

# Chapter 3

# Metabolic Labeling of Ras with Tritiated Palmitate to Monitor Palmitoylation and Depalmitoylation

**Frederick D. Tsai, Joseph P. Wynne, Ian M. Ahearn, and Mark R. Philips**

## Abstract

Metabolic labeling with tritiated palmitate is a direct method for monitoring posttranslational modification of Ras proteins with this fatty acid. Advances in intensifying screens have allowed for the easy visualization of tritium without the need for extended exposure times. While more energetic radioisotopes are easier to visualize, the lack of commercial source and need for shielding make them more difficult to work with. Since radiolabeled palmitate is directly incorporated into Ras, its loss can be monitored by traditional pulse-chase experiments that cannot be accomplished with the method of acyl-exchange chemistry. As such, tritiated palmitate remains a readily accessible and direct method for monitoring the palmitoylation status of Ras proteins under a multitude of conditions.

**Key words** Ras, Palmitoylation, Pulse-chase, Tritium, Transcreen

## 1 Introduction

Ras proteins must associate with cellular membranes in order to regulate signaling events [1]. Ras undergoes a series of posttranslational modifications that result in a lipidated protein with affinity for phospholipid bilayers. A CAAX sequence at the C-terminus of Ras proteins signals for these modifications. The CAAX sequence of nascent Ras proteins is modified in three steps. First, a farnesyl lipid is added to the CAAX cysteine by farnesyltransferase. Next, the C-terminal AAX peptide is cleaved by Rce1. Finally, the newly C-terminal farnesyl cysteine is methyl esterified by Icmt [2]. CAAX processing produces relatively weak membrane affinity and is insufficient to properly localize Ras to the plasma membrane [3]. A second signal immediately upstream of the CAAX sequence affords added affinity. In the case of K-Ras4B, the second signal consists of a polybasic sequence that can interact with the negatively charged headgroups of the phospholipids in the inner leaflet of the plasma membrane. For all other Ras isoforms, including N-Ras, H-Ras, and K-Ras4A, the second signal consists of palmitoylation

Lorenza Trabalzini and Saverio Francesco Retta (eds.), *Ras Signaling: Methods and Protocols*,
Methods in Molecular Biology, vol. 1120, DOI 10.1007/978-1-62703-791-4_3, © Springer Science+Business Media, LLC 2014

of one or two cysteines in proximity to the farnesyl cysteine. Whereas CAAX processing is irreversible, the second signals can be modulated. In the case of the polybasic region of K-Ras4B, phosphorylation of a nearby serine partially neutralizes the positive charge [4]. For the palmitoylated Ras isoforms, the palmitates can be readily removed by hydrolysis under physiologic conditions [5, 6]. The dynamic nature of palmitoylation and its necessity for oncogenic function have made it an active area of investigation in Ras biology.

The most direct way to monitor protein palmitoylation is by metabolic labeling with palmitic acid that incorporates a radioisotope. Because tritiated [$^{3}$H]palmitate is commercially available, this has been the reagent of choice for more than three decades. Following metabolic labeling, Ras can be readily immunoprecipitated and analyzed by SDS-PAGE and the radiolabeled protein observed by fluorography using a scintillation reagent [7]. The challenge of working with [$^{3}$H]palmitate has been the low energy of the β particle that is emitted when tritium naturally decays into $^{3}$He. The low energy is insufficient to allow the β particle to pass out of the polyacrylamide gel or to penetrate the emulsion of X-ray film, which necessitates impregnating the gel with a fluor such as 2,5-diphenyloxazole in order to convert the β particles into photons and thereby allow for exposure of film. While this method reliably produces images of labeled proteins, it often requires long exposures of up to several weeks to attain a clear image. Alternatively, the gel can be cut into slices and digested to release the labeled proteins that can be quantified by scintillation counting [8]. This method affords direct quantification but is cumbersome and time-consuming. Some groups have solved the problem of low signal by labeling Ras with 16-[$^{125}$I] iodohexadecanoic acid (IC), an analog of palmitate [9, 10]. The gamma rays emitted upon decay of $^{125}$I are much more penetrating, allowing for direct autoradiography of dried gels with overnight exposures. However, the lack of a commercial source of 16-[$^{125}$I] IC makes its use much more complicated since it must be synthesized by the investigator. Moreover, organic synthesis involving sodium [$^{125}$I] iodide requires far more precautions than does working with tritium.

Recently, several new methods have been developed to analyze protein acylation without using radioisotopes. The most widely used of these is known as acyl-exchange chemistry. This method is applied ex vivo to lysates of cells or tissues. It entails a three-step procedure that includes blocking free thiols with *N*-ethylmaleimide (NEM), hydrolyzing thioester-linked lipids (e.g., palmitate) with hydroxylamine, and labeling the free sulfhydryls produced with a sulfhydryl-reactive reagent that is biotinylated, thereby allowing for purification and/or detection with streptavidin [11]. The great advantage of this method is the ability to assay the level of palmitoylation in samples that were not or could not be labeled with radioactive palmitate [12]. Perhaps the most powerful use of this method has been to survey global acylation of proteins in complex lysates of cells [13, 14]. However, the disadvantages of this method include the need to optimize each step to obtain reproducible signals and the inability to conduct pulse-chase experiments since the method is always ex vivo. Another recently developed method utilizes bioorthogonal chemical ligation, or "click chemistry." This two-step process involves metabolically labeling cells with alkynyl-analogues of palmitic acid; the labeled proteins are then reacted with azido-modified detection markers, such as fluorescent tags for visualization or biotin for visualization and/or purification [15]. The applicability of this method for examining palmitoylation in vivo, e.g., with pulse-chase experiments, and improvements in detection and use with newer generations of chemical reporters may result in click chemistry supplanting radioisotope labeling in the future [16].

In the meantime, we have taken advantage of the advances in X-ray film sensitivity and specialized intensifying screens for tritium to optimize conventional metabolic labeling with [$^{3}$H]palmitate. Our method consists of [$^{3}$H]palmitate labeling, immunoprecipitation, SDS-PAGE, transfer onto PVDF membranes, and high sensitivity fluorography using the Kodak BioMax Transcreen-LE intensifying screen and BioMax MS film. Our method allows for visualization of fluorograms in 1–3 days, depending on the Ras isoform being investigated. Moreover, since the exposed substrate is bound to PVDF membrane, protein levels can also be directly assessed by immunoblots quantified with a LiCor imaging system. This has allowed us to monitor changes in the palmitoylation status of H-Ras caused by peptidyl–prolyl isomerase inhibitors [17]. The method also allows for pulse-chase experiments that are crucial for determining how alterations to the steady-state palmitoylation of Ras are achieved. In this chapter, we outline our protocol for such assays.

## 2 Materials

### 2.1 Cell Culture

1. Labeling medium: Dulbecco's Modified Eagle Medium (DMEM) + 10 % dialyzed fetal bovine serum (FBS), 5 mM sodium pyruvate (to inhibit beta oxidation of palmitate), 3.6 mg/mL fatty acid-free bovine serum albumin (BSA). Make fresh prior to experiment. For 10 mL, weigh 36 mg of fatty acid-free BSA. Add 1 mL of dialyzed FBS, 0.5 mL of sodium pyruvate (from 100 mM stock), 8.5 mL of DMEM, and mix. Sterile-filter using 0.45 μm filter (*see* **Note 1**).
2. Chase medium: DMEM + 10 % dialyzed FBS, 5 mM sodium pyruvate, 3.6 mg/mL fatty acid-free BSA, 200 μM palmitic

acid. Prepare medium as indicated for labeling media, then add 200 μM palmitic acid (1:1,000 dilution from 200 mM stock in ethanol).

3. Palmitic acid, [9,10-$^{3}$H(N)]- (Perkin Elmer, Waltham, MA, USA) (tritiated).
4. Palmitic acid (Sigma Aldrich, St. Louis, MO, USA) (untritiated): make 200 mM stock in absolute ethanol.
5. Transfection reagent: Lipofectamine 2000 (Invitrogen).

### 2.2 Immunoprecipitation

1. RIPA buffer: 20 mM Tris–HCl, pH 7.5, 137 mM NaCl, 2 mM EDTA, 0.5 % sodium deoxycholate, 0.1 % SDS, 1 % Triton X-100, 10 % glycerol, protease inhibitors (Roche Complete, Mini, EDTA-free protease inhibitor cocktail tablets), 2 mM Pefabloc SC. Add protease inhibitors fresh before use; store at 4 °C.
2. Immunoprecipitation of Ras: Y13-259 pan-Ras conjugated antibody (Santa Cruz Biotechnology).

### 2.3 SDS-PAGE Analysis

1. SDS-PAGE running buffer: 0.3 % Tris (w/v), 1.44 % glycine (w/v), 0.1 % SDS (w/v).
2. Laemmli buffer (SDS-PAGE sample buffer): 120 mM Tris–HCl, pH 6.8, 4 % SDS (w/v), 20 % glycerol, 0.02 % bromophenol blue (w/v).
3. Dithiothreitol (DTT).
4. 14 % acrylamide gel.
5. Towbin buffer (Tris–glycine transfer buffer): 0.3 % Tris (w/v), 1.44 % glycine (w/v).
6. Semidry transfer: PVDF membrane, semidry transfer apparatus, Whatman filter paper.
7. Immunoblot detection of Ras: RAS10 (Calbiochem), anti-GFP (Invitrogen).
8. Secondary antibodies: LiCor IRDye 680/800 conjugated goat anti-mouse/rabbit IgG.
9. LiCor Odyssey Imaging System for development and quantification of immunoblot.

### 2.4 Tritium Fluorography

X-ray film cassette.

Kodak BioMax Transcreen-LE intensifying screen.

Kodak BioMax MS Film, Maximum Sensitivity—Radioisotope.

Kodak X-Omat or other film developer system.

## 3 Methods

### 3.1 Steady-State Monitoring of H-Ras Palmitoylation

1. Seed $10^6$ COS-1 cells in each well of a 6-well dish.
2. The next day, transfect cells with GFP-H-Ras using Lipofectamine 2000, according to the manufacturer's instructions.
3. To prepare tritiated labeling medium for six samples, in a sterile tissue culture hood, add 1.2 mCi of [$^3$H]palmitic acid to a 10 cm plate (for a final concentration of 0.2 mCi/mL). Allow ethanol solvent to evaporate completely (*see* **Note 2**).
4. Add 6 mL of labeling medium on top of the 10 cm plate containing the [$^3$H]palmitate and incubate at 37 °C for 5 min.
5. Change medium for the transfected cells using the prepared labeling medium containing [$^3$H]palmitate. If applicable, apply the appropriate concentration of pharmacologic agent, e.g., 1 μM FK506, 50 ng/mL cycloheximide, 500 nM rapamycin, or 25 μM 2-BP. Incubate at 37 °C overnight (*see* **Note 3**).
6. The next day, place cells on ice. Manually remove medium from cells and wash once with ice-cold PBS. Treat all media and washes as liquid radioactive waste.
7. Lyse cells in ice-cold RIPA buffer for 5 min on ice.
8. Collect cells using a cell scraper into microfuge tubes and clarify lysate by spinning at top speed (~20,000 × *g*) in a 4 °C pre-chilled tabletop centrifuge for 10 min.
9. Immunoprecipitate Ras from the clarified lysate by adding clarified lysate to a fresh tube containing 10 μL of Y13-259 beads (anti-pan-Ras). Rotate for 1 h at 4 °C.
10. Spin down beads at 2,500 × *g* for 5 min at 4 °C in a pre-chilled tabletop centrifuge.
11. Manually remove supernatant and wash the beads once with ice-cold RIPA buffer (*see* **Note 4**).
12. Manually remove supernatant and resuspend beads in 20 μL of 2× Laemmli sample buffer + 5 mM DTT (*see* **Note 5**).
13. Perform SDS-PAGE on samples, followed by transfer onto PVDF membrane using a semidry transfer apparatus (*see* **Note 6**).
14. Block membrane with 5 % milk in PBS for 20 min, then blot for Ras or GFP with the antibodies described above for 1 h. Wash 3× with PBS + 1 % Triton X-100 for 5 min each, then blot with LiCor IRDye 680/800 conjugated secondary antibodies for 1 h. Visualize and quantify Ras by LiCor Odyssey Imaging.
15. After immunoblotting, dry membranes overnight on Whatman paper.

16. Affix membranes with tape to the Kodak BioMax intensifying screen that is mounted in an X-ray film cassette.
17. In a darkroom, place Kodak BioMax MS film into the cassette containing the membrane. This must be done in complete darkness.
18. Place cassette in a −80 °C freezer for 24–48 h (the cold temperature increases the sensitivity of the film).
19. When ready to develop film, remove cassette from −80 °C freezer and allow to thaw to room temperature (roughly half an hour or more). Wipe away condensation from outside of cassette to prevent water from getting onto film or membrane.
20. In a darkroom (again, in complete darkness), develop film using a Kodak X-Omat film developer or alternative. If exposure is too low, additional pieces of film can be put down for longer exposures (2 weeks or less is usually sufficient).
21. The bands representing palmitoylated proteins on the film can be scanned using a flatbed scanner with a transillumination mode and their optical density quantified using ImageJ software (Fig. 1).

### 3.2 Pulse-Chase Depalmitoylation Assay

1. Seed $10^6$ COS-1 cells in each well of a 6-well dish.
2. The next day, transfect cells with GFP-H-Ras using Lipofectamine 2000, according to the manufacturer's instructions.
3. On the following day (post-transfection), prepare [$^3$H]palmitic acid labeling medium for the pulse. For six samples, in a sterile tissue culture hood, add 6 mCi of [$^3$H]palmitic acid to a 10 cm plate (for a final concentration of 1 mCi/mL). Allow ethanol solvent to evaporate completely before adding 6 mL of labeling medium on top of the [$^3$H]palmitate in the 10 cm plate. Incubate at 37 °C for 5 min (*see* **Note 7**).
4. To begin pulse, add 500 μL of labeling medium containing [$^3$H]palmitate to the well corresponding to the maximum chase timepoint (=60 min). If applicable, add pharmacologic agents. Incubate for 5 min (*see* **Note 8**).
5. Remove [$^3$H] medium and wash cells twice with room-temperature PBS. Add 1 mL of chase medium and incubate for 60 min.
6. 10 min after **step 3**, aspirate media from cells in the adjacent well and repeat **steps 2** and **3**, but incubate in chase medium for 45 min.
7. Repeat **step 4** for each successive well, decreasing length of chase by 15 min. Timepoints may be adjusted as desired (*see* **Note 9**).

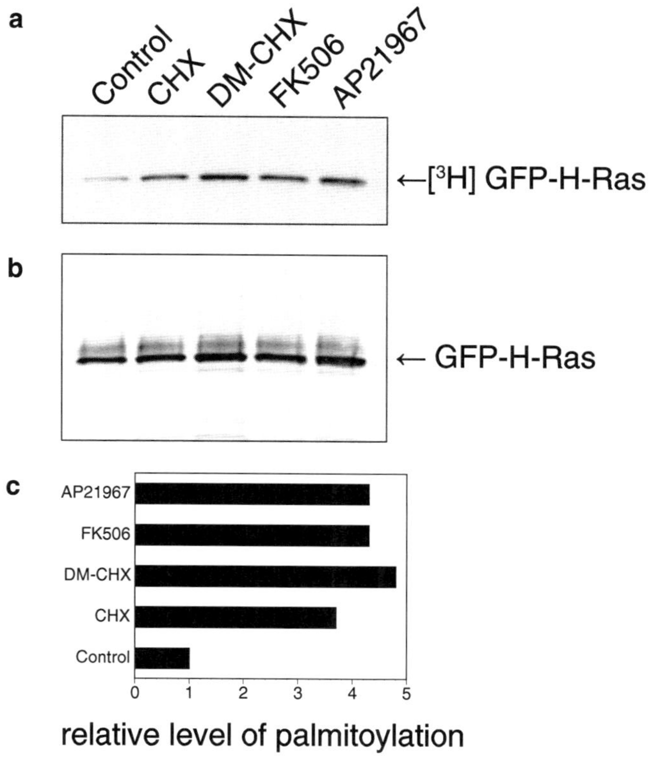


**Fig. 1** Steady-state metabolic labeling of GFP-H-Ras with various drug treatments. COS-1 cells were transfected with GFP-H-Ras and labeled with [$^3$H]palmitate while being treated either with DMSO (control), 50 ng/mL cyclohexamide (CHX), 10 μM DM-CHX, 1 μM FK506, or 2 μM AP21967. Cells were lysed the following day in RIPA buffer. Cell lysates were immunoprecipitated for Ras and subjected to SDS-PAGE analysis. (**a**) Fluorogram for [$^3$H]palmitate labeling, 1 day exposure. (**b**) Immunoblot for Ras. (**c**) [$^3$H]GFP-H-Ras bands were quantified using ImageJ software (for fluorogram) and LiCor software (for immunoblot) and are plotted as normalized ratio. Levels indicate amount of GFP-H-Ras incorporating [$^3$H]palmitate relative to total protein amount

8. Once all timepoints are complete, place cells on ice and manually remove medium. Wash once with ice-cold PBS and proceed to lyse samples for SDS-PAGE, Western analysis, and film exposure as described above starting from **step** 7 of steady-state labeling protocol (Subheading 3.1).

## 4 Notes

1. We have found that to avoid confusion between labeling and chase media, it is advantageous to use phenol red-free DMEM in making labeling medium.

2. Treat this 10 cm plate, pipette tips, and all other [$^{3}$H]palmitate-containing materials as solid waste. When working with radioisotopes, care must be taken to minimize both solid and liquid waste. All forms of waste must be carefully contained and monitored to prevent contamination of lab materials and surfaces. This is particularly important due to the long half-life of tritium (12.3 years). Monitoring potential tritium contamination should be performed by wipe testing and scintillation counting for all lab surfaces after every experiment and decontamination should be performed promptly, if necessary.
3. Treatment with pharmacologic agents must be titrated for both dose and duration to allow for sufficient pharmacologic activity while at the same time minimizing cell toxicity. For the pharmacologic agents used here to inhibit peptidyl–prolyl isomerization, overnight treatments are suitable for steady-state labeling, while much shorter treatments are used for pulse-chase labeling.
4. To facilitate removal of supernatant from beads, use flat-orifice gel loading tips.
5. Dithiothreitol (DTT) is used at 5 mM as reducing agent for SDS-PAGE sample preparation instead of β-mercaptoethanol to avoid cleavage of thioester linkages. Samples in Laemmli buffer may be stored at −80 °C prior to SDS-PAGE analysis.
6. PVDF affords a greater affinity for hydrophobic and lipidated proteins than nitrocellulose and, as such, is the preferred membrane to use; however, nitrocellulose membranes can sufficiently bind and be used to detect palmitoylated Ras species. A semidry transfer system is used to minimize production of liquid radioactive waste.
7. A greater concentration of [$^{3}$H]palmitate is required for pulse-chase analysis than for steady-state labeling. Here we increase the concentration fivefold.
8. Our pulse-chase assay utilizes a pulse of only 5 min. Previous efforts to monitor the latency of H-Ras palmitoylation used pulse intervals of 1 h or more [18], but these methods are confounded by the kinetics of palmitate metabolism. The rate of accumulation of exogenous palmitic acid in cells is controlled by saturable fatty acid transporters. Once unlabeled palmitic acid has gained access to the cytosol, it must be conjugated with Coenzyme A in order to compete for the labeling of proteins with the [$^{3}$H]palmitoyl Co-A that was produced during the pulse. Thus, an instantaneous chase using even a great excess of unlabeled palmitate is not possible. Further confounding the ability to achieve a rapid and complete chase is the fact that labeled fatty acids can be recycled [19] such that a pool of [$^{3}$H] palmitoyl Co-A can be produced even after [$^{3}$H]palmitate is

removed from the growth medium. In our experience, a characteristic increase in the incorporation of the label was routinely seen between 15 and 30 min following chase, necessitating both a shorter pulse and longer chase timepoints [17].

9. In our pulse-chase protocol, we perform timepoints of 15 min intervals, which corresponds to the time elapsed between treatment of each well (**steps 2** and **3**). Performing each chase timepoint counting backwards allows us to harvest cells at the end of each chase duration simultaneously, minimizing timing errors in our pulse-chase. The length of time between **steps 3** and **4** can be adjusted to allow for different chase intervals, e.g., 5 or 10 min differences, as required.

## References


1. Willumsen BM, Norris K, Papageorge AG et al (1984) Harvey murine sarcoma virus p21 ras protein: biological and biochemical significance of the cysteine nearest the carboxy terminus. EMBO J 3:2581–2585
2. Wright LP, Philips MR (2006) Thematic review series: lipid posttranslational modifications. CAAX modification and membrane targeting of Ras. J Lipid Res 47:883–891
3. Choy E, Chiu VK, Silletti J et al (1999) Endomembrane trafficking of ras: the CAAX motif targets proteins to the ER and Golgi. Cell 98:69–80
4. Bivona TG, Quatela SE, Bodemann BO et al (2006) PKC regulates a farnesyl-electrostatic switch on K-Ras that promotes its association with Bcl-XL on mitochondria and induces apoptosis. Mol Cell 21:481–493
5. Goodwin JS, Drake KR, Rogers C et al (2005) Depalmitoylated Ras traffics to and from the Golgi complex via a nonvesicular pathway. J Cell Biol 170:261–272
6. Smotrys JE, Linder ME (2004) Palmitoylation of intracellular signaling proteins: regulation and function. Annu Rev Biochem 73:559–587
7. Sefton BM, Trowbridge IS, Cooper JA et al (1982) The transforming proteins of Rous sarcoma virus, Harvey sarcoma virus and Abelson virus contain tightly bound lipid. Cell 31: 465–474
8. Hancock JF (1995) Prenylation and palmitoylation analysis. Methods Enzymol 55:237–245
9. Peseckis SM, Deichaite I, Resh MD (1993) Iodinated fatty acids as probes for myristate processing and function. Incorporation into pp 60v-src. J Biol Chem 268:5107–5114
10. Liang X, Nazarian A, Erdjument-Bromage H et al (2001) Heterogeneous fatty acylation of Src family kinases with polyunsaturated fatty acids regulates raft localization and signal transduction. J Biol Chem 276:30987–30994
11. Drisdel RC, Green WN (2004) Labeling and quantifying sites of protein palmitoylation. Biotechniques 36:276–285
12. Drisdel RC, Alexander JK, Sayeed A et al (2006) Assays of protein palmitoylation. Methods 40:127–134
13. Kang R, Wan J, Arstikaitis P et al (2008) Neural palmitoyl-proteomics reveals dynamic synaptic palmitoylation. Nature 456:904–909
14. Roth AF, Wan J, Bailey AO et al (2006) Global analysis of protein palmitoylation in yeast. Cell 125:1003–1013
15. Charron G, Zhang MM, Yount JS et al (2009) Robust fluorescent detection of protein fatty-acylation with chemical reporters. J Am Chem Soc 131:4967–4975
16. Yang YY, Ascano JM, Hang HC (2010) Bioorthogonal chemical reporters for monitoring protein acetylation. J Am Chem Soc 132:3640–3641
17. Ahearn IM, Tsai FD, Court H et al (2011) FKBP12 binds to acylated H-ras and promotes depalmitoylation. Mol Cell 41:173–185
18. Veit M, Ponimaskin E, Schmidt MF (2008) Analysis of S-acylation of proteins. Methods Mol Biol 446:163–182
19. Qanbar R, Bouvier M (2004) Determination of protein-bound palmitate turnover rates using a three-compartment model that formally incorporates [3H]palmitate recycling. Biochemistry 43:12275–12288

# Chapter 4

# Ubiquitination of Rac1 by Inhibitors of Apoptosis (IAPs)

## Tripat Kaur Oberoi-Khanuja and Krishnaraj Rajalingam

## Abstract

Ubiquitination of proteins has emerged as a vital posttranslational modification at the crux of numerous signalling pathways, regulating them in various ways. Most members of the small GTPase family including Ras and Rho proteins are regulated by GEFs, GAPs, and RhoGDIs that modulate their cycling between the active and inactive states. Ubiquitination has added another layer to the regulation of small GTPases. Recently, we have uncovered that inhibitors of apoptosis (IAPs) function as direct E3 ubiquitin ligases for Rho GTPase Rac1 and target it for proteasomal degradation. Here, we describe in vitro and in vivo ubiquitination assays for detecting the conjugation of ubiquitin to Rac1 by XIAP and cIAP1.

**Key words** IAP, XIAP, cIAP1, In vitro ubiquitination, In vivo ubiquitination, Endogenous ubiquitination, Ni-NTA-His ubiquitination assay, RhoGTPase, Rac1

## 1 Introduction

Ubiquitination is one of the most versatile forms of posttranslational modifications, where a small, 8 kDa ubiquitin moiety is covalently conjugated to substrate proteins in several combinations of monomers and chains. This might lead to a variety of consequences including proteasomal degradation, altered subcellular localization, and/or assemblage of protein complexes [1]. Ubiquitination is a reversible modification as ubiquitin conjugated to proteins can be cleaved off by ubiquitin isopeptidases called deubiquitinases (DUBs). Proteins conjugated with K48- and K11-linked polyubiquitin chains are often targeted to proteasomes for their degradation [2].

RhoGTPases are small evolutionarily conserved proteins predominantly known for their crucial role in the regulation of actin cytoskeleton. Rac1 is a ubiquitously expressed, well-studied Rho GTPase, known to control numerous cellular processes ranging from cell polarity, cell adhesion, migration to proliferation and phagocytosis [3]. Previous studies have shown that bacterial toxins deamidate Rac1 leading to its persistent activation and proteasomal degradation. However, the molecular machinery driving Rac1

Lorenza Trabalzini and Saverio Francesco Retta (eds.), *Ras Signaling: Methods and Protocols*,
Methods in Molecular Biology, vol. 1120, DOI 10.1007/978-1-62703-791-4_4, © Springer Science+Business Media, LLC 2014

ubiquitination and proteasomal degradation was not known for a long time. Recent studies have revealed that inhibitors of apoptosis (IAPs) and HACE1 can function as direct E3 ubiquitin ligases of Rac1 targeting it for proteasomal degradation [4, 5]. IAPs are RING domain containing E3 ubiquitin ligases and they bind to Rac1 in a nucleotide-independent manner [6]. Loss of XIAP or cIAP1 led to an increase in basal Rac1 levels. Further, loss of XIAP prevents the degradation of Rac1 upon cytotoxic necrotizing factor 1 (CNF1) treatment or by RhoGDI1 depletion. IAPs directly conjugate ubiquitin to lysine 147 of Rac1 both in vitro and in vivo. Here, we present step-by-step protocols for the detection of Rac1 polyubiquitination by IAPs at endogenous levels in mammalian cells, using 293T overexpression system and Ni-NTA pulldowns, and in vitro by employing bacterially purified components.

## 2 Materials

### 2.1 In Vitro Ubiquitination Assay

1. Ubiquitination assay buffer (1×): 50 mM Tris–HCl, pH 7.5, 100 mM NaCl, 2.5 mM $MgCl_2$, 1 mM DTT (*see* **Note 1**). To prepare 10× ubiquitination buffer, weigh 6.057 g of Tris, 5.844 g of NaCl, and 238 mg of $MgCl_2$ and transfer to a glass bottle. Add water to a volume of 90 ml. Mix and adjust to pH 7.5 with HCl. Make up the volume to 100 ml with water. Aliquot it into 1.5 ml Eppendorf tubes and freeze at −20 °C. Make a 50 mM stock solution of dithiothreitol (DTT) by dissolving 771.25 mg of DTT in 100 ml of water. Make 1.5 ml aliquots and store at −20 °C for extended periods of time. Always add fresh DTT to 1 mM final concentration (1 μl of 50 mM DTT for 50 μl of reaction mix).
2. Inorganic pyrophosphatase: Prepare stocks at the concentration of 100 U/ml in water. Make 1 ml aliquots and store at −20 °C. Final working concentration is 1 U (10 μl) in 50 μl reaction mix.
3. $Mg^{2+}$/ATP: Use 2.5 μl (0.1 M stock) in 50 μl reaction mix. Final working concentration is 5 mM.
4. E1 enzyme: Use 2.5 μl (stock concentration: 5 μM) in 50 μl reaction mix. Final working concentration is 250 nM.
5. E2 enzyme: Use 2.5 μl (stock concentration: 50 μM stock) in 50 μl reaction mix. Final working concentration is 2.5 μM.
6. $His_6$-ubiquitin: Dissolve powdered ubiquitin in 1× ubiquitination buffer to obtain a 2.083 mM solution (20 μg/μl). Make 10 μl aliquots and store at −20 °C. Use 1 μl of this ubiquitin stock solution for 50 μl reaction (final working concentration is 41.6 μM or 0.4 μg/μl).
7. Recombinant human cIAP1 and XIAP: Use cIAP1 and XIAP proteins at 100 nM final concentration (stock concentration

varies according to the batch). Use bacterially purified cIAP1 and XIAP.

8. Recombinant human Rac1Q61L: Purify using pGEX-2T plasmid (*see* **Note 2**) and remove the GST tag using thrombin (*see* **Note 3**). Final concentration of Rac1 Q61L used in the reaction is approximately 3 μM.
9. 5× Laemmli buffer: 250 mM Tris–HCl, pH 6.8, 5 % β-mercaptoethanol, 10 % SDS, 30 % glycerol, 0.02 % bromophenol blue. In a graduated cylinder, add 12.5 ml of 2 M Tris–HCl, pH 6.8, 10 g of SDS, 5 ml of β-mercaptoethanol, 20 mg of bromophenol blue, and water to 50 ml while stirring continuously to dissolve SDS. Finally, add 30 ml of glycerol and make the final volume to 100 ml with water.

### 2.2 Ni-NTA-His *In vivo* Ubiquitination Assay

1. 293T cells: Culture cells in DMEM media supplemented with 10 % FCS and 0.2 % penicillin (100 U/ml)/streptomycin (100 μg/ml).
2. GeneJuice® transfection reagent (Merck Millipore): Perform transfections using GeneJuice® transfection reagent in accordance with manufacturer's instructions. Other plasmid transfection reagents can also be employed.
3. Plasmids: pRK5 myc-Rac1 Q61L (active Rac1 mutant), pRK5 myc-Rac1 Q61L K147R (Rac1 ubiquitination mutant), CMV Flag-cIAP1, CMV Flag-cIAP1 H588A (cIAP1 RING mutant), pCS2 His-ubiquitin (*see* **Note 4**).
4. MG132: Make 10 mM stock solution of proteasomal inhibitor, MG132 (solubilize in DMSO). Prepare 1 ml aliquots and store at −20 °C. Use 1 μl of MG132 stock solution per ml of media in transfected 10 cm dishes (final working concentration: 10 μM) (*see* **Note 5**).
5. Ni-NTA resin: Use 50 μl of 50 % Ni-NTA resin per transfection.
6. 1× Phosphate buffer saline (PBS): 137 mM NaCl, 2.7 mM KCl, 10 mM $Na_2HPO_4$, 2 mM $KH_2PO_4$, pH 7.4. Dissolve the following in 800 ml of distilled $H_2O$: 8 g of NaCl, 10.2 g of KCl, 1.44 g of $Na_2HPO_4$, 0.24 g of $KH_2PO_4$. Adjust pH to 7.4 with HCl. Adjust volume to 1 l with additional distilled $H_2O$. Sterilize by autoclaving and refrigerate at 4 °C.
7. Buffer A: 6 M guanidine HCl, 0.1 M $Na_2HPO_4/NaH_2PO_4$, 10 mM imidazole, pH 8.0. Weigh 143.3 g of guanidine HCl and transfer to a glass bottle. Add 25 ml of 1 M $Na_2HPO_4/NaH_2PO_4$ and 2.5 ml of 1 M imidazole. Dissolve in 200 ml of water and adjust the pH to 8.0 with NaOH. Make up the final volume to 250 ml with water (*see* **Note 6**).
8. Buffer TI: 25 mM Tris–HCl, pH 6.8, 20 mM imidazole. Add 5 ml of 1 M Tris–HCl, pH 6.8, and 4 ml of 1 M imidazole to a glass bottle and make up volume to 200 ml with water.

9. Buffer A/TI: Add 1 volume of Buffer A to 3 volumes of Buffer TI.
10. 2× Laemmli–imidazole: 100 mM Tris–HCl, pH 6.8, 2 % β-mercaptoethanol, 4 % SDS, 12 % glycerol, 0.02 % bromophenol blue, 0.2 M imidazole. In a graduated cylinder, add 5 ml of 2 M Tris–HCl, pH 6.8, 4 g of SDS, 2 ml of β-mercaptoethanol, 20 mg of bromophenol blue, and water to 50 ml while stirring continuously to dissolve SDS. Finally add 30 ml of glycerol and make up the final volume to 100 ml with water. Mix 200 μl of this 2× Laemmli buffer with 20 μl of 2 M imidazole (*see* **Note 6**).

### 2.3 Endogenous in vivo Ubiquitination Assay

1. HeLa cells: Culture cells in RPMI media supplemented with 10 % FCS and 0.2 % penicillin (100 U/ml)/streptomycin (100 μg/ml).
2. siRNAs: (a) XIAP3′UTR siRNA: Target sequence: CTGAC TGATCTAATTGTATTA (Qiagen), (b) RhoGDI1 siRNA: 5′-UCAAUCUUGACGCCUUUCCTT-3′ (Sigma Genosys), and (c) Control siRNA: Catalog No. D-001210-01-20 (Dharmacon). Transfect all siRNAs at a final concentration of 60 nM each.
3. HiPerFect transfection reagent: Perform transfections using HiPerFect transfection reagent in accordance with manufacturer's instructions. Other siRNA transfection reagents can also be used.
4. MG132: Make 10 mM stock solution of proteasomal inhibitor, MG132 (solubilize in DMSO). Prepare 1 ml aliquots and store at −20 °C. Use 1 μl of MG132 stock solution per ml of media in transfected 10 cm dishes (final working concentration: 10 μM) (*see* **Note 5**).
5. 1× Phosphate buffer saline (PBS): *See* Subheading 2.2, **item 6**.
6. Anti-Rac1 antibodies: Use 15 μl of anti-Rac1 rabbit polyclonal antibody (C11; Santa Cruz) for immunoprecipitation per sample and 1:500 dilution of anti-Rac1 mouse monoclonal antibody (BD Biosciences, #610651) for Rac1 detection by Western blotting (*see* **Note 7**).
7. Agarose A and G beads: Use 30 μl each of agarose A and G beads per sample.
8. RIPA buffer: 50 mM Tris–HCl, pH 7.5, 250 mM NaCl, 10 % glycerol, 1 % Triton X-100, 1.5 mM $MgCl_2$, 25 mM NaF, 1 mM PMSF, 1 mM $Na_3VO_4$, 0.01 % β-mercaptoethanol, and 1× Protease Inhibitor Cocktail (Set I; Calbiochem). Weigh 3.03 g of Tris and 7.3 g of NaCl and transfer them to a glass bottle. Add 300 ml of water and adjust the pH to 7.5. Add 50 ml of glycerol and 5 ml of Triton X-100 with constant stirring. Finally make up the volume to 500 ml with water. Store at 4 °C. All the other components of RIPA buffer are freshly added (*see* **Note 8**).

## 3 Methods

Prepare all solutions in sterile, deionized/distilled water and with analytical grade reagents. Prepare all cell culture reagents under sterile conditions, autoclave, or sterile filter. Store all reagents at room temperature unless otherwise stated. Aliquot all proteins and enzymes and snap freeze, and when required, thaw on ice.

### 3.1 In Vitro Ubiquitination Assay

1. Each in vitro ubiquitination reaction requires three negative controls: first, ubiquitination mix without the E1 enzyme (first enzyme of the ubiquitination cascade); second, ubiquitination mix without the E3 ligase (enzyme that executes substrate ubiquitination; cIAP1 or XIAP in this case); and third, ubiquitination mix without $Mg^{2+}$/ATP (essential for covalent catalysis). Therefore, label the Eppendorf tubes accordingly (*see* **Note 9**).
2. Thaw all the components of ubiquitination reaction on ice. Make a table for the volume of components needed for each reaction according to the final concentration required in 50 μl. For example, in this case, 5 μl of 10× ubiquitination buffer, 2.5 μl of 20× $Mg^{2+}$/ATP solution, 10 μl of 100 U/ml stock (final concentration 1 U), and so forth. Make up the final volume with sterile double-distilled water (DD water). An example of a typical ubiquitination mix with its components is shown in Table 1.

**Table 1**
**Ubiquitination assay mix used for in vitro ubiquitination**

| | 1 | 2 | 3 | 4 | 5 | 6 | 7 | 8 | 9 | 10 |
|---|---|---|---|---|---|---|---|---|---|---|
| 10× Ubiquitination buffer | 5 | 5 | 5 | 5 | 5 | 5 | 5 | 5 | 5 | 5 |
| Inorganic pyrophosphatase | 10 | 10 | 10 | 10 | 10 | 10 | 10 | 10 | 10 | 10 |
| Dithiothreitol (DTT) | 1 | 1 | 1 | 1 | 1 | 1 | 1 | 1 | 1 | 1 |
| $Mg^{2+}$/ATP | 2.5 | 2.5 | – | 2.5 | 2.5 | 2.5 | 2.5 | – | 2.5 | 2.5 |
| E1 | 2.5 | 2.5 | 2.5 | – | 2.5 | 2.5 | 2.5 | 2.5 | – | 2.5 |
| UbcH5a (E2) | 2.5 | 2.5 | 2.5 | 2.5 | 2.5 | 2.5 | 2.5 | 2.5 | 2.5 | 2.5 |
| XIAP (E3) | – | 0.3 | 0.3 | 0.3 | 0.3 | – | – | – | – | – |
| cIAP1 (E3) | – | – | – | – | – | – | 1 | 1 | 1 | 1 |
| $His_6$-ubiquitin | 1 | 1 | 1 | 1 | 1 | 1 | 1 | 1 | 1 | 1 |
| Rac1 Q61L | 5 | 5 | 5 | 5 | – | 5 | 5 | 5 | 5 | – |
| DD water | 11.5 | 11.2 | 13.7 | 13.7 | 16.2 | 11.5 | 10.5 | 13.0 | 13.0 | 15.5 |

This mix was used to generate Fig. 1. All the components listed here are added in μl amounts, and the final volume is made up to 50 μl with water. The conditions used for performing this experiment are as follows: (1) control reaction without an IAP, (2) Rac1 ubiquitination with XIAP, (3) control reaction without $Mg^{2+}$/ATP, (4) control reaction without E1, (5) control reaction without Rac1 Q61L (to detect XIAP autoubiquitination), (6) control reaction without cIAP1, (7) Rac1 ubiquitination with cIAP1, (8) control reaction without $Mg^{2+}$/ATP, (9) control reaction without E1, and (10) control reaction without Rac1 Q61L (cIAP1 autoubiquitination)

3. After adding all the components in the reaction mix, vortex briefly to mix them and then centrifuge in a microcentrifuge at the maximum speed for a few seconds to collect liquid from the sides of the tube.
4. Incubate the reaction mix for 30 min to 1 h at 37 °C in a heating block (*see* **Note 10**). After incubation, briefly centrifuge again as above for a few seconds to collect the liquid condensed in the cap.
5. Add 20 μl of 5× Laemmli buffer and vortex to uniformly mix the solution. Heat the samples at 100 °C for 2 min (*see* **Note 11**) and then load on to a 10 % SDS-PAGE gel to separate the proteins. Then, proceed further for Western blotting and detection using anti-Rac1 antibody (*see* **Note 7**) (Fig. 1).

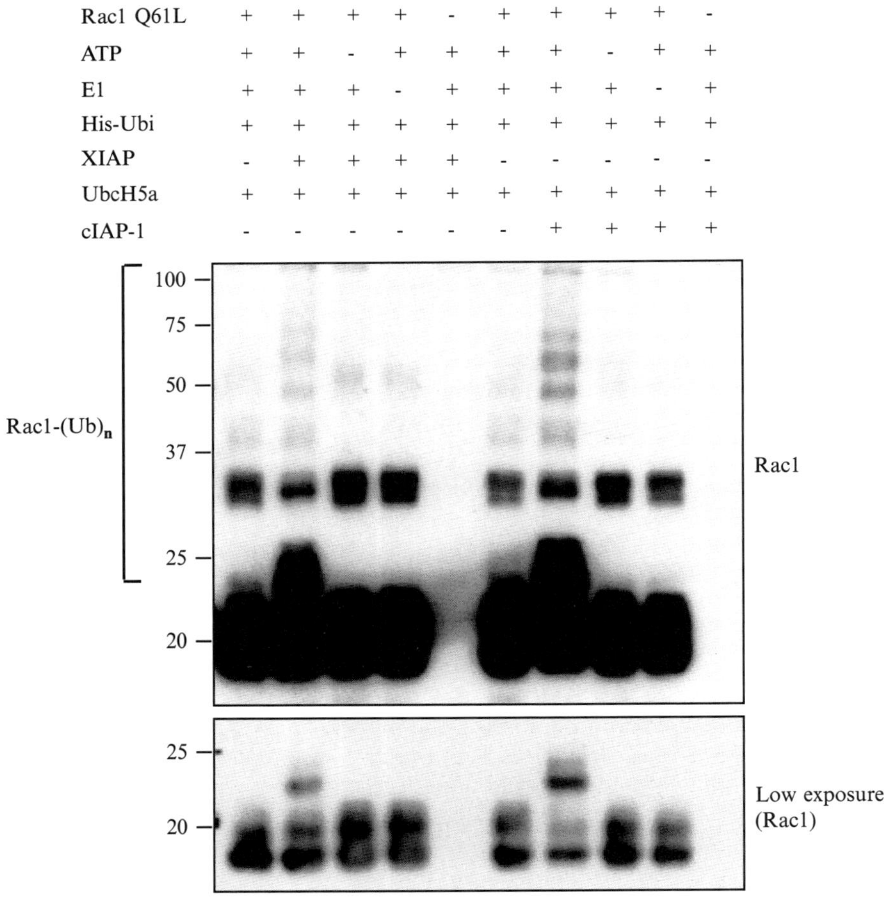


**Fig. 1** Direct ubiquitination of Rac1 by XIAP and cIAP1. Rac1Q61L was purified from bacteria and the in vitro ubiquitination was performed as mentioned in the methods section. The samples were resolved by SDS-PAGE and the conjugation of polyubiquitin chains to Rac1 was monitored by Western blots with Rac1 antibody. Two negative controls were included where E1 or Mg-ATP is excluded from the reaction mix to prevent ubiquitination

### 3.2 Ni-NTA-His Ubiquitination Assay

1. On the first day, seed 293T cells in 10 cm dishes and 24 h later (second day), transfect the cells with different plasmids (around 50 % confluent culture). There are generally four different conditions per experiment. First, Rac1 Q61L with ubiquitin and empty vector; second, Rac1 Q61L with ubiquitin and cIAP1; third, Rac1 Q61L with ubiquitin and cIAP1 H588A; and finally, Rac1 Q61L K147R with ubiquitin and cIAP1. Perform transfections using GeneJuice® reagent in accordance with manufacturer's instructions (*see* **Note 4**).
2. 24 h post-transfection (third day), exchange culture media in transfected cells.
3. On the fourth day, lyse cells for Ni-NTA pulldown. 6 h prior to lysis, treat cells with MG132 at a final concentration of 10 μM (1 μl of stock solution/ml of media).
4. For each transfection, 50 μl of 50 % Ni-NTA resin is needed. For four transfections, transfer 2 × 150 μl of resin into two Eppendorf tubes. Centrifuge at high speed (max. speed of microcentrifuge) for 7 s at room temperature. Remove supernatant and wash each tube of resin thrice with 1 ml of Buffer A. Resuspend both samples in 150 μl of Buffer A and pool in one tube (*see* **Note 12**).
5. 6 h after MG132 treatment, harvest the cells for pulldown. Remove the media from cells and wash them once with 10 ml of ice-cold PBS. Aspirate the liquid completely (*see* **Note 13**).
6. Add 1 ml of ice-cold PBS to each dish, scrap the cells, and collect them in a labelled tube. Keep them on ice.
7. Centrifuge the cells at high speed for 45 s and aspirate the supernatant. Carefully resuspend the cells in 500 μl of PBS by tapping the tube. Transfer 20 μl of each suspension to an Eppendorf tube containing 50 μl of 2× Laemmli. After vortexing, boil the samples for 10 min before freezing them at −20 °C. Use these samples for testing the expression of the different plasmids.
8. Centrifuge the remaining cells in microfuge at the maximum speed for 45 s and aspirate PBS carefully so as not to remove any cells. Resuspend cells in 1 ml of Buffer A. Mix carefully by pipetting (*see* **Note 14**).
9. Sonicate the suspension on ice (50 % power, 8 pulses of 1 s each). Take care in washing the sonicator tip while changing the samples.
10. Add 50 μl of washed and equilibrated Ni-NTA agarose suspension to each tube. Carefully resuspend uniformly before adding to the tubes. Put the samples on a rotator for 3 h at room temperature. The rest of the procedure is carried out at room temperature.

11. Centrifuge the tubes containing Ni-NTA agarose for 10 s at high speed and discard the supernatant carefully (*see* **Note 15**).
12. Add 1 ml of Buffer A to each sample tube and rotate the tubes back and forth five to six times. Centrifuge and repeat the washing with Buffer A.
13. Repeat the washing twice with Buffer A/TI and finally wash once with Buffer TI. Aspirate the supernatant carefully.
14. Remove all the liquid from the Ni-NTA resin using an insulin syringe. The resin turns white.
15. Resuspend resin in 50 μl of 2× Laemmli/imidazole. Vortex to mix. Boil the samples for 10 min and load them on a 10 % SDS-PAGE gel. Analyze 20 μl of lysates collected before Ni-NTA pulldown and 25 μl of post-pulldown samples by SDS-PAGE and Western blotting.

### 3.3 Endogenous Rac1-Ubiquitination Assay

1. Seed HeLa cells in 12-well plates for transfection. Rac1 ubiquitination is detected at endogenous levels upon RhoGDI1 depletion [7]. So the following conditions are tested. First, cells transfected with scrambled control siRNA; second, cells transfected with RhoGDI1 siRNAs to observe Rac1 degradation; and third, cells transfected with XIAP and RhoGDI1 siRNAs together to prevent Rac1 degradation by depletion of the E3 ligase (XIAP). Therefore, transfect six wells of a 12-well plate per condition (total 18 wells transfected) (*see* **Note 16**).
2. 24 h post-transfection, trypsinize the transfected cells (6 wells/condition), resuspend in fresh media, and pool them into three separate falcon tubes. Make up the volume of each tube to 10 ml with fresh RPMI media and seed the pooled cells onto 10 cm dishes (1 dish/condition; total of 3 dishes).
3. After 2 days post-transfection, lyse the cells for immunoprecipitation. Six hours prior to lysis, treat the cells with MG132 at a final concentration of 10 μM (1 μl of stock solution/ml of media) (*see* **Note 5**).
4. Prepare 3 ml of RIPA buffer; 1 ml per sample is required for lysis. Wash the cells with 10 ml of ice-cold 1× PBS and aspirate it completely (*see* **Note 13**).
5. Add 1 ml of RIPA to each of the 10 cm dishes, scrap and collect the cells in labelled Eppendorf tubes. Incubate the tubes for 30 min on ice with vortexing every 10 min (*see* **Note 17**).
6. Centrifuge the lysates at top speed (20,000 ×*g*), at 4 °C for 20 min, and collect the supernatants in labelled Eppendorf tubes. From these supernatants, transfer 30 μl to tubes containing 30 μl of 5× Laemmli, mix thoroughly, and freeze at −20 °C as control lysates.

7. To the rest of the lysates, add 15 μl of Rac1 antibody and incubate them at 4 °C, in Eppendorf tubes overnight on a rotator (*see* **Note 7**).
8. Next day, add 30 μl each of agarose A and G beads to the lysates and incubate at 4 °C for 2 h on a rotator.
9. Centrifuge the lysates at 650 ×*g* for 1 min in a table top centrifuge. Discard the supernatants.
10. Wash the beads with RIPA buffer without the inhibitors. Aspirate supernatants carefully. Repeat the washing step twice.
11. Remove all the liquid from the beads using an insulin syringe. The beads turn white.
12. Add 50 μl of 5× Laemmli to the beads. Vortex to mix. Boil the samples for 10 min and load on a 10 % SDS-PAGE gel. Analyze 15 μl of control lysates and 20 μl of immunoprecipitated samples by SDS-PAGE and Western blotting.

## 4 Notes

1. It is more convenient to have a 10× ubiquitination buffer, as the volumes of enzymes and proteins to be added for ubiquitination reaction seem to vary from batch to batch due to differences in their concentrations. So it is easier to make up the final volume to 50 μl with water.
2. Rac1 Q61L is used instead of Rac1 WT because it has been shown in various instances, and we have also observed that active Rac1 (Rac1 Q61L or Rac1-GTP) is ubiquitinated much better than either Rac1 WT or Rac1 T17N (dominant negative) [4]. Further, ubiquitination reaction can also be performed using Rac1-GDP and Rac1-GTPγS proteins. To the purified GST-Rac1 protein in 100 μl of buffer, add 2 μl of 0.5 M EDTA, pH 8.0 (final concentration 10 mM), and vortex shortly. Add 1 μl of 10 mM GTPγS (final concentration 0.1 mM) or 1 μl of 100 mM GDP (final concentration 1 mM) and vortex shortly. Incubate the mixture at 30 °C for 15 min with constant agitation. Terminate the reaction by placing the sample on ice and adding 6.5 μl of 1 M $MgCl_2$ (final concentration 60 mM) and vortex the sample. GSH beads can be used to isolate GST-Rac1-GDP or GST-Rac1-GTPγS from the sample, from which GST can be cleaved off as described in **Note 3** and used further for ubiquitination reaction.
3. Perform thrombin cleavage using cleavage buffer: 20 mM Tris–HCl, pH 8.4, 150 mM NaCl, 2.5 mM $CaCl_2$, 1 mM DTT, 10 % glycerol, and 1 % Triton X-100. Amount of thrombin used for cleavage as well as temperature for thrombin

cleavage varies with proteins used. In the present case, 1 l bacterial culture (BL21 CodonPlus) of GST-Rac1 was purified using 1 ml of GSH beads. 10 μl of GST-Rac1-loaded beads is cleaved using 100 μl of cleavage buffer and 10 U of thrombin, rotated overnight at room temperature. Protein is eluted and 100 μM PMSF is added to inactivate thrombin.

4. pRK5 myc-Rac1 Q61L plasmid is obtained from Addgene, and the K147R mutant is generated by site-directed mutagenesis. CMV Flag-cIAP1 plasmid is a kind gift from Dr. Domagoj Vucic, and the H588A mutant is generated by site-directed mutagenesis. pCS2 His-ubiquitin plasmid is a kind gift from Prof. Michael Rape. K147 is used as a negative control as it is the major ubiquitination site on Rac1, and mutation of this site impedes ubiquitination of Rac1. Further, another negative control is the use of cIAP1 mutant H588A. This mutant disrupts the activity of cIAP1 RING domain and, hence, prevents Rac1 ubiquitination.
5. It is very important to use a proteasomal inhibitor like MG132 in all in vivo ubiquitination assays meant for detecting degradative ubiquitination. If MG132 is not used in these assays, all the proteins getting ubiquitinated would be recognized by the proteasomal machinery and would be degraded. Thus, it would be quite difficult to detect polyubiquitin-conjugated proteins by Western blotting in the absence of proteasomal inhibitors.
6. Imidazole is included in the binding buffers to minimize the binding of untagged or contaminating proteins and increase the purity of the His-tagged protein. Amount of imidazole can be varied depending on the amount of contaminants to target protein ratio. Imidazole is also added to the elution buffers as it competes with the target protein for binding to the beads and thus promotes elusion. For this reason, it is also added to Laemmli buffer.
7. All primary antibodies used for detection in case of Rac1 ubiquitination are prepared in TBST containing 3 % BSA. Ubiquitin (dilution 1:500; Invitrogen), Rac1 (dilution 1:500; BD Transduction Laboratories), myc (dilution 1:500; 9E10, Santa Cruz Biotechnology), Flag (dilution 1:8,000, Sigma), XIAP (dilution 1:500; BD Transduction Laboratories), Actin (dilution 1:1,000; Sigma), and RhoGDI1 (dilution 1:1,000; A 20, Santa Cruz Biotechnology). Rac1 is immunoprecipitated using rabbit polyclonal Rac1 antibody (Santa Cruz), and mouse monoclonal Rac1 antibody is used for Western blot detection as it prevents the detection of heavy chain and light chain bands on the blots. This is important as the size of Rac1 coincides with the size of antibody light chain bands.

8. The components freshly added to RIPA buffer just before lysis are PMSF (stock solution 333 mM), $Na_2VO_4$ (stock solution 100 mM), NaF (stock solution 1 M), $MgCl_2$ (stock solution 1 M), 0.01 % β-mercaptoethanol, and Protease Inhibitor Cocktail (stock solution 100×; Protease Inhibitor Cocktail Set I, Calbiochem). For example, to prepare 6 ml of RIPA buffer, amounts of freshly added components are as follows: 0.6 μl β-mercaptoethanol, 18 μl PMSF, 9 μl $MgCl_2$, 150 μl NaF, 60 μl $Na_2VO_4$, and 60 μl of protease inhibitors.
9. Using IAPs as E3 ligases, we can have an internal positive control as IAPs can autoubiquitinate themselves utilizing the same E2 (UbcH5a) that is required for ubiquitination of Rac1. So we have one sample with cIAP1 or XIAP with all components of the ubiquitination reaction, without any Rac1. Autoubiquitination of IAPs can be checked using specific antibodies against them (cIAP1, R&D Systems, and XIAP, BD Biosciences). This serves to determine whether the ubiquitination reaction works per se in case we do not observe any Rac1 ubiquitination. Also, it can serve as a negative control when the blot is probed with Rac1 antibody to ensure that the ubiquitination chains observed are specific for Rac1 and not as a result of IAP ubiquitination.
10. The reaction time can be varied depending on the protein. We can observe ubiquitin chain synthesis on Rac1 quite well even in less than 1 h and so we usually terminate the reaction by that time. However, most of the reaction should happen in less than 30 min.
11. The in vitro ubiquitination samples are heated at 100 °C for 2 min instead of 5 min that we normally use for sample preparation for SDS-PAGE. It blocks nonspecific chain cleavage and helps in better ubiquitin chain detection in our hands.
12. Always use Ni-NTA resin for one extra sample than is required for pulldown as we lose some resin in the process. Also, the resin is distributed into two tubes to facilitate centrifugation.
13. Before starting the pulldown, label two sets (three sets for endogenous Rac1-IP) of Eppendorf tubes for the samples and add 50 μl of Laemmli to one set of tubes for collecting control lysates. Also check beforehand if there are adequate amounts of ice-cold PBS.
14. The cell suspension is quite viscous so check carefully to remove every drop of the suspension from the pipette tip before discarding.
15. Ni-NTA resin is quite indistinguishable from the liquid portion. So it is advised to leave nearly 50–100 μl of supernatant in the tube during each washing step to avoid losing the resin.

16. Since it is an endogenous immunoprecipitation, more cells are needed to observe pulldown of the proteins so six wells per condition are transfected. Cells are not directly seeded on 10 cm dishes as it is observed that the transfection efficiency is better when cells are transfected in 12-well plates in our hands.
17. The lysates can also be sonicated (50 % power, 2 pulses of 1 s each) if they are very viscous and difficult to handle only by centrifugation.

## Acknowledgements

The work discussed here is supported through an ENP grant RA1739/1-1 to KR from the DFG. KR is a PLUS3 fellow of the Boehringer Ingelheim Foundation.

### References

1. Haglund K, Dikic I (2005) Ubiquitylation and cell signaling. EMBO J 24:3353–3359
2. Zhang N, Wang Q, Ehlinger A et al (2009) Structure of the s5a:k48-linked diubiquitin complex and its interactions with rpn13. Mol Cell 35:280–290
3. Vega FM, Ridley AJ (2008) Rho GTPases in cancer cell biology. FEBS Lett 582: 2093–2101
4. Oberoi TK, Dogan T, Hocking JC et al (2011) IAPs regulate the plasticity of cell migration by directly targeting Rac1 for degradation. EMBO J 31:14–28
5. Torrino S, Visvikis O, Doye A et al (2011) The E3 ubiquitin-ligase HACE1 catalyzes the ubiquitylation of active Rac1. Dev Cell 21:959–965
6. Vaux DL, Silke J (2005) IAPs, RINGs and ubiquitylation. Nat Rev Mol Cell Biol 6:287–297
7. Boulter E, Garcia-Mata R, Guilluy C et al (2010) Regulation of Rho GTPase crosstalk, degradation and activity by RhoGDI1. Nat Cell Biol 12:477–483

# Chapter 5

# Ras GTPases Are Both Regulators and Effectors of Redox Agents

**Elisa Ferro, Luca Goitre, Eva Baldini, Saverio Francesco Retta, and Lorenza Trabalzini**

## Abstract

Redox agents have been historically considered pathological agents which can react with and damage many biological macromolecules including DNA, proteins, and lipids. However, a growing number of reports have suggested that mammalian cells can rapidly respond to ligand stimulation with a change in intracellular ROS thus indicating that the production of intracellular redox agents is tightly regulated and that they serve as intracellular signaling molecules being involved in a variety of cell signaling pathways. Numerous observations have suggested that some members of the Ras GTPase superfamily appear to regulate the production of redox agents and that oxidants can function as effector molecules for the small GTPases, thus contributing to their overall biological function. In addition, many of the Ras superfamily small GTPases have been shown to be redox sensitive, thanks to the presence of redox-sensitive sequences in their primary structure. The action of redox agents on these redox-sensitive GTPases is similar to that of guanine nucleotide exchange factors in that they perturb GTPase nucleotide-binding interactions that result in the enhancement of the guanine nucleotide exchange of small GTPases.

Thus, Ras GTPases may act both as upstream regulators and downstream effectors of redox agents.

Here we overview current understanding concerning the interplay between Ras GTPases and redox agents, also taking into account pathological implications of misregulation of this cross talk and highlighting the potentiality of these cellular pathways as new therapeutical targets for different pathologies.

**Key words** Ras GTPases, Redox agents, Reactive oxygen species, Reactive nitrogen species, Ras/ROS interplay, Rac, Rho

## 1 Introduction

Signal transduction through Ras proteins occurs by reversible binding of GTP, while the inactive form is bound to GDP. Switching between these two states is regulated by guanine nucleotide exchange factors (GEFs), that catalyze the exchange of GDP with GTP to promote Ras activation, and GTPase-activating proteins (GAPs), that deactivate the Ras protein by stimulating hydrolysis of bound GTP to GDP. Deactivation can also be achieved by

Lorenza Trabalzini and Saverio Francesco Retta (eds.), *Ras Signaling: Methods and Protocols*,
Methods in Molecular Biology, vol. 1120, DOI 10.1007/978-1-62703-791-4_5, 

association with guanine nucleotide dissociation inhibitors (GDIs), which prevent membrane association, and GDP dissociation. All of these regulatory proteins are themselves affected by diverse upstream signals, which serve to activate or inactivate Ras GTPase signaling pathways [1]. Besides modulatory factors such as GAPs and GEFs, reactive small molecule oxidants have also been demonstrated to regulate Ras activity [2–4]. Many of the Ras superfamily small GTPases have been shown to be redox sensitive, and their known conserved redox-sensitive sequences have been termed the NKCD, GXXXXGK(S/T)C, and CGNKXD motifs. The action of redox agents on these redox-sensitive GTPases is similar to that of GRFs in that they perturb GTPase nucleotide-binding interactions that result in the enhancement of the guanine nucleotide exchange of small GTPases [5–7].

For many years, the generation of intracellular redox agents such as reactive oxygen species (ROS) and reactive nitrogen species (RNS) was viewed solely as the unregulated by-product of aerobic metabolism and other enzymatic processes, and redox agents have been historically considered pathological agents which can react with and damage many biological macromolecules including DNA, proteins, and lipids. However, over the last years, a growing number of reports have suggested that mammalian cells can rapidly respond to ligand stimulation with a change in intracellular ROS indicating that the production of intracellular redox agents is tightly regulated. Thus redox agents may serve as intracellular signaling molecules being involved in a variety of cell signaling pathways [8–13], to fulfill a wide range of essential biological processes, including cell adhesion, migration, proliferation, differentiation, and survival [14–17]. Numerous observations have suggested that the Ras GTPases regulate the production of redox agents and that oxidants may function as effector molecules for the small GTPases, thus contributing to their overall biological function [18].

Thus, Ras GTPases may act both as upstream regulators and downstream effectors of redox agents.

Here we discuss current understanding concerning the cross talk between GTPases and redox agents. The discussion also takes into account pathological implications of alterations of both ROS regulation by small GTPases and small GTPases regulation by ROS and highlights the potentiality of these cellular pathways as new therapeutical targets for different pathologies.

## 2 Metabolism and Signaling of Redox Agents

ROS are highly reactive oxygen-containing molecules generated as common by-products of oxidative metabolism or in response to the activation of several oxidative enzyme complexes.

The key determinant of the overall effects of ROS is the superoxide anion ($O_2^{\bullet -}$). $O_2^{\bullet -}$ is generated by a number of sources located throughout the cell via the incomplete, one-electron reduction of molecular oxygen ($O_2$); even though it has a short half-life, $O_2^{\bullet -}$ is the precursor of all other major reactive oxygen species found in biological systems, including the highly reactive hydroxyl radical ($^{\bullet}OH$), hydrogen peroxide ($H_2O_2$), and peroxynitrite ($OONO^-$) [14–17].

Mitochondria are a major intracellular source of ROS, whose generation is inversely related to the efficiency of the electron transport chain. Any compromise in the transport of electrons facilitates the leakage of an electron onto $O_2$, thus generating $O_2^{\bullet -}$. This leakage of electrons takes place mainly at complex I (NADH: ubiquinone oxidoreductase) and III (ubiquinol: cytochrome c oxidoreductase) that convert up to 5 % of molecular $O_2$ to $O_2^{\bullet -}$ [19].

In addition, $O_2^{\bullet -}$ is produced by the activity of NAD(P)H oxidase (NOX), xanthine oxidase (XO), cytochrome p450 (CYP), uncoupled NO synthase (NOS), myeloperoxidase (MPO), lipoxygenase (LOX), and cyclooxygenase (COX) [14–17], which can be induced by a variety of chemical and physical stimuli, including integrins, growth factors, G protein-coupled receptor agonists, cytokines, neurotransmitters, metabolic factors, shear stress, ischemia/reperfusion, chemotherapeutics and ionizing radiations, as well as aging [14, 15, 20, 21]. Conversely, $O_2^{\bullet -}$ is rapidly removed by distinct superoxide dismutase (SOD) isoenzymes, located in the mitochondria (SOD2), cytoplasm (SOD1), and extracellular (SOD3) compartments, which catalyze the dismutation of $O_2^{\bullet -}$ into $H_2O_2$ and $O_2$. In turn, $H_2O_2$ is reduced to $H_2O$ by the catalase and glutathione peroxidase enzymes. In addition, $O_2^{\bullet -}$ can be converted to hydroxyl radical ($^{\bullet}OH$) by the Fenton or Haber–Weiss reactions, or to the nonradical oxidant, peroxynitrite anion ($ONOO^-$), by reacting with nitric oxide ($^{\bullet}NO$) [6, 22, 23] (Fig. 1).

Within the body, $^{\bullet}NO$ is generated by three main NOS isoforms which produce it at different levels in response to numerous stimuli such as $Ca^+$ and calmodulin, posttranslational modifications, and inflammatory agents [24]. Nitric oxide, itself, is a very poor oxidant, but is the parent compound for the number of reactive nitrogen species.

Another major route for RNS production begins with the reaction between $^{\bullet}NO$ and $O_2$ producing nitrogen dioxide, $^{\bullet}NO_2$. $^{\bullet}NO_2$ is known to react and modify a variety of biomolecules resulting in lipid peroxidation [25, 26], guanine base nitration [27, 28], tyrosine side chain nitration [29, 30], and oxidation of thiols and thiyl radicals [31]. Thiyl radical intermediates, whether protein or small molecule based (i.e., glutathione), act as a precursor for radical-based S-nitrosation events. This process is considered extremely important for activation of Ras GTPases [6]. In addition, $^{\bullet}NO_2$ has been shown

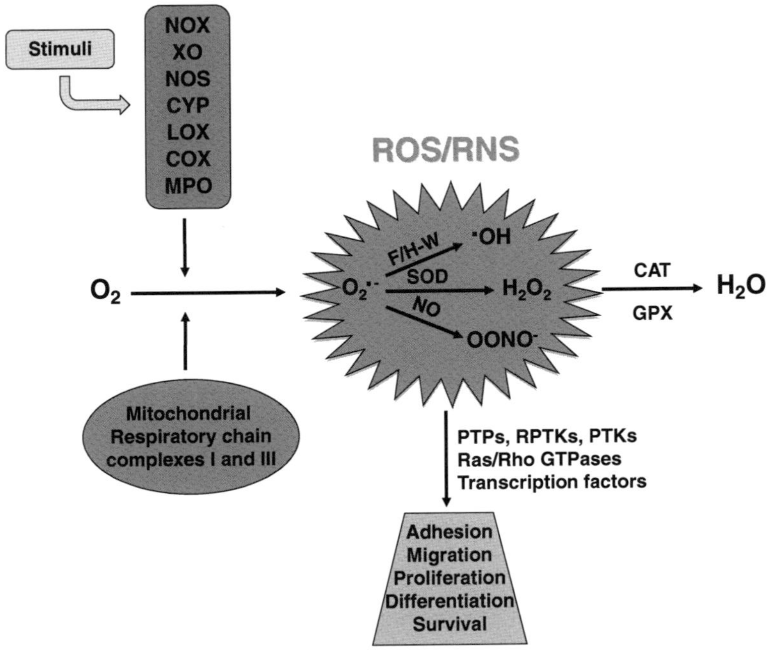


**Fig. 1** ROS metabolism and signaling. The superoxide anion ($O_2^{\bullet-}$) is generated via incomplete, one-electron reduction of molecular oxygen ($O_2$) by the redox complexes I and III of the mitochondrial electron transport chain and by the activity of NAD(P)H oxidase (NOX), xanthine oxidase (XO), uncoupled NO synthase (NOS), cytochrome p450 (CYP), lipoxygenase (LOX), cyclooxygenase (COX), and myeloperoxidase (MPO), which can be induced by a variety of chemical and physical stimuli. $O_2^{\bullet-}$ is removed by superoxide dismutase (SOD) isoenzymes, which catalyze the dismutation of $O_2^{\bullet-}$ into $H_2O_2$ and $O_2$. In turn, $H_2O_2$ is reduced to $H_2O$ by the catalase (CAT) and glutathione peroxidase (GPX) enzymes. In addition, $O_2^{\bullet-}$ can be converted to hydroxyl radical ($^{\bullet}OH$) by the Fenton (F) or Haber–Weiss (H-W) reactions or to peroxynitrite ($OONO^-$) by reacting with nitric oxide (NO). At physiological concentrations, ROS are involved in the redox-dependent regulation of multiple signal transduction pathways to fulfill a wide range of essential biological processes

to induce cis–trans isomerization of arachidonic acid leading to increased levels of trans-fatty acids in human plasma [32].

The reactivity of $^{\bullet}NO_2$ toward other free radical species contrasts its reactions with small organics (lipids) and large macromolecules (protein thiols). $^{\bullet}NO_2$ reacts with $^{\bullet}NO$ to produce nonradical dinitrogen trioxide, $N_2O_3$ [33, 34], a potent S-nitrosation agent for small molecule thiols (i.e., glutathione, cysteine), and protein thiols (i.e., cysteine side chains) [35, 36]. In addition to its reaction with $^{\bullet}NO$, $^{\bullet}NO_2$ also dimerizes to form an additional oxidizing species, dinitrogen tetroxide, $N_2O_4$ [6, 33].

It is now well established that physiological concentrations of ROS are endowed with essential signaling properties, which are mainly due to the reversible oxidation of redox-sensitive molecular targets, thereby functioning as signaling molecules. However, at high levels, ROS are known to exert very damaging effects through oxidative stress. This is caused by an imbalance between the production of ROS and the ability of cellular antioxidant mechanisms to readily detoxify the reactive intermediates.

Importantly, as $O_2^{\bullet-}$ can spontaneously react with $^{\bullet}NO$ to form $OONO^-$ at a rate three times faster than $O_2^{\bullet-}$ dismutation by SOD, modest increases of $O_2^{\bullet-}$ can result in a great reduction of $^{\bullet}NO$ bioavailability and increased formation of $OONO^-$, a very strong oxidant with the potential to produce multiple cytotoxic effects [6, 37–42]. In addition, $OONO^-$ can also trigger feedforward mechanisms that further amplify $O_2^{\bullet-}$ generation and oxidative stress, including the uncoupling of NO synthase (NOS) which produces $O_2^{\bullet-}$ instead of $^{\bullet}NO$, thus amplifying the risk of cellular dysfunction and oxidative injury [14]. The maintenance of highly regulated mechanisms to control ROS levels and functional specificity is therefore essential for normal cellular homeostasis and proper response to environmental stimuli.

NOX are major sources of ROS and have been demonstrated to play a fundamental role in the compartmentalization of ROS production and redox signaling [43]. Besides NOX, an important role in the spatiotemporal regulation of ROS production is also played by enzymes involved in arachidonic acid (AA) metabolism, such as phospholipase $A_2$ (PLA2), lipoxygenases (LOX), and cyclooxygenases (COX), suggesting that a complex regulatory network may take place for proper modulation of redox signaling [44].

Accumulating evidence points to the protein tyrosine phosphatases (PTPs) as the major redox-sensitive molecular targets of ROS [45]. This protein family is indeed characterized by the presence in the active site of a highly conserved sequence motif containing a Cys residue that is essential for catalysis and very susceptible to reversible inactivating oxidation by ROS. In turn, oxidative inactivation of PTPs promotes phosphorylation-dependent downstream signaling events.

In addition to PTPs, other important signaling proteins have been shown to act as endogenous redox sensors for mediating ROS signaling, including receptor protein tyrosine kinases (RPTKs), cytoplasmic kinases (protein tyrosine kinases, PTKs), transcription factors, and, as it will be reviewed here, members of the Ras superfamily of small GTPases [43, 45, 46] (Fig. 1).

## 3 Ras GTPases Regulate the Production of Redox Agents

The NOX complex was originally identified in phagocytic leukocytes as an enzymatic defense system against infections required for the oxidative burst-dependent microbial killing [47, 48]. The active complex is composed of membrane-associated cytochrome b558, comprising the catalytic gp91$^{phox}$ (also known as NOX2) and regulatory p22$^{phox}$ subunits, and four cytosolic regulatory components, including p40$^{phox}$, p47$^{phox}$, p67$^{phox}$, and the small GTPase Rac1 [48]. Neutrophils express two different Rac isoforms, including the phagocyte specific Rac2 and the more ubiquitously expressed Rac1. Detailed molecular analysis has revealed that Rac proteins function as a necessary switch for ROS generation and that the protein is recruited to the membrane following neutrophil activation where it can bind to both p67$^{phox}$ and gp91$^{phox}$ [49].

Certain aspects of neutrophil biology appear to be conserved in the ROS signaling of nonphagocytic cells. In particular, homologues of the NOX complex were found in vascular endothelial cells and smooth muscle cells, as well as in other normal or transformed cells such as colon cancer or melanoma [50]. Several isoforms of the catalytic NOX2 protein were identified, including NOX1, NOX3, NOX4, and NOX5, and shown to localize in proximity of specific redox-sensitive molecular targets within discrete subcellular compartments, thereby facilitating the compartmentalization of redox signaling [43]. In addition, the expression of a constitutively activated form of Rac1 was noted to increase the basal level of hydrogen peroxide in immortalized fibroblasts [51] as well as in certain transformed cell lines [52], while the expression of a dominant negative form of Rac1 was shown to inhibit the production of ROS following addition of various ligands [51]. These data suggest that a Rac-regulated oxidase exists in a wide range of cell types and participates in normal signal transduction.

It has been shown that NOX1 constitutively binds the RacGEF βPIX, and the interaction is caused by growth factor stimulation [53]. This and previous studies [54] also support a pathway where ligand addition results in the sequential activation of phosphatidylinositol 3-kinase (PI3K), which in turn generates lipid products that can activate GEFs through the PH (pleckstrin homology) domain present within the exchange factors. Activation of the GEF leads to increased Rac activity that is presumed to directly stimulate NOX [18].

The small GTPase Rac1 is also thought to be implicated in the control of NOX4 function in endothelial cells and mesangial cells [55, 56]. NOX4 is one of the most abundant and extensively expressed NOX isoform with constitutive ROS-generating activity [57, 58] that, in contrast to the other NOXs, exists in the immediate environment of the nucleus.

Several evidence support the role of ROS as specific effector molecules that act downstream of Rac. In a recent paper, it has been shown a role of Rac-regulated ROS in the cross talk between G protein-coupled receptors (GPCRs) and the JAK/STAT pathway [59], while different studies support a role of Rac1 as a crucial, common upstream mediator of ROS production in integrin-mediated outside-in signaling [11–13, 44, 60].

ROS have been implicated in the integration of signals from VEGF and Rac to regulate the integrity of the endothelial barrier [51, 61–65]. Further studies demonstrated that the VEGF-dependent phosphorylation of VE-cadherin and β-catenin are dependent on Rac and ROS and result in decreased junctional integrity and enhanced vascular permeability [66, 67].

In addition to NOX, Rac1 has been demonstrated to act upstream of AA-metabolizing enzymes, such as $PLA_2$ [68, 69], 5-LOX [13, 59, 60], and COX-2 [70], whereas many reports show that AA metabolism modulates NOX and mitochondrial ROS production [44].

An important aspect of Rac-regulated oxidant signaling is the demonstration of redox-dependent cross talk between different small GTPase family members. ROS production is apparently an essential component in signaling cascades that mediate Rac1/p190RhoGAP-induced downregulation of RhoA and concomitant formation of membrane ruffles and integrin-mediated cell spreading. A cellular pathway linking generation of ROS to downregulation of Rho involves inhibition of the low-molecular-weight protein tyrosine phosphatase (LMW-PTP) and a consequent increase in the activation by phosphorylation of the Rho inhibitor p190Rho-GAP [71]. It has been shown that ROS production causes p190RhoGAP translocation to the adherens junctions (AJs), where it binds p120ctn, and subsequently inhibits local Rho activity [72]. It thus plays a role in the stabilization of cell–cell contacts [67]. These findings suggest that Rac1 downregulates Rho and stress fiber formation in a redox-dependent manner and define a mechanism that relates changes in cellular redox state to the control of actin cytoskeleton rearrangements by Rho GTPases.

In addition to Rac, the production of ROS by nonphagocytic cell types stimulated by growth factors or cytokines includes the participation of p21Ras [51]. Fibroblasts expressing constitutively active mutants of both Rac and Ras produce high levels of ROS associated with a high rate of proliferation. In the same study experimental evidence was provided suggesting that Rac is positioned downstream to Ras. Similar overexpression of Ras in other cell types such as keratinocytes [73] and epithelial cells [74] also demonstrated an increase in basal ROS levels.

In NIH3T3 cells, oncogenic H-Ras activates DNA repair through the Ras/PI3K/Rac1/NADPH oxidase-dependent pathway; the increased ROS production via this signaling pathway is required for enhancement of the DNA repair capacity induced by oncogenic H-Ras [54]. Studies in human thyroid cells led to senescence demonstrated that the NADPH oxidase NOX4 is a critical mediator in oncogenic H-Ras-induced DNA damage response and subsequent senescence [75, 76]. A further mechanism insight into redox regulation of Ras-induced premature senescence has been provided by recent data showing that the Ras-oncogene, through the Ras/MEK pathway, upregulates the expression of NOX1 and NOX4, thus leading to an increase of intracellular ROS and DNA damage [77].

In other cell types the source of Ras-induced ROS appears to be linked to the mitochondria [78]. Mitochondria have the highest levels of antioxidants in the cell and play an important role in the maintenance of cellular redox status, thereby acting as ROS and redox sink and limiting NOX activity. However, in many mammalian cells the mitochondrial electron transport chain is a major source of ROS that can serve as signaling molecules in other cellular compartments and under certain conditions may stimulate NOXs. Many findings indicate the existence of a bidirectional signaling cross talk between mitochondria and NOX, where small GTPases can orchestrate a complex web of regulation for ROS production [79–81].

Indeed, in integrin signaling, the regulation of mitochondria by both Rac and RhoA appears to be related to their ability to alter intracellular ROS [12].

It has been shown that nerve growth factor (NGF)-induced differentiation of PC12 cells is mediated by significant alteration of mitochondrial metabolism by reducing mitochondrial-produced ROS and stabilizing the electrochemical gradient. This is accomplished by stimulation of mitochondrial manganese superoxide dismutase (MnSOD) via Ki-Ras and ERK1/2 [82]. Thus ROS produced by small GTPases could regulate mitochondrial properties, including the overall metabolic rate and the generation of mitochondrial oxidants with important signaling functions within the cell [18].

Using both knockout mice and *Drosophila* models, it has been recently demonstrated that members of the Rit subfamily GTPases (mammalian Rit and Rin and the *Drosophila* Ric homologue) play an evolutionarily conserved role in governing survival in response to oxidative stress through the regulation of a p38 MAPK cascade [83]. The described signaling link between the Rit GTPases and the p38 MAPK cascade represents a novel survival mechanism. The identification of p38 as a major target of Rit signaling, particularly in response to oxidative stress, suggests that Rit-p38 signaling may be critical in determining whether ROS-dependent p38 signaling results in cell death or recovery and establishes Rit as the central regulator of a p38 signaling cascade that functions as a critical survival mechanism in response to stress [83, 84].

## 4 Ras GTPases Are Regulated by Redox Agents

Although several studies implicate Ras GTPases in the production and regulation of intracellular ROS, many evidences indicate that Ras proteins can also be direct targets of ROS. Similar to the action of GEFs, various redox agents, including both ROS and RNS, have been shown to stimulate Ras guanine nucleotide dissociation in vitro and upregulate Ras function in vivo [6, 7].

A correlation between RNS and Ras activation was first discovered by Lander and coworkers. This group showed for the first time that NO can activate Ras by promoting RasGDP dissociation in vitro, GTP binding to Ras in vivo, and stimulation of pathways downstream to Ras [85–92]. The increased Ras activity was correlated to S-nitrosation of a single solvent-exposed cysteine, $Cys^{118}$, which is located in the nucleotide-binding NKCD motif [87, 88, 92, 93]. Further studies indicated that the nitrogen dioxide $\cdot NO_2$, produced by the reaction of NO with $O_2$, reacts with the Ras $Cys^{118}$ through a radical-based mechanism leading to stimulation of nucleotide exchange on Ras [3, 94]. In addition to NO, the superoxide anion $O_2\cdot^-$ was able to enhance guanine nucleotide dissociation from Ras and the Ras-related GTPase Rap1A through a molecular mechanism similar to that of the $NO/O_2$-mediated guanine nucleotide dissociation [95]. The redox-sensitive NKCD motif has been found within the Ras subfamily of GTPases such as H-, N-, K-, and E-Ras, as well as in Rap1A [5]. In addition to $Cys^{118}$, three C-terminal cysteines in Ras showed to be modified by RNS/ROS donors both in vitro and in mouse fibroblasts [96]. As these cysteines undergo lipid modification and membrane association, modification of these residues could potentially be an RNS-/ROS-mediated mechanism of downregulation of Ras by preventing proper membrane localization, in contrast to the upregulation of Ras by RNS/ROS modification of $Cys^{118}$ [6].

In addition to the members of the Ras subfamily, redox-active motifs were afterwards found to be present in other Ras superfamily GTPases, suggesting that redox regulation of GTPase signaling is more widespread that previously envisioned [4].

Members of the Rho family contains the GXXXXGK(S/T)C redox-sensitive motif, located in the phosphoryl-binding loop important for redox-mediated regulation of guanine nucleotide exchange activity in vitro. This motif contains a redox-sensitive cysteine ($Cys^{18}$, Rac1 numbering) at the C-terminus, and it is conserved in almost half of Rho family GTPases such as Rac1 (and its isoforms Rac2 and 3), Cdc42, and RhoA (and its isoforms RhoB and C) [4, 97]. The radical-based molecular mechanism of Rho GTPase guanine nucleotide exchange is similar to the mechanism characterized for Ras GTPases. RhoA showed to be able to be directly activated by ROS in cells by oxidative modification of critical Cys residues within the redox-active motif, and ROS-mediated activation of RhoA induced cytoskeletal rearrangement, thus supporting the hypothesis that cytoskeletal dynamics may be affected by a novel mechanism of regulation of RhoA signaling, independent to classical regulation by GEFs and GAPs [98].

A number of Rab proteins also have the GXXXXGK(S/T)C motif (Rab1B, Rab2A/B, Rab4A/B, Rab14, Rab15, Rab19, and Sec4). Intriguingly, many Rab GTPases (Rab1A, Rab8A/B, Rab10, and Rab13) possess both the NKCD and GXXXXGK(S/T) C motifs, whereas some Rab proteins (Rab3A/B/C/D, Rab 7, Rab22, and Rab38) possess only the NKCD motif [5].

A CGNKXD redox-sensitive motif was found in Ran protein [2]; this motif contains a redox-sensitive cysteine, $Cys^{120}$, at the N-terminal. In addition to this CGNKXD motif, Ran possesses and additional redox-sensitive cysteine $Cys^{85}$ (Ran numbering). This type of redox center is also conserved in Dexras1 and Rhe proteins, as well as in some Rab GTPases [2]. Although redox regulation of the members of Rab and Ran families has been recently discovered, its physiological relevance and pathological consequences linked to the misregulation of redox signaling associated with these redox-sensitive small GTPases have not yet been explored [5].

## 5 The Interplay Between Small GTPases and ROS: Pathological Aspects and Therapeutical Implications

Over the last years, a growing number of evidence suggest that ROS play an important role in the regulation of physiological processes like cell differentiation, proliferation, migration, and vasodilatation. On the other hand, production of ROS "in the wrong place at the wrong time" results in oxidative stress leading to cellular dysfunction and apoptosis, which contributes to different pathologies like atherosclerosis, heart failure, hypertension, ischemia/reperfusion injury, cancer, aging, and neurodegeneration.

It is well known that activated Ras signaling contributes to oncogenic transformation by providing molecular signals that promote cell proliferation, obstruct cell death, inhibit cellular differentiation, and induce angiogenesis [99]. Signaling pathways starting from activated Ras and resulting in mitochondrial ROS production and downstream signaling regulation have been the subject of several recent interesting studies, and different mechanisms have been proposed to elucidate the role of mitochondrial respiration in cancer.

It has been shown that the activation of K-Ras(G12V) leads to mitochondrial dysfunction and a metabolic switch from oxidative phosphorylation to glycolysis finalized to support growth under hypoxic conditions and leading to increased generation of ROS [100]. The major source of ROS generation required for K-Ras-induced anchorage-independent growth is the $Q_o$ site of mitochondrial complex III [101]. Thus mitochondrial dysfunction appears to be an important mechanism by which K-Ras(G12V) causes metabolic changes and ROS stress in cancer cells and promotes tumor development [100].

Mitochondrial dysfunction and ROS production mediated by activation of Ras, Myc, and p53 produce downstream signaling pathways (e.g., NFkappaB, STAT3) that are crucial in cancer-related inflammation. Different inflammation-associated cancers resulting from signaling pathways coordinated at the mitochondrial level have been identified that may be useful for developing innovative strategies for both cancer prevention and cancer treatment [102].

Beside cellular metabolism and bioenergetics, several recent studies have identified a number of biological mechanisms and pathways involved in oncogenic transformation and maintenance of the cancer phenotype, such as redox regulation of cell fate decisions, autophagy induction as a survival signal, and how all these cellular processes interplay with oncogene-induced transformation.

Cell fate regulation is a function of diverse cell signaling pathways that promote cell survival and or inhibit cell death. In this regard, the Bcl-2 family proteins play a crucial role in maintaining a tight balance between cell death and cell proliferation. Recent studies provided a novel mechanism for death regulation by the Bcl-2 family based on the evidence that Bcl-2 displayed a pro-oxidant behavior as its overexpression triggered an increase in intracellular $O_2^{\bullet -}$, which was linked to its ability to block apoptosis. The interaction of Bcl-2 with Rac1 and other members of the Ras superfamily appeared to be involved in the pro-oxidant effect of Bcl-2 [103, 104]. This mechanism supports a model where overexpression of Bcl-2 and its interaction with Ras superfamily proteins confer survival advantage to cancer cells by creating a pro-oxidant milieu. The converging role of small GTPases and Bcl-2 in promoting the pro-oxidant state of cancer cells through physical interaction opens up a new horizon for future redox-based therapeutical designs [104].

Several studies suggest that autophagy may be important in the regulation of cancer development and progression and in determining the response of tumor cells to anticancer therapy [103, 105]. It has been recently reported that autophagy is an important mechanism for sustaining glycolysis during Ras-mediated oncogenic transformation. In addition, the connection between glucose metabolism and autophagy was necessary to promote anchorage-independent cell growth [106]. According to that, it has been shown that autophagy is associated with the malignant transformation of mammalian cells induced by K-Ras and that ROS are involved as signaling molecules in K-Ras(G12V)-induced autophagy. The increase in intracellular ROS produced in response to oncogenic K-Ras involves p38 MAPK signaling and leads to JNK activation. JNK acts downstream of ROS and plays a causal role in autophagy induction through upregulation of autophagy-specific genes 5 and 7 (ATG5 and ATG7) [107]. As mitochondria sustain viability of Ras-expressing cells in starvation, autophagy is required to maintain the pool of functional mitochondria necessary to support growth of Ras-driven tumors [108]. These and other recent findings reviewed by Bellot and coworkers [103] testify that ROS, autophagy, mitochondrial metabolism, and Ras-driven oncogenic progression are interlinked through a fine balance. This cross talk presents multiple potential targets for intervention for an effective anticancer drug design.

Oncogenic activation of the *H-Ras* gene has been found in more than 35 % of patients with urothelial carcinomas [109]. It has been recently shown that in addition to tumorigenic ability, oncogenic H-Ras possesses a novel proapoptotic activity to facilitate the induction of apoptosis by histone deacetylase inhibitors (HDACI), a new class of anticancer agents characterized by high cytotoxicity toward transformed cells [110]. Expression of oncogenic H-Ras in human bladder tumor J82 cells and treatment of cells with the HDACI, FK228, synergistically induce the ERK pathway, resulting in differentially increased NOX-1 elevation and ROS production, leading to differential activation of caspases and cell death [111–113]. Thus, in addition to its well-known role in mediating mitogenic signals for cell proliferation and transformation, the ERK pathway plays an essential role in mediating apoptotic signals induced by HDACIs through induction of NOX-1 elevation to ROS production and caspase activation for inducing cell death. In addition, expression of oncogenic H-Ras in J82 cells also results in an increased susceptibility to exogenous $H_2O_2$ for inducing caspase activation and apoptosis [111]. Further studies revealed that FK228 combined with exogenous $H_2O_2$ cooperatively induces activation of MEK1/2 and ERK1/2 to increase NOX-1 elevation, intracellular ROS production, caspase activation, and cell death. Expression of oncogenic H-Ras significantly increases these FK288- and exogenous $H_2O_2$-induced effects. Oncogenic H-Ras-increased susceptibility to FK228 could be alternatively achieved by additional treatment with exogenous $H_2O_2$. These findings have important and useful implications as combined use of HDACIs with ROS-generating agents may apply to therapeutical strategies to preferentially kill malignant cells with or without oncogenic H-Ras activation [114].

A cellular model to mimic chronic induction of human breast carcinogenesis associated with accumulated exposures to low doses of environmental carcinogens was recently used to clarify the carcinogenic activity of 2-amino-1-methyl-6-phenylimidazo[4,5-b]pyridine (PhIP) in inducing progressive carcinogenesis of non-cancerous human breast epithelial MCF10A cells from a non-cancerous stage to premalignant and malignant stages. Biological changes in cells induced by chronic exposure to PhIP were accompanied by biochemical and molecular changes, including upregulated H-Ras gene expression, ERK pathway activation, NOX-1 expression, and ROS elevation, suggesting that the Ras-ERK-NOX-ROS pathway played an important role in both initiation and in maintenance of cellular carcinogenesis induced by PhIP. Antioxidant agents like the green tea catechin components, at non-cytotoxic doses, were capable of suppressing PhIP-induced cellular carcinogenesis and tumorigenesis thus contributing to the identification of non-cytotoxic preventive agents effective in targeted intervention of carcinogenesis as well as tumorigenesis associated with chronic exposure to environmental carcinogens [115].

Cancer is one of the most prevalent disorders caused by deregulation of Ras activity by a redox agent. Numerous studies show that cancers, to a large extent, are induced by deregulation of Ras redox signaling combined with an alteration of Ras downstream cellular transduction cascades. As with cancers, many cardiovascular and neuronal disorders appear to be the result of deregulation of various cellular signaling events via the redox-sensitive Ras [5]. The deregulation of the redox signaling of Ras with its downstream cascades has also been linked to various disorders related to immune and inflammation processes.

Ras and the downstream MAPK and PI3K pathways have crucial functions in T-cell activation. NO binding to the $Cys^{118}$ of Ras showed to trigger Ras activation and recruitment of PI3K and Ras-dependent activation of MAPKs which include ERKs, JNK, and p38 MAP kinase in T lymphocytes. In addition, NO activation resulted in association of Ras with the effector Raf-1 and consequent increase in Raf-1 kinase activity, MAPKs activity, and phosphorylation of the transcription factor Elk-1. Moreover, NO triggered an increase in mRNA levels of the proinflammatory cytokine, TNF-α, which was ERK dependent [116]. This data strongly suggest that this NO-activated Ras-mediated signaling pathway may play a critical role in Elk-1-induced transcriptional activation of T lymphocytes, host defense, and inflammation.

Proinflammatory cytokines such as TNF-α and IL-1β enhance degradation of cartilage-specific, type II collagen by matrix metalloproteinase-13 (MMP-13). By using pharmacologic, Ras genetic knockdown and antioxidant approaches, it has been shown that H-Ras and NOX mediate TNF-α and IL-1β induction of MMP-13, and it has been suggested that these targets could be utilized for therapeutical inhibition of MMP-13-driven cartilage erosion by Ras and NOX inhibitors and antioxidants [117].

Systemic sclerosis (scleroderma) is a connective tissue disorder characterized by extensive fibrosis of the skin and visceral organs, by vascular abnormalities and immunological manifestations. Recent evidences suggest that the mechanisms involved in the progression of scleroderma fibrosis include an autoamplification circuit linking ROS, Ras, and ERK 1–2 which in turn amplifies and maintains the autocrine loop made up by cytokines, growth factors, and their cognate receptors [118].

There is a vast body of literature that links vascular ROS production to cardiovascular disease [119]. Vascular ROS production, as well as Rac1 activation, has been associated with hypertrophy and smooth muscle cell proliferation and endothelial dysfunction as well as endothelial cell migration, hypertension inflammation, and atherosclerosis [120–123]. Vascular hypertrophy has been ascribed to the effects of various receptor agonists, including angiotensin II (Ang II), which induces ROS production in VSMCs in a Rac1-dependent fashion [124]. Recent studies showed that this Ang II-induced ROS production also requires the membrane adapter caveolin, which is involved in Rac1 activation [125, 126], and the lipid kinase PI3K-γ [124, 127].

Ischemia/reperfusion (I/R) injury is also associated with ROS production. This is a clinically relevant problem occurring as damage to the myocardium following blood restoration after a critical period of coronary occlusion. It is well known that immediately following the reinstitution of oxygenated blood into ischemic tissue, there is a rapid burst of ROS, but the molecular basis and source of this process are not yet convincingly identified [17, 18]. However, both in vitro and in vivo experiments [128, 129] have suggested that Rac1 plays a dominant role in ROS generation after I/R, and it activates the nuclear factor NF-κB and stimulates mRNA expression of several inflammatory genes, such as TNF-α and iNOS in the liver, leading to massive hepatocyte necrosis. Thus, efforts aimed at inhibiting Rac protein function could be useful therapeutical strategies in a variety of clinical settings in which there is concern about the potential harmful effects of I/R injury [128, 129].

Data from the literature suggest that RhoA activation and ROS are associated to airway smooth muscle contractility [130–132]; it has been shown that oxidative stress with $H_2O_2$ leads to airway smooth muscle contraction mediated by increases in intracellular $Ca^{2+}$ concentration and the Rho/Rho kinase pathway [132].

Both ROS and Rho/Rho kinase have been suggested to play important roles in vasoconstriction and may contribute to the pathogenesis of hypertension in experimental animals and humans. Jin and coworkers demonstrated the direct activation of the Rho/Rho kinase signaling pathway by ROS in rat aorta, suggesting an important role for ROS-mediated Rho/Rho kinase activation in vasoconstriction [133].

As previously discussed, a novel mechanism for the regulation of RhoA in cells by ROS has been proposed that allows predicting that ROS may directly activate Rho signaling in smooth muscle and in the endothelium thus affecting vascular permeability [98]. This mechanism of regulation, which is independent of classical regulatory proteins, may be particularly relevant in pathological conditions where ROS are generated and the cellular redox-balance altered, such as in asthma and I/R injury.

## 6 Concluding Remarks

Members of the Ras GTPase superfamily are able to regulate the production of ROS and RNS, and this is an important aspect of the function of these monomeric G proteins. In addition, the functional cross talk between some different Ras GTPases, such as Rac1 and RhoA, appears strictly related to redox signaling. Finally, many of the Ras superfamily small GTPases have been shown to be targets of regulation by ROS and RNS, due to the presence of conserved redox-sensitive sequences in their primary structure.

Thus, Ras GTPases may act both as upstream regulators and downstream effectors of redox agents, strongly contributing to their overall biological function and playing a key role in various cellular signaling events. Deregulation of small GTPases by redox agents or deregulation of redox signaling by small GTPases may significantly alter these cellular signaling pathways and lead to the pathological state.

As both Ras GTPases and redox agents play an important role in the regulation of fundamental processes like cell differentiation, proliferation, migration, adhesion, and survival, the comprehension of molecular mechanisms of the interplay between small GTPases and ROS/RNS may strongly help to clarify how this cross talk contributes to physiological and pathological cellular events.

In this review we summarized and illustrated some recent experimental evidence indicating the interplay between Ras GTPases and ROS/RNS. The discussion of the studies reviewed here was mainly aimed to describe the pathological consequences of alterations of both redox agents regulation by Ras GTPases and Ras GTPase regulation by redox agents. More interestingly, we tried to emphasize the fact that fully understanding of these cellular pathways may provide novel potential targets for the design of strategic therapies for different pathologies where both small GTPases and oxidative stress play a role.

### References


1. Herrmann C (2003) Ras-effector interactions: after one decade. Curr Opin Struct Biol 13:122–129
2. Heo J (2008) Redox regulation of Ran GTPase. Biochem Biophys Res Commun 376:568–572
3. Heo J, Campbell SL (2004) Mechanism of p21Ras S-nitrosylation and kinetics of nitric oxide-mediated guanine nucleotide exchange. Biochemistry 43:2314–2322
4. Heo J, Campbell SL (2005) Mechanism of redox-mediated guanine nucleotide exchange on redox-active Rho GTPases. J Biol Chem 280:31003–31010
5. Heo J (2011) Redox control of GTPases: from molecular mechanisms to functional significance in health and disease. Antioxid Redox Signal 15:689–724
6. Davis MF, Vigil D, Campbell SL (2011) Regulation of Ras proteins by reactive nitrogen species. Free Rad Biol Med 51:565–575
7. Mitchell L, Hobbs GA, Aghajanian A et al (2013) Redox regulation of Ras and Rho GTPases: mechanism and function. Antioxid Redox Signal 20:250–258
8. Sundaresan M, Yu ZX, Ferrans VJ et al (1995) Requirement for generation of $H_2O_2$ for platelet-derived growth factor signal transduction. Science 270:296–299
9. Bae YS, Kang SW, Seo MS et al (1997) Epidermal growth factor (EGF)-induced

generation of hydrogen peroxide. Role in EGF receptor-mediated tyrosine phosphorylation. J Biol Chem 272:217–221
10. Leto TL, Geiszt M (2006) Role of Nox family NADPH oxidases in host defense. Antioxid Redox Signal 8:1549–1561
11. Kheradmand F, Werner E, Tremble P et al (1998) Role of Rac1 and oxygen radicals in collagenase-1 expression induced by cell shape change. Science 280:898–902
12. Werner E, Werb Z (2002) Integrins engage mitochondrial function for signal transduction by a mechanism dependent on Rho GTPases. J Cell Biol 158:357–368
13. Chiarugi P, Pani G, Giannoni E et al (2003) Reactive oxygen species as essential mediators of cell adhesion: the oxidative inhibition of a FAK tyrosine phosphatase is required for cell adhesion. J Cell Biol 161:933–944
14. Chrissobolis S, Faraci FM (2008) The role of oxidative stress and NADPH oxidase in cerebrovascular disease. Trends Mol Med 14: 495–502
15. Fortuño A, San José G, Moreno MU et al (2005) Oxidative stress and vascular remodelling. Exp Physiol 90:457–462
16. Miller AA, Drummond GR, Sobey CG (2006) Reactive oxygen species in the cerebral circulation: are they all bad? Antioxid Redox Signal 8:1113–1120
17. Valko M, Leibfritz D, Moncol J et al (2007) Free radicals and antioxidants in normal physiological functions and human disease. Int J Biochem Cell Biol 39:44–84
18. Finkel T (2006) Intracellular redox regulation by the family of small GTPases. Antioxid Redox Signal 8:1857–1863
19. Turrens JF (2003) Mitochondrial formation of reactive oxygen species. J Physiol 552: 335–344
20. Girouard H, Iadecola C (2006) Neurovascular coupling in the normal brain and in hypertension, stroke, and Alzheimer disease. J Appl Physiol 100:328–335
21. Iadecola C, Park L, Capone C (2009) Threats to the mind: aging, amyloid, and hypertension. Stroke 40:S40–S44
22. Faraci FM, Didion SP (2004) Vascular protection: superoxide dismutase isoforms in the vessel wall. Arterioscler Thromb Vasc Biol 24:1367–1373
23. Blough NV, Zafiriou OC (1985) Reaction of superoxide with nitric oxide to form peroxynitrite in alkaline aqueous solution. Inorg Chem 24:3502–3504
24. Hill BG, Dranka BP, Bailey SM et al (2010) What part of NO don't you understand? Some answers to the cardinal questions in nitric oxide biology. J Biol Chem 285: 19699–19704
25. Pryor WA, Lightsey JW (1981) Mechanisms of nitrogen dioxide reactions: initiation of lipid peroxidation and the production of nitrous acid. Science 214:435–437
26. Thomas HV, Mueller PK, Lyman RL (1968) Lipoperoxidation of lung lipids in rats exposed to nitrogen dioxide. Science 159:532–534
27. Misiaszek R, Crean C, Geacintov NE et al (2005) Combination of nitrogen dioxide radicals with 8-oxo-7,8-dihydroguanine and guanine radicals in DNA: oxidation and nitration end-products. J Am Chem Soc 127: 2191–2200
28. Niles JC, Wishnok JS, Tannenbaum SR (2006) Peroxynitrite-induced oxidation and nitration products of guanine and 8-oxoguanine: structures and mechanisms of product formation. Nitric Oxide 14:109–121
29. Bian K, Gao Z, Weisbrodt N et al (2003) The nature of heme/iron-induced protein tyrosine nitration. Proc Natl Acad Sci U S A 100: 5712–5717
30. Thomas DD, Espey MG, Vitek MP et al (2002) Protein nitration is mediated by heme and free metals through Fenton-type chemistry: and alternative to the $NO/O_2$-reaction. Proc Natl Acad Sci U S A 99:12691–12696
31. Bonini MG, Fernandes DC, Augusto O (2004) Albumin oxidation to diverse radicals by the peroxidase activity of Cu, Zn-superoxide dismutase in the presence of bicarbonate or nitrite: diffusible radicals produce cysteinyl and solvent-exposed and -unexposed tyrosyl radicals. Biochemistry 43:344–351
32. Jiang H, Kruger N, Lahiri DR et al (1999) Nitrogen dioxide induces cis–trans-isomerization of arachidonic acid within cellular phospholipids. Detection of trans-arachidonic acids in vivo. J Biol Chem 274: 16335–16341
33. Augusto O, Bonini MG, Amanso AM et al (2002) Nitrogen dioxide and carbonate radical anion: two emerging radicals in biology. Free Radic Biol Med 32:841–859
34. Ford E, Hughes MN, Wardman P (2002) Kinetics of the reactions of nitrogen dioxide with glutathione, cysteine and uric acid at physiological pH. Free Radic Biol Med 32: 1314–1323
35. Herold S, Rock G (2005) Mechanistic studies of S-nitrosothiol formation by $NO^{\bullet}/O_2$ and by $NO^{\bullet}$/methemoglobin. Arch Biochem Biophys 436:386–396
36. Keszler A, Zhang Y, Hogg N (2010) Reaction between nitric oxide, glutathione, and oxygen in the presence and absence of protein: how are S-nitrosothiols formed? Free Radic Biol Med 48:55–64
37. Beckman JS, Beckman TW, Chen J et al (1990) Apparent hydroxyl radical production

by peroxynitrite-implications for endothelial injury from nitric-oxide and superoxide. Proc Natl Acad Sci U S A 87:1620–1624
38. Beckman JS, Koppenol WH (1996) Nitric oxide, superoxide, and peroxynitrite: the good, the bad, and the ugly. Am J Physiol Cell Physiol 40:C1424–C1437
39. Pacher P, Beckman JS, Liaudet L (2007) Nitric oxide and peroxynitrite in health and disease. Physiol Rev 87:315–424
40. Radi R, Peluffo G, Alvarez MN et al (2001) Unraveling peroxynitrite formation in biological systems. Free Radic Biol Med 30: 463–488
41. Squadrito GL, Pryor WA (1998) Oxidative chemistry of nitric oxide: the roles of superoxide, peroxynitrite, and carbon dioxide. Free Radic Biol Med 25:392–403
42. Girouard H, Park L, Anrather J et al (2007) Cerebrovascular nitrosative stress mediates neurovascular and endothelial dysfunction induced by angiotensin II. Arterioscler Thromb Vasc Biol 27:303–309
43. Ushio-Fukai M (2009) Compartmentalization of redox signaling through NADPH oxidase-derived ROS. Antioxid Redox Signal 11: 1289–1299
44. Gregg D, de Carvalho DD, Kovacic H (2004) Integrins and coagulation: a role for ROS/redox signaling? Antioxid Redox Signal 6: 757–764
45. Monteiro HP, Arai RJ, Travassos LR (2008) Protein tyrosine phosphorylation and protein tyrosine nitration in redox signaling. Antioxid Redox Signal 10:843–889
46. Liu H, Colavitti R, Rovira II et al (2005) Redox-dependent transcriptional regulation. Circ Res 97:967–974
47. Babior BM (1978) Oxygen-dependent microbial killing by phagocytes. N Engl J Med 298:659–668, 721–725
48. Babior BM (2004) NADPH oxidase. Curr Opin Immunol 16:42–47
49. Groemping Y, Rittinger K (2004) Activation and assembly of the NADPH oxidase: a structural perspective. Biochem J 386:401–416
50. Moldovan L, Mythreye K, Goldschmidt-Clermont PJ et al (2006) Reactive oxygen species in vascular endothelial cell motility. Roles of NAD(P)H oxidase and Rac1. Cardiovas Res 71:236–246
51. Sundaresan M, Yu ZX, Ferrans VJ et al (1996) Regulation of reactive-oxygen-species generation in fibroblasts by Rac1. Biochem J 318: 379–382
52. Sulciner DJ, Irani K, Yu ZH et al (1996) Rac1 regulates a cytokine-stimulated, redox-dependent pathway necessary for NF-kappaB activation. Mol Cell Biol 16:7115–7121
53. Park HS, Lee SH, Park D et al (2004) Sequential activation of phosphatidylinositol 3-kinase, beta Pix, Rac1 and Nox1 in growth factor-induced production of $H_2O_2$. Mol Cell Biol 24:4384–4394
54. Cho HJ, Jeong HG, Lee JS et al (2002) Oncogenic H-Ras enhances DNA repair through the Ras/phosphatidylinositol 3-kinase/Rac1 pathway in NIH3T3 cells. Evidence for association with reactive oxygen species. J Biol Chem 277:19358–19366
55. Gorin Y, Ricono JM, Kim NH et al (2003) Nox4 mediates Angiotensin II-induced activation of Akt/protein kinase B in mesangial cells. Am J Physiol Ren Physiol 285:F219–F229
56. Chai D, Wang B, Shen L et al (2008) RXR agonists inhibit high-glucose-induced oxidative stress by repressing PKC activity in human endothelial cells. Free Radic Biol Med 44:1334–1347
57. Brown DI, Griendling KK (2009) Nox proteins in signal transduction. Free Radic Biol Med 47:1239–1253
58. Miller FJ Jr (2009) NADPH oxidase 4: walking the walk with Poldip2. Circ Res 105: 209–210
59. Pelletier S, Duhamel F, Coulombe P et al (2003) Rho family GTPases are required for activation of JAK/STAT signaling by G protein-coupled receptors. Mol Cell Biol 23: 1316–1333
60. Taddei ML, Parri M, Mello T et al (2007) Integrin-mediated cell adhesion and spreading engage different sources of reactive oxygen species. Antioxid Redox Signal 9:469–481
61. van Wetering S, van Buul JD, Quik S et al (2002) Reactive oxygen species mediate Rac-induced loss of cell–cell adhesion in primary human endothelial cells. J Cell Sci 115: 1837–1846
62. Abid MR, Kachra Z, Spokes KC et al (2000) NADPH oxidase activity is required for endothelial cell proliferation and migration. FEBS Lett 486:252–256
63. Abid MR, Tsai JC, Spokes KC et al (2001) Vascular endothelial growth factor induces manganese-superoxide dismutase expression in endothelial cells by a Rac1-regulated NADPH oxidase-dependent mechanism. FASEB J 15:2548–2550
64. Yeh LH, Park YJ, Hansalia RJ et al (1999) Shear-induced tyrosine phosphorylation in endothelial cells requires Rac1-dependent production of ROS. Am J Physiol 276: C838–C847
65. Sohn HY, Keller M, Gloe T et al (2000) The small G-protein Rac mediates depolarization-induced superoxide formation in human endothelial cells. J Biol Chem 275:18745–18750

66. Monaghan-Benson E, Burridge K (2009) The regulation of vascular endothelial growth factor-induced microvascular permeability requires Rac and reactive oxygen species. J Biol Chem 284:25602–25611
67. Beckers CM, van Hinsbergh VW, van Nieuw Amerongen GP (2010) Driving Rho GTPase activity in endothelial cells regulates barrier integrity. Thromb Haemost 103:40–55
68. Peppelenbosch MP, Qiu RG, de Vries-Smits AM et al (1995) Rac mediates growth factor-induced arachidonic acid release. Cell 81: 849–856
69. Woo CH, Lee ZW, Kim BC et al (2000) Involvement of cytosolic phospholipase A2, and the subsequent release of arachidonic acid, in signalling by rac for the generation of intracellular reactive oxygen species in rat-2 fibroblasts. Biochem J 348:525–530
70. Wu R, Coniglio SJ, Chan A et al (2007) Up-regulation of Rac1 by epidermal growth factor mediates COX-2 expression in recurrent respiratory papillomas. Mol Med 13:143–150
71. Nimnual AS, Taylor LJ, Bar-Sagi D (2003) Redox-dependent downregulation of Rho by Rac. Nat Cell Biol 5:236–241
72. Wildenberg GA, Dohn MR, Carnahan RH et al (2006) p120-catenin and p190RhoGAP regulate cell–cell adhesion by coordinating antagonism between Rac and Rho. Cell 127: 1027–1039
73. Yang JQ, Li S, Domann FE et al (1999) Superoxide generation in v-Ha-ras-transduced human keratinocyte HaCaT cells. Mol Carcinog 26:180–188
74. Yang JQ, Buettner GR, Domann FE et al (2002) v-Ha-ras mitogenic signalling through superoxide and derived reactive oxygen species. Mol Carcinog 33:206–218
75. Weyemi U, Lagente-Chevallier O, Boufraqech M et al (2012) ROS-generating NADPH oxidase NOX4 is a critical mediator in oncogenic HRas-induced DNA damage and subsequent senescence. Oncogene 31:1117–1129
76. Weyemi U, Dupuy C (2012) The emerging role of ROS-generating NADPH oxidase NOX4 in DNA-damage responses. Mut Res 751:77–81
77. Kodama R, Kato W, Furuta S et al (2013) Ros-generating oxidases NOX1 and NOX4 contribute to oncogenic Ras-induced premature senescence. Genes Cells 18:32–41
78. Lee AC, Fenster BE, Ito H et al (1999) Ras proteins induce senescence by altering the intracellular levels of reactive oxygen species. J Biol Chem 274:7936–7940
79. Lee SB, Bae IH, Bae YS et al (2006) Link between mitochondria and NADPH oxidase 1 isozyme for the sustained production of reactive oxygen species and cell death. J Biol Chem 281:36228–36235
80. Daiber A (2010) Redox signaling (cross-talk) from and to mitochondria involves mitochondrial pores and reactive oxygen species. Biochim Biophys Acta 1797:897–906
81. Dikalov S et al (2011) Cross talk between mitochondria and NADPH oxidases. Free Radic Biol Med 51:1289–1301
82. Cassano S, Agnese S, D'Amato V et al (2010) Reactive oxygen species, Ki-Ras, and mitochondrial superoxide dismutase cooperate in nerve growth factor-induced differentiation of PC12 cells. J Biol Chem 285:24141–24153
83. Cai W, Rudolph JL, Harrison SMW et al (2011) An evolutionary conserved Rit GTPase-p38 MAPK signaling pathway mediates oxidative stress resistance. Mol Biol Cell 22:3231–3241
84. Cai W, Shi G-X, Andres D (2013) Rit, p38 MAPK and oxidative stress. Commun Integr Biol 1:e22297
85. Baker T, Booden M, Buss J (2000) S-Nitrosocysteine increases palmitate turnover on Ha-Ras in NIH 3T3 cells. J Biol Chem 275:22037–22047
86. Dawson TM, Sasaki M, Gonzalez-Zulueta M et al (1998) Regulation of neuronal nitric oxide synthase and identification of novel nitric oxide signaling pathways. Prog Brain Res 118:3–11
87. Lander HM, Hajjar DP, Hempstead BL et al (1997) A molecular redox switch on p21ras. Structural basis for the nitric oxide-p21Ras interaction. J Biol Chem 272:4323–4326
88. Lander HM, Milbank AJ, Tauras JM et al (1996) Redox regulation of cell signaling. Nature 381:380–381
89. Lander HM, Ogiste JS, Pearce SF et al (1995) Nitric oxide-stimulated guanine nucleotide exchange on p21ras. J Biol Chem 270: 7017–7020
90. Lander HM, Sehajpal PK, Novogrodsky A (1993) Nitric oxide signaling: a possible role for G proteins. J Immunol 151:7182–7187
91. Saavedra AP, Tsygankova OM, Prendergast GV et al (2002) Role of cAMP, PKA and Rap1A in thyroid follicular cell survival. Oncogene 21:778–788
92. Mittar D, Sehajpal PK, Lander HM (2004) Nitric oxide activates Rap1 and Ral in a Ras-independent manner. Biochem Biophys Res Commun 322:203–209
93. Lander HM, Tauras JM, Ogiste JS et al (1997) Activation of the receptor for advanced glycation end products triggers a p21ras-dependent mitogen-activated protein kinase pathway regulated by oxidant stress. J Biol Chem 272:17810–17814
94. Heo J, Prutzman KC, Mocanu V et al (2005) Mechanism of free radical nitric oxide-mediated Ras guanine nucleotide dissociation. J Mol Biol 346:1423–1440

95. Heo J, Campbell SL (2005) Superoxide anion radical modulates the activity of Ras and Ras-related GTPases by a radical-based mechanism similar to that of nitric oxide. J Biol Chem 280:12438–12445
96. Mallis RJ, Buss JE, Thomas JA (2001) Oxidative modification of H-Ras: S-thiolation and S-nitrosylation of reactive cysteines. Biochem J 355:145–153
97. Heo J, Raines KW, Mocanu V et al (2006) Redox regulation of RhoA. Biochemistry 45:14481–14489
98. Aghajanian A, Wittchen ES, Campbell SL et al (2009) Direct activation of RhoA by reactive oxygen species requires a redox-sensitive motif. PLoS One 4:e8045
99. Hanahan D, Weinberg RA (2000) The hallmarks of cancer. Cell 100:57–70
100. Hu Y, Lu W, Chen G et al (2012) K-ras(G12V) transformation leads to mitochondrial dysfunction and a metabolic switch from oxidative phosphorylation to glycolysis. Cell Res 22:399–412
101. Weinberg F, Hamanaka R, Wheaton WW et al (2010) Mitochondrial metabolism and ROS generation are essential for Kras-mediated tumorigenicity. Proc Natl Acad Sci U S A 107:8788–8793
102. Kamp DW, Shacter E, Weitzman SA (2011) Chronic inflammation and cancer: the role of mitochondria. Oncology 25:400–410
103. Bellot GL, Liu D, Pervaiz S (2012) ROS, autophagy, mitochondria and cancer: Ras, the hidden master? Mitochondrion 13:155–162
104. Kang J, Pervaiz S (2013) Crosstalk between Bcl-2 family and Ras family small GTPases: potential cell fate regulation? Front Oncol 2: 206
105. Nippert MM, O'Toole PS, Thorburn A (2006) Autophagy in cancer: good, bad, or both? Cancer Res 66:9349–9351
106. Lock R, Roy S, Kenific CM et al (2011) Autophagy facilitates glycolysis during Ras-mediated oncogenic transformation. Mol Biol Cell 22:165–178
107. Guo JY, Chen H-Y, Fan J et al (2011) Activated Ras requires autophagy to maintain oxidative metabolism and tumorigenesis. Genes Dev 25:460–470
108. Kim M-J, Woo S-J, Yoon C-H et al (2011) Involvement of autophagy in oncogenic K-Ras-induced malignant cell transformation. J Biol Chem 286:12924–12932
109. Buyru N, Tigli H, Ozcan F et al (2003) Ras oncogene mutations in urine sediments of patients with bladder cancer. J Biochem Mol Biol 36:399–402
110. Choudhary S, Wang HCR (2007) Proapoptotic ability of oncogenic H-Ras to facilitate apoptosis induced by histone deacetylase inhibitors in human cancer cells. Mol Cancer Ther 6:1099–1111
111. Choudhary S, Wang HCR (2009) Role of reactive oxygen species in proapoptotic ability of oncogenic H-Ras to increase human bladder cancer cell susceptibility to histone deacetylase inhibitor for caspase induction. J Cancer Res Clin Oncol 135:1601–1613
112. Choudhary S, Rathore K, Wang HCR (2011) Differential induction of reactive oxygen species through Erk1/2 and Nox-1 by FK288 for selective apoptosis of oncogenic H-Ras-expressing human urinary bladder cancer J82 cells. J Cancer Res Clin Oncol 137:471–480
113. Choudhary S, Rathore K, Wang HC (2010) FK228 and oncogenic H-Ras synergistically induce Mek1/2 and Nox1 to generate reactive oxygen species for differential cell death. Anticancer Drugs 21:831–840
114. Choudhary S, Wang KKA, Wang HCR (2011) Oncogenic H-Ras, FK228, and exogenous $H_2O_2$ cooperatively activated the ERK pathway in selective induction of human urinary bladder cancer J82 cell death. Mol Carcinogen 50:215–219
115. Choudhary S, Sood S, Donnel RL et al (2012) Intervention of human breast cell carcinogenesis chronically induced by 2-amino-1-methyl-6-phenylimidazo[4,5-b]pyridine. Carcinogenesis 33:876–885
116. Deora AA, Hajjar DP, Lander HM (2000) Recruitment and activation of Raf-1 kinase by nitric oxide-activated Ras. Biochemistry 39: 9901–9908
117. Ahmad R, Sylvester J, Ahmad M et al (2011) Involvement of H-Ras and reactive oxygen species in proinflammatory cytokine-induced matrix metalloproteinase-13 expression in human articular chondrocytes. Arch Biochem Biophys 507:350–355
118. Gabrielli A, Svegliati S, Moroncini G et al (2012) New insights into the role of oxidative stress in scleroderma fibrosis. Open Rheumatol J 6:87–95
119. Hordijk OL (2006) Regulation of NADPH oxidases: the role of Rac proteins. Circ Res 98:453–462
120. Harrison D, Griendling KK, Landmesser U et al (2003) Role of oxidative stress in atherosclerosis. Am J Cardiol 91:7A–11A
121. Lassegue B, Clempus RE (2003) Vascular NAD(P)H oxidases: specific features, expression, and regulation. Am J Physiol Regul Integr Comp Physiol 285:R277–R297
122. Madamanchi NR, Vendrv A, Runge MS (2005) Oxidative stress and vascular disease. Arterioscler Thromb Vasc Biol 25:29–38
123. Gregg D, Rauscher FM, Goldschmidt-Clermont PJ (2003) Rac regulates cardiovascular superoxide through diverse molecular

interactions: more than a binary GTP switch. Am J Physiol Cell Physiol 285:C723–C734
124. Seshiah PN, Weber DS, Rocic P et al (2002) Angiotensin II stimulation of NAD(P)H oxidase activity: upstream mediators. Circ Res 91:406–413
125. Zuo L, Ushio-Fukai M, Ikeda S et al (2005) Caveolin-I is essential for activation of Rac1 and NAD(P)H oxidase after angiotensin II type I receptor stimulation in vascular smooth muscle cells. Role in redox signaling and vascular hypertrophy. Arterioscler Thromb Vasc Biol 25:1824–1830
126. del Pozo MA, Balasubramanian N, Alderson NB et al (2005) Phospho-caveolin-1 mediates integrin-regulated membrane domain internalization. Nat Cell Biol 7:901–908
127. Vecchione C, Patrucco E, Marino G et al (2005) Protection from angiotensin II-mediated vasculotoxic and hypertensive response in mice lacking PI3Kgamma. J Exp Med 201:1217–1228
128. Kim K-S, Takeda K, Sethi R et al (1998) Protection from reoxygenation injury by inhibition of Rac1. J Clin Invest 101:1821–1826
129. Harada N, Iimuro Y, Nitta T et al (2003) Inactivation of the small GTPase Rac1 protects the liver from ischemia/reperfusion injury in the rat. Surgery 134:480–491
130. Andreadis AA, Hazen SL, Comhair SA et al (2003) Oxidative and nitrosative events in asthma. Free Radic Biol Med 35:213–225
131. Rabe KF, Dent G, Magnussen H (1995) Hydrogen peroxide contracts human airways in vitro: role of epithelium. Am J Physiol 269: 332–338
132. Kojima K, Kume H, Ito S et al (2007) Direct effects of hydrogen peroxide on airway smooth muscle tone: roles of $Ca^{2+}$ influx and Rho-kinase. Eur J Pharmacol 556:151–156
133. Jin L, Ying Z, Webb RC (2004) Activation of Rho/Rho kinase signaling pathway by reactive oxygen species in rat aorta. Am J Physiol Heart Circ Physiol 287:1495–1500

# Chapter 6

# Biophysical and Proteomic Characterization Strategies for Cysteine Modifications in Ras GTPases

G. Aaron Hobbs, Harsha P. Gunawardena, and Sharon L. Campbell

## Abstract

Cysteine is one of the most reactive amino acids and is modified by a number of oxidants. The reactivity of cysteines is dependent on the thiol p$K_a$; however, measuring cysteine p$K_a$ values is nontrivial. Ras family GTPases have been shown to contain a free cysteine that is sensitive to oxidation, and free radical-mediated oxidation of this cysteine has been shown to be activating. Here, we present a new technique that allows for measuring cysteine p$K_a$ values using a fluorescent detection system with the molecule 4-fluoro-7-aminosulfonylbenzofurazan (ABD-F). In addition, we also describe how to generate several oxidants. Lastly, we describe several mass spectrometry-based experiments and the necessary adjustments to the experiments to detect cysteine oxidation.

**Key words** Cysteine p$K_a$, Oxidation, Ras GTPases, Mass spectrometry, Reactive oxygen and nitrogen species

## 1 Introduction

One of the most reactive amino acids in proteins is cysteine, which can undergo a variety of different posttranslational modifications and several different types of oxidative reactions [1]. In fact, thiol oxidation plays a key role in regulating redox homeostasis and protecting the cell during oxidative stress [2–4]. Moreover, many genes that respond to redox stress show altered regulation by reactive oxygen and nitrogen species (ROS and RNS) [5]. Some of the common end products of cysteine oxidation are disulfide bond formation, mixed disulfide bond formation with glutathione (glutathiolation), and nitrosation, as well as sulfenic, sulfinic, and sulfonic acid [6]. Many redox-sensitive proteins contain cysteine residue(s) that has an altered p$K_a$. The microenvironment surrounding the cysteine determines the p$K_a$ of the thiol side chain. However, cysteine p$K_a$ values are difficult to reliably predict as several factors influence p$K_a$, including solvent exposure, hydrogen bond formation, and charge-charge interactions [7]. The p$K_a$ of L-cysteine in

Lorenza Trabalzini and Saverio Francesco Retta (eds.), *Ras Signaling: Methods and Protocols*,
Methods in Molecular Biology, vol. 1120, DOI 10.1007/978-1-62703-791-4_6, © Springer Science+Business Media, LLC 2014

water has been measured to be ~8.5–9, and the thiolate form ($RS^-$) is approximately tenfold more reactive than the thiol form (RSH) [8]. Thus, redox-regulated proteins often contain reactive cysteines with altered $pK_a$ values that are close to or lower than physiological pH.

Critical cellular enzymes that regulate redox homeostasis include peroxiredoxins (reduce peroxide), thioredoxins (reduce oxidized peroxiredoxin), and glutathione peroxidases (reduce peroxide using the cellular glutathione pool), which act through mechanisms involving cysteine oxidation. The $pK_a$ of the active-site cysteines have been shown to be markedly reduced in each of these classes of enzymes; for example, the active-site cysteine $pK_a$ has been shown to be approximately 5–6 in 2-Cys peroxiredoxins [9], the active-site cysteine in *Escherichia coli* thioredoxin was measured to be between 7.1 [10] and 7.5 [11], and the $pK_a$ of the active-site cysteine in glutathione peroxidase has been estimated to be 7.2 [12].

The work in our lab is centered on the redox regulation of Ras superfamily GTPases. We have shown that the activity of a subset of Ras and Rho GTPases can be regulated through redox-sensitive cysteines [13–15]. Ras GTPases, in particular, have received a great deal of interest in the field of redox biology. There are four distinct Ras genes in the human genome, H-, K- (1A and 1B), and N-Ras, which differ primarily in their carboxyl-terminal regions. They encode small, 21-kDa guanine nucleotide-binding proteins that function as molecular switches to modulate signaling pathways that control cell growth, differentiation, and apoptosis [16]. This is achieved by cycling between the inactive ("off") GDP-bound and active ("on") GTP-bound states. Given the high affinity interaction between Ras and its nucleotide ligands (GDP and GTP) as well as the slow intrinsic rate of GDP dissociation and GTP hydrolysis, two classes of modulatory proteins regulate the activation state of Ras proteins. One class of proteins that activate Ras is guanine nucleotide exchange factors (GEFs), which promote exchange of GDP for GTP. Analogous to guanine nucleotide exchange factors, redox agents have been shown to stimulate nucleotide exchange and alter the activity of Ras proteins through reaction with cysteine 118 ($Cys^{118}$), which is located in a conserved guanine nucleotide-binding motif [17]. We have found that only redox agents capable of thiyl radical formation, such as $NO_2^{\bullet}$, can modulate Ras activity [18]. Thiyl radical formation at $Cys^{118}$ facilitates guanine nucleotide exchange in Ras by promoting oxidation and the subsequent dissociation of the bound guanine base [19]. In cells, where the GTP/GDP ratio is in excess, this can lead to exchange of GDP for GTP and result in Ras activation. Importantly, this mechanism of Ras regulation has been shown to play a role in Ras-mediated tumorigenesis and tumor maintenance [20].

Rho GTPases are members of the Ras superfamily, and like Ras GTPases, function as molecular switches to regulate cellular growth. However, they also regulate cell motility and oxidant regulation [21]. We have previously shown that a subset of Rho GTPases can be regulated through redox modification [22]. These GTPases contain a distinct, reactive cysteine that is conserved in ~40 % of Rho GTPases and renders them sensitive to oxidation and oxidative modification in cells [23].

Although several methods to determine cysteine p$K_a$ values have been employed in the field [7], we have recently developed a fluorescence-based method to measure cysteine p$K_a$ values [24]. Using 4-fluoro-7-aminosulfonylbenzofurazan (ABD-F), a compound that specifically reacts with the thiolate form of the cysteine side chain, we have been able to measure the p$K_a$ of reactive thiols in Ras and Rho family GTPases. In Ras, Cys$^{118}$ is the only solvent accessible cysteine in the core catalytic domain (Ras residues 1–166). Thus, Ras provides an excellent system to demonstrate the use of the ABD-F strategy for measuring p$K_a$ values of cysteine thiol side chains.

When coupled with mass spectrometry (MS), this method provides an analytical platform for the detection and quantification of redox modifications in Ras family GTPases and other proteins. The bottom-up MS method we've employed [25] requires proteolytic digestion with trypsin followed by LC-MS analysis by electrospray ionization (ESI). The accurate identification of peptides has become a routine practice with the availability of a vast number of programs and search engines that assist in assigning peptide sequences to mass spectra using probabilistic prediction algorithms [26]. In addition, the intensity, or spectral count information, can be used for the relative or absolute quantification of peptides associated with site-specific protein modifications.

Herein, we describe in detail biophysical and proteomic approaches to detect cysteine modifications in Ras GTPases, as well as the preparation and handling of ROS and RNS. These methods include:

1. Cysteine p$K_a$ determination using 4-fluoro-7-aminosulfonylbenzofurazan (ABD-F).
2. Generation of cysteine-modifying redox-active compounds.
3. $NO_2^{\bullet}$ generation and NONOates.
4. Quantitative mass spectrometry including

   Isotope-coded affinity tag-labeling for relative cysteine quantification.

   Quantification of the cysteine oxidation.

   Differential quantification of cysteine oxidation.

   Differential thiol trapping to determine reversible oxidation of cysteine residues.

   Filter-aided sample preparation.

## 2 Materials

### 2.1 ABD-F Buffers and Reagents

1. Components: 4-fluoro-7-aminosulfonylbenzofurazan (ABD-F), black, flat-bottom, non-coated plates, and protein spin concentrators.
2. Reducing buffer: 15 mM 4-(2-hydroxyethyl)-1-piperazine-ethanesulfonic acid (HEPES), pH 8.0, 5 mM $MgCl_2$, 30 mM NaCl, 200 μM diethylene triamine pentaacetic acid (DTPA), and 5 mM dithiothreitol (DTT; added day of use).
3. ABD-F modification buffer: 15 mM 2-(*N*-morpholino)ethanesulfonic acid (MES), pH 6.5; (*see* **Note 1**), 5 mM $MgCl_2$, 30 mM NaCl, and 200 μM DTPA.
4. pH screen buffer (for a pH range between 5.5 and 8.5): 100 mM MES, 100 mM HEPES, 5 mM $MgCl_2$, and 200 μM DTPA (titrate each pH value individually).

### 2.2 Nitrosation Reagent (CysNO/GSNO) Generation

1. Solution I: 50 mM L-cysteine (or reduced glutathione) in 120 mM HCl.
2. Solution II: 50 mM $NaNO_2$.
3. Solution III: 40 mM ammonium sulfamate.
4. CysNO dilution buffer: 100 mM HEPES, pH 7.5, 5 mM $MgCl_2$, and 1 mM DTPA.

### 2.3 Nitric Oxide-Releasing Agents (NONOates)

1. All NONOate compounds (available from Cayman Chemicals).
2. 4,5-Diaminofluorescein (DAF-2).

### 2.4 Mass Spectrometry Techniques

#### 2.4.1 Isotope-Coded Affinity Tag (ICAT)-Labeling for Relative Cysteine Quantification

1. Strong cation-exchange (SCX) PolySULFOETHYL A column.
2. Isotope-coded affinity tag (ICAT) reagents, including a cation-exchange cartridge and avidin affinity cartridge.
3. Micro-concentrator/spin-filter device.
4. C18 spin column.

#### 2.4.2 Quantification of the Cysteine Oxidation

1. Ammonium bicarbonate (ABC) buffer: 200 mM $NH_4HCO_3$.
2. ABC + Iodoacetamide (IAM): 200 mM $NH_4HCO_3$ and 5 mM IAM.
3. Phosphate buffer: 100 mM $Na_2HPO_4$, pH 7.4, 50 mM NaCl, and 5 mM $MgCl_2$.
4. Phosphate-buffered saline (PBS): 10 mM $Na_2HPO_4$, 2 mM $KH_2PO_4$, 137 mM NaCl, and 2.7 mM KCl.
5. BIAM-elution buffer: 50 $Na_2HPO_4$, pH 7.2, 2 mM D-biotin, and 150 mM NaCl.

*2.4.3 Differential Quantification of Cysteine Oxidation*

1. Tris buffer: 50 mM Tris, pH 8.5, 25 mM NaCl, and 5 mM $MgCl_2$.
2. SCX buffer A: 10 mM $KH_2PO_4$, pH 3.0, and 20 % (vol/vol) acetonitrile (ACN).
3. SCX buffer B: 10 mM $KH_2PO_4$, pH 3.0, 350 mM KCl, and 20 % (vol/vol) ACN.

*2.4.4 Differential Thiol Trapping to Determine Reversible Oxidation of Cysteine Residues*

Denaturing alkylation buffer (DAB): 200 mM Tris, pH 8.5, 8 M urea, 0.5 % (wt/vol) SDS, and 10 mM EDTA.

*2.4.5 Filter-Aided Sample Preparation (FASP)*

1. Denaturing buffer: 0.1 M Tris, pH 8.5, and 8 M urea.
2. Iodoacetamide (IAM) buffer (should be made day of use): 0.1 M Tris, pH 8.5, 50 mM IAM, and 8 M urea.

## 3 Methods

### *3.1 Cysteine p$K_a$ Determination Using 4-Fluoro-7-aminosulfonyl-benzofurazan (ABD-F)*

1. Determine the amount (in moles) of protein that will be required to perform the assay. The assay as described is performed in a 96-well plate and requires fluorescence detection at 513 nm. Each well should contain approximately 5 or 10 μM protein and be performed in triplicate at each pH value (technical repeats). To determine the p$K_a$ accurately, a pH range must be performed with adequate coverage. As ABD-F fluorescence intensity is sensitive to the site of modification, it is important to ensure similar buffer conditions across the pH range selected.
2. Calculate the quantity of protein required; for example, 12 pH values performed in triplicate requires approximately 72 nmol of protein to perform the assay (200 μl/well; 10 μM/well × (36 wells × 200 μl) = 72 nmol required).
3. Reduce the protein with dithiothreitol (DTT) or dithiobutylamine (DTBA) (*see* **Note 2**) to increase the yield of protein that can react with ABD-F. For adequate reduction, exchange the protein into a buffer containing 5 mM (or greater) DTT. Prepare the reducing buffer (minimum pH of 8.0) and exchange the protein into the buffer (using Amicon Ultra concentrators). As a subset of oxidative reactions (sulfenic and sulfonic acid) will not be reversed by DTT (*see* **Note 3**), this step serves to increase the percentage of protein that can be modified by ABD-F and enhances the signal-to-noise ratio of the assay.
4. While the protein is being buffer exchanged, prepare the ABD-F modification buffer. To remove the dissolved oxygen

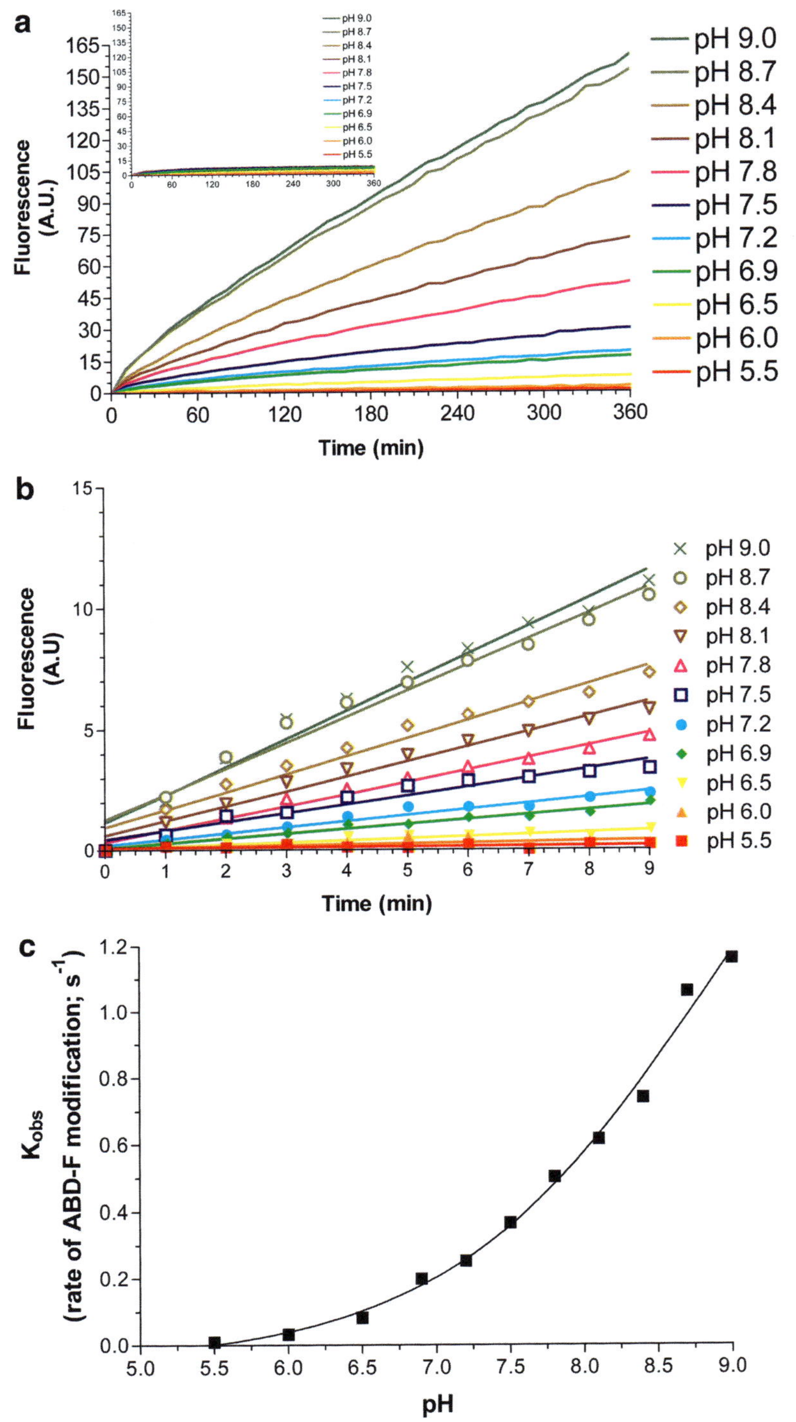


**Fig. 1** ABD-F modification of H-Ras$^{WT}$ over a selected pH range. (**a**) H-Ras$^{WT}$ (10 μM) was reacted with ABD-F (1 mM) over a period of 6 h at the indicated pH values. Data was collected every minute, and each reaction was performed in triplicate (error bars removed for clarity). Absolute fluorescence values are presented. In systems containing one modifiable cysteine, the total fluorescence

(*see* **Note 4**), sparge the ABD-F modification buffer with an inert gas, such as $N_2$, for 30 min.

5. Incubate the protein in reducing buffer for at least 60 min on ice for complete reduction.
6. Buffer exchange the protein into the sparged ABD-F modification buffer using Amicon Ultra concentrators (use appropriately sized membrane pores). This step removes reducing agents from the sample. It is critical that at least three rounds of exchange are performed to efficiently remove reducing agent from the sample (*see* **Note 5**).
7. Determine the protein concentration.
8. Set up the 96-well plates. The excitation wavelength for ABD-F is 389 nm and emission wavelength is 513 nm. Set up a "dummy tray" that contains half of the volume of the final reaction (100 μl). In the "dummy tray," add ABD-F to a final concentration that is double the intended concentration (add 2 mM ABD-F in 100 μl). As the reaction proceeds through second-order kinetics, it is critical that all wells contain exactly the same amount of protein and ABD-F; therefore, the rate of modification is related to the thiol/thiolate state of the cysteine. In a separate 96-well plate, replicate the pH screen of the "dummy tray," but add 20 μM protein in 100 μl to each well (*see* **Note 6**).
9. Using a multichannel pipette, quickly remove the solution from the "dummy tray" and place the solution into the respective wells of the test plate to initiate the ABD-F reaction. Quickly start the plate reader for data collection.
10. Data analysis (*see* **Note 7**). Use the linear portion of the initial slopes to calculate the rate of fluorescence generation over time (Fig. 1a, b). Plot the determined rates of fluorescence generation versus pH and fit the resulting sigmoidal curve to a

**Fig. 1** (continued) values will plateau at a similar value if given sufficient time. The inset shows the relative reactivity of H-Ras$^{C118S}$, which was performed under identical conditions as H-Ras$^{WT}$. In this graph, it is apparent that little to no modification of cysteines occurred in Ras$^{C118S}$. (**b**) The initial rates of ABD-F modification are used for p$K_a$ determination. Here, the initial data are presented (zoomed in from **a**), and the initial rate was determined using GraphPad Prism; linear curve fitting was performed. Error bars are removed for clarity. (**c**) The observed rates of modification are plotted against pH. Each data point represents the rate of modification determined from the curve fitting in (**b**). The data were fit to a Boltzmann Sigmoidal distribution to determine the p$K_a$. Control data (see inset in **a**) show that only Ras$^{C118}$ is modified by ABD-F as the Ras$^{C118S}$ control shows no reactivity. However, Ras$^{Cys118}$ does not have an altered p$K_a$. Hence, the maximal rate of modification could not be obtained as Ras becomes unstable at high pH values. Therefore, the p$K_a$ is an estimate

Boltzman sigmoidal equation (GraphPad Prism). This equation is used to calculate the p$K_a$ for a single site-specific modification (*see* **Note 8**; Fig. 1c).

11. After the fluorescence data has been collected, determine the site(s) modified by ABD-F (*see* **Note 9**). A straightforward method is to employ liquid chromatography-mass spectrometry (LC-MS) on trypsin-digested samples. We have also found that cysteine-to-serine mutants can be used to confirm modification sites if the mutation does not alter the protein structure and/or ABD-F reactivity.

### 3.2 Generation of Cysteine-Modifying Redox-Active Compounds (CysNO/GSNO Generation)

As the reactions of ROS and RNS with proteins are diverse and often complex, this section will focus on the generation and quantification of various redox-active compounds. Subheading 3.4 will describe mass spectrometry (MS) approaches to quantify and determine the sites of modification.

1. Prepare solution I and solution II on the day of use (*see* **Notes 10** and **11**).
2. Mix 100 μl of solution I with 100 μl of solution II and place into a foil-covered 1.5-ml Eppendorf tube. Allow the reaction mixture to sit in the dark for 10 min. These reactions are highly light sensitive; therefore, care should be maintained to keep all reactions in the dark.
3. Add 20 μl of 40 mM ammonium sulfamate to remove unreacted nitrate and allow the reaction to proceed in the dark for an additional 2 min. This protocol assumes an ~85 % efficiency of the reaction (although published reports suggest an efficiency as high as ~90 % [27]) and adds a sufficient concentration of sulfamate to remove all remaining $NO_2$.
4. Dilute the thiol-NO (i.e., CysNO or GSNO) with 880 μl (fourfold) of CysNO dilution buffer. While this buffer can be altered to fit the needs of the individual researcher, the high concentration of the buffer component prevents changes in pH, reduces metal contaminants, and increases the half-life of the thiol-NO in solution [28, 29].
5. Measure the concentration of the generated thiol-NO using absorbance at 336 nm. The molar absorptivity for CysNO and GSNO is 900 $M^{-1}$ $cm^{-1}$. Alternatively, the molar absorptivity of 16.8 $M^{-1}$ $cm^{-1}$ at 543 nm can be used for concentration determination (*see* **Note 12**).

### 3.3 $NO_2^{\bullet}$ Generation and NONOates

While CysNO represents a good nitrosation agent that can modify protein thiols through non-radical oxidative (two-electron) chemistry, the activity of Ras GTPases are uniquely sensitive to free radical oxidation. Free radical agents, such as $NO_2^{\bullet}$, can react with a solvent accessible thiol in Ras to produce a thiyl radical.

Through electron transfer, the bound guanine nucleotide becomes oxidized, which results in enhanced guanine nucleotide dissociation and enhanced guanine nucleotide exchange (activation) [13]. Thus, Ras activity is modulated by free radical agents capable of thiyl radical formation but not through non-radical oxidation. However, this radical-mediated mechanism may not be unique to Ras GTPases. Increasing evidence suggests that in many cases, non-radical-mediated thiol oxidation reactions may be too slow to be biologically relevant [30]. Thus, in this section, we describe the preparation and handling of $NO_2^•$ for use in radical-mediated protein modification.

#### 3.3.1 $NO_2^•$ Gas Generation

1. $NO_2^•$ gas can be purchased from various companies; however, it can just as easily be generated. Obtain a small reaction vial and add a small piece of copper wire (~100 mg or less) and cover the vial with a rubber stopper that is airtight.
2. Deplete the oxygen from the vial using an inert gas, such as $N_2$.
3. Inject in a small volume of nitric acid. Allow the reaction to proceed for 5–10 min. Nitrogen dioxide radicals are formed by the following equation:

$$Cu + 4HNO_3 \rightarrow Cu(NO_3)_2 + 2^{•}NO_2 + 2H_2O \quad (1)$$

The reaction proceeds by the above reaction pathway if a brown gas is generated. This reaction should be performed in a hood as $NO_2^•$ gas is highly toxic.

4. Estimate the concentration of $NO_2^•$ gas that will dissolve in solution using Henry's Law. Henry's Law states that the amount of a gas that will dissolve in solution is directly proportional to the partial pressure of the gas in equilibrium with the liquid. The coefficient ($k$) for $NO_2^•$ is needed for these calculations. From a table containing these coefficients, a range of 1,800–2,500 mol/l/atm can be obtained [31, 32]. As the density for $NO_2^•$ gas (2.62 g/cm$^3$) is heavier than $N_2$ gas (0.808 g/cm$^3$), one can assume that the relative gas concentration is 100 % $NO_2^•$ at the bottom of the vial. The coefficient needs to be converted to units of atm/mol/l by taking the reciprocal of the coefficient. Thus, a range of $1.2–3.4 \times 10^{-2}$ atm/mol/l is obtained.
5. Prepare a reaction vial with the protein. Seal the reaction vial and deplete the oxygen using $N_2$ or another inert gas. The head-space remaining will be critical; thus, the volume remaining in the reaction vial after the addition of the protein sample needs to be determined.
6. Add $NO_2^•$ gas to the reaction vial. The amount of gas added to the sealed reaction vial will not likely affect the pressure in the reaction vial significantly; therefore, the pressure is estimated to be 1 atm. However, the percent of $NO_2^•$ added to the vial

relative to the remaining headspace must be calculated to determine the percent of $NO_2^•$ gas in the gas mixture. Thus, if 100 μl of pure $NO_2^•$ gas is added to a reaction vial with 1 ml head space (inert $N_2$ gas), then the total concentration of $NO_2^•$ gas will be 10 % (thus, $0.10 \times 1.00$ atm = 0.10 atm $NO_2^•$ gas in this example).

7. Using the equation $c = p/k$, where $c$ is the concentration, $p$ is the partial pressure (from **step 8**, 0.10 atm), and $k$ is the inverse of Henry's law coefficient (from **step 4**, using an average of $2.3 \times 10^{-2}$ atm/mol/l), approximately 4.35 mol/l $NO_2^•$ gas will dissolve in liquid in a sealed reaction vial.

### 3.3.2 NONOates

NONOates, such as the compounds listed in Table 1, are compounds that release $NO^•$ in solution in a pH and time-dependent manner over seconds, minutes, or hours.

There are 8 NONOate compounds available. All NONOates are relatively stable at alkaline pH (approximately 24 h at 0 °C) and are water soluble. However, at pH 5, the release of $NO^•$ is nearly instantaneous. In addition, each NONOate has a characteristic UV absorbance value. Relevant information on NONOates is listed in Table 1.

1. Suspend the desired amount of NONOate in 10 mM NaOH to minimize $NO^•$ release (*see* **Note 13** and Fig. 2).
2. To measure $NO^•$ release in the selected buffer, 4,5-diaminofluorescein (DAF-2) can be used. DAF-2 specifically

**Table 1**
**Specifications and use of NONOate compounds**

| Name | Cas# | Efficiency (mol $NO^•$ per NONOate) | Half-life at 37 °C; 22–25 °C | $\lambda_{max}$ (nm) | $\varepsilon$ ($M^{-1}cm^{-1}$) |
|---|---|---|---|---|---|
| DETA NONOate | 146724-94-9 | 2 | 20 h; 56 h | 252 | 7,640 |
| Spermine NONOate | 136587-13-8 | 2 | 39 min; 230 min | 252 | 8,500 |
| Proli NONOate | N/A | 2 | 1.8 s; N/A | 252 | 8,400 |
| DPTA NONOate | 146724-95-0 | 2 | 3 h; 5 h | 252 | 7,860 |
| DEA NONOate | 372965-00-9 | 1.5 | 2 min; 16 min | 250 | 6,500 |
| PAPA NONOate | 146672-58-4 | 2 | 15 min; 77 min | 252 | 8,100 |
| MAHMA NONOate | 146724-86-9 | 2 | 1 min; 2.7 min | 250 | 7,250 |
| Sulpho NONOate | 61142-90-3 | 0 | 7 min; 77 min | 252 | 8,100 |
| [54–56] | | | | | |

All NONOate compounds currently available and the CAS# (Proli NONOate currently has no CAS#) are listed. The efficiency represents the number of moles $NO^•$ released per mol of NONOate compound. The half-life information is provided at both temperatures. The $\lambda_{max}$ is the wavelength where concentration of the compounds can be taken using a UV spectrophotometer, and $\varepsilon$ is the corresponding extinction coefficient

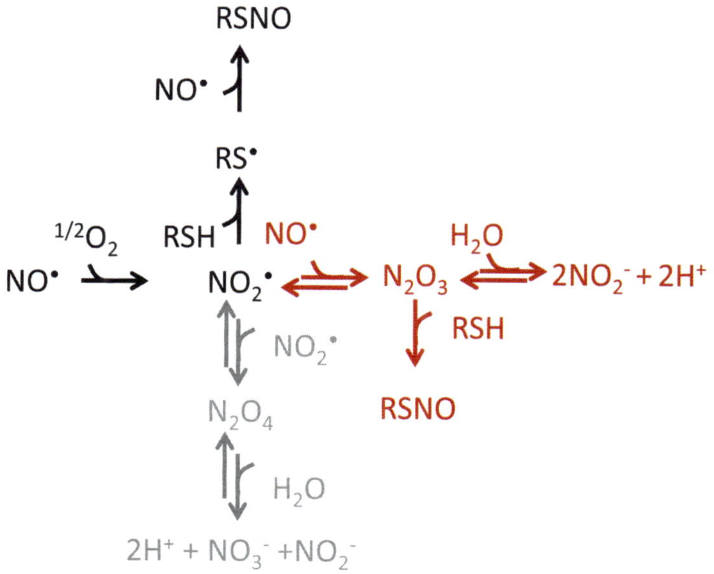


**Fig. 2** The reaction profile of NO• generated from NONOates. The two most important reactions with $NO_2^{\bullet}$ are with free thiols (*black* pathway) and free NO• (*red* pathway). The rate for $NO_2^{\bullet}$ with free thiol (RSH) is $\geq 2 \times 10^7\ M^{-1}\ s^{-1}$ [51, 52]. For the reaction of $NO_2^{\bullet}$ with NO• to form $N_2O_3$, the rate is $1.1 \times 10^9\ M^{-1}\ s^{-1}$ [51, 53]. While this reaction is readily reversible, the reaction kinetics in vitro will favor the nitrosation of thiols through the reaction $N_2O_3 + RSH \rightarrow RSNO$ (in *red*, rate of $1.2 \times 10^7\ M^{-1}\ s^{-1}$ and an autohydrolysis rate with water of $4.75 \times 10^7\ s^{-1}$). However, in an in vivo system, the pathway in *black* will likely be the major route of protein oxidation [13]. While the end product of the pathway in *red* is identical to the pathway in *black* (free radical mediated), the reactions involving thiyl radical species result in unique regulation in Ras family GTPase activity. The gray reaction pathway is less likely to occur because it relies on three bimolecular reactions ($2(NO^{\bullet} + NO^{\bullet} \rightarrow NO_2^{\bullet})$ and $NO_2^{\bullet} + NO_2^{\bullet} \rightarrow N_2O_4$), whereas the *red* and *black* pathways require only two bimolecular reactions

binds to NO• and undergoes a change in fluorescence upon reaction. DAF-2 has excitation and emission wavelengths of 485 and 538 nM, respectively. Approximately 10 μM DAF-2 is recommended. DAF-2 has a detection limit of 5 nM at neutral pH (*see* **Note 14**).

### 3.4 Quantitative Mass Spectrometry

The cysteine labeling procedure known as isotope-coded affinity tag (ICAT) [33] was the first stable isotope-based chemical labeling method used in quantitative proteomics. The ICAT approach was designed primarily for the semiquantitation of proteins expressed in two cellular states by labeling proteins with heavy and light cysteine-reactive tags. The trypsin-digested peptides that correspond to the differential labels are identified using MS, and the relative signal intensities are used for quantitative comparison (*see* Fig. 3).

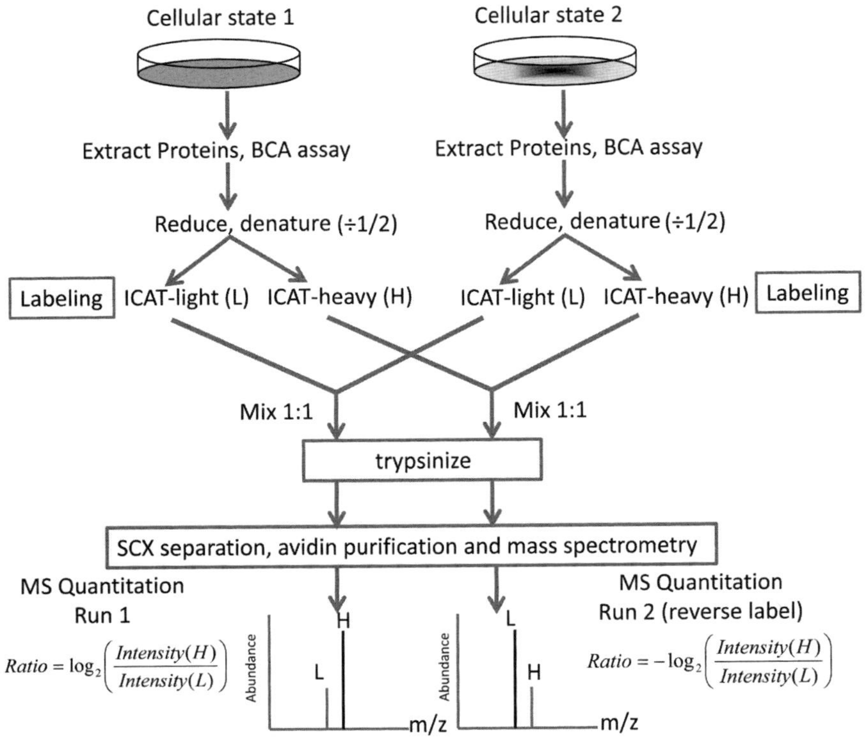


**Fig. 3** Generalized ICAT-based relative quantification schematic for measuring cysteine peptides. ICAT allows for the relative quantification of samples from two separate conditions, such as quantification of protein expression levels under different cellular conditions. However, if one omits the reducing steps and adds an oxidation step to one of the conditions, then this approach can be adapted to study oxidation of proteins in cellular tissue (described in detail in the acid-cleavable ICAT with $H_2O_2$ approach). Each sample can be subjected to a technical replicate by reverse labeling to ensure reproducibility. Furthermore, ICAT reagents can be swapped out for other stable isotope-labeled tags, such as BIAM or Cys-tandem mass tag reagents

Proteomics-based methodologies have recently been developed to study redox regulation in proteins, also known as the redoxome [34, 35]. A number of stable isotope labeling methods exist for measuring cysteine oxidation by MS. The majority of these methods fall into two main categories: (1) cysteine oxidation measured directly by comparing the loss of oxidized peptide as a function of oxidants by the relative signal intensities of cysteine-containing peptides labeled with light and heavy stable isotope reagents, such as ICAT, biotinylated-ICAT [36, 37], or biotinylated iodoacetamide (BIAM) [35], and (2) cysteine oxidation measured indirectly by thiol-blocking, selective reduction, and reversible modification using thiol probes that detect a gain in signal due to thiol oxidation, which is defined as oxICAT (*see* Fig. 4) [38]. In the context of measuring redox modifications, ICAT and other types of cysteine-reactive tags allow for the accurate quantification of cysteine-containing peptides [39]. Here, we describe new

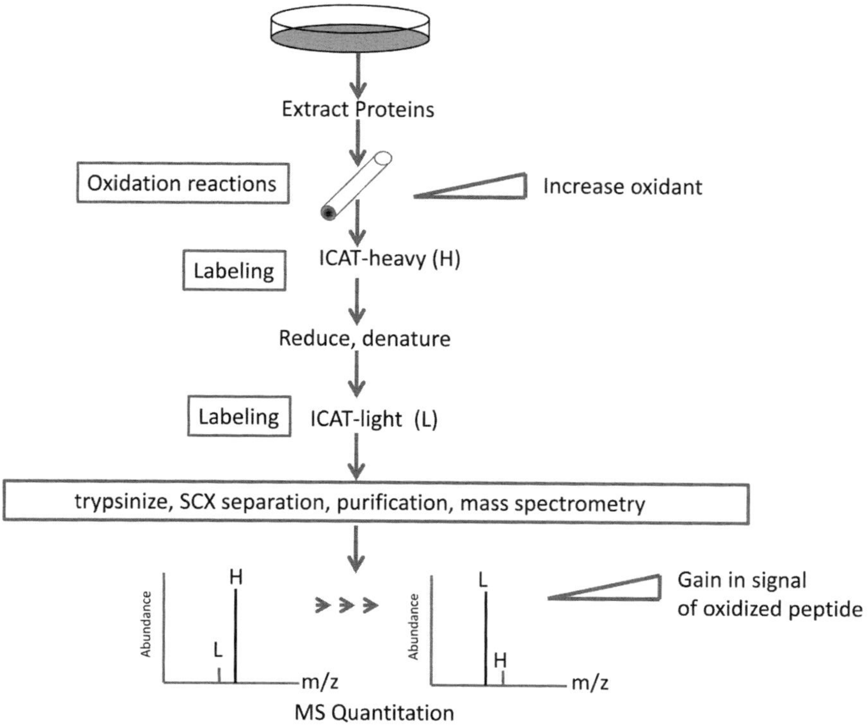


**Fig. 4** The oxICAT schematic for measuring cysteine oxidation. This method measures reversible cysteine oxidation. Here, the protein is denatured and labeled in one step, which labels all non-oxidized and buried cysteines with the heavy label, and is completed by reducing the oxidized species and labeling with the light isotope. Thus, if several experiments are performed by varying levels of oxidant exposure, one can estimate the level of oxidation (relative reactivity) of cysteines in a protein. A gain in signal intensity of the light isotope-labeled signal is an indirect measure of reversible thiol oxidation. Note that a technical or biological replicate analysis can be performed by reversing the order of the ICAT labels with the heavy isotope representing the oxidized species

techniques for detecting and quantifying cysteine oxidation using mass spectrometry-based methods, including isotope-coded affinity tag-labeling for relative cysteine quantification, quantification of the cysteine oxidation, differential quantification of cysteine oxidation, differential thiol trapping to determine reversible oxidation of cysteine residues, and filter-aided sample preparation (*see* **Note 15**).

#### *3.4.1 Isotope-Coded Affinity Tag (ICAT)-Labeling for Relative Cysteine Quantification*

1. Generate proteins from cellular extracts and measure total protein using the bicinchoninic acid (BCA) assay.
2. Supplement the buffer with 0.1 % sodium dodecyl sulfate (SDS) before modification with the ICAT reagents (*see* **Note 16**).
3. Incubate the modified and unmodified samples at 37 °C with the acid-cleavable $^{12}C$ (light) or $^{13}C$ (heavy) ICAT reagents in the absence of reducing agent using the protocol supplied by the manufacturer.

4. After 2 h of incubation, mix the light and heavy ICAT-labeled protein samples at a stoichiometric ratio of 1:1, and trypsin digest the mixture by incubating at 37 °C overnight in ABC buffer.
5. Purify the trypsin-digested peptides using a cation-exchange cartridge to remove excess labeling reagent. The desalted peptides are affinity purified using the avidin affinity cartridge. Dry the peptides and suspend in the cleavage reagent to release the peptides from the acid-cleavable linker by incubating at 37 °C for 2 h. Dry the acid-cleaved peptides and suspend in 0.1 % formic acid for LC-MS analysis (*see* **Notes 17** and **18** for data analysis, *see* **Notes 19** and **20** for alternative approaches).

#### 3.4.2 Quantification of the Cysteine Oxidation

The biotinylated iodoacetamide (BIAM) approach can be used to study oxidation of exposed cysteines under controlled redox conditions by quantifying differential BIAM labeling. The redox chemistry includes the reversible oxidation of cysteine residues by the addition of $H_2O_2$ or other oxidants as well as reduction with reducing agents to reduce oxidized cysteines.

1. Generate oxidized proteins.
2. Dissolve the proteins in ABC buffer containing 5 mM iodoacetamide (IAM) and incubate at room temperature in the dark for 30 min to complete the modification of accessible Cys residues with IAM.
3. Centrifuge the samples at 15,000 × *g* for 1 h, dialyze the supernatant against phosphate buffer, and divide the sample into two equal parts.
4. Dilute the two equal samples to 5 ml with phosphate buffer.
5. Add 250 μl of 100 mM DTT. Similarly, incubate the second control sample with 250 μl phosphate buffer.
6. Incubate both samples at room temperature for 30 min and treat with 275 μl of 100× BIAM for 30 min in the dark at room temperature. Remove excess BIAM by overnight dialysis against ABC buffer.
7. Load the sample on an avidin affinity cartridge followed by incubation for 20 min at room temperature.
8. Wash the column with 1–4 volumes of PBS, collect the flow-through, and wash the column further with PBS until the absorbance at 280 nm returns to baseline.
9. Elute the BIAM-modified proteins with four bed volumes of BIAM-elution buffer.
10. Trypsin digest the proteins overnight at 37 °C. Purify peptides using a PepClean desalting column according to the manufacturer's protocol and analyze the peptides by LC-MS analysis.

*3.4.3 Differential Quantification of Cysteine Oxidation*

1. Incubate oxidized and reduced protein samples (200 μg) in 100 μl of Tris buffer with light and heavy acid-cleavable ICAT reagents, respectively, at 37 °C for 3 h in the absence of reducing agents.
2. Volumetrically mix the light and heavy-labeled proteins at a 1:1 ratio and buffer exchange the samples against ABC buffer.
3. Digest the samples with trypsin by overnight incubation at 37 °C.
4. Lyophilize the digested peptides and suspend in either affinity loading buffer for direct avidin affinity purification or fractionate by reconstituting the dried peptides in SCX buffer A.
5. Fractionate the peptides using HPLC and a strong cation-exchange (SCX) PolySULFOETHYL A column with a step gradient of SCX buffer A to SCX buffer B.
6. Mix the individual fractions from SCX with equivalent amounts of affinity loading buffer and load onto an avidin affinity cartridge. Dry the avidin affinity purified peptides using a lyophilizer and suspend in the cleavage reagent to release the ICAT-labeled peptides from the acid-cleavable linker by incubating at 37 °C for 2 h.
7. Dry the peptides obtained by acid cleavage and suspend in 0.1 % formic acid for LC-MS/MS analyses.

*3.4.4 Differential Thiol Trapping to Determine Reversible Oxidation of Cysteine Residues*

1. After purifying the oxidized protein sample, precipitate the protein with 10 % trichloroacetic acid (TCA).
2. Centrifuge the TCA precipitates (13,000 × *g*, 4 °C, 30 min) and wash the pellet with 500 μl of ice-cold 10 % TCA and 200 μl of ice-cold 5 % TCA.
3. Dissolve the pellet in 80 μl of denaturing alkylation buffer (DAB) and the contents of one vial of cleavable heavy ICAT reagent dissolved in 20 μl of ACN.
4. Incubate the sample at 900 rpm for 1 h at 37 °C in the dark. To remove the light ICAT reagent, precipitate the proteins with 400 μl of chilled (20 °C) acetone for 4 h at 20 °C. After centrifugation (13,000 × *g*, 4 °C, 30 min), wash the pellet twice with 500 μl of chilled acetone.
5. Dissolve the protein pellet in a mixture of 80 μl of DAB, 1 μl of 100 mM DTT, or other selected reducing agents.
6. Add the contents of one vial of cleavable light ICAT reagent dissolved in 20 μl of ACN.
7. Incubate the sample at 900 rpm for 1 h at 37 °C in the dark.
8. Perform trypsin digestion of the ICAT-labeled peptides, enrichment on streptavidin columns, and cleavage of the

biotin tag according to the description in Subheadings 3.4.1 and 3.4.2.

9. Subject the sample to LC-MS/MS analysis.

#### 3.4.5 Filter-Aided Sample Preparation (FASP)

There are two major strategies for converting proteins extracted from biological material to generating peptides suitable for MS-based proteome analysis: SDS-PAGE separation and in-gel digestion or in-solution digestion. The SDS-PAGE separation method reduces sample complexity by removing sample contaminants, such as detergents, salts, DNA, and other nonprotein compounds. The in-gel digestion approach is generally less efficient compared to in-solution digestion; however, when sample amounts are not limiting, this technique has shown wider usage amongst biologists for generating peptides. More recently, the filter-aided sample preparation (FASP) method has been introduced as it removes high levels of salts, SDS, and other contaminants prior to in-solution digestion [40–43]. In the FASP protocol, all protein cleanup steps, enrichment, and tryptic digestions are performed in a single micro-concentrator/spin-filter device. The technique can be quite useful for performing fast sample cleanup of in vitro-purified proteins where no further separation is required.

1. Mix 30 μl of sample with 200 μl of denaturing buffer in a YM10 or YM30 (depending on protein size) spin column and centrifuge for 15 min at 14,000 ×*g*.
2. Add an additional 200 μl of denaturing buffer to a YM10 or YM30 spin column and centrifuge for 15 min at 14,000 ×*g* (*see* **Note 21**).
3. Discard the flow-through.
4. Add 100 μl of IAM buffer and mix on a shaker at 600 rpm for 1 min.
5. Incubate for 20 min at 25 °C (*see* **Note 22**).
6. Centrifuge for 10 min at 14,000 ×*g*.
7. Add 100 μl of denaturing buffer to the spin column and spin for 15 min at 14,000 ×*g*. Repeat this step two additional times.
8. Add 100 μl of ABC buffer and spin as in **step** 7. Repeat this step 2 additional times.
9. Add 16 μl of 0.1 μg/μl trypsin solution (trypsin should be diluted with ABC buffer) to the micro-concentrator/spin-filter device (*see* **Note 23** for an alternative approach).
10. Remove old collection tubes from the spin column and replace with fresh collection tubes before centrifuging for 10 min at 14,000 ×*g*.
11. Add 40 μl of ABC buffer and centrifuge for 10 min at 14,000 ×*g* to collect peptides.

## 4 Notes

1. pH 6.5 was selected because low pH values help maintain cysteine residues in the thiol (–SH) state. Higher pH buffers (>8.0) allow cysteines to populate the thiolate ($–S^-$) state, which makes them more susceptible to air oxidation.
2. To modify proteins with redox-active reagents, it is important that the thiol moieties of the protein be in the reduced state. ABD-F will not react with oxidized cysteines and will result in reduced fluorescence yield. Further, oxidation of a thiol to sulfenic acid (-SOH) can result in disulfide bond formation. Thus, disulfide bonds and further oxidation states can readily occur when the protein is in the sulfenic acid state regardless of the thiol $pK_a$. If the protein is in the sulfinic ($-SO_2H$) or sulfonic ($-SO_3H$) state, the protein cannot be reduced and will be less reactive to ROS and RNS. By maintaining the protein in a reduced state, the reactivity will be dependent on the thiol/thiolate state of the protein.
3. DTT works most efficiently at pH values greater than 8.0. This is because the reducing agent is effective when one of the thiol moieties is in the thiolate form. The $pK_a$ of the DTT thiols are ~9.2 and 10.1 [44]. Mercaptoethanol (β-ME) is not optimal as it can stably modify thiols, which will reduce reactivity with ABD-F. Tris(2-carboxyethyl)phosphine (TCEP), which is a phosphine-reducing agent, appears to be efficient at reducing disulfides; however, other oxidation states, such as sulfenic acid (-SOH), are not efficiently reduced by TCEP according to our observations. A new reducing agent, dithiobutylamine (DTBA), functions similarly to DTT and has a thiol $pK_a$ that is approximately one unit lower [45], which increases the reactivity of DTBA at physiological pH values. However, this reducing agent needs to be generated in house and is currently unavailable for purchase.
4. The buffer should have all oxygen removed as this will aid in preventing air oxidation over time, especially when studying thiols with altered $pK_a$ values, as the thiolate ($RS^-$) is more reactive than a thiol (RSH).
5. ABD-F reacts with free thiols in solution. Thus, reducing agents that act through a free thiol (e.g., DTT, β-ME, and DTBA) will generate fluorescence signal (false reactivity) and interfere with $pK_a$ measurements. Therefore, these reagents cannot be present during data collection.
6. A final concentration of 10 μM protein is sufficient to give a good signal-to-noise ratio for most proteins. In general, a 100:1 ABD-F:protein ratio is sufficient.

7. Although the curve-fitting approach outlined quantifies the initial linear slopes, collecting data at longer times will aid in verifying whether equivalent amounts of protein and ABD-F were added to each well (*see* Fig. 1a).
8. This method is most effective when using a protein that only has one solvent accessible cysteine because the reactivity of other cysteines will complicate data analysis.
9. If the protein of interest contains multiple solvent accessible cysteines, the assay can be modified to determine whether the protein has a cysteine with an altered p$K_a$. Using a substitution mutant of the suspected redox-sensitive cysteine (Cys-to-Ser (or Ala) mutation), perform the assay essentially as described but at one pH value (pH 6.5 is recommended such that non-redox-sensitive cysteines that do not have altered p$K_a$ values do not result in high background signal). Thus, the cysteine mutant serves as a control and provides the background level of ABD-F modification in the protein.
10. As reported in Grossi et al. [27], the pH of the reaction is critical for generating optimum nitroso-thiol content; therefore, after solution I and II are mixed, the solution pH should be approximately 1.7–2.0.
11. In general, any volume and concentration of nitroso-thiol is possible. The protocol explained in detail here is scaled to meet the needs of our experiments. Thus, this protocol will yield approximately 1,100 μl of 8 mM thiol-NO.
12. An alternative is to use the Saville assay [46] to measure thiol-NO concentration. However, this assay detects $NO_2^-$ as well as the thiol-NO, which could lead to error if the free $NO_2^-$ in solution has not been reacted with the thiol or sulfamate. In our experience, the absorbance at 336 nm is reliable.
13. An inherent disadvantage to using NONOates is that they generate $NO^•$. In the case of Ras, $NO_2^•$ is required for thiyl radical production. However, $NO^•$ can react with $O_2$ to form $NO_2^•$ as well as other reaction products. A further disadvantage is that once $NO_2^•$ is generated, it can react with $NO^•$ released by the NONOate to produce other reaction products. A common reaction product is $N_2O_3$, which can nitrosate cysteines at an approximate rate of $1.2 \times 10^7\ M^{-1}\ s^{-1}$ (Fig. 2) [47]. In Ras reactions with DEA NONOate, we observed that only a fraction of Ras reacts with $NO_2^•$ to induce nucleotide dissociation and a sizable fraction of Ras can be nitrosated by $N_2O_3$. However, nitrosation of Ras through a non-radical pathway does not alter Ras activity [48, 49]. Modification by $N_2O_3$ is only likely under the reaction conditions used in vitro with purified proteins; however, in cells, this oxidation pathway is not favored.

14. One can verify the dependence of the NO• species generated on protein activity when using NONOates by adding in PTIO (2-Phenyl-4,4,5,5-tetramethylimidazoline-1-oxyl 3-oxide), which is a selective NO• scavenger. This control will allow the user to verify whether free radical byproducts cause the observed effects, as opposed to a competing reaction with the NONOate breakdown products.

15. All of the strategies described in Subheading 3.4 rely on the loss of the analytical signal upon cysteine modification (oxidation). Most cysteine residues are maintained in their reduced state in cells, measuring glutathiolation directly using chemical labeling significantly impacts the sensitivity and dynamic range of (oxidation) detection.

16. Buffers with low concentrations of SDS or urea allow for more complete labeling of the unfolded protein by the ICAT reagents.

17. All buffers should be prepared in HPLC grade water for optimum performance. All reversed-phase LC-MS buffers should be prepared in LC-MS grade water.

18. All peptides containing cysteine residues will contain heavy and light pairs of ICAT labels, and their relative intensities are measured using a full MS scan. The relative intensity, ratio = [$L^{int}$/ $H^{int}$], is used to determine the relative levels of modification.

19. A similar strategy to ICAT has been demonstrated to measure cysteine oxidation due to NO• through the use of light and heavy *n*-ethylmaleimide (NEM) reagents that are generally used to protect sulfhydryls. We believe that this strategy could be adapted to study glutathiolation as well because the ICAT reagents can be easily replaced by NEM and d5-NEM. More recently, other stable isotope labeling reagents have become available, such as Cys-TMT (Thermo Fisher Scientific; San Jose CA). The use of Cys-TMT allows for an analytical workflow to quantitatively assay free cysteine residues and determine the extent of modification under six biologically distinct conditions in a single mass spectrometry experiment. The multiplexing is particularly useful for studies involving time-course measurements.

20. An alternative strategy compared to the methods described in Subheading 3.4 is to introduce a reduction step to remove reversible cysteine modifications. Protein modifications can be measured indirectly by the following steps: (a) alkylation of free sulfhydryl groups on cysteine residues; (b) reduction of glutathione adducts by glutaredoxin (Grx), which does not affect other cysteine modifications; and (c) blocking all nascent sulfhydryl groups with an irreversible labeling reagent, such as IAM or NEM. The advantage of this method is that glutathione-specific

cysteines will be quantitatively assayed. This methodology is more sensitive due to a net increase in the analytical signal for measuring glutathiolation, which is advantageous because most cysteine residues have low levels of glutathiolation in cells.

21. A YM30 filter (MW cut off of 30 kDa) is used to improve filtration speeds, whereas denaturants assist on-filter retention of proteins less than 30 kDa. However, the use of YM10 filters would be a conservative approach for ensuring the retention of lower molecular weight proteins/polypeptides (approximately 5–8 kDa range).
22. As we are studying cysteine oxidation in the sample prior to alkylation, the reducing step is omitted. Alkylation is performed to prevent disulfide exchange from scrambling the oxidative modifications to other cysteine residues that are exposed upon trypsin digest.
23. Incubate overnight at 37 °C or incubate for 1–5 min using microwave irradiation as described in the GOFAST method [50]. An alternative approach uses endoprotease Lys-C (1:100 enzyme/protein ratio) for 4 h at 37 °C prior to trypsin digestion to improve the overall peptide digestion efficiency.

## Acknowledgements

We would like to thank Dan Isom for his technical insights. The research efforts described herein were supported by NIH RO1GM75431 and RO1CA089614 to SLC, and GAH was partially funded by the Program in Molecular and Cellular Biophysics (NIH T32GM008570).

## References

1. Netto LE, de Oliveira MA, Monteiro G et al (2007) Reactive cysteine in proteins: protein folding, antioxidant defense, redox signaling and more. Comp Biochem Physiol C Toxicol Pharmacol 146:180–193
2. Martinez-Ruiz A, Lamas S (2007) Signalling by NO-induced protein S-nitrosylation and S-glutathionylation: convergences and divergences. Cardiovasc Res 75:220–228
3. Mieyal JJ, Chock PB (2012) Posttranslational modification of cysteine in redox signaling and oxidative stress: focus on s-glutathionylation. Antioxid Redox Signal 16:471–475
4. Wang Y, Yang J, Yi J (2012) Redox sensing by proteins: oxidative modifications on cysteines and the consequent events. Antioxid Redox Signal 16:649–657
5. Adler V, Yin Z, Tew KD et al (1999) Role of redox potential and reactive oxygen species in stress signaling. Oncogene 18:6104–6111
6. Reddie KG, Carroll KS (2008) Expanding the functional diversity of proteins through cysteine oxidation. Curr Opin Chem Biol 12:746–754
7. Roos G, Foloppe N, Messens J (2013) Understanding the pK(a) of redox cysteines: the key role of hydrogen bonding. Antioxid Redox Signal 18:94–127
8. Bulaj G, Kortemme T, Goldenberg DP (1998) Ionization-reactivity relationships for cysteine thiols in polypeptides. Biochemistry 37: 8965–8972
9. Nelson KJ, Parsonage D, Hall A et al (2008) Cysteine pK(a) values for the bacterial peroxiredoxin AhpC. Biochemistry 47:12860–12868
10. Dyson HJ, Jeng MF, Tennant LL et al (1997) Effects of buried charged groups on cysteine thiol ionization and reactivity in Escherichia coli thioredoxin: structural and functional characterization of mutants of Asp 26 and Lys 57. Biochemistry 36:2622–2636

11. Chivers PT, Prehoda KE, Volkman BF et al (1997) Microscopic pKa values of Escherichia coli thioredoxin. Biochemistry 36:14985–14991
12. Tosatto SC, Bosello V, Fogolari F et al (2008) The catalytic site of glutathione peroxidases. Antioxid Redox Signal 10:1515–1526
13. Davis MF, Vigil D, Campbell SL (2011) Regulation of Ras proteins by reactive nitrogen species. Free Radic Biol Med 51:565–575
14. Lander HM, Hajjar DP, Hempstead BL et al (1997) A molecular redox switch on p21(ras). Structural basis for the nitric oxide-p21(ras) interaction. J Biol Chem 272:4323–4326
15. Mitchell L, Hobbs GA, Aghajanian A et al (2013) Redox regulation of ras and rho GTPases: mechanism and function. Antioxid Redox Signal 18:250–258
16. Takai Y, Sasaki T, Matozaki T (2001) Small GTP-binding proteins. Physiol Rev 81: 153–208
17. Lancaster JR Jr (2008) Protein cysteine thiol nitrosation: maker or marker of reactive nitrogen species-induced nonerythroid cellular signaling? Nitric Oxide 19:68–72
18. Heo J, Campbell SL (2006) Ras regulation by reactive oxygen and nitrogen species. Biochemistry 45:2200–2210
19. Heo J, Campbell SL (2004) Mechanism of p21Ras S-nitrosylation and kinetics of nitric oxide-mediated guanine nucleotide exchange. Biochemistry 43:2314–2322
20. Lim KH, Ancrile BB, Kashatus DF et al (2008) Tumour maintenance is mediated by eNOS. Nature 452:646–649
21. Jaffe AB, Hall A (2005) Rho GTPases: biochemistry and biology. Annu Rev Cell Dev Biol 21:247–269
22. Heo J, Raines KW, Mocanu V et al (2006) Redox regulation of RhoA. Biochemistry 45:14481–14489
23. Aghajanian A, Wittchen ES, Campbell SL et al (2009) Direct activation of RhoA by reactive oxygen species requires a redox-sensitive motif. PLoS One 4:e8045
24. Isom DG, Marguet PR, Oas TG et al (2011) A miniaturized technique for assessing protein thermodynamics and function using fast determination of quantitative cysteine reactivity. Proteins 79:1034–1047
25. Chait BT (2006) Chemistry. Mass spectrometry: bottom-up or top-down? Science 314: 65–66
26. Yates JR, Ruse CI, Nakorchevsky A (2009) Proteomics by mass spectrometry: approaches, advances, and applications. Annu Rev Biomed Eng 11:49–79
27. Grossi L, Montevecchi PC (2002) S-nitrosocysteine and cystine from reaction of cysteine with nitrous acid. A kinetic investigation. J Org Chem 67:8625–8630
28. Moore KP, Mani AR (2002) Measurement of protein nitration and S-nitrosothiol formation in biology and medicine. Methods Enzymol 359:256–268
29. Gu J, Lewis RS (2007) Effect of pH and metal ions on the decomposition rate of S-nitrosocysteine. Ann Biomed Eng 35:1554–1560
30. Jones DP (2008) Radical-free biology of oxidative stress. Am J Physiol Cell Physiol 295:C849–C868
31. Berdniko VM, Bazhin NM (1970) Oxidation-Reduction Potentials of Certain Inorganic Radicals in Aqueous Solutions. Russ J Phys Ch Ussr 44:395–398
32. Chameides WL (1984) The Photochemistry of a Remote Marine Stratiform Cloud. J Geophys Res-Atmos 89:4739–4755
33. Gygi SP, Rist B, Gerber SA et al (1999) Quantitative analysis of complex protein mixtures using isotope-coded affinity tags. Nat Biotechnol 17:994–999
34. Chiappetta G, Ndiaye S, Igbaria A et al (2010) Proteome screens for Cys residues oxidation: the redoxome. Methods Enzymol 473:199–216
35. Marino SM, Li Y, Fomenko DE et al (2010) Characterization of surface-exposed reactive cysteine residues in Saccharomyces cerevisiae. Biochemistry 49:7709–7721
36. Sethuraman M, Clavreul N, Huang H et al (2007) Quantification of oxidative posttranslational modifications of cysteine thiols of p21ras associated with redox modulation of activity using isotope-coded affinity tags and mass spectrometry. Free Radic Biol Med 42:823–829
37. Sethuraman M, McComb ME, Huang H et al (2004) Isotope-coded affinity tag (ICAT) approach to redox proteomics: identification and quantitation of oxidant-sensitive cysteine thiols in complex protein mixtures. J Proteome Res 3:1228–1233
38. Leichert LI, Gehrke F, Gudiseva HV et al (2008) Quantifying changes in the thiol redox proteome upon oxidative stress in vivo. Proc Natl Acad Sci U S A 105:8197–8202
39. Chouchani ET, James AM, Fearnley IM et al (2011) Proteomic approaches to the characterization of protein thiol modification. Curr Opin Chem Biol 15:120–128
40. Wisniewski JR, Ostasiewicz P, Mann M (2011) High recovery FASP applied to the proteomic analysis of microdissected formalin fixed paraffin embedded cancer tissues retrieves known colon cancer markers. J Proteome Res 10:3040–3049
41. Wisniewski JR, Zielinska DF, Mann M (2011) Comparison of ultrafiltration units for proteomic and N-glycoproteomic analysis by the filter-aided sample preparation method. Anal Biochem 410:307–309

42. Wisniewski JR, Zougman A, Mann M (2009) Combination of FASP and Stage Tip-based fractionation allows in-depth analysis of the hippocampal membrane proteome. J Proteome Res 8:5674–5678
43. Wisniewski JR, Zougman A, Nagaraj N et al (2009) Universal sample preparation method for proteome analysis. Nat Methods 6: 359–362
44. Houk J, Singh R, Whitesides GM (1987) Measurement of thiol-disulfide interchange reactions and thiol pKa values. Methods Enzymol 143:129–140
45. Lukesh JC 3rd, Palte MJ, Raines RT (2012) A potent, versatile disulfide-reducing agent from aspartic acid. J Am Chem Soc 134: 4057–4059
46. Saville B (1958) A scheme for the colorimetric determination of microgram amounts of thiols. Analyst 83:670–672
47. Keshive M, Singh S, Wishnok JS et al (1996) Kinetics of S-nitrosation of thiols in nitric oxide solutions. Chem Res Toxicol 9:988–993
48. Hobbs GA, Bonini MG, Gunawardena HP et al (2013) Glutathiolated Ras: characterization and implications for Ras activation. Free Radic Biol Med 57:221–229
49. Williams JG, Pappu K, Campbell SL (2003) Structural and biochemical studies of p21Ras S-nitrosylation and nitric oxide-mediated guanine nucleotide exchange. Proc Natl Acad Sci U S A 100:6376–6381
50. Yu Y, Xie L, Gunawardena HP et al (2012) GOFAST: an integrated approach for efficient and comprehensive membrane proteome analysis. Anal Chem 84:9008–9014
51. Ford E, Hughes MN, Wardman P (2002) Kinetics of the reactions of nitrogen dioxide with glutathione, cysteine, and uric acid at physiological pH. Free Radic Biol Med 32:1314–1323
52. Raines KW, Bonini MG, Campbell SL (2007) Nitric oxide cell signaling: S-nitrosation of Ras superfamily GTPases. Cardiovasc Res 75:229–239
53. Augusto O, Bonini MG, Amanso AM et al (2002) Nitrogen dioxide and carbonate radical anion: two emerging radicals in biology. Free Radic Biol Med 32:841–859
54. Keefer LK, Nims RW, Davies KM et al (1996) "NONOates" (1-substituted diazen-1-ium-1,2-diolates) as nitric oxide donors: convenient nitric oxide dosage forms. Methods Enzymol 268:281–293
55. Maragos CM, Morley D, Wink DA et al (1991) Complexes of •NO with nucleophiles as agents for the controlled biological release of nitric oxide. Vasorelaxant effects. J Med Chem 34:3242–3247
56. Maragos CM, Wang JM, Hrabie JA et al (1993) Nitric oxide/nucleophile complexes inhibit the in vitro proliferation of A375 melanoma cells via nitric oxide release. Cancer Res 53:564–568

# Chapter 7

# Use of the Yeast Two-Hybrid Technology to Isolate Molecular Interactions of Ras GTPases


Elisa Ferro, Eva Baldini, and Lorenza Trabalzini



## Abstract

Since its original description, the yeast two-hybrid system has been extensively used to identify protein–protein interactions from many different organisms, thus providing a convenient mean to both screen for proteins that interact with a protein of interest and to characterize the known interaction between two proteins. In these years the technique has improved to overcome the limitations of the original assay, and many efforts have been made to scale up the technique and to adapt it to large-scale studies. In addition, variations have been introduced to enlarge the range of proteins and interactors that can be assayed by hybrid-based approaches. Several groups studying molecular mechanisms that underlie signal transduction pathways regulated by Ras GTPases have successfully used the yeast two-hybrid system or related methods to isolate and identify new binding partners of Ras proteins.

Here we describe the basic protocol for a yeast two-hybrid library screening and for a small-scale yeast two-hybrid assay.



**Key words** Yeast two-hybrid, Protein–protein interactions, Library screening, Ras GTPases


## 1 Introduction

The yeast two-hybrid (Y2H) system was originally described by Fields and Song in 1989 [1] and is based on the observation that eukaryotic transcription factors have a modular structure, consisting of a DNA-binding domain (BD) and a transactivation domain (AD). The BD recognizes and binds a specific DNA sequence in the promoter region of a gene, and the AD contacts other proteins of the transcriptional machinery to enable transcription to occur. These two domains are physically separable and can function in proximity to each other without covalent binding.

In the original Y2H system, two putative interacting proteins, X and Y, are fused to the BD and AD of a yeast transcriptional activator. The two hybrid proteins BD-X and AD-Y are co-expressed in a yeast strain containing the transcriptional activator's specific DNA-binding site (the upstream activating sequence, UAS) upstream of a


Lorenza Trabalzini and Saverio Francesco Retta (eds.), *Ras Signaling: Methods and Protocols*,
Methods in Molecular Biology, vol. 1120, DOI 10.1007/978-1-62703-791-4_7, © Springer Science+Business Media, LLC 2014

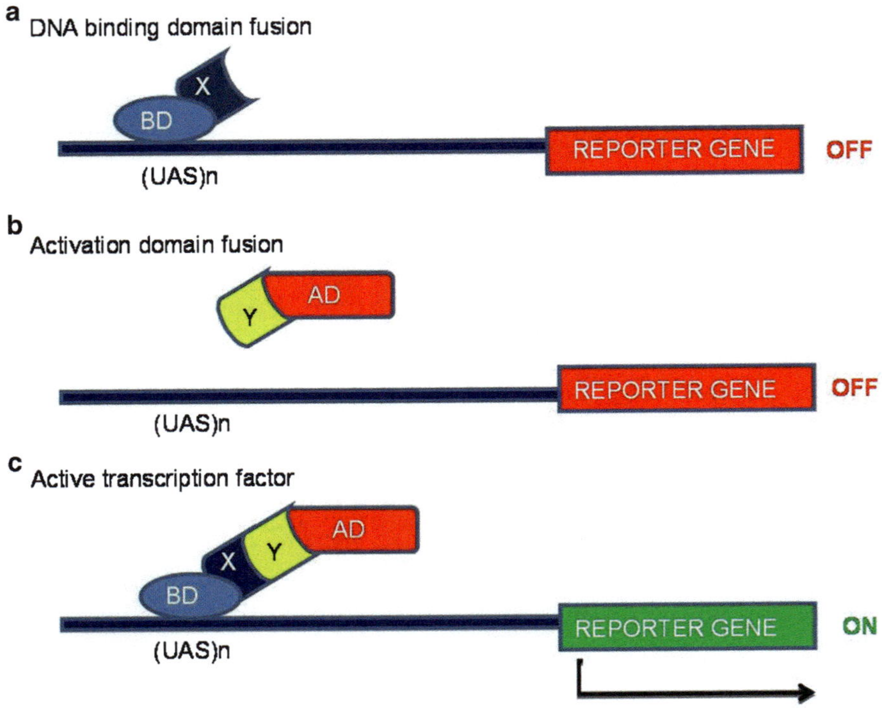


**Fig. 1** The two-hybrid system. DNA-binding domain (BD) and transcription activation domain (AD) fusions are expressed in yeast. (**a**) In the absence of AD, the DNA-BD, expressed as a fusion with protein X, is capable of binding to the upstream activator sequences (UAS) but is not itself capable of activating transcription. (**b**) The transcription AD, expressed as a fusion with protein Y, is similarly incapable of activating transcription in the absence of a DNA-BD. (**c**) Interaction between the two fusion proteins (by virtue of an interaction between X and Y) results in reconstitution of an active transcription factor and subsequent transcription of a reporter gene (e.g., *LACZ* or *HIS3*) providing a means to assay the interaction between the two fusion proteins

gene (reporter gene) whose expression can easily be detected by observing cell growth or by a quick enzyme assay. Physical association between X and Y in yeast cells brings the BD and AD into proximity, thereby reconstituting the transcription factor and driving transcription of the reporter gene, resulting in an easily detectable phenotype. The method is summarized in Fig. 1.

A typical Y2H assay consists of the following steps: (1) genes coding X and Y proteins are cloned in two distinct plasmids in frame with BD or AD coding sequences, respectively; (2) the two constructs are cotransformed in a yeast strain bearing appropriate reporter genes; and (3) interaction between X and Y is verified by analysis of the yeast phenotype.

The most used reporter genes are nutritional markers, such as genes encoding amino acid biosynthetic enzymes, as *HIS3* or *LEU2*, or enzymatic reporters as *LACZ* or *MEL1*, encoding β- or α-galactosidase, respectively.

Although the Y2H system may be very useful to test the interaction between two known proteins, the main and more useful application of the Y2H assay is the screening of cDNA libraries to isolate new putative partners for a protein of interest. The latter, fused to the BD, acts as a "bait"; the library, fused to the AD, is the source of putative "preys." Thanks to its ability to select many putative interacting partners at the same time, this method has been adopted by numerous laboratories and converted to the kit-based format that is popular with molecular biologists.

Two different approaches can be used to perform library screening: (a) cotransformation of "bait" and library in the same yeast strain and (b) the "mating protocol" that will be described in this chapter.

The development of haploid yeast strains of opposite mating type (MATα and MATa) to be used in the Y2H system has proved to be very useful, especially for high-throughput two-hybrid approaches. Bait and prey constructs are initially transformed into two different haploid yeast strains with opposite mating types (MATa and MATα) in order to separately express the DNA-BD and the AD fusion proteins. Then, bait and prey strains are mated to generate diploids that contain both constructs, and the diploids are tested for reporter activation as in a conventional Y2H experiment. It has been suggested that yeast mating is more likely to detect interactions than in standard cotransformation screens because in the mating procedure the bait and prey proteins are already expressed at a steady state level before mating. In contrast, in the cotransformation procedure expression begins after transformation [2].

The Y2H system presents three major advantages over alternative assays for gene identification. First, since it is based on a powerful genetic selection, it is able to assay very high numbers of potential coding sequences in a relatively simple experiment. Second, as it relies on an assay performed in vivo, it is not limited by the artificial conditions of in vitro assays. Finally, since it is based on a physical binding assay, a wide variety of protein–protein interactions can be detected and characterized following one single commonly used protocol [3]. However, limitations of the system have been reported too, including the possibility to isolate very large numbers of clones with no biological relevance ("false positives") or to miss expected interactions ("false negatives").

The following events can cause the isolation of false positives: (1) the bait and the prey do indeed interact in the context of the Y2H assay, but not in the normal in vivo context, where the two proteins are not expressed at the same time or in the same tissue or subcellular district; these false positives are difficult to eliminate without extensive biological knowledge of the tested protein. (2) Transcriptional activity occurs independently of any protein–protein interaction, due to aspecific activation by BD or AD fusion proteins, plasmid rearrangements, or copy number changes that generate such auto-activators, alterations at a reporter gene that result in constitutive expression [4, 5]. (3) A third partner can bridge the bait and the prey that are not able to interact directly. Such a false positive can be either biologically relevant or artificial. All these events need to be carefully considered to avoid misinterpretation of the results.

To increase the specificity of the Y2H assay and to decrease the number of false positives, additional selection markers have been introduced in the most recent generation of the Y2H system. These additional reporter genes are engineered downstream of distinct and unrelated UAS and TATA box sequences. Library proteins that interact with unrelated sequences flanking or within the UAS (i.e., false positives) are automatically screened out with this selection system. An alternative strategy consists of reducing the expression levels of the two-hybrid proteins by using yeast centromeric vectors, which are maintained at low copies in the cell, and truncated versions of promoters such ADH1 [3, 6].

False negatives may be due to (1) incorrect folding and/or instability of a DB or AD fusion protein that could affect the ability to interact with its interacting partner and (2) toxicity of some fusion proteins that could affect the viability of transformed cells.

The quality of the library is a key parameter for the success of a screening and the assessment of the data. In cDNA libraries, mostly used in Y2H screenings, the relative representation of each cDNA reflects the endogenous expression level of the corresponding gene. Thus, it could happen that putative interacting proteins are underrepresented if their mRNA is expressed at relatively low levels [3]. In addition, some encoding genes might be missing due to defects in library construction. Both these events can cause the loss of potential interactors. To date, several premade cDNA libraries from different sources and suitable for Y2H screenings are commercially available. All these libraries are guaranteed to contain a high number of independent clones. Some of them have been normalized in order to increase the level of underrepresented cDNAs; other libraries, the most used, are already pretransformed in a yeast strain to guarantee a high efficiency of transformed yeast cells.

As protein–protein interactions, identified by the Y2H system, are inferred indirectly through activation of reporter genes in the host cell, they need to be validated by alternative techniques, in non-yeast-based assays. Approaches conventionally used include in vitro "pull-down" and binding assay, co-immunoprecipitation of endogenous proteins, co-immunolocalization, or gradient sedimentation.

Since its first setup, in the first 1990s, the Y2H approach has been extensively used to identify novel Ras interactors, thus giving a great contribution to the knowledge of signal transduction pathways regulated by Ras GTPases [7–16]. Thanks to their low molecular weight, the full length of Ras family proteins has been directly used as a bait in several library screenings, thus enforcing the significance of isolated interactions. The use of Y2H solved out many questions on Ras interactions, including the identification of regions and residues directly involved in the interaction. In particular, deletion analysis coupled with Y2H has been useful to identify the Ras-binding domain (RBD) of several Ras effectors [17–22]; RBD, which mediates specific interaction with the active GTP-bound form of several Ras GTPases, is common to different proteins with different functions [23, 24]. The Y2H combined with mutation analysis has been also used to determine the contribution to the interaction with Ras of single amino acid residues in the RBD [18, 25, 26].

Several variants of the original two-hybrid system described by Fields and Song are available. The commonly used systems rely upon either the GAL4 or LexA DNA-binding domains in combination with either a GAL4, VP16, or B42 transcription activation domain. Several of these alternatives are interchangeable (e.g., the GAL4 DNA-BD can be used with the GAL4, VP16, or B42 AD) [27].

Here we will describe a system based on GAL4, reporting a typical Y2H library screening by using the mating protocol and a pretransformed cDNA library. The protocol for a small-scale two-hybrid assay will be also detailed.

## 2 Materials

All media and solutions must be sterile (*see* **Note 1**).

### 2.1 General Use Materials

1. Sterile tubes and flasks for cell culture and transformation.
2. Sterile glass rods, plastic loops, and/or bent Pasteur pipettes for streaking and/or spreading cells on plates.
3. 100- and 150-mm diameter sterile Petri plates.

### 2.2 Plasmids and Yeast Strains

1. cDNA(s) of protein(s) of interest.
2. GAL4 DNA-BD and GAL4 AD vectors and suitable yeast host strains: choose any of commercially available Y2H system kits. The different systems also contain DNA-BD and AD constructs to be used as positive and negative controls (*see* **Note 2**).
3. Only for library screening: cDNA library constructed in the GAL4 AD vector.

Depending on your protein of interest, you can choose among a great number of cDNA libraries from different sources to be used in the Y2H screening and commercially available. cDNA libraries cloned into a GAL4 AD vector and already transformed into the yeast strain (pretransformed cDNA library) are also available and very easy to use.

#### 2.3 Growth and Maintenance of Yeasts (See Note 3)

1. Dextrose stock solution (40 %): to prepare 1 l, weigh 400 g of dextrose and add slowly to water until complete solubilization. Filter sterilize or autoclave and store at room temperature (*see* **Note 4**).
2. YPD medium: 20 g/l peptone, 10 g/l yeast extract. To prepare 1 l of YPD, weigh components as indicated and solubilize in 950 ml of $H_2O$, adjust the pH to 5.8, and autoclave. Allow medium to cool to ~55 °C and then add 50 ml of 40 % dextrose. Store at 4 °C (*see* **Note 5**).
3. YPD plates: prepare YPD broth as above, add 20 g/l agar and autoclave. Cool to 55 °C, add dextrose as above, and pour plates. Store at 4 °C.
4. YPDA medium: to 1 l of YPD medium add 15 ml of filter-sterilized 0.2 % adenine hemisulfate to a final concentration of 0.003 %.

#### 2.4 Selection of Yeasts

1. 10× dropout solution (DO): 300 mg/l L-isoleucine, 1,500 mg/l L-valine, 200 mg/l L-adenine hemisulfate, 200 mg/l L-arginine HCl, 200 mg/l L-histidine HCl, 1,000 mg/l L-leucine, 300 mg/l L-lysine HCl, 200 mg/l L-methionine, 500 mg/l L-phenylalanine, 2,000 mg/l L-threonine, 200 mg/l L-tryptophan, 300 mg/l L-tyrosine, 200 mg/l L-uracil.

   Omit the appropriate component(s) to prepare selective DO supplements to be added to SD to produce synthetic minimal media lacking one or more specific nutrients.

   Autoclave or filter sterilize and store at 4 °C for up to 1 year (*see* **Note 6**).
2. SD medium: 6.7 g/l Yeast Nitrogen base without amino acids. To prepare 1 l of SD, solubilize 6.7 g of Yeast Nitrogen base w/o amino acids in 850 ml of $H_2O$, adjust the pH to 5.8, and autoclave. Allow medium to cool to ~55 °C. Add 50 ml of 40 % dextrose and 100 ml of the appropriate sterile 10× DO.
3. SD plates: prepare SD broth as above, add 20 g/l agar, and autoclave. Cool to 55 °C, add dextrose and DO as above, and pour plates. Store at 4 °C.
4. X-α-Gal stock solution: dissolve X-α-Gal at 20 mg/ml in dimethylformamide (DMF). Store in the dark at −20 °C.
5. SD/X-α-Gal plates: prepare as SD plates and add X-α-gal stock solution to a final concentration of 80 mg/l before pouring plates. Store at 4 °C.

#### 2.5 Yeast Transformation

1. 10× TE stock solution: 0.1 M Tris–HCl, 10 mM EDTA, pH 7.5. Autoclave and store at room temperature.
2. 10× LiAc stock solution: 1 M lithium acetate. Adjust to pH 7.5 with diluted acetic acid, autoclave, and store at room temperature.

3. 50 % polyethylene glycol (PEG) 4000 stock solution. Filter sterilize or autoclave. Store at room temperature.
4. 100 % dimethyl sulfoxide (DMSO).
5. 1× TE/LiAc solution: to prepare 10 ml, add 1 ml of 10× TE and 10× LiAc to 8 ml of $H_2O$. Make fresh from stock solutions immediately prior to use.
6. PEG/LiAc solution: to prepare 10 ml, mix 8 ml of 50 % PEG, 1 ml of 10× TE 10×, and 1 ml of 10× LiAc. Make fresh from stock solutions immediately prior to use.
7. 10 mg/ml sonicated salmon sperm carrier DNA (commercially available): boil in water bath for 20 min, cool in ice, and vortex immediately before use (*see* **Note 7**).

### 2.6 β-Galactosidase Filter Assay

1. Z buffer: 16.1 g/l $Na_2HPO_4{\cdot}7H_2O$, 5.5 g/l, $NaH_2PO_4{\cdot}H_2O$, 0.75 g/l KCl, 0.246 g/l $MgSO_4{\cdot}7H_2O$, pH 7. Autoclave and store at room temperature.
2. X-Gal stock solution: dissolve 5-bromo-4-chloro-3-indolyl-beta-D-galactoside (X-Gal) at 20 mg/ml in DMF. Store in the dark at −20 °C.
3. Z buffer/X-Gal solution: 100 ml of Z buffer, 0.27 ml of β-mercaptoethanol, 1.67 ml of X-Gal stock solution. Make fresh immediately prior to use.
4. Whatman #1 filter papers: prepare 75-mm diameter filters for use with 100-mm diameter Petri plates. Wrap them in aluminum foil and sterilize by autoclaving.

### 2.7 Extraction of Plasmid DNA from Yeast

1. Yeast lysis solution (YLS): 2 % Triton X-100, 1 %, SDS, 100 mM NaCl, 1 mM EDTA, 10 mM Tris-HCl, pH 8.0.
2. Phenol/chloroform/isoamyl alcohol (25:24:1) mix, commercially available.
3. Acid-washed glass beads (425–600 μm).
4. 70 % and 95 % ethanol.

### 2.8 Transformation of E. coli with Plasmid DNA

1. Mg stock solution: 1 M $MgSO_4$, 1 M $MgCl_2$. Filter sterilize. Store at room temperature.
2. Ampicillin stock solution: 50 mg/ml in sterile $H_2O$. Filter sterilize, aliquot, and store at −20 °C.
3. SOB medium: 20 g/l Bacto tryptone, 5 g/l Bacto yeast extract, 0.58 g/l NaCl, 0.186 g/l KCl. To prepare 1 l of SOB, weigh components as indicated and solubilize in 990 ml of $H_2O$, adjust the pH to 7.5, and autoclave. Allow medium to cool to ~55 °C and then add 10 ml of Mg stock solution. Store at room temperature.
4. SOC medium: add 90 μl of dextrose stock solution (*see* Subheading 2.3) to 10 ml of SOB. Prepare fresh prior to use.

5. LB/amp plates: 10 g/l Bacto tryptone, 5 g/l Bacto yeast extract, 5 g/l NaCl, 18 g/l agar. Solubilize Bacto tryptone, Bacto yeast extract, and NaCl in $H_2O$, adjust pH to 7.0, then add agar and autoclave. Cool to 50 °C and add ampicillin to 100 μg/ml. Pour plates and store at 4 °C.

### 2.9 Preparation of Protein Extracts (See Note 8)

1. Protease inhibitor stock solutions.
   - 1 mg/ml Pepstatin A in DMSO.
   - 10.5 mM Leupeptin in $H_2O$.
   - 200 mM Benzamidine in $H_2O$.
   - 2.1 mg/ml Aprotinin in $H_2O$.

   Store these individual stock solutions as directed on the labels and follow label precautions.
2. Protease inhibitor cocktail: prepare 688 μl by mixing 66 μl of Pepstatin A, 2 μl of Leupeptin, 500 μl of Benzamidine, and 120 μl of Aprotinin.

   Always prepare this solution fresh just before using. Place on ice to prechill.
3. Phenylmethyl-sulfonyl fluoride (PMSF) stock solution: dissolve 0.1742 g in 10 ml of isopropanol. Wrap tube in foil and store at room temperature (*see* **Note 9**).
4. Acid-washed glass beads (425–600 μm).
5. 1 M Tris–HCl, pH 6.8 stock solution.
6. 0.5 M EDTA stock solution.
7. Cracking buffer stock solution. 8 M urea, 5 % w/v SDS, 40 mM Tris–HCl, pH 6.8, 0.1 mM EDTA, 0.4 mg/ml bromophenol blue. To prepare 100 ml: 48 g of urea, 5 g of SDS, 4 ml of 1 M Tris-HCl, 20 μl of 0.5 M EDTA, 40 mg of bromophenol blue, and deionized $H_2O$ to a final volume of 100 ml.
8. Cracking buffer (complete). The following recipe is sufficient for one protein extract. Scale up recipe as required. Prepare only the volume you need just before use (*see* **Note 10**).

To prepare 1.13 ml of complete cracking buffer: 1 ml of cracking buffer stock solution, 10 μl of β-mercaptoethanol, 70 μl of prechilled protease inhibitor cocktail, 50 μl of PMSF stock solution.

## 3 Methods

*Preparing for Y2H* (Fig. 2).

### 3.1 Construct Fusion Genes

1. For library screening, prepare the BAIT by cloning the cDNA of the protein of interest in the GAL4 DNA-BD vector.
2. For small-scale Y2H assay, prepare both the BAIT and the PRAY by cloning the cDNA of the two candidate interaction

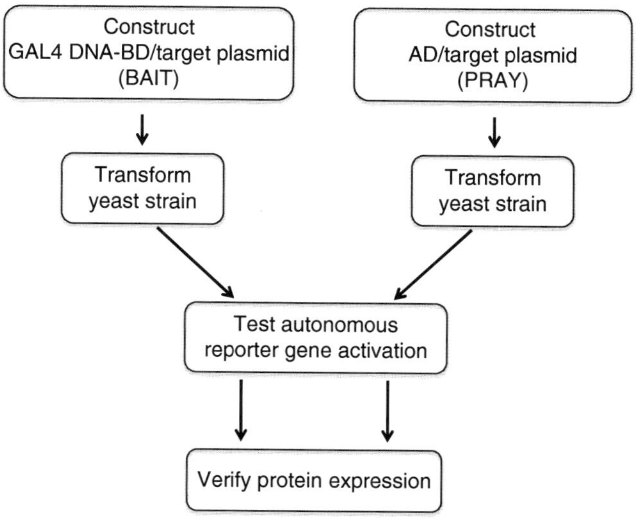


**Fig. 2** Preparing for the Y2H screening. Overview of the procedure

partners, X and Y, in the GAL4 DNA-BD and in the AD vector, respectively. For a more accurate interaction analysis, either cDNA can be cloned in each vector, and the two different combinations of constructs DNA-BD-X/AD-Y and DNA-BD-Y/AD-X may be used in the small-scale assay (*see* **Note 11**).

### 3.2 Prepare Yeast Working Stock Plates

1. Streak a small portion of the frozen glycerol stock of your yeast strain onto a 100 mm YPD agar plate.
2. Incubate the plate at 30 °C until yeast colonies reach ~2 mm in diameter (this takes 3–5 days). Use these colonies as your working stock. You can store the plate at 4 °C up to 1 month.

### 3.3 Test Autonomous Reporter Gene Activation

1. For library screening, transform your DNA-BD construct into AH109 yeast strain.
2. For small-scale Y2H, independently transform yeast cells with your DNA-BD and AD constructs (*see* Subheading 3.5 for the transformation procedure).
3. Assay the transformants for *MEL1* activation by spreading 100 μl of a 1/10 dilution and a 1/100 dilution of the transformation mix on SD/-Trp/X-α-Gal plates (for DNA-BD constructs) or SD/-Leu/X-α-Gal plates (for AD constructs). Perform positive control with pCL1 and negative control with either DNA-BD or AD empty vectors. Incubate plates at 30 °C and let them grow for 3 days.
4. If the transformant colonies are white, there is no autonomous reporter gene activation, and you can keep your transformants. Prepare stock plates and liquid cultures for freezing.

If the transformant colonies are blue, your fusion protein may have a transcriptional activation domain. This problem can be solved by creating deletion mutants of the protein and retesting the deletion constructs. Note that such deletions may also eliminate a potentially interacting domain.

### 3.4 Verify Protein Expression

*Preparation of yeast cultures for protein extraction*:

1. Transform your fusion construct(s) into AH109 strain (*see* Subheading 3.5 for the transformation procedure).
2. For each transformed yeast strain you wish to assay, use a single isolate colony (1–2 mm in diameter, no older than 4 days) to prepare a 5 ml overnight culture in the SD selection medium appropriate for your system and plasmids. Also prepare a 10-ml culture of an untransformed yeast colony in YPD as a negative control.
3. Vortex the overnight cultures for 0.5–1 min to disperse cell clumps. For each clone to be assayed (and the negative control), separately inoculate 50 ml aliquots of YPD medium with the entire overnight culture.
4. Incubate at 30 °C with shaking (220–250 rpm) until the $OD_{600}$ reaches 0.4–0.6 (depending on the fusion protein, this will take 4–8 h). Multiply the $OD_{600}$ (of a 1 ml sample) by the culture volume (i.e., 55 ml) to obtain the total number of $OD_{600}$ units; this number will be used in the next section (Subheading 2.9, **step 1**) (e.g., 0.6 × 55 ml = 33 total $OD_{600}$ units) (*see* **Note 12**).
5. Quickly chill the culture by pouring it into a prechilled 100 ml centrifuge tube halfway filled with ice.
6. Immediately place tube in a prechilled rotor and centrifuge at 1,000 × *g* for 5 min at 4 °C.
7. Pour off supernatant and resuspend the cell pellet in 50 ml of ice-cold $H_2O$.
8. Recover the pellet by centrifugation at 1,000 × *g* for 5 min at 4 °C.
9. Immediately freeze the cell pellet by placing the tube on dry ice or in liquid nitrogen.

Immediately use the cells or store at −70 °C until you are ready to prepare the protein extract.

*Preparation of protein extracts* (*see* **Note 8**):

Unless otherwise stated, keep protein samples on ice.

1. Prewarm complete cracking buffer to 60 °C. Use 100 μl of cracking buffer per 7.5 $OD_{600}$ units of cells (e.g., for 33 total $OD_{600}$ units of cells, use 0.44 ml of cracking buffer).
2. Quickly thaw cell pellets by separately resuspending each one in the prewarmed cracking buffer. If cell pellets are not

immediately thawed by the prewarmed cracking buffer, place the tubes briefly at 60 °C to hasten melting. To avoid risk of proteolysis, do not leave them longer than 2 min at 60 °C.

3. Transfer each cell suspension to a 1.5-ml screw-cap microcentrifuge tube containing 80 μl of glass beads per 7.5 $OD_{600}$ units of cells. The volume of the glass beads can be measured using a graduated 1.5-ml microcentrifuge tube.
4. Heat samples at 70 °C for 10 min (*see* **Note 13**).
5. Vortex vigorously for 1 min.
6. Pellet debris and unbroken cells in a microcentrifuge at the maximum speed for 5 min at 4 °C.
7. Transfer the supernatants to fresh 1.5 ml screw-cap tubes and place on ice (*first supernatants*).
8. Treat the pellets as follows:
   (a) Place tubes in a 100 °C (boiling) water bath for 3–5 min.
   (b) Vortex vigorously for 1 min.
   (c) Pellet debris and unbroken cells in a microcentrifuge as above.
   (d) Combine each supernatant (*second supernatant*) with the corresponding *first supernatant* (from **step 7**).
   If no supernatant is obtained, add more cracking buffer (50–100 μl) and repeat (b) and (c).
9. Boil the samples briefly and use them for SDS-PAGE analysis and Western blot. Probe the blots with antibodies to the epitope tags (c-Myc or HA in our system) or to the GAL4 DNA-BD or AD. Use untransformed cell protein extracts as controls.

Alternatively, samples may be stored on dry ice or in a –70 °C freezer until you are ready to perform analysis.

### 3.5 Small-Scale Yeast Transformation

You can use this protocol both to transform yeast cells with only one plasmid, like described in this section, and to simultaneously cotransform yeast with two plasmids (DNA-BD and AD constructs) when you want to verify or detect an interaction between two proteins.

1. Inoculate 1 ml of YPD with a large yeast colony from a fresh working plate (*see* Subheading 3.2) (*see* **Note 14**).
2. Vortex vigorously for 5 min to disperse any clumps.
3. Transfer into a flask containing 100 ml of YPD.
4. Incubate at 30 °C for 16–18 h with shaking at 250 rpm. Measure $OD_{600}$ of the culture, it should be >1.5 (*see* **Note 15**).
5. Transfer a suitable volume of overnight culture to a flask containing 300 ml of YPD to obtain an $OD_{600}$ 0.2–0.3.

6. Incubate at 30 °C for 3 h with shaking (250 rpm) or until $OD_{600}$ 0.4–0.6 (*see* **Note 16**).
7. Place cells in 50 ml tubes and centrifuge at 1,000 × *g* for 5 min at room temperature.
8. Discard the supernatant and thoroughly resuspend the cell pellet in sterile TE or distilled $H_2O$. Pool the cells into one tube (final volume 25–50 ml).
9. Centrifuge at 1,000 × *g* for 5 min at room temperature.
10. Decant the supernatant.
11. Resuspend the cell pellet in 1.5 ml of freshly prepared, sterile 1× TE/1× LiAc. You have prepared competent yeast cells ready to be transformed.
12. Add 0.4–0.6 μg plasmid DNA and 0.1 mg of carrier DNA to a fresh 1.5-ml tube and mix.
13. Add 0.1 ml of yeast competent cells to each tube and mix well by vortexing.
14. Add 0.6 ml of sterile PEG/LiAc solution to each tube and vortex to mix.
15. Incubate at 30 °C for 30 min with shaking at 200 rpm.
16. Add 70 μl of DMSO. Mix well by gentle inversion. Do not vortex.
17. Heat shock for 15 min in a 42 °C water bath.
18. Chill cells on ice for 1–2 min.
19. Centrifuge cells for 30 s in a microfuge at the maximum speed at room temperature. Remove the supernatant. Resuspend cells in 0.5 ml of sterile 1× TE buffer and centrifuge for 30 s as above (*see* **Note 17**).
20. Discard super and resuspend in 0.2 ml of TE 1× buffer.
21. Plate 100 μl of cell suspension on 100 mm SD selective plates (SD/-Trp for DNA-BD constructs and SD/-Leu for AD constructs).
22. Incubate plates at 30 °C until colonies appear (generally, 2–4 days).

### 3.6 Library Screening: Mating Protocol

The protocol described here is suitable for the screening of a cDNA library cloned in the pGADT7 vector and already transformed in the Y187 (*MATα*) yeast strain, that can be mated with the AH109 (*MATa*) yeast strain pretransformed with the BAIT (*see* **Note 2** and Table 1) as described in Subheading 3.5.

*Determination of the library titer* (*see* **Note 18**):

1. Transfer a 10 μl aliquot of the pretransformed library to complete medium and plate five appropriate dilutions of this suspension on 10 mm SD/-Leu plates.

**Table 1**
**Two-hybrid vectors provided by the Matchmaker™ GAL4 Two-Hybrid System 3 (Clontech)**

| | Fusion | Epitope | Yeast selection | Bacterial selection |
|---|---|---|---|---|
| *Cloning vectors* | | | | |
| pGBKT7 | DNA-BD/BAIT | c-Myc | *TRP1* | Kanamycin |
| pGADT7 | AD/library or AD/PREY | HA | *LEU2* | Ampicillin |
| *Control vectors* | | | | |
| pGBKT7-53 | DNA-BD/p53 | c-Myc | *TRP1* | Kanamycin |
| pGADT7-T | AD/T-antigen | HA | *LEU2* | Ampicillin |
| pCL1 | GAL4 | | *LEU2* | Ampicillin |

The table reports the two-hybrid vectors provided by the commercial kit mentioned in this chapter
Cloning vectors: pGBKT7, express proteins as fusion products with the GAL4 DNA-BD domain. Contains the *Kanr* gene for selection in *E. coli* and the nutritional marker *TRP1* for selection in yeast. pGADT7, express proteins as fusion products with the GAL4 AD domain. Contains the *Ampr* gene for selection in *E. coli* and the nutritional marker *LEU2* for selection in yeast
Control vectors: pGBKT7-53 and pGADT7-T encode a DNA-BD/murine p53 fusion protein and an AD/SV40 large T-antigen fusion protein, respectively. As the two proteins show a strong interaction in a Y2H assay, they are generally used as positive controls for the interaction [28]
pCL1 encodes the full-length, wild-type GAL4 protein and provides a positive control for α-galactosidase and β-galactosidase assays
*HA* hemagglutinin

2. Incubate plates at 30 °C until colonies appear (3–5 days).
3. Calculate the library titer as follows:

$$\frac{\text{N° colonies}}{\text{Plating volume (ml)} \times \text{dilution factor}} = \text{cfu / ml}.$$

*Mating* (Fig. 3):

1. Inoculate one fresh, large colony of your bait strain (AH109 transformed with the BAIT) into 50 ml of SD/-Trp liquid medium. Incubate shaking (250–270 rpm) at 30 °C until the $OD_{600}$ reaches 0.8 (16–20 h).
2. Centrifuge to pellet the cells (1,000 ×*g* for 5 min) and discard the supernatant.
3. Resuspend the pellet to a cell density of $>1 \times 10^8$ cells/ml in 4–5 ml of SD/-Trp.
4. Combine 1 ml of library strain with the 5 ml of bait strain in a sterile 2 l flask. Add 45 ml of 2× YPDA liquid medium with 50 mg/ml kanamycin.
5. Incubate at 30 °C for 20–24 h, slowly shaking (30–50 rpm) (*see* **Note 19**).

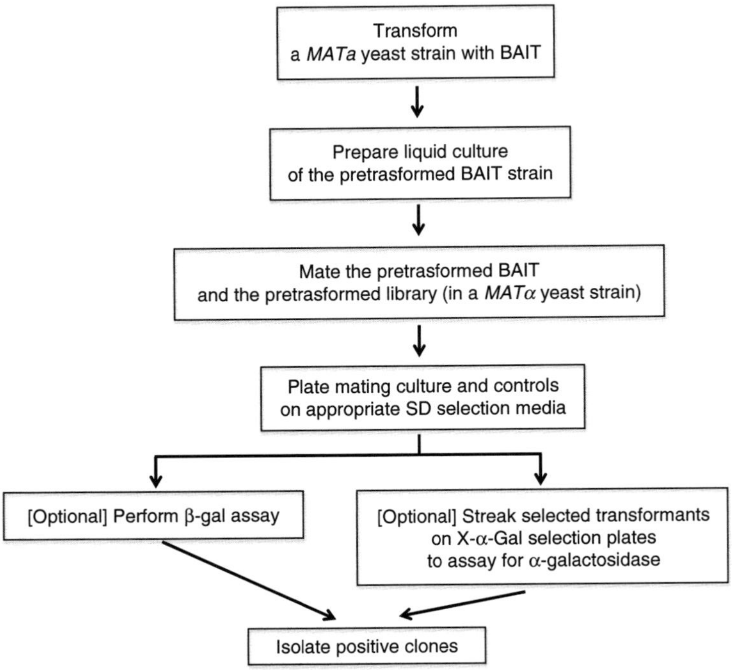


**Fig. 3** Y2H screening using a pretransformed cDNA library and mating. Overview of the procedure

6. Centrifuge to pellet the cells at 1,000 × *g* for 10 min at room temperature.
7. Meanwhile rinse the 2 l flask twice with 50 ml of 0.5× YPDA with 50 μg/ml kanamycin, combine the rinses, and use them to resuspend the pelleted cells.
8. Centrifuge to pellet the cells at 1,000 g for 10 min at room temperature and discard the supernatant.
9. Resuspend cells in 10 ml of 0.5× YPDA/Kan. Measure the total volume of cells + medium.
10. To calculate the number of clones screened, prepare 1/10, 1/100, 1/1,000, and 1/10,000 dilutions of the cell suspension and spread 100 μl of each dilution on the following 100 mm selective plates and incubate at 30 °C for 3–5 days.

    SD/-Trp (negative control).

    SD/-Leu (negative control).

    SD/-Trp/-Leu (diploids cells containing both DNA-BD and AD constructs will grow on this plate).
11. Plate 200 μl aliquots of the remainder of the culture, on 150 mm SD/-Trp/-Leu/-His plates (you will need 50–55 plates). Incubate at 30 °C for 3–8 days. Ignore small pale colonies that may appear after 2 days but never grow to >1 mm in diameter.

True His+ colonies are robust and can grow to >2 mm. They will be tested for expression of the other reporter genes.

12. Calculate the number of screened clones (diploids) by counting the colonies from the SD/-Leu/-Trp plates (**step 10**) after 3–5 days.

    Number of screened clones = cfu/ml of diploids × resuspension volume (ml).

    It is imperative that at least one million diploids are screened. A lower amount will result in a lower chance of detecting genuine interactions (*see* **Note 20**).

13. Peak Trp+Leu+His+ colonies with a plastic loop and streak on 100 mm selective SD/-Trp/-Leu/-His/-Ade to test the activation of the *ADE2* reporter gene. Let them grow at 30 °C.

14. Streak Trp+Leu+His+Ade+ colonies on 100 mm SD/-Trp/-Leu/-His/X-α-GAL plates to test the activation of the *MEL1* reporter gene (colonies turning blue) (*see* **Note 21**).

Clones able to grow in the absence of His and Ade and expressing α-galactosidase activity may be considered true positives of the screening and further analyzed. If the number of true positives is very high, the clones can be further selected by verifying activation of the *LACZ* reporter gene (β-galactosidase filter assay).

### 3.7 β-Galactosidase Filter Assay

1. For best results use fresh colonies grown on SD/-Trp/-Leu/-His plates (*see* **Note 22**).
2. For each plate, presoak a sterile filter by placing it in 2.5–5 ml of Z buffer/X-gal solution in a clean 100-mm diameter plate.
3. Using forceps, place a clean, dry filter over the surface of the plate of colonies to be assayed. Gently rub the filter with the side of the forceps to help colonies cling to the filter.
4. Carefully lift the filter off the agar plate with forceps and transfer it (colonies facing up) to a pool of liquid nitrogen. Using the forceps, completely submerge the filters in the liquid nitrogen.
5. After the filter has frozen completely (~10 s), remove it from the liquid nitrogen and allow it to thaw at room temperature. This freeze/thaw treatment permeabilizes the cells.
6. Carefully place the filter, colony side up, on the presoaked filter (from **step 2**). Avoid trapping air bubbles under or between the filters.
7. Incubate the filters at room temperature and check periodically for the appearance of blue colonies (*see* **Note 23**).
8. Identify the β-galactosidase-producing colonies. Pick the corresponding positive colonies from the original plates to fresh medium and keep them for further analysis.

### 3.8 Analysis of Positive Clones (Fig. 4)

This phase of the screening is aimed to isolate and identify the AD construct contained in each positive clone and encoding putative interactors of the bait protein. To accomplish this aim, total plasmid DNA is first extracted from yeast cells and then used to transform bacterial cells. AD constructs are finally rescued by bacterial cells by taking advantage of the different selection markers of DNA-BD and AD vectors.

*Extraction of plasmid DNA from yeast cells* (*see* **Note 24**):

1. For each positive clone, pick the colony, inoculate into 2 ml of SD/-Leu/-Trp medium, and incubate overnight at 30 °C with shaking.
2. Pellet the cells in a 1.5 ml microcentrifuge tube and remove the supernatant.

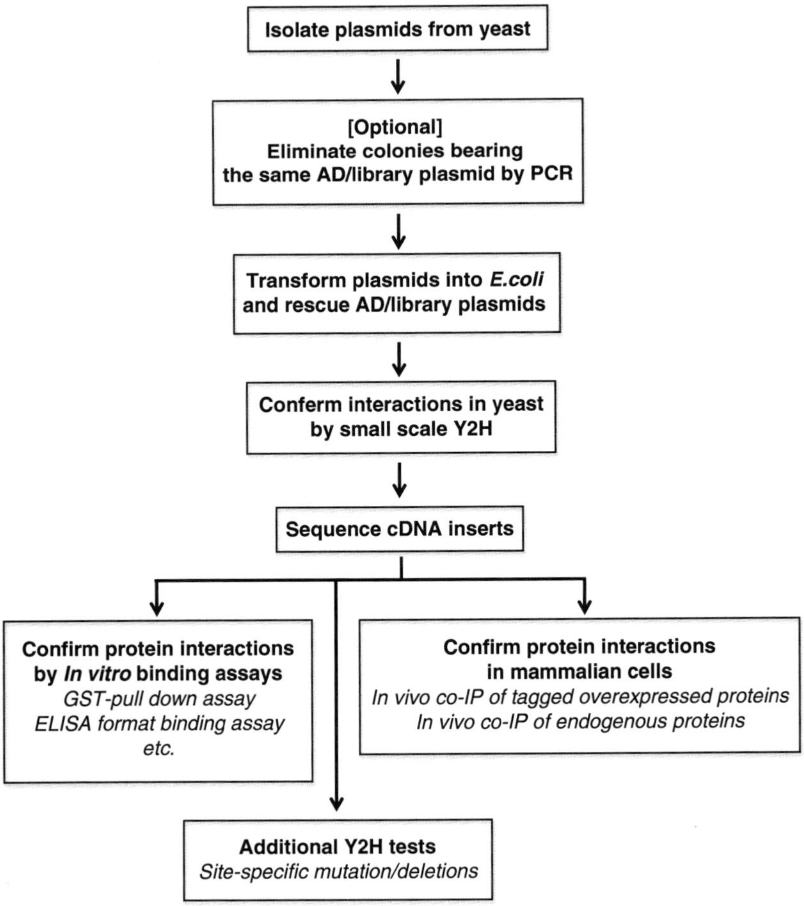


**Fig. 4** Analysis of positive clones. Overview of the general strategy

3. Add 200 μl of YLS and gently resuspend the cell pellet with a micropipette tip.
4. Add 400 μl of glass beads and 200 μl of phenol/chloroform/isoamyl alcohol.
5. Vortex vigorously for 2 min and centrifuge in microfuge at maximum speed for 1 min.
6. Transfer the aqueous phase to a fresh tube and precipitate the nucleic acids by adding 20 μl of sodium acetate and 500 μl of 95 % ethanol.
7. Collect the precipitate by centrifugation at maximum speed for 5 min and wash the nucleic acid pellet once with 500 μl of 70 % ethanol.
8. Air-dry the pellet for 5 min at room temperature.
9. Dissolve the pellet in 25 μl of TE buffer.
10. Use this DNA extract to transform *E. coli* cells.

*Selection of plasmids in bacterial cells* (*see* ***Note 25***):

1. For each positive clone, add 5 μl of the plasmid DNA extract to competent bacterial cells and incubate in ice for 30 min.
2. Heat shock cells by incubating at 42 °C for 45 s and then in ice for 3 min.
3. Add 900 μl of SOC medium and incubate at 37 °C for 1 h under shaking.
4. Pellet cells in the microfuge, remove the most of the supernatant, and leave 150–200 μl.
5. Resuspend cells, plate on LB/amp, and incubate overnight at 37 °C to select bacterial clones containing the AD constructs.

*Extraction of AD constructs from bacterial cells*:

1. For each positive clone, inoculate one amp+ colony into 5 ml on LB/amp and incubate overnight at 37 °C with shaking.
2. Purify the AD construct from transformed bacterial cellular suspension using any standard mini-prep DNA purification protocol or kit.

*Analysis of AD constructs*:

1. Verify the presence and the size of the insert in AD constructs isolated from positive clones by enzymatic digestion and analysis by agarose gel electrophoresis.
2. If the number of positive clones is high, you may need to amplify AD/library inserts by PCR and characterize PCR products by digesting with a frequent cutter restriction enzyme, such as AluI or HaeIII. Analyze fragment size by agarose gel electrophoresis to check for multiple AD/library plasmids and eliminate multiple copies.

*Bait dependency test*:

Use each AD vector containing an insert in a small-scale two-hybrid assay to verify the interaction with the bait protein. For each AD construct, cotransform AH109 cells with the following couples of plasmids:

BD-bait/AD-library insert.

Empty BD vector/AD library insert.

Only the AD/library insert.

Cotransform also with the following couples of plasmids as controls:

Empty BD vector/empty AD vector (negative control).

pGBKT7-53/pGADT7-T (positive control).

1. Perform cotransformation as described in Subheading 3.5 using 0.4–0.6 μg of each plasmid DNA.
2. Plate cotransformed cells onto SD/-Trp/-Leu plates and incubate at 30 °C until colonies appear.
3. Streak Trp+/Leu+ transformants on SD/-Trp/-Leu/-His and SD/-Trp/-Leu/-His/X-α-GAL and incubate at 30 °C.
4. Select clones that exhibit a bait-dependent HIS3/MEL1-positive phenotype (Table 2) and sequence by using a GAL4AD sequencing primer.
5. Identify putative interacting clones by using bioinformatics tools.

*Follow-up*:

1. Validate protein–protein interaction isolated by Y2H screening by using alternative techniques, in non-yeast-based assays. Approaches conventionally used include in vitro binding assays, in vitro "pull-down" assay, co-immunoprecipitation of endogenous proteins, and co-immunolocalization.

**Table 2**
**Bait dependency test**

| Plasmid 1 (DNA-BD) | Plasmid 2 (AD) | Expected Y2H results for a true positive |
|---|---|---|
| None | AD/library insert | Negative |
| DNA-BD only (empty vector) | AD/library insert | Negative |
| DNA-BD/BAIT | AD/library insert | Positive |

AH109 cells are cotransformed with combinations of DNA-BD and AD plasmid as indicated and plated on SD/-Trp/-Leu selective plates. Protein–protein interactions are verified on the basis of the ability of Trp+/Leu+ transformants to grow on SD/-Trp/-Leu/-His plates (activation of HIS3 reporter gene) and to turn blue on SD/-Trp/-Leu/-His/X-α-GAL plates (activation of MEL1 reporter gene). The table shows expected results for bait-dependent interactions

2. You may further characterize the isolated interaction by generating site-specific mutations and/or deletions of the binding partners and using them in small-scale Y2H assays in order to map regions or residues critical for the interaction.

## 4 Notes

1. It is not necessary to handle yeast under laminar flood, but take care to work always in "clean" environment, because bacterial contaminations are worst enemies for this assay.
2. We use the Matchmaker™ GAL4 Two-Hybrid System 3 (Clontech), and we will refer to the components of this system in this chapter. As other systems commercially available, this system contains the GAL4 DNA-BD and GAL4 AD cloning vectors and DNA-BD and AD constructs to be used as positive and negative controls. The ones that will be mentioned in this chapter are summarized in Table 1. The vectors contain c-Myc and HA tags to facilitate detection of fusion proteins and distinct bacterial selection markers (AD-*Amp*$^r$, DNA-BD-*Kan*r) to facilitate selective rescue of plasmids from bacterial cells.

   The kit provides also suitable yeast strains, and in particular the AH109 yeast strain, *MATa*, auxotroph for Trp and Leu, and containing four genes under the control of distinct GAL4 UASs and TATA boxes: *ADE2*, *HIS3*, *MEL1*, and *LACZ*. This strain can be used as the host strain both for the screening of an AD/library and for small-scale assays. Protein–protein interaction may be deduced by the activation of any of these reporter genes. It can be mated with *MATα* strains (such as Y187) pretransformed with cDNA libraries. The Y187 yeast strain is auxotroph for Trp and Leu and contains the *LACZ* reporter gene under the control of GAL4. It can be used to test for an interaction between two known proteins or as a mating partner to verify protein interactions. In addition to cloning vectors, control plasmids, and yeast strains, commercial kits may contain additional components like oligos to be used as sequence primers in sequence analysis of clones and primary antibodies specific for DNA-BD, AD, or the different tags contained in the Y2H vectors, for detection of fusion proteins. A detailed user manual is generally enclosed to the kit or in alternative can be downloaded by the website of the company.
3. Yeast strains can be stored indefinitely in YPD medium with 25 % glycerol at −70 °C. Transformed yeast strains are best stored in the appropriate SD dropout medium to keep selective pressure on the plasmid. A small amount of frozen cells can be scraped from the stock with a loop and streaked onto YPD plates (or SD selective plates) to recover the strain.

4. Due to the high concentration of dextrose, solubilization may be difficult at room temperature. Put about half of the volume of water in a beaker with a magnetic anchor, warm up to 45–50 °C and add slowly the glucose keeping stirring until complete solubilization. Add water to 1 L.
5. It is fundamental to carefully check the pH of media used for yeasts. The pH 5.8 is the optimal value for yeast growth. Small variations of the pH may affect the growth; in addition, very small increase of the pH strongly favors bacterial contaminations.
6. Solubilize in water by adding each component one-by-one under stirring. It may be necessary to slightly warm the water up to improve solubilization.
7. Use only high-quality carrier DNA; nicked calf thymus DNA is not recommended.
8. Yeast cells are protected by rigid cell walls, which are hard to break. Thus one of the challenges regarding the preparation of protein extracts from yeast is the efficiency of the procedure. There are different protocols available; the results (i.e., protein yield and quality) will vary depending on the protein and may be more successful with one protocol than with the other. The protocol described here (urea/SDS method) is the one we have been used in our lab. The two most challenging aspects of isolating proteins from yeast are disrupting yeast cell walls and inhibiting the many endogenous yeast proteases. Yeast cell walls are tough and must be disrupted by a combination of physical and chemical means. Endogenous proteases must be counteracted with a cocktail of strong protease inhibitors. If you know your protein of interest is susceptible to a protease not inhibited by the cocktail described here, add the appropriate inhibitor before using the mixture. You may also wish to add other inhibitors such as sodium fluoride to prevent dephosphorylation, if that is appropriate for your protein.
9. Caution: PMSF is hazardous. Wear gloves. Handle with care and read label precautions.
10. As PMSF has a short half-life (~7 min) in aqueous solutions, you may need to add additional aliquots of PMSF during the preparation of protein extracts.
11. Use classic protocols for DNA cloning [29]. You can easily generate a fusion gene if compatible restriction sites are present in the test gene and the corresponding vector. If not, generate the gene fragment by PCR with useful restriction sites incorporated into the primers. Briefly (a) purify the gene fragment, whether generated by restriction digestion or PCR; (b) digest the DNA-BD and/or the AD vector with the appropriate restriction enzyme(s), treat with phosphatase

when necessary, and purify; (c) ligate the appropriate vector and the insert; (d) transform ligation mix into *E. coli*; (e) identify insert-containing plasmids by restriction analysis; and (f) check orientation and reading frame of the junctions by using appropriate sequencing primers. Sequence the entire insert if produced by PCR.

12. It is fundamental to stop cell culture growth when the $OD_{600}$ is 0.4–0.6. During late log phase, the ADH1 promoter shuts down and the level of endogenous yeast proteases increases.
13. This initial incubation at 70 °C is important because frees membrane-associated proteins. If you skip this step, membrane-associated proteins will be removed from the sample at **step 6** (high-speed centrifugation).
14. It is important to use young colonies. Some yeast strains, like AH109 and Y187, carry the *ade2-101* mutation and when they become old their color turns to dark pink. You should not pick these colonies.
15. Shake well the flask before measuring the $OD_{600}$, because yeast cells can stick on the bottom of the flask.
16. If the cell suspension takes more than 3 h to reach an $OD_{600} = 0.4–0.6$, something is wrong with the culture. It is advisable to stop the experiment and start again with a fresh culture.
17. This washing step is important. It can be noted that after the second centrifugation, pellets have a different color and look. If the pellet is too sticky and it is difficult to resuspend, there's something wrong (possible bacterial contamination).
18. The titer of pretransformed libraries is generally around $5 \times 10^7$ cfu/ml, while the number of independent clones is guaranteed to be at least $1 \times 10^6$. Before starting the screening, it is worthy to determine the titer of the transformed cDNA library and the number of independent clones that will be analyzed in the screening.
19. A low shaking speed prevents the cells from settling on the base of the flask. Vigorous shaking can reduce the mating efficiency.
20. Example of calculation:

    Resuspension volume = 11.5 ml.

    Plating volume = 100 μl.

    50 colonies grew on the 1/1,000 dilution on SD/-Trp/-Leu plates.

    Number of clones screened = $50 \times 11.5 \times 10 \times 1{,}000$ = 5.75 millions.

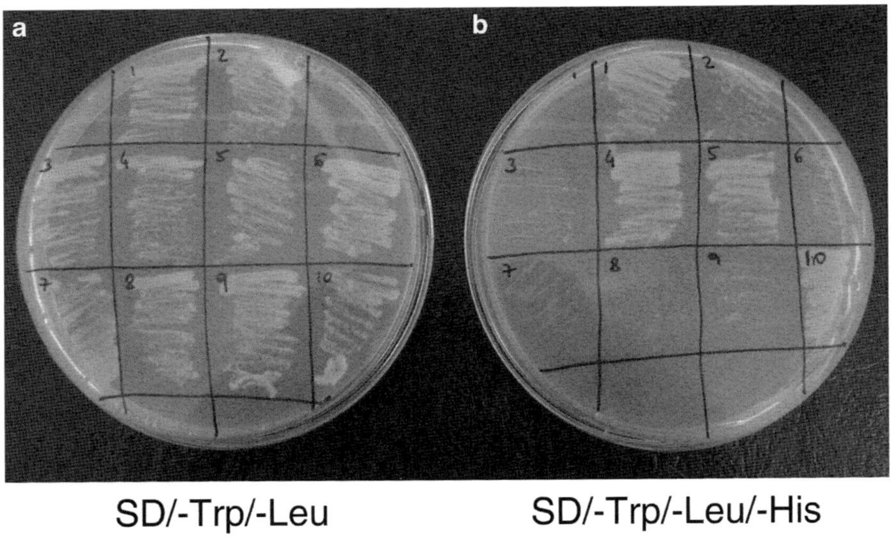


**Fig. 5** Selection of transformants. In the example showed in the figure, AH109 yeast cells have been cotransformed with different combinations of DNA-BD and AD constructs. Cotransformants have been streaked both on a SD/-Trp/-Leu (**a**) and on a SD/-Trp/-Leu/-His (**b**) selective plate. A grid has been drawn on the bottom of each plate in order to streak several colonies in the same plate. The colonies have been numbered to trace them during selection. Colonies growing on plate (**a**) contain both the DNA-BD (*trp1*) and the AD (*leu2*) constructs, while colonies growing on plate (**b**) activate the *HIS3* reporter gene, indicating interaction between the DNA-BD and AD fusion proteins. Clones #1, 4, 5, and 10 in the (**a**) plate express interacting partners

21. **Steps 13** and **14**: it is not necessary to remove the whole colony, just pick a small amount of it and streak it on the new plate. Streaking a big amount of cell may produce aspecific growth generating a false positive. Draw a grid on the bottom of each plate in order to streak several colonies in the same plate. Number the colonies to trace them during selection (Fig. 5).
22. Streak His+Ade+Mel+ transformants on SD/-Trp/-Leu/-His plates and let them grow at 30 °C for 3–5 days. Streak yeast cotransformed with pGBKT7-53 and pGADT7-T as positive control and yeast cotransformed with pGBKT7 and pGADT7 empty vectors as negative control. In this experiment the age of the colonies is crucial: colonies need to be big enough to be lifted but still metabolically active, thus don't let them get too old.
23. The time it takes colonies expressing β-galactosidase to turn blue typically varies from 30 min to 8 h in a library screening. Prolonged incubation (>10 h) may give false positives. After 12 h of incubation, all the colonies, including the negative control, turn to a greenish color. Most yeast reporter strains cotransformed with the positive controls for a two-hybrid interaction give a positive blue signal within 60 min. If the controls do not behave as expected, check the reagents and repeat the assay.

24. The plasmid DNA isolated from each positive yeast colony will be a mixture of the DNA-BD/BAIT plasmid and at least one type of the AD/library plasmid. The extraction of plasmid DNA from yeast cells is not trivial, primarily because of the tough cell wall. Furthermore the relatively large size (>6 kb) and low copy number of some yeast plasmids result in very low DNA yields, regardless of the plasmid isolation method used. Besides, plasmid DNA isolated from yeast is often contaminated by genomic DNA because yeast contains ~3× as much genomic DNA as *E. coli* and the isolation method breaks the yeast chromosomes and releases them from cellular material. There are several yeast plasmid isolation procedures currently in use. The various protocols differ primarily in the method used to break the cell walls. The protocol described here is the one that we use in our lab and is suitable to recover AD plasmids from the positive clones isolated by Y2H screening.
25. It is recommended to use bacterial cells able to yield a transformation efficiency of at least $10^7$ cfu/μg. The strong contamination by yeast genomic DNA in the plasmid DNA extracted from yeast cells strongly reduces the transformation efficiency.

## References


1. Fields S, Song O (1989) A novel genetic system to detect protein–protein interactions. Nature 340:245–246
2. Uetz PH, Hughes RE, Fields S (1998) The two hybrid system: finding likely partners for lonely protein. Focus 20(363):62–64
3. Vidal M, Legrain P (1999) Yeast forward and reverse 'n'-hybrid systems. Nucl Acid Res 27:919–929
4. Vidalain P, Boxem M, Ge H et al (2004) Increasing specificity in high-throughput yeast two-hybrid experiments. Methods 32:363–370
5. Fields S (2005) High-throughput two-hybrid analysis. The promise and the peril. FEBS J 272:5391–5399
6. Chevray P, Nathans D (1992) Protein interaction cloning in yeast: identification of mammalian proteins that react with the leucine zipper of Jun. Proc Natl Acad Sci U S A 89: 5789–5793
7. Vojtek AB, Hollenberg SM, Cooper JA (1993) Mammalian Ras interacts directly with the serine/threonine kinase Raf. Cell 74:205–214
8. Spaargaren M, Bischoff JR (1994) Identification of the guanine nucleotide dissociation stimulator for Ral as a putative effector molecule of R-ras, H-ras, K-ras, and Rap. Proc Natl Acad Sci U S A 91:12609–12613
9. Hofer F, Fields S, Schneider C et al (1994) Activated Ras interacts with the Ral guanine nucleotide dissociation stimulator. Proc Natl Acad Sci U S A 91:11089–11093
10. Kikuchi A, Demo SD, Ye ZH et al (1994) ralGDS family members interact with the effector loop of ras p21. Mol Cell Biol 14:7483–7491
11. Wolthuis RMF, Bauer B, van't Veer LJ et al (1996) RalGDS-like factor (Rlf) is a novel Ras and Rap1A-associating protein. Oncogene 13:353–362
12. Peterson SN, Trabalzini L, Brtva TR et al (1996) Identification of a novel RalGDS-related protein as a candidate effector for Ras and Rap1. J Biol Chem 271:29903–29908
13. Shao H, Andres D (2000) A novel RalGEF-like protein, RGL3, as a candidate effector for rit and Ras. J Biol Chem 275:26914–26924
14. Nancy V, Wolthuis RM, de Tand MF et al (1999) Identification and characterization of potential effector molecules of the Ras-related GTPase Rap2. J Biol Chem 274:8737–8745
15. Janoueix-Lerosey I, Pasheva E, de Tand MF et al (1998) Identification of a specific effector of the small GTP-binding protein Rap2. Eur J Biochem 252:290–298
16. Janoueix-Lerosey I, Jollivet F, Camonis J et al (1995) Two-hybrid system screen with the small GTP-binding protein Rab6. Identification

of a novel mouse GDP dissociation inhibitor isoform and two other potential partners of Rab6. J Biol Chem 270:14801–14808
17. Spaargaren M, Martin GA, McCormick F et al (1994) The Ras-related protein R-ras interacts directly with Raf-1 in a GTP-dependent manner. Biochem J 300:303–307
18. Ferro E, Magrini D, Guazzi P et al (2008) G-protein binding features and regulation of the RalGDS family member, RGL2. Biochem J 415:145–154
19. Van Aelst L, Barr M, Marcus S et al (1993) Complex formation between RAS and RAF and other protein kinases. Proc Natl Acad Sci U S A 90:6213–6217
20. Jaitner BK, Becker J, Linnemann T et al (1997) Discrimination of amino acids mediating Ras binding from noninteracting residues affecting Raf activation by double mutant analysis. J Biol Chem 272:29927–29933
21. Chan EY, Stang SL, Bottorff DA et al (2002) Mutations in conserved regions 1, 2, and 3 of Raf-1 that activate transforming activity. Mol Carcinogen 33:189–197
22. Huang L, Weng X, Hofer F et al (1997) Three-dimensional structure of the Ras-interacting domain of RalGDS. Nat Struct Biol 4:609–614
23. Ponting CP, Benjamin DR (1996) A novel family of Ras-binding domains. Trends Biochem Sci 21:422–425
24. Wohlgemuth S, Kiel C, Kramer A et al (2005) Recognizing and defining true Ras binding domains I: biochemical analysis. J Mol Biol 348:741–758
25. Fabian JR, Vojtek AB, Cooper JA et al (1994) A single amino acid change in Raf-1 inhibits Ras binding and alters Raf-1 function. Proc Natl Acad Sci U S A 91:5982–5986
26. Mosteller RD, Han J, Broek D (1994) Identification of residues of the H-ras protein critical for functional interaction with guanine nucleotide exchange factors. Mol Cell Biol 14:1104–1112
27. Stephens DJ, Banting G (2000) The use of yeast two-hybrid screens in studies of protein:protein interactions involved in trafficking. Traffic 1:763–768
28. Li B, Fields S (1993) Identification of mutations in p53 that affect its binding to SV40 T antigen by using the yeast two hybrid system. FASEB J 7:957–963
29. Sambrook J, Fritsch EF, Maniatis T (1989) Molecular cloning: a laboratory manual. Cold Spring Harbor Laboratory, Cold Spring Harbor, NY

# Chapter 8

# Screening for MAPK Modulators Using an In-Cell Western Assay

**Simon Schnaiter, Beatrix Fürst, Johannes Neu, Frigyes Wáczek, László Őrfi, György Kéri, Lukas A. Huber, and Winfried Wunderlich**

## Abstract

Several essential cellular functions are critically influenced by the Ras-Raf-MEK-ERK cascade, and pathological problems have been implicated with loss as well as gain of its activity. Therefore, there is a great demand for chemical probes that act as modulators of the cascade in order to correct medical problems associated with its malfunctions.

Here we describe the use of the In-Cell Western assay, a fast and easy method for the detection of ERK activity. This method is useful for the screening of chemical libraries for compounds that modulate the intensity and duration of growth factor-induced MAPK activity. Examples are provided from a screen for MAPK modulators in Vichem Chemie Research´s Nested Chemical Library™.

**Key words** Ras, Raf, MEK, ERK, In-Cell Western, Screening

## 1 Introduction

The Raf-MEK-ERK signaling cascade is one of the central transducers of Ras-mediated growth factor signals. It is consisting of three sequentially activated protein kinases: a MAP kinase kinase kinase or MAP3K (A, B, and C-Raf) activates the MAP kinase kinases or MAPK2Ks MEK1 and MEK2 by phosphorylation on two serines. These activated MAP2Ks are in turn activating the MAP kinases or MAPKs ERK1 and ERK2 by means of their dual specificity protein kinase activity through phosphorylation on threonine and tyrosine [1].

Downstream of ERKs are more than 100 substrates involved in pleiotropic cellular functions such as survival and apoptosis, proliferation and differentiation, metabolism and motility [2–4]. Such diverse and partially opposing effects demand complex regulatory mechanisms [5,6]. One of these mechanisms is temporal regulation, i.e., the regulation of intensity and duration of ERK activity. In order to identify molecules that interfere with temporal regulation, it is necessary to reliably monitor ERK activity over time.

Lorenza Trabalzini and Saverio Francesco Retta (eds.), *Ras Signaling: Methods and Protocols*,
Methods in Molecular Biology, vol. 1120, DOI 10.1007/978-1-62703-791-4_8, © Springer Science+Business Media, LLC 2014

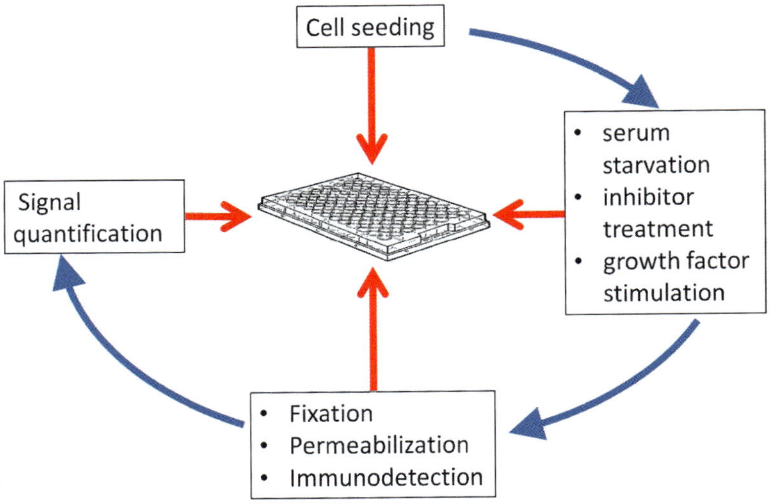


**Fig. 1** Schematic representation of the In-Cell Western assay

The dual phosphorylation of ERK (ppERK) correlates with its enzymatic activity, and several antibodies have been generated to detect this event. There is a variety of methods that make use of these antibodies to measure ERK phosphorylation such as Western blotting, intracellular flow cytometry, or immunocytochemistry. The problem of these methods is that they involve many steps that need the attention of a researcher and thus they are incompatible with a screening approach.

We describe the use of an In-Cell Western (ICW) assay for the identification of MAPK modulators in Vichem Chemie Research´s Nested Chemical Library™ [7]. ICW involves the stimulation of a confluent monolayer of cells with growth factor followed by a fixation-immunodetection protocol finished by infrared fluorescence-based quantification (depicted in Fig. 1). It is a fast method with limited hands-on-time and as such amenable to automation, making it a straightforward approach for compound library screening.

## 2 Materials

### *2.1 Preparation of Target Cells*

1. HeLa cell line.
2. Dulbecco's modified Eagle medium (DMEM).
3. Fetal Bovine Serum Gold (FBS).
4. Phosphate-buffered saline (PBS): 137 mM sodium chloride, 2.7 mM potassium chloride, 10 mM disodium hydrogen phosphate, 1.8 mM potassium dihydrogen phosphate.

5. Trypsin EDTA (1:250) (10×): dilute this stock solution with PBS to give a 1× working solution.
6. 96-well plates (Sarstedt).
7. 50 ml Reagent reservoirs.

### 2.2 Treatment of Target Cells

1. Dimethyl sulfoxide (DMSO); when used for cell treatment prepare a 10× stock solution by diluting DMSO to 10 % (v/v) with serum-free DMEM (SFD).
2. Test compounds: prepare 10 mM solutions in DMSO. Use these solutions to prepare 1,000 and 10 μM stock solutions in SFD (10×).
3. MEK inhibitor U0126: prepare 10 mM solutions in DMSO. Use this solution to prepare a 100 μM stock solution in SFD (10×).
4. Murine epidermal growth factor (EGF): dissolve at 100 μg/ml in 1 % (w/v) bovine serum albumin in PBS and store as 20 μl aliquots at −80 °C. For cell treatment, dilute this solution with SFD to obtain a 100 ng/ml stock solution.

### 2.3 Detection and Quantification

1. Fixation solution: 3.7 % formaldehyde solution (w/v) in PBS diluted 1:10 from a 37 % (w/v) solution (*see* **Notes 1** and **2**). Store at 4 °C.
2. Washing solution 1: 0.1 % Triton X-100 (v/v) in PBS (*see* **Notes 1** and **3**). Store at 4 °C.
3. Odyssey blocking buffer (LI-COR). Store at 4 °C.
4. Alexa Fluor® 680 Phalloidin (Life Technologies). Dissolved in methanol. Store at −20 °C protected from light.
5. Anti-ppERK antibody and IRDye 800CW Protein Labeling Kit—High Molecular Weight (LI-COR). Store at 4 °C protected from light (*see* **Note 4**).
6. Antibody master mix: mix 10 ml of Odyssey blocking buffer with 0.5 μl of anti-ppERK (1:20,000) and 6.7 μl of phalloidin-680 dye (1:1,500) (*see* **Notes 5** and **6**). Try to avoid any exposure to light and use lightproof tubes or wrap transparent tubes with aluminum foil. Store the solution at 4 °C.
7. Washing solution 2: 0.1 % Tween-20 (v/v) in PBS (*see* **Note 3**).
8. Aluminum foil.
9. Parafilm®.
10. LI-COR 9201 ODYSSEY Infrared Imaging System.
11. LI-COR ODYSSEY 2.1 Software.
12. In-Cell Western Plug-in Software.

# 3 Methods

### 3.1 Preparation of Target Cells

*Cell seeding* (first day morning)

1. Harvest HeLa cells using 1× Trypsin EDTA in PBS.
2. Suspend the cells in DMEM supplemented with 10 % FBS to reach a concentration of 300,000 cells/ml (*see* **Note 7**).
3. Pipette 100 μl of the suspension per well of a 96-well plate (cell plate) (*see* **Note 8**).
4. Incubate the plate at 37 °C, in a 5 % $CO_2$ atmosphere for approximately 6 h until cells are properly attached (*see* **Note 9**).

*Serum starvation* (first day afternoon)

1. Check if the cells are firmly attached (*see* **Note 10**).
2. Carefully remove growth medium.
3. Wash cells carefully twice with 100 μl of SFD.
4. Starve the cells overnight (*see* **Note 11**) with SFD (100 μl/well) at 37 °C, in a 5 % $CO_2$ atmosphere.

### 3.2 Treatment of Target Cells (Second Day Morning)

*Inhibitor treatment*

1. Prepare an extra 96-well plate (EGF plate): dispense 100 μl of the EGF 100 ng/ml stock solution into columns 2–6 and 8–12.
2. Remove the starved cells from the incubator (cell plate).
3. Add 10 μl of each 10× stock in SFD:
   (a) Test compounds: the 1,000 and 10 μM stock solutions.
   (b) DMSO (negative control): the 10 % stock solution.
   (c) U0126 (positive control): the 100 μM stock solution to the respective wells of both plates (EGF plate and cell plate) according to the schedule in Fig. 2 (10 μl of SFD is added to the wells A1–6 and B1–6).
4. Incubate the cell plate at 37 °C in a 5 % $CO_2$ atmosphere for 1 h.

| | 1 | 2 | 3 | 4 | 5 | 6 | 7 | 8 | 9 | 10 | 11 | 12 |
|---|---|---|---|---|---|---|---|---|---|---|---|---|
| A | background control | | | | | | positive control | | | | | |
| B | untreated control | | | | | | negative control | | | | | |
| C | Test compound 1, concentration 1 | | | | | | Test compound 1, concentration 1 | | | | | |
| D | Test compound 1, concentration 2 | | | | | | Test compound 1, concentration 2 | | | | | |
| E | Test compound 2, concentration 1 | | | | | | Test compound 2, concentration 1 | | | | | |
| F | Test compound 2, concentration 2 | | | | | | Test compound 2, concentration 2 | | | | | |
| G | Test compound 3, concentration 1 | | | | | | Test compound 3, concentration 1 | | | | | |
| H | Test compound 3, concentration 2 | | | | | | Test compound 3, concentration 2 | | | | | |
| min | 0 | 5 | 10 | 25 | 35 | 60 | 0 | 5 | 10 | 25 | 35 | 60 |

**Fig. 2** Schedule for inhibitor treatment and growth factor stimulation

*Growth factor stimulation*

Perform the next three steps consecutively for each column at the respective time point, i.e. start at $t = 0$ min with treatment of columns 6 and 12 and put the plate back into the incubator. Then 25 min later treat columns 5 plus 11, at $t = 35$ min treat columns 4 plus 10, 50 min after starting treat columns 3 and 9 and at 55 min treat columns 2 plus 8:

(a) Remove medium from the cell plate using a multichannel pipette.

(b) Add 100 µl from the EGF plate to the cell plate using a multichannel pipette.

(c) Incubate the cell plate at 37 °C in a 5 % $CO_2$ atmosphere.

### 3.3 Detection and Quantification

*Fixation of cells*

1. Remove medium carefully.
2. Fix the cells with fixation solution (100 µl/well) and incubate in a fume hood for 20 min (*see* **Notes 2** and **8**).
3. Remove fixation solution (*see* **Note 12**) and wash the preparations five times for 5 min each using washing solution 1. The washing should be conducted with gentle shaking.

*Blocking and antibody incubation*

1. Remove the washing solution 1 (*see* **Note 12**) and add 50 µl of blocking buffer per well followed by gentle rotation for 90 min.
2. Remove the blocking buffer (*see* **Note 12**) and add the antibody master mix (50 µl/well). *Don't forget to make a background control by adding just blocking buffer without antibodies to wells A1–6.* Seal the well plate light and evaporation proof (i.e. use aluminum foil and ideally Parafilm®). Incubate the preparation overnight on a shaker under gentle rotation.

*Washing of cells* (third day morning)

1. Remove the antibody solution (*see* **Note 12**) and wash the preparations five times for 5 min each using 100 µl of washing solution 2 per well. Protect from light!
2. Shake all liquids off and wrap the well plate lightproof with aluminum foil. Carry on with scanning immediately (*see* **Note 13**).

*Scanning and evaluation of the data*

1. Set the LI-COR scanner as pictured in Fig. 3 (*see* **Note 14**).
2. Execute scan (*see* **Note 15**). The scanner generates data for the intensities of both 700 nm (red pseudo color) and 800 nm (green pseudo color) wavelengths (*see* Fig. 4, upper panels).

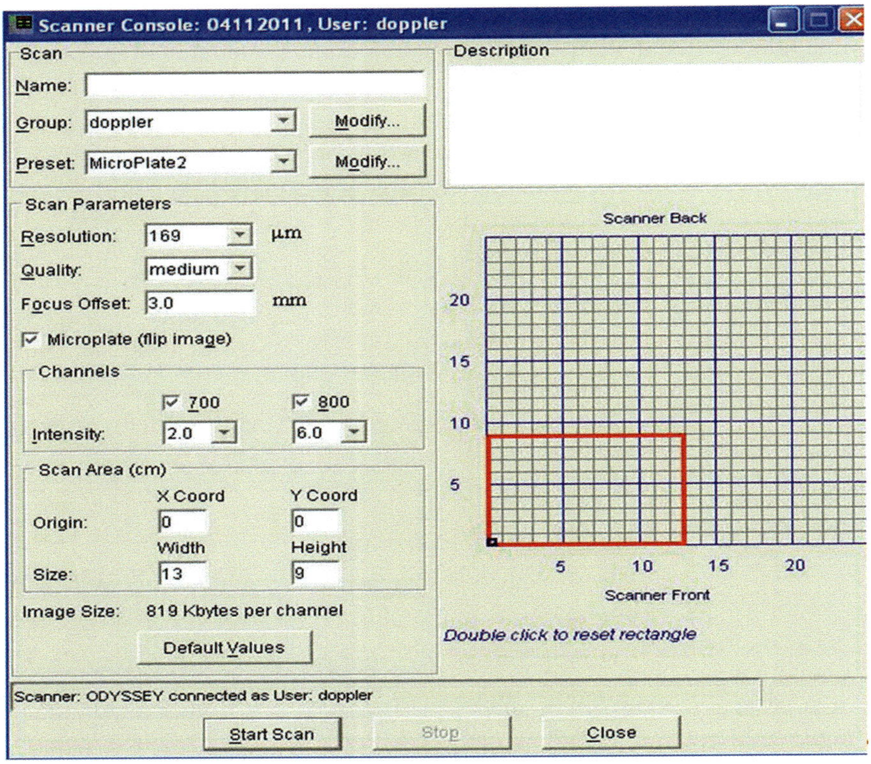


**Fig. 3** Set the LI-COR scanner as pictured here before starting the scan (*see* **Note 16**)

3. Evaluate the data using the In-Cell Western plug in (*see* **Notes 15** and **16**). Thereby both pseudo colors are matched (*see* Fig. 4, lower panel) and the values of ppERK are normalized to those of phalloidin. The background control is subtracted automatically.
4. Export the data into an Excel file (*see* **Note 17**).
5. Normalize all values to the respective zero minute time points.
6. Conduct statistics and prepare graphs that ideally fit your experiment. We used a line diagram (*see* Fig. 5).

## 4 Notes

1. Prepare fresh each time for best results.
2. Wear a mask, gloves, and an adequate eye protection during the handling procedures and perform the fixation process in a fume hood. Store the liquid waste separately and let formaldehyde evaporate from paper towel in the fume hood for a day.
3. For handling reasons, prepare a solution of 10 % (v/v) in PBS and use it as a stock. Store at 4 °C.

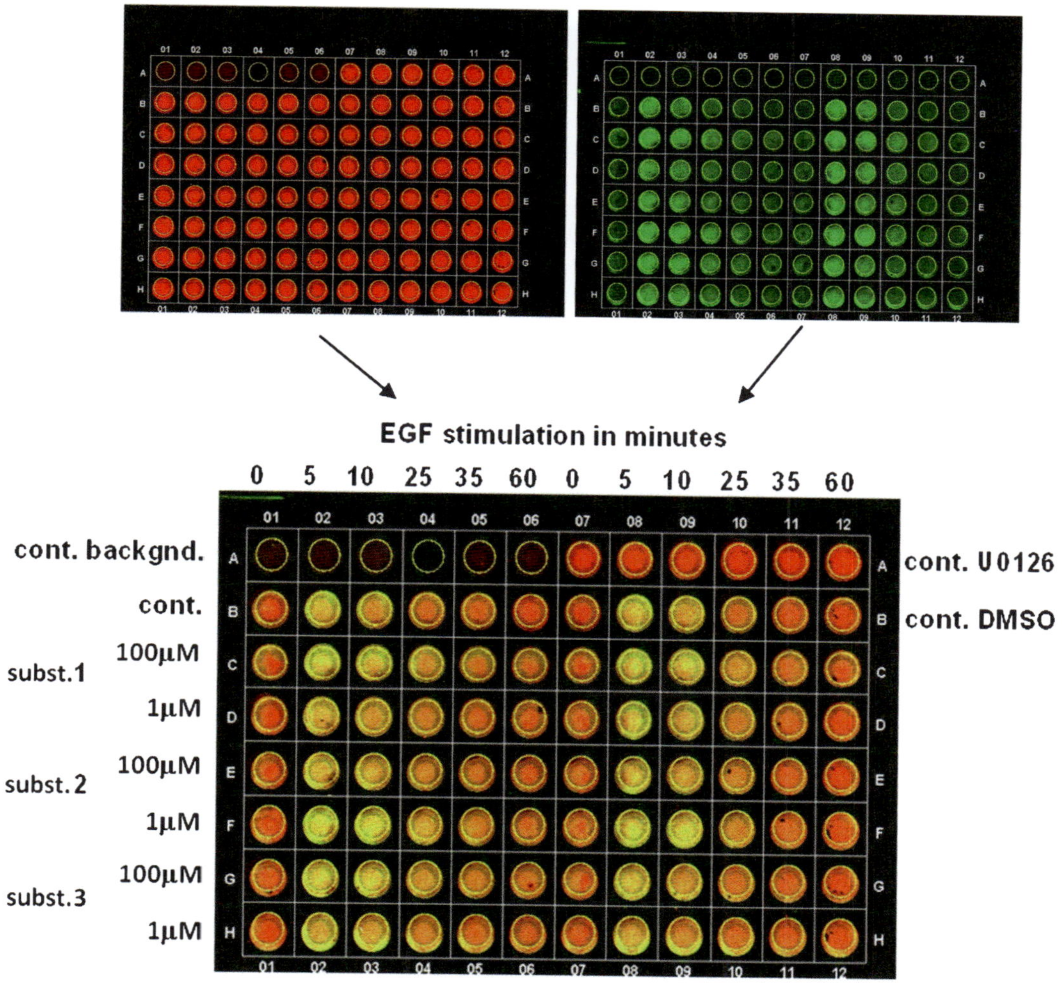


**Fig. 4** Example images created by the LI-COR infrared scanner. The *upper left part* of the figure shows the 700 nm channel (phalloidin, *red pseudocolor*); the 800 nm wavelength is illustrated in the *upper right* (anti-ppERK, *green pseudocolor*). The two images are overlayed on the bottom, ready for evaluation using the In-Cell Western Plug-in

4. Rabbit mAb anti-phospho-p44/42 MAPK (Erk1/2) (Thr202/Tyr204) (197G2) (Cell Signaling, Cat. No.: 4377, carrier free and custom formulation) was used for the demonstrated experiments. For the labeling reaction follow the manufacturer's instructions. In brief, 1 mg antibody was dissolved in a mixture of 0.9 ml PBS + 0.1 ml of 1 M potassium phosphate buffer ($K_2HPO_4$, pH 9) and then labeled with 7.3 µl labeling dye for 2 h at 20 °C. Subsequently the labeled antibody was separated from free dye using the desalting spin columns.
5. Antibody concentrations may vary when using other cells lines or older reagents.

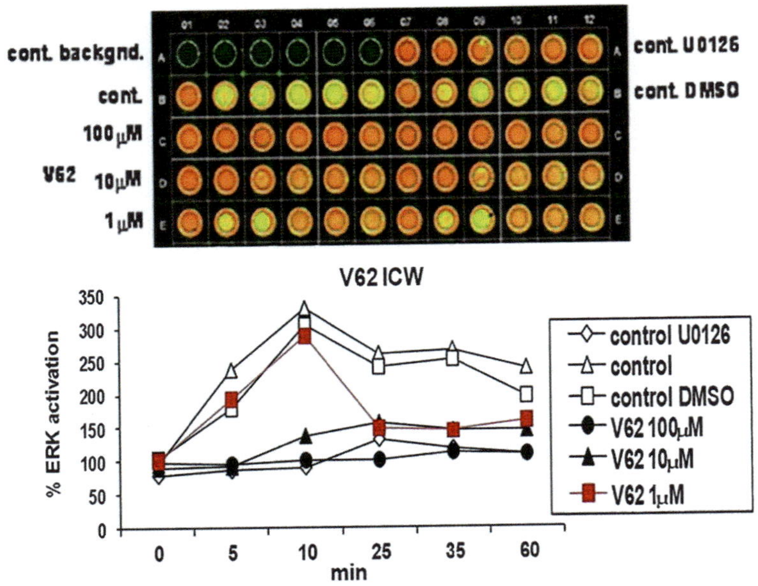


**Fig. 5** A typical time curve for EGF induced ERK activation and its modulation by the test compound V62 (see Note 18). The *upper part* of the figure shows an overlay as described in Fig. 4. The *bottom part* illustrates the calculated values as *line chart*

6. Prepare the blocking buffer first and then add the antibodies.
7. 300,000 cells/ml provided reproducible results for HeLa cells. The optimal density may vary when using other cell lines.
8. This step and further steps in which the same solution is distributed over a 96-well plate is optimally performed using a multichannel pipette with multi-dispense-function.
9. A confluent cell layer covering each well evenly should be present.
10. The cells need to be firmly attached so they are not getting lost during the manipulations.
11. 14 h of serum starvation is optimal for HeLa cells as the level of un-stimulated ppERK is at its minimum after this time. The starvation time needs to be optimized for each cell type.
12. The following method was used: pour out liquid into a tray and cover the well plate with a paper towel, invert it and shake it firmly 2–3 times to remove practically all liquids.
13. The object can be stored (dry) at 4 °C for short time periods (several days). However, there will be a loss of signal intensity.
14. Readjust several parameters if microplates from other suppliers than Sarstedt or other cell lines are used.

15. Follow the instructions described in the In-Cell Western Plug-in manual provided by LI-COR.
16. Align the In-Cell Western Grid exactly to obtain unbiased data. The circles (measurement areas) must not contact the well's walls to avoid interfering reflections.
17. Equivalent programs can be used alternatively.
18. Typical results achieved by screening through Vichem Chemie Research´s Nested Chemical Library™ [7].

## Acknowledgements

This work was supported by the European Union FP6 project "Growthstop," the Austrian Science Fund (SFB021 "Cell Proliferation and Cell Death in Tumors"), and the COMET Center ONCOTYROL, which is funded by the Austrian Federal Ministries BMVIT/BMWFJ (via FFG) and the Tiroler Zukunftsstiftung/ Standortagentur Tirol (SAT).

### References

1. Roskoski R Jr (2012) MEK1/2 dual-specificity protein kinases: structure and regulation. Biochem Biophys Res Commun 417:5–10
2. Osborne JK, Zaganjor E, Cobb MH (2012) Signal control through Raf: in sickness and in health. Cell Res 22:14–22
3. Wortzel I, Seger R (2011) The ERK cascade: distinct functions within various subcellular organelles. Genes Cancer 2:195–209
4. Geest CR, Coffer PJ (2009) MAPK signaling pathways in the regulation of hematopoiesis. J Leukoc Biol 86:237–250
5. Murphy LO, Blenis J (2006) MAPK signal specificity: the right place at the right time. Trends Biochem Sci 31:268–275
6. Teis D, Huber LA (2003) The odd couple: signal transduction and endocytosis. Cell Mol Life Sci 60:2020–2033
7. Keri G, Szekelyhidi Z, Banhegyi P et al (2005) Drug discovery in the kinase inhibitory field using the Nested Chemical Library technology. Assay Drug Dev Technol 3: 543–551

# Chapter 9

# Behavioral Methods for the Study of the Ras–ERK Pathway in Memory Formation and Consolidation: Passive Avoidance and Novel Object Recognition Tests

**Raffaele d'Isa, Riccardo Brambilla, and Stefania Fasano**

## Abstract

Memory is a high-level brain function that enables organisms to adapt their behavioral responses to the environment, hence increasing their probability of survival. The Ras–ERK pathway is a key molecular intracellular signalling cascade for memory consolidation. In this chapter we will describe two main one-trial behavioral tests commonly used in the field of memory research in order to assess the role of Ras–ERK signalling in long-term memory: passive avoidance and object recognition.

Passive avoidance (PA) is a fear-motivated instrumental learning task, designed by Jarvik and Essman in 1960, in which animals learn to refrain from emitting a behavioral response that has previously been associated with a punishment. We will describe here the detailed protocol and show some examples of how PA can reveal impairments or enhancements in memory consolidation following loss or gain of function genetic manipulations of the Ras–ERK pathway. The phenotypes of global mutants as Ras-GRF1 KO, GENA53, and ERK1 KO mice, as well as of conditional region-specific mutants (striatal K-CREB mice), will be illustrated as examples. Novel object recognition (NOR), developed by Ennaceur and Delacour in 1988, is instead a more recent and highly ecological test, which relies on the natural tendency of rodents to spontaneously approach and explore novel objects, representing hence a useful non-stressful tool for the study of memory in animals without the employment of punishments or starvation/water restriction regimens. Careful indications will be given on how to select the positions for the novel object, in order to counterbalance for individual side preferences among mice during the training. Finally, the methods for calculating two learning indexes will be described. In addition to the classical discrimination index (DI) that measures the ability of an animal to discriminate between two different objects which are presented at the same time, we will describe the formula of a new index that we present here for the first time, the recognition index (RI), which quantifies the ability of an animal to recognize a same object at different time points and that, by taking into account the basal individual preferences displayed during the training, can give a more accurate measure of an animal's actual recognition memory.

**Key words** Ras–ERK pathway, Memory, Passive avoidance, Object recognition, Recognition index (RI), Ras-GRF1 KO, GENA 53, Ras-GRF1 OE, ERK1 KO, Mice

Lorenza Trabalzini and Saverio Francesco Retta (eds.), *Ras Signaling: Methods and Protocols*,
Methods in Molecular Biology, vol. 1120, DOI 10.1007/978-1-62703-791-4_9, © Springer Science+Business Media, LLC 2014

# 1 Introduction

### 1.1 Ras–ERK Signalling in Memory Consolidation

The ability to modify behavioral responses to external stimuli is essential for biological organisms in terms of evolutionary fitness, as adaptation of an organism's behavior to the environment sharply increases its probability of survival. Memory, technically defined in cognitive neuroscience as the faculty to encode, store, and retrieve information, is the high-level brain function that allows this adaptive behavioral plasticity. Memory can be categorized into two main phases: short-term memory (STM) and long-term memory (LTM) [1–4]. STM, whose duration ranges from minutes to hours, does not require protein synthesis and relies merely on phosphorylating already-existing proteins. On the other hand, the formation of LTM, which can persist for days, months, or even decades (in humans), depends instead on gene expression and de novo protein synthesis [5, 6].

The Ras-controlled mitogen-activated protein kinase (MAPK)–extracellular-regulated kinase (ERK) pathway is one of the key molecular intracellular cascades implicated in the formation of LTM [7, 8]. Although the Ras–ERK signalling pathway was originally described in the context of cancer research as a regulator of cell proliferation and differentiation [9–13], its involvement in synaptic plasticity and memory consolidation has become increasingly evident since the mid-1990s.

In the brain, this cascade provides a link between ionotropic, metabotropic, and neurotrophin receptors to cytosolic (regulation of ion channels and protein translation) and nuclear events, leading to gene transcription, de novo protein synthesis, and changes in synaptic remodelling, plasticity, and memory formation. Once activated by neurotransmitter receptors through GTP/GDP exchange factors, the small GTPases belonging to the Ras class (p21 H-, K- and N-Ras) stimulate sequentially the cascade of protein kinases consisting of the Raf (c-Raf and B-Raf), the MEK1/2, and the ERK1/2 proteins. In particular, ERK1 and ERK2, the two major MAPK in the brain, are able to translocate into the nucleus and activate either directly or indirectly (via the kinases of the MSK and RSK families) transcription factors like CREB, serum response element (SRE)-binding protein like Elk-1, and chromatin remodelling by phosphorylating histone H3. English and Sweatt were the first, in 1996, to provide evidence that an increase of ERK phosphorylation is associated with a form of plasticity called long-term potentiation (LTP) in area CA1 of hippocampus [14] and the following year they were the first to demonstrate that ERK activation is necessary for LTP induction in this brain region [15]. Long-term potentiation (LTP) and long-term depression (LTD) are two main forms of long-term plasticity, which are considered as valid cellular correlates of memory formation. LTP is a persistent increase in synaptic strength as measured by the amplitude of the excitatory postsynaptic potential (EPSP), which can be produced by high-frequency stimulation. LTD, on the other hand, is an opposite phenomenon associated to a prolonged reduction of synaptic transmission.

In 1997, the first genetic evidence became available demonstrating that ablation in the mouse of Ras-GRF1, a neuronal specific activator (guanine nucleotide exchange factor, GEF) of Ras proteins, is both necessary for LTP induction in the amygdala, a brain region necessary for emotional memory and for LTM consolidation [16]. The following year Sweatt and collaborators demonstrated that hippocampus-dependent memory consolidation can be blocked using SL327, a MEK inhibitor which can cross the blood–brain barrier [17]. Indeed, it is now widely accepted that Ras–ERK activation is required for both synaptic plasticity and memory formation in a variety of systems [8, 18–23].

In the past 15 years, several pharmacological and genetic manipulations have highlighted the involvement of the Ras–ERK pathway in many distinct types of memory. Injection of the MEK inhibitor PD098059, for instance, in entorhinal cortex resulted in impairments in passive avoidance fear memory [24] and Morris water maze spatial memory [25], whereas its injection in insular cortex brought to attenuated conditioned taste aversion [26]. When administered intracerebroventricularly, PD098059 affected conditioned taste aversion [27] and object recognition memory [28]. Furthermore, U0126, another MEK inhibitor, injected in amygdala, led to deficits in Pavlovian fear conditioning [29]. Finally, as already mentioned above, intraperitoneal administration of SL327 produced impaired fear conditioning [17] and Morris water maze spatial memory [30].

Genetic manipulations corroborate these findings. MEK1 KO mice, for example, exhibit a reduction of fear memory [31], whereas, conversely, mice carrying a homozygous null mutation for RIN1, which negatively regulates Ras, display an enhanced fear conditioning [32]. *Mrsk2*_KO mice, an animal model of Coffin-Lowry syndrome, a X-linked syndrome associated with mental retardation, are loss of function mutants for the gene that codes for RSK2, a direct downstream target of ERK1/2, and reveal a poorer performance in the acquisition of water maze memory and in its recall in the 9-day delay memory test [33]. On the other hand, mice lacking RGS14, which seems to act as a negative regulator of the Ras–ERK pathway, have recently been reported to display enhanced long-term potentiation in hippocampal area CA2 (which is blocked by MEK inhibition) and increased object recognition memory, in addition to a superior performance in the learning curve of Morris water maze training [34]. In any case, loss of function and gain of function mutations of components of the Ras–ERK cascade do not always lead to such linear effects on memory function.

Examples of more complicated and counterintuitive outcomes are ERK1 KO mice [35–37] and the neurofibromatosis type 1 (NF1) mutants [38–41], demonstrating that aberrant alterations of ERK signalling in the brain can lead either to a memory enhancement (as in the case of ERK1 KO mice) or to a memory impairment (as in the NF1 mutants) (*see* for review [23]).

In this chapter we will describe in detail the protocols of two main one-trial behavioral tests adopted in the Ras–ERK field, in their version for mice: passive avoidance and novel object recognition.

### 1.2 Passive Avoidance

Avoidance conditioning occurs when an organism learns a behavioral response that leads to avoidance of an aversive stimulus. Avoidance tasks are traditionally subdivided into active avoidance tasks and passive avoidance tasks, according to the terminology introduced by the American psychologist Orval Hobart Mowrer in the context of his two-factor theory of avoidance learning [42]. Although this distinction has been questioned many times, since it seems that passive avoidance is not actually a so passive process but more likely an active inhibitory process [43–45], the categorization still remains useful for didactic purposes. Basically, in active avoidance an animal learns to perform a certain response in order to avoid a punishment. In passive avoidance, also called inhibitory avoidance, an animal learns instead NOT to perform a certain response to avoid the punishment. Based on the nature of the behavior to suppress, passive avoidance can be divided into approach–avoidance tasks and escape–avoidance tasks. In the first kind of tasks, animals are initially trained to perform an approach response (like running or pressing a lever) in order to obtain a reward (e.g., food or water). Subsequently the approach response is punished and the animals have to learn to avoid it. In escape–avoidance tasks instead, untrained natural responses, which are likely to be performed by a determined species, are utilized. Typically animals are placed in a fearful or distressful situation, which elicits spontaneous escape behavior, and the attempts to withdraw from it are punished. Currently, tests commonly referred to as passive avoidance are almost always escape–avoidance tasks. Two main variants exist: step-down passive avoidance and step-through passive avoidance. The step-down version, the earlier of the two, was originally designed by Jarvik and Essman in 1960, with the aim at realizing a simple one-trial passive avoidance procedure for mice [46]. Animals are located on an elevated platform in the middle of an exposed field and are given an electric foot shock when they step off the platform in the attempt to reach the perimeter. In the test session the animals that have learnt will suppress their natural phobia for open elevated spaces and will avoid descending from the platform. Latency to step off serves as an index of memory. Since the original method revealed to be susceptible to a certain degree of variability due to genetic strain, sex, age, individual emotionality, seasonal changes, and daily activity/rest cycles, the authors introduced some modifications to the test [47], and few years later they proposed an improved one-trial passive avoidance procedure based on a two-compartment step-through situation [48], which they then described analytically in a dedicated methodological paper [49]. In this version animals are placed in an illuminated chamber connected to a dark compartment by means of a guillotine door. Due to their natural photophobia, mice and rats tend to leave the highly lit compartment to enter the dark one, where they are punished with an electric shock. In the retention test session, animals that remember will refrain from leaving the lit chamber and latencies to step-through are taken as an index of memory.

Passive avoidance proved to be fundamental for the study of the timing of memory consolidation processes, since the fact that the training was represented by a single trial (unlike preexisting multiple-trial tests) allowed experimental manipulations interfering with memory to be performed at different temporal points after the training session, in order to understand the boundaries of the post-acquisition processes stabilizing the memory trace. Indeed, immediately the year after their first methodological paper presenting the test, Jarvik and Essman used their protocol to investigate the anterograde amnesia inducing effects of ether administered shortly after the training and published their results directly on Science [50]. Few months later McGaugh adopted the test in his milestone paper on the amnesic effects of post-training electroconvulsive shock (ECS) treatment [51] and established passive avoidance as the golden standard test for the great number of studies on ECS that followed in the subsequent years. In the successive decade a series of articles by McGaugh and colleagues featuring passive avoidance, all published in Science [52–57], largely contributed to the popularity of this behavioral test. As can be seen in Fig. 1, the number of publications listed in PubMed for passive avoidance sees a sharp increase in the 1970s and since then keeps growing at a constant rate till the present day. In 2012 the top record was reached, with 206 publications (*see* **Note 1**).

As for its neural requirements, a variety of brain structures have been shown to be implicated in inhibitory avoidance memory, including hippocampus [58–60], amygdala [61–64], and striatum [58–68]. More recently, the quest for the molecular bases of avoidance memory began. The first paper in which passive avoidance was used to investigate the role of Ras–ERK signalling in memory through genetic manipulations was published by Brambilla and colleagues in 1997 [16]. Since then, the test has been used several times by our laboratory and others to evaluate the effects of abnormal Ras–ERK function in mutant mice. We will report here, in a concise review, some examples of the results obtained by our laboratory by using this test.

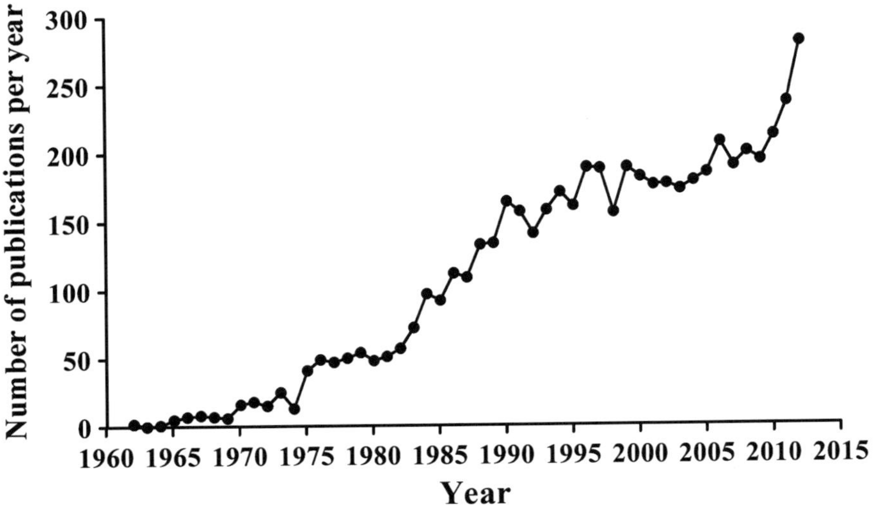


**Fig. 1** Number of publications per year listed by PubMed for passive avoidance (1960–2012)

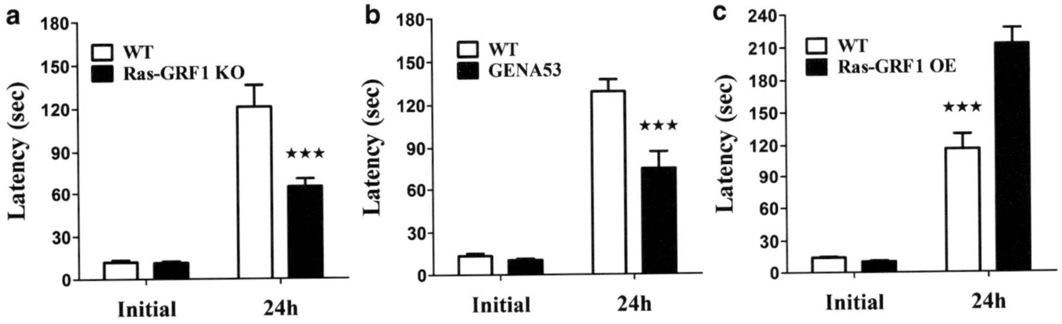


**Fig. 2** Long-term (24 h) memory in passive avoidance for Ras-GRF1 KO (**a**), GENA53 (**b**), and Ras-GRF1 OE mutant mice (**c**)

The input level of the Ras–ERK cascade is provided by Ras, a small G-protein whose activity depends on its binding to GDP or GTP [69]. When it is bound to GDP (Ras-GDP), it is inactive, while the binding with GTP brings it into its active form (Ras-GTP). Ras-guanine nucleotide-releasing factor 1 (Ras-GRF1) is a guanine nucleotide exchange factor (GEF), which promotes the conversion of Ras-GDP into Ras-GTP, acting as a turn-on switch for the MAPK pathway [70].

In Fig. 2 the effects of loss of function and gain of function mutations of Ras-GRF1 on passive avoidance memory are compared. In panel (a) we can see that Ras-GRF1 KO mice [16] have impaired 24 h long-term memory compared with their wild-type littermates (independent-samples $t$-test, $t(57) = -3.398$; $p = 0.0005$), whereas the initial latencies are absolutely comparable (independent-samples $t$-test, $t(57) = -0.395$; $p = 0.694$), demonstrating

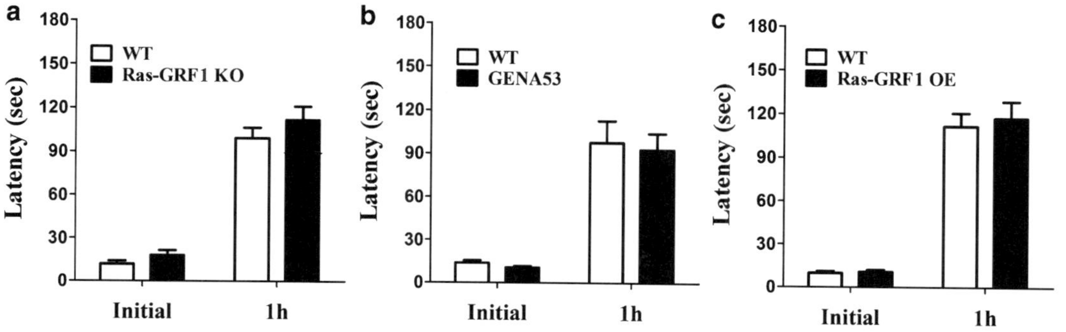


**Fig. 3** Short-term (1 h) memory in passive avoidance for Ras-GRF1 KO (**a**), GENA53 (**b**), and Ras-GRF1 OE mutant mice (**c**)

absence of sensorimotor abnormalities in the mutants. In panel (b) we observe that GENA53 mice, transgenic mutant animals in which an ENU-induced point mutation inactivates the Ras-GRF1 gene with minimum disturbance to the genome [71], have a similar deficit, showing impaired memory compared with their wild-type siblings (independent-samples *t*-test, $t(42) = -3.692$; $p = 0.001$), but normal initial latencies (independent-samples *t*-test, $t(42) = -1.630$; $p = 0.111$). As recently demonstrated in fear conditioning [72], Ras-GRF1 KO and GENA53 mice, although obtained by two different genetic techniques, show convergent evidence of the effect of the lack of Ras-GRF1 on fear memory and reveal a deficit of comparable entity (independent-samples *t*-test, $t(51) = 0.727$; $p = 0.473$). Finally, Ras-GRF1 overexpressing (OE) mice [73, 74] exhibit the opposite phenotype that is a memory enhancement (Fig. 2c). Compared to the wild-type mice, their memory is significantly higher, by 84 % (independent-samples *t*-test, $t(60) = 4.813$; $p < 0.0001$).

Interestingly, in 1 h short-term memory (Fig. 3a–c), none of the three mutants shows any abnormality (independent-samples *t*-tests versus respective wild-type littermates, Ras-GRF1 KO: $t(25) = 1.088$, $p = 0.287$; GENA53: $t(42) = -0.249$, $p = 0.805$; Ras-GRF1 OE: $t(44) = 0.384$, $p = 0.703$). This dissociation between long-term and short-term memory confirms that the Ras–ERK pathway is selectively implicated in memory consolidation, that is, the protein synthesis requiring process by which memories are stabilized after the initial acquisition [4].

The cAMP response element-binding protein (CREB) is one of the main transcription factors activated by ERK1/2 after its translocation to the nucleus and has largely been implicated in long-term synaptic plasticity and in the consolidation process converting short-term memory to long-term memory [75–79]. Figure 4 illustrates the performance in inhibitory avoidance of str-KCREB mice, a transgenic mouse line which reversibly expresses

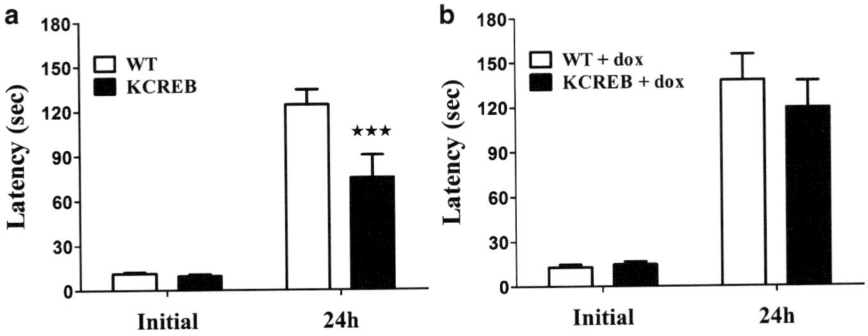


**Fig. 4** Long-term passive avoidance memory in striatal KCREB mutants, in absence (**a**) and presence (**b**) of doxycycline in the daily ration of food

killer CREB, a potent negative dominant inhibitor of CREB, in the striatum [80, 81]. As one can observe in Fig. 4a, str-KCREB mutants display impaired 24 h long-term memory (independent-samples *t*-test, $t(58) = -5.644$; $p < 0.0001$), in absence of alteration of the initial latencies (independent-samples *t*-test, $t(58) = -1.015$; $p = 0.314$). Notably, if the transgene expression of KCREB is turned off by administration of doxycycline in food (at 40 mg/kg) for 2 weeks (Fig. 4b), the deficient phenotype is completely rescued. Transgenic mice treated with doxycycline are undistinguishable from their wild-type controls (independent-samples *t*-test, $t(22) = -0.739$; $p = 0.468$) and their performance is significantly higher than the mutants which didn't receive doxycycline (independent-samples *t*-test, $t(40) = 2.615$; $p = 0.013$).

The data presented above support the idea that loss of function mutations of the Ras–ERK signalling pathway (Ras-GRF1 KO, GENE53, str-KCREB) lead to deficits in memory consolidation while gain of function mutations (Ras-GRF1 OE) promote LTM. However, a more complex scenario is found when evaluating avoidance memory of ERK1 KO mice. Indeed, due to the crosstalk between ERK1 and ERK2, mice lacking ERK1 (ERK1 KO) present abnormal biochemical responses that are linked to a compensatory upregulation of ERK2 signalling and manifest in the end, in a counterintuitive manner, a memory enhancement [35–37]. We show here for the first time the results of memory assessment in a double mutant carrying both Ras-GRF1 KO and ERK1 KO mutations (Fig. 5). The rationale of this experiment is that, by combining a loss of function mutation (Ras-GRF1 KO) with a functional gain of function mutant (ERK1 KO), the two conditions could neutralize each other. As one can see, initial latencies are all comparable across genotypes (one-way ANOVA, $F(3,182) = 1.018$; $p = 0.386$), but in the 24 h long-term memory test a significant effect of genotype is detected (one-way ANOVA, $F(3,182) = 10.169$; $p < 0.0001$). Bonferroni's post hoc test shows that Ras-GRF1 KO single mutants (Ras-GRF1 KO/ERK1 WT mice) reveal impaired memory

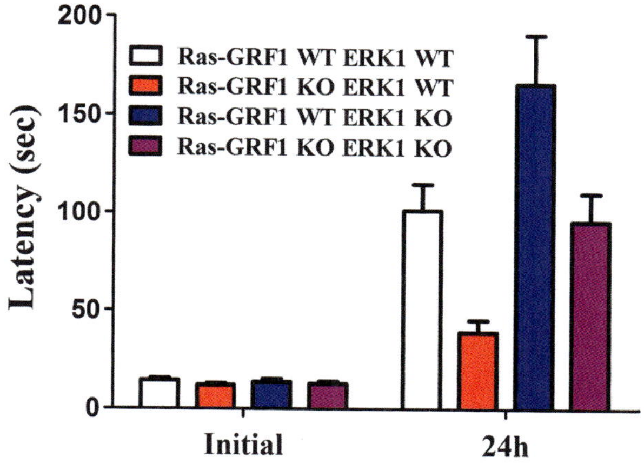


**Fig. 5** Long-term passive avoidance memory in Ras-GRF1 KO/ERK1 KO double mutant mice, compared with the performance of their single mutant (Ras-GRF1 KO/ERK1 WT and Ras-GRF1 WT/ERK1 KO) and wild-type littermate controls

compared to Ras-GRF1 WT/ERK1 WT mice ($p=0.003$), while ERK1 KO single mutants display enhanced memory ($p=0.035$). Noteworthy, double mutants (Ras-GRF1 KO/ERK1 KO) demonstrate instead a complete reversion of the behavioral phenotype to normal memory function, being totally undistinguishable from their wild-type siblings ($p=1$). The two mutations appear to have mutually eliding effects, rescuing the memory alterations observed in the single mutants. Indeed, compared to the single Ras-GRF1 KO mice, the double mutants exhibit a significant increase of memory ($p=0.018$), while, compared against the single ERK1 KO mice, they prove a significant reduction ($p=0.023$).

### 1.3 Novel Object Recognition

One-trial object recognition is a behavioral test originally developed by Ennaceur and Delacour in 1988 [82, 83]. It is based on the natural tendency of rodents to approach and explore novel objects that have never been encountered before, even if these have no natural significance for them and they have never been associated with a reward. Their instinctive search for novelty leads mice and rats to have an innate preference for a novel object over a familiar one. This simple principle of rodent behavior gives the possibility to set up a test for memory. Indeed, if in an initial phase two identical objects are presented to the animal (the sample phase) and in a subsequent session one of the two now familiar objects is replaced with a novel one (the test session), rodents will be able to display their natural preference only if they remember having seen the objects of the training session. If an animal doesn't remember the first exposure to the objects, during the second exposure it will perceive both objects as equally novel and hence it will show no preference between them. On the contrary, rodents that have memory of the first encounter will show a strong bias for the novel object.

This task has been demonstrated to rely on different brain structures, mainly perirhinal cortex [84–91] and hippocampus [92–94].

Object recognition has many advantages compared to the majority of tests classically used in behavioral studies. First of all, it exploits a natural behavior and for this reason is considered an ecological test. The response measured belongs to the natural animal's repertoire, not an artificially induced behavior that rodents learn to perform only in a specific laboratory setting. Secondly, since the behavior is spontaneous, there is no necessity to motivate the animals to act. This avoids the employment of food- or water-restricted protocols (which rely on hunger or thirst to motivate, respectively), of painful punishments like electrical shocks, and of stressful situations that induce escape behaviors (as coerced immersion in water in the Morris water maze). Besides the relevance from the ethical point of view (animals should experience the minimum level of stress if experimental alternatives are available), the reduction of stress during the training actually enhances the reliability and reproducibility of the results. Moreover, no extensive training is required, so the test is time saving, and the equipment is not expensive (the scoring can be done also manually without a video-tracking software in case of necessity). Finally, the delay between sample and test sessions can be chosen as desired, allowing testing of short-term and long-term memory at variable intervals, and, remarkably, the test is repeatable on the same animals, since the memory for the objects is labile and completely vanishes within a week, contrarily to what happens in, for instance, a different one-trial test like passive avoidance. Altogether, these characteristics make object recognition an ideal example of new generation behavioral tests to investigate the effects of genetic and pharmacological manipulations.

## 2 Materials

### 2.1 Passive Avoidance

1. Passive avoidance cage for mice (Panlab, Barcelona, Spain, model LE872). Size: 25 × 25 × 24 cm for the white compartment; 19.5 × 10.8 × 12 cm for the dark compartment.
2. Avoidance programmer and integrated shock generator with scrambler, 0–2 mA output (Panlab, Barcelona, Spain, model LE 10026).
3. Ethanol 30 %.
4. At least 10–12 mice per group.

### 2.2 Object Recognition

1. A square open-field arena (45 × 45 × 45 cm) made of dark gray polyvinyl chloride (PVC).
2. Two iron parallelepipeds 3 cm long, 3 cm wide, and 5 cm high (*see* **Note 2**).

3. Two glass cylindrical vials filled with water, with a circular basis of 2.5 cm in diameter and 6 cm high.
4. A digital video camera.
5. A personal computer.
6. Video-tracking software SMART version 2.0 (Panlab, Barcelona, Spain).
7. Two stopwatches.
8. 30 % ethanol.
9. At least 13–15 mice per group (since some mice are likely to be excluded due to insufficient exploration, especially if using 129Sv and C57BL/6 strains, it is better to start with more animals than the ones required for passive avoidance).

## 3 Methods

### 3.1 Housing

1. If animals are obtained from a commercial source, another research institute or an internal breeding facility that is separated from the animal behavior testing structure, then ensure a period of at least 1 week for acclimatization and recovery from shipment and transportation stress. This kind of stress generates a variety of effects on animal physiology (including changes in cardiovascular, immune, endocrine, central nervous, and reproductive systems) and behavior [95–101]. With some genetic strains a longer period can be required. BALB/c mice, for instance, have been shown to have high glucocorticoid levels, a common index of hypothalamus–pituitary–adrenal (HPA) axis activation, even 15 days after shipment [102]. An insufficient acclimatization period does not only put at risk animal welfare but also represents a confounding factor which can alter the outcome of the experiments, menacing result reliability.

   Cage the animals in groups to avoid social isolation stress, which can lead to enhanced anxiety and neophobia, locomotor hyperactivity, hyperaggressivity, impaired sensorimotor gating, lack of cognitive flexibility, and reduced short-term and long-term memory [103–106]. In any case groups should be composed of no more than 4–5 mice per cage, in order to avoid overcrowding stress, which, although often underrated, also can cause enhancement of corticosteroid levels and anxiety ([107], *see* **Note 3**).
2. Mark each mouse so it can be individually identified. There are many different methods to mark rodents, which include acting on ears, coat, tail, or paws. The most common methods are ear notching, ear tags, dye spots on the coat, tail marking, toe clipping, tattooing, and implantation of subcutaneous transponders [108].

3. Provide food and water ad libitum (unless animals are on a food-restricted and/or water-restricted regimen as a result of a specific experimental manipulation).
4. House mice in transparent plastic cages of at least 33 cm in length and 15 cm in width, so that free horizontal movement is allowed. Height should be ~13 cm, to permit complete vertical rearing. A metal wire grid as top lid allows climbing and wire-gnawing behavior. Endow the cages with atoxic sawdust bedding. Aspen wood is recommended, while pine and cedar woods, which contain aromatic phenols, are to avoid due to respiratory toxicity [109], alteration of liver size and function [110–112], and even reduction of sleeptime induced by hypnotic drugs like pentobarbital [110, 113]. Moreover, when given a free choice, mice exhibit a strong natural aversion for pine and cedar woods and a preference for aspen wood as nesting material ([112], *see* **Note 4**).
5. The environment in which the animals are housed should be kept temperature controlled (21.5 ± 1 °C) and humidity controlled (40 ± 10 %). Set up a fixed light/dark cycle of 12 h:12 h with lights on at 9:00 am, in order to ensure constant circadian rhythms (*see* **Note 5**).
6. Change cages weekly, in order to maintain them clean, but do not fix a higher frequency since cage change is a stress for the animals, which have all their odor clues/marks and nesting arrangements destroyed each time. Never change cages prior to a behavioral training or testing day, so mice can have at least 48 h to habituate to their new home cages.

### 3.2 Passive Avoidance

#### 3.2.1 Experimental Procedures

We will here describe in detail a protocol for step-through passive avoidance in mice. The apparatus is divided into two compartments separated by a small guillotine door: an open and strongly lit (with a 60 W lamp) one and a dark and closed one.

1. *Setting*. Choose a quiet and dimly lit room. Set the required parameters on the avoidance programmer: 1 s as interval between the animal's entry and the release of the shock; 360 s as maximum trial duration; 2 s as shock duration; 0.20 mA as shock intensity. Before starting, switch on the calibration mode and control on the amperometer that the actual intensity of the shock released corresponds to desired one.
2. *Training*. In the training session put the mouse in the lit compartment, facing away from the dark chamber, and allow it to explore for 30 s. After that, raise the guillotine door and leave the animal free to exit. Due to their natural photophobia, mice tend to enter the dark compartment in a few seconds (generally 15–20). Measure the latency to step-through. Once the animal has entered with all four paws, the door will slide down

and one single, continuous, 0.20 mA shock will be delivered through the metal grid floor for 2 s. Thirty seconds after the punishment, remove the mouse from the dark compartment and return it to its home cage. At the end of all the training trials, clean the apparatus with ethanol 30 %, paying particular attention to the shock chamber (if the metal grid is not maintained perfectly clean, the shock intensity may be altered).

3. *Test*. Testing sessions can be carried out at time points selected at choice. Commonly, 1 and 24 h intervals are chosen to evaluate, respectively, short-term and long-term memory. No shock is released during the testing session, so before starting remember to set to zero the value of the shock intensity on the avoidance programmer. For the retention session, introduce again the animal into the lit compartment, with its head facing away the door. After 10 s open the door and allow the mouse to freely move to the other side. Measure the latency to enter the dark box as an index of memory. Mice that retain memory will avoid entering the punished compartment. Cut-off time is set at 360 s, after which the guillotine door automatically closes and the animal can be removed from the apparatus.

#### 3.2.2 Data Analysis

Compare latencies to step-through in the test session across groups by means of independent-samples *t*-test or one-way ANOVA followed by the appropriate post hoc tests.

Check differences in the initial latencies in the same manner, in order to control for abnormalities in the sensorimotor or motivational domains. A further learning test can be performed by comparing a group's latency in the test trial with the initial latency exhibited in the training trial, by means of a paired-samples *t*-test. A significant increase in the latencies, with respect to the basal levels, indicates a significant learning.

Perform a categorical meta-analysis of the latency data by setting a threshold for determinate categories, for example learners (>55 s) and non-learners (<55 s), or bad learners (between 55 and 90 s) and good learners (>90 s). Frequencies of mice in the different categories can be compared through a chi-square analysis in crosstabs.

### 3.3 Novel Object Recognition

#### 3.3.1 Setting Preparation and Validation

1. Before starting, arrange the experimental setting. Choose a quiet and dimly lit room. The arena should be located with its center under the lens of the video camera, in order to minimize perspective distortion effects. Use your video-tracking software to create the images that will be used as masks superimposed on the scene to allow the scoring. The first image will outline the thigmotaxis zone, the boundaries of whom are represented by the perimeter of the base of the arena and by a line running 4 cm from the borders of the walls. The second image will depict two circular areas around the objects and will define

the exploration zones. First of all, choose the positioning points for the objects, which will be maintained constant for all the experiment. Draw them at 21 cm along the diagonal of the square arena (starting from two opposite corners at choice). In this way, the two objects will be at 20 cm between them and equidistant from the center (10 cm), their respective corners (21 cm), and the walls. Once the points are traced (mark them also physically on the arena with an indelible pen so you can see where to place the objects during the training and testing sessions), draw with your video-tracking software two circles that have, respectively, the left and the right positioning point as center and 4 cm in radius. In this way, when a mouse enters with the extremity of its head inside the circle, considering that the object length from the center is ~1.5 cm, there will be a distance of 2.5 cm between the mouse's nose and the object. Indeed, this information will be useful during the scoring, since the experimenter, as explained further, when evaluating the object explorations, will have to know not only when the animal is in contact with the object but also when it is pointing the object at a distance of 2.5 cm or less.

2. Before performing the experiment for the first time, the setting should be validated. The left and the right sides of the arena need to be equally illuminated to avoid side preference effects. On the other hand, there must not be a spontaneous preference for one of the two different objects. For example if a vial with a diameter of 3 cm and a plastic cube with a side of 15 cm are chosen as experimental objects, the latter will obviously be more salient. The objects proposed in this protocol have already been validated, but details for a complete validation are described in **Notes 6** and 7.

*3.3.2 Experimental Procedure*

1. *Habituation (Day 1)*. Place a single mouse, with its head oriented towards the walls of a corner, in the empty arena and let it explore for 5 min in order to familiarize with the testing apparatus. Select as releasing point one of the two corners that do not face an object zone and keep the releasing point, from which exploration is started, constant across days and mice. Thigmotaxis (the tendency to remain close to the walls avoiding open spaces) is measured as an index of anxiety. Record mice's performance with a digital video-camera, so it can be viewed on a monitor and scored by using the superimposed images created with the video-tracking software. Use a stopwatch to manually score the time spent inside the thigmotaxis area (i.e., within 4 cm from the walls). A mouse is to be considered inside the area if more than half of its body is so. After each trial is finished, return the animal in its house cage and clean the arena with ethanol 30 %.

2. *Training (Day 2)*. Introduce two identical non-displaceable objects (e.g., the two 3 × 3 × 5 cm iron parallelepipeds) in the open field and allow the mice (one at a time) to explore them for 10 min. The objects must be positioned, facing one another, at opposite sides symmetrically along the diagonal, at 20 cm of distance between them, 10 cm from the center, and 21 cm from the corner (follow the already-traced positioning points). Manually measure the time spent exploring each object, considering as object exploration touching with at least one forepaw, licking, sniffing, or orienting the head towards the object and pointing it within 2.5 cm of distance. Direct contacts that are judged to be accidental should not be counted as an exploratory behavior (e.g., backing into an object, tail-only contact, bumping into an object with a side while passing by). In addition, sitting on the object should not be scored as exploration as well. Animals with less than 8 s of total exploration time are excluded from the training session. Wash the arena and the objects with 30 % ethanol after each mouse in order to equalize olfactory cues.
3. *Choice of the position for the novel object*. Although mice should not exhibit group side preferences, individual preferences are often observed. The position of the novel object must be randomized and balanced for preference and side. First, calculate the percentage of exploration for each of the two identical objects (100 × time of exploration of the object/total time of exploration of both objects). You will obtain a list of percentages as in Table 1. Then evaluate the performance of each mouse and sign individual preferences for the left or right side, if present. Preferences are classified as follows: exploring 58 % or more of the time one of the two identical objects is classified as side preference (Left or Right); exploring 70 % or more demonstrates a strong preference (Left! or Right!); finally, animals with more than 75 % of preference are considered biased and have to be excluded from the experiment. Utilize these preferences to randomize the choice of the object to replace (for each treatment, half of the mice will have the novel object in the non-preferred side and half in the preferred one; for each group, if it is odd in number; for the spare preference always put the novel object in the non-preferred side). Animals with a preference of 50–57.9 % are considered to have a null preference (=). For mice without a preference, the position of the novel object could merely be randomized for side (left/right). Indeed, the best way to randomize the positions is to simply rank in decreasing order all the preferences and assign alternatively non-preferred (NP) and preferred (P), starting with NP for the highest value, to all the animals. Then continue doing the same for the mice without a preference. The result of such a procedure is showed in Table 2.

**Table 1**
**Percentages of exploration of the two identical objects during the training session for C57BL/6 WT mice**

| Mouse | Expl. left obj. (%) | Expl. right obj. (%) |
|---|---|---|
| 1 | 52 | 48 |
| 2 | 64 | 36 |
| 3 | 39 | 61 |
| 4 | 49 | 51 |
| 5 | 54 | 46 |
| 6 | 51 | 49 |
| 7 | 54 | 46 |
| 8 | 56 | 44 |
| 9 | 49 | 51 |
| 10 | 74 | 26 |
| 11 | 59 | 41 |
| 12 | 50 | 50 |

**Table 2**
**Assignment of side preferences and choice of the positions for the new object (L: left; R: right; NP: non-preferred side; P: preferred side)**

| Mouse | Expl. left obj. (%) | Expl. right obj. (%) | Preference | Position of novel |
|---|---|---|---|---|
| 10 | 74 | 26 | L! | R (NP) |
| 2 | 64 | 36 | L | L (P) |
| 3 | 39 | 61 | R | L (NP) |
| 11 | 59 | 41 | L | L (P) |
| 8 | 56 | 44 | = | R (NP) |
| 7 | 54 | 46 | = | L (P) |
| 5 | 54 | 46 | = | R (NP) |
| 1 | 52 | 48 | = | L (P) |
| 6 | 51 | 49 | = | R (NP) |
| 4 | 49 | 51 | = | R (P) |
| 9 | 49 | 51 | = | L (NP) |
| 12 | 50 | 50 | = | R |

4. *Test (Day 3)*. Mice are tested for their 24 h long-term memory. Place again the animals in the arena with two objects, but this time replace one of them with a novel object (e.g., a glass cylindrical vial filled with water, with a circular basis of 2.5 cm in diameter and 6 cm high), to assess object recognition memory. Follow the indications given at the previous point for the positioning of the new object. Manually record the time spent exploring each object. A minimum threshold of 6.5 s of total exploration is required to evaluate memory. Animals with less than 6.5 s are excluded from the testing session. Wash the arena and the objects with 30 % ethanol after each mouse.

#### 3.3.3 Data Analysis

1. *Thigmotaxis*. Calculate the percentages of thigmotaxis by dividing times spent in the thigmotaxis area (×100) by the total time (in this case 300 s). Percentages can be compared across groups by means of an independent-sample *t*-test (if groups are two) or one-way ANOVA followed by an appropriate post hoc test (if groups are three or more). A thigmotaxis index (TI) can be calculated by dividing the percentage of thigmotaxis by the percentage of area that the thigmotaxis zone represents with respect to the total area of the arena (in our case, the thigmotaxis area represents 35 % of the arena; your video-tracking software will provide you with this value when drawing the zones). The percentage of thigmotaxis can be compared against chance level (35 %), but since normal wild-type mice are already thigmotactic, this test will yield little information. Various genetic or pharmacological manipulations can make the animals more or less thigmotactic. These effects can be detected by comparing the TIs by means of a *t*-test or a one-way ANOVA with its related post hoc tests. The TIs, being an absolute value, are particularly useful when comparing the thigmotaxis of animals tested in arenas with different dimensions.
2. *Exploration*. Calculate the total time of exploration of the two identical objects presented during the training (left object + right object) and compare it across groups through *t*-test or a one-way ANOVA. This variable measures reactivity to novel discrete stimuli in naïve animals. Tendency to explore can be altered in many mutants. Hyper-exploration is observed, for example, in 5-$HT_{5A}$ receptor knockout mice [114] and dopamine transporter knockout mice [115]. Interestingly, higher or lower levels of exploration are not necessarily associated with locomotor hypo- or hyperactivity. Generally hyper-exploration is associated with an impulsive phenotype, while hypo-exploration with a depressive or anxious one.
3. *Memory*. Absolute times spent exploring the novel and the familiar objects can be compared between them by means of a

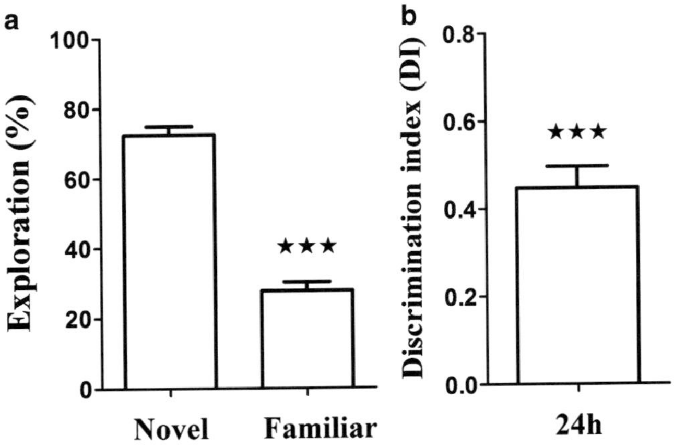


**Fig. 6** Long-term (24 h) object recognition memory in C57BL/6 WT mice, evaluated through the DI method. Panel (**a**) represents the percentages of exploration of the novel and the familiar objects during the test day, while panel (**b**) shows the corresponding DI value

paired-samples *t*-test. A significant difference indicates a significant learning. Compute the percentage of exploration of the novel [100 × time exploring novel/(time exploring novel + familiar)] and the familiar [100 × time exploring familiar/(time exploring novel + familiar)] objects. These two percentages can be compared against chance level (50 %) but can also be compared across groups through an independent-samples *t*-test or a one-way ANOVA. Mice with a reduction (versus controls) in the percentage of exploration of the novel object have a memory deficit; on the other hand, mice with a higher percentage reveal a memory enhancement.

4. *Learning indexes.* Two learning indexes can be calculated. The first is the classical discrimination index (DI), which quantifies the ability of an animal to discriminate between two different objects that are presented at the same time. It is computed as follows: [(time spent exploring novel object) − (time exploring familiar object)]/[(time spent exploring novel object) + (time exploring familiar object)]. The DI can assume values between −1 and +1. A DI significantly higher than zero (chance level) reveals preference for the novel object, indicating object recognition memory. In some cases a DI significantly lower than zero can be found, which demonstrates a preference for the familiar object and typically represents a perseverative/repetitive phenotype [116, 117]. DIs are compared among groups through an independent-samples *t*-test or a one-way ANOVA. Figure 6 illustrates the results of the 24 h memory test of the mice shown in the previous tables, evaluated through the DI method. Animals show a significant memory, as the percentage of time spent exploring the novel object is significantly higher than the one for

the familiar object (Fig. 6a; paired-samples $t$-test, $t(8)=9.068$; $p<0.0001$). The DI results $0.45 \pm 0.05$, significantly above chance level (Fig. 6b; one-sample $t$-test, $t(8)=9.039$; $p<0.0001$).

The second learning index is new and we present it here for the first time. The recognition index (RI) measures the ability of an animal to recognize a same object at different time points. It is based on the fact that if an animal remembers the familiar object, its percentage of exploration for that object will decrease when the animal is exposed to the same object for a second time. The percentages of exploration of the familiar object on day 2 versus day 3 can been compared by means of a paired-samples $t$-test. A significant difference proves a significant learning. But, in order to compare the percentages correctly, we have to know which of the two identical familiar objects to select as a baseline. Since on the test day the position of the familiar object is variable, for the comparison we have to select the familiar object that on the training day was in the same position as the familiar object of the test day, that is, the non-displaced object of the training trial must be selected as a baseline. Use the list of the positions of the familiar objects on day 3 as a guide and pair the percentages of exploration of the non-displaced familiar object during the training with the ones during the test. The RI is calculated as follows: [(percentage of exploration of the non-displaced familiar object during the training) − (percentage of exploration of the non-displaced familiar object during the test)]/percentage of exploration of the non-displaced familiar object during the training. Since the RI contains also information regarding the basal level of preference displayed by the animals, it is actually more reliable than the DI. Let's consider, for instance, an animal that exhibits, during the test, a preference of 70 % for the novel object, that is positioned on the left side, and 30 % for the familiar, which is placed on the right side. On day 2, during the training, the animal already manifested a strong preference, 70 %, for the left side (hence already had from the beginning only 30 % of preference for the non-displaced familiar object, which is on the right side). The DI would return a value of 0.40 [(70–30)/(70 + 30)], erroneously making us think the animal has a good memory. The RI, on the other hand, adjusting for the basal individual preference, would result 0 [(30–30)/30], correctly showing that the training had no effect on that animal. DI and RI will be very similar when the animals have very little individual preferences (when these are zero, the indexes coincide). Indeed, if individual preferences are present, the RI will be a more reliable index of learning. RIs can be compared against chance level (zero) through a one-sample $t$-test and across groups by means of an independent-samples $t$-test or a one-way ANOVA. Figure 7 represents the results of the 24 h memory test of the mice

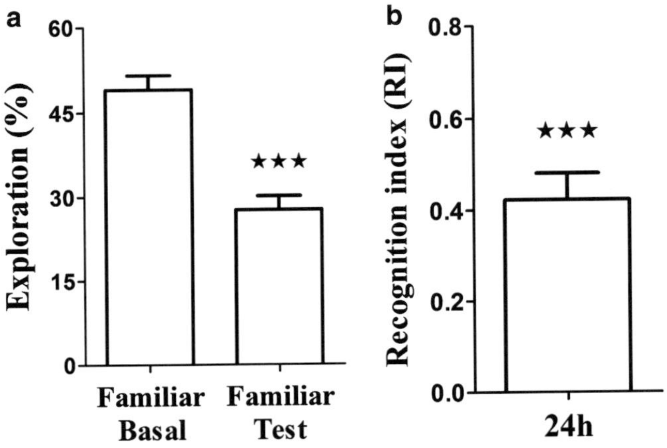


**Fig. 7** Long-term (24 h) object recognition memory in C57BL/6 WT mice, evaluated through the RI method. Panel (**a**) illustrates the percentages of exploration of the familiar object on the day of the training and on the day of the test, while panel (**b**) shows the corresponding RI value

shown in the previous tables, evaluated through the RI method. Animals show a significant memory, as the percentage of time spent exploring the familiar object on the test day is significantly decreased compared with the basal percentage of exploration observed on the day of the training (Fig. 7a; paired-samples $t$-test, $t(8) = -5.782$; $p = 0.0004$). The RI results $0.42 \pm 0.06$, significantly above chance level (Fig. 7b; one-sample $t$-test, $t(8) = 7.262$; $p < 0.0001$).

## 4 Notes

1. Since the test has two alternative names, the total number of publications were counted by summing the matches for "passive avoidance" in title/abstract with the matches for "inhibitory avoidance" AND NOT "passive avoidance," to avoid including twice the same papers. Up to December 2013 the total resulted to be 5,772 matches (4,538 for passive avoidance plus 1,234 for inhibitory avoidance). Interestingly, this total is higher than those obtained for behavioral tests like radial maze (1,401), active avoidance (1,713), rotarod (2,038), forced swimming (2,214), fear conditioning (3,343), elevated plus maze (4,764), and even Morris water maze (4,929).
2. The choice of the two different objects should be made according to certain criteria. More specifically, both objects should be heavy enough to be non-displaceable. Moreover, they should be comparable in size but different in shape, texture, and material. Affordance properties (i.e., the perceptual properties connecting to action) should instead be comparable between the objects. Heyser and Chemero have recently shown that objects

which can be climbed on, compared against objects which can only be touched, are explored more in the training phase and even lead to a greater discrimination index in the memory test session [118].

3. The American ethologist John Calhoun is famous for having performed pioneering studies on the effects of overcrowding on rat behavior [119–122]. His article "Population Density and Social Pathology," in which Calhoun reported the results he obtained at the National Institute of Mental Health (NIMH), publishing them on a 1962 edition of Scientific American [120], has arrived to be cited up to 150 times a year and has been included as one of the "Forty Studies that Changed Psychology," along with the works of figures like Freud, Pavlov, Skinner, Watson, Rorschach, and Milgram [123]. In this paper Calhoun showed how rats living in a controlled laboratory setting in which food, water, and bedding were provided ad libitum and protection from external predators and diseases was guaranteed, initially seemed to live in a "rodent utopia," rapidly breeding. But when the overcrowding led to lack of space, the behavior of the animals began to degenerate. The increased frequency of undesired social contact brought to increased stress and aggressivity. Violence became out of control. Cannibalism and infanticide were observed. Males became hypersexual and homosexual. Calhoun named this escalation "behavioral sink." The number of the population dramatically fell down, to an extent that put the colony at risk of extinction. At the end of the experiment, the only rats survived were asexual and socially withdrawn. Even when reintroduced in normal rat communities, these animals continued to manifest an autistic phenotype till death.

4. Odynets and colleagues studied the behavior of mice in a free choice test for selection of nesting material, in which animals were placed in a system of five communicating cages, with the central cage empty and the other four containing beddings from different woods, namely, aspen, pine, spruce, and birch [112]. C57BL/6 mice chose aspen in 70 % of the cases and spruce in 30 %. BALB/c chose aspen in 100 % of the cases. For both genetic strains pine and birch were rejected. Wild *Mus musculus* mice were tested as well and their order of preference was comparable with that of the laboratory mice (aspen: 62.5 %; spruce: 25 %; birch: 12.5; pine: 0 %).

5. Mice are a crepuscular species (i.e., they mainly initiate their active phase at dusk). Since their circadian rhythms are different from those of humans, in certain conditions, like, for example, for the study of spontaneous behavior with an ecological approach, inversion of the 12 h:12 h light/dark cycle may be considered (lights off at 9:00 am). But in this case

animals should be born in the inverted cycle. If mice come from a structure where the regular cycle was kept, do not change the rhythm they have lived in from birth, as the transition would generate great distress in the animals.

6. In order to validate illumination and context use at least 20 wild-type mice and control for side preference effects. Expose each animal to the empty arena (for 5 min) and measure the time spent in the two exploration zones by utilizing the specifically designed filter image. The mean time spent in the left zone and the one spent in right zone should not be statistically different. If there is a significant difference, the problem could be due to nonhomogenous illumination of the arena. Change or relocate the lightning sources, and retry. Alternatively, the difference could be due to the presence of an object which elicits fear or attraction in the mice in one of the two sides of the room. On day 2, repeat the test by exposing the animals to the first couple of identical objects (for 10 min). On day 3, repeat it with the second couple of objects instead (again 10 min). As for the empty arena, also when mice are confronted with the objects, there should be no group preference for the left or the right identical objects.
7. Validation of the objects requires at least 20, possibly naïve, wild-type animals. Perform this test only after side preference effects have been checked (as described in **Note 6**). To verify that the two different objects are equally salient, carry out a habituation session of 5 min on day 1 and the following day expose the mice directly to the two different objects (for 10 min). As eventual side preferences have already been ruled out, if one of the two objects is preferred in a statistically significant manner, then one of the two objects is perceived as more stimulating (or fearful) by the animals. Change the objects and retry.

## References

1. Squire LR, Knowlton B, Musen G (1993) The structure and organization of memory. Annu Rev Psychol 44:453–495
2. Milner B, Squire LR, Kandel ER (1998) Cognitive neuroscience and the study of memory. Neuron 20:445–468
3. Squire LR (2004) Memory systems of the brain: a brief history and current perspective. Neurobiol Learn Memory 82:171–177
4. D'isa R, Solari N, Brambilla R (2011) Biological memory in animals and in man. In: Campardo G, Tiziani F, Iaculo M (eds) Memory mass storage. Springer, Berlin, pp 417–441
5. Davis HP, Squire LR (1984) Protein synthesis and memory: a review. Psychol Bull 96: 518–559
6. Mcgaugh JL (2000) Memory – a century of consolidation. Science 287:248–251
7. Blum S, Dash PK (2009) MAP kinase signaling in learning and memory. In: Squire LR (ed) Encyclopedia of neuroscience, vol 5. Elsevier, San Diego, CA, pp 657–662
8. Peng S, Zhang Y, Zhang J et al (2010) ERK in learning and memory: a review of recent research. Int J Mol Sci 11:222–232
9. Boulton TG, Nye SH, Robbins DJ et al (1991) ERKs: a family of protein-serine/threonine

kinases that are activated and tyrosine phosphorylated in response to insulin and NGF. Cell 65:663–675
10. Crews CM, Alessandrini A, Erikson RL (1992) Erks: their fifteen minutes has arrived. Cell Growth Differ 3:135–142
11. Marshall C (1995) The Ras/Raf/ERK pathway. In: Zerial M, Huber LA (eds) Guidebook to the small GTPases. Oxford University Press, New York, pp 65–73
12. Malumbres M, Barbacid M (2003) RAS oncogenes: the first 30 years. Nat Rev Cancer 3:459–465
13. Fernandez-Medarde A, Santos E (2011) Ras in cancer and developmental diseases. Genes Cancer 2:344–358
14. English JD, Sweatt JD (1996) Activation of p42 mitogen-activated protein kinase in hippocampal long term potentiation. J Biol Chem 271:24329–24332
15. English JD, Sweatt JD (1997) A requirement for the mitogen-activated protein kinase cascade in hippocampal long term potentiation. J Biol Chem 272:19103–19106
16. Brambilla R, Gnesutta N, Minichiello L et al (1997) A role for the Ras signalling pathway in synaptic transmission and long-term memory. Nature 390:281–286
17. Atkins CM, Selcher JC, Petraitis JJ et al (1998) The MAPK cascade is required for mammalian associative learning. Nat Neurosci 1:602–609
18. Orban PC, Chapman PF, Brambilla R (1999) Is the Ras-MAPK signalling pathway necessary for long-term memory formation? Trends Neurosci 22:38–44
19. Mazzucchelli C, Brambilla R (2000) Ras-related and MAPK signalling in neuronal plasticity and memory formation. CMLS 57:604–611
20. Adams JP, Sweatt JD (2002) Molecular psychology: roles for the ERK MAP kinase cascade in memory. Annu Rev Pharmacol Toxicol 42:135–163
21. Thomas GM, Huganir RL (2004) MAPK cascade signalling and synaptic plasticity. Nat Rev Neurosci 5:173–183
22. Davis S, Laroche S (2006) Mitogen-activated protein kinase/extracellular regulated kinase signalling and memory stabilization: a review. Genes Brain Behav 5(Suppl 2):61–72
23. Fasano S, Brambilla R (2011) Ras-ERK signaling in behavior: old questions and new perspectives. Front Behav Neurosci 5:1–6
24. Walz R, Roesler R, Quevedo J et al (1999) Dose-dependent impairment of inhibitory avoidance retention in rats by immediate post-training infusion of a mitogen-activated protein kinase kinase inhibitor into cortical structures. Behav Brain Res 105:219–223
25. Hebert AE, Dash PK (2002) Extracellular signal-regulated kinase activity in the entorhinal cortex is necessary for long-term spatial memory. Learn Mem 9:156–166
26. Berman DE, Hazvi S, Rosenblum K et al (1998) Specific and differential activation of mitogen-activated protein kinase cascades by unfamiliar taste in the insular cortex of the behaving rat. J Neurosci 18:10037–10044
27. Swank MW (2000) Pharmacological antagonism of tyrosine kinases and MAP kinase in brainstem blocks taste aversion learning in mice. Physiol Behav 69:499–503
28. Kelly A, Laroche S, Davis S (2003) Activation of mitogen-activated protein kinase/extracellular signal-regulated kinase in hippocampal circuitry is required for consolidation and reconsolidation of recognition memory. J Neurosci 23:5354–5360
29. Schafe GE, Atkins CM, Swank MW et al (2000) Activation of ERK/MAP kinase in the amygdala is required for memory consolidation of pavlovian fear conditioning. J Neurosci 20:8177–8187
30. Selcher JC, Atkins CM, Trzaskos JM et al (1999) A necessity for MAP kinase activation in mammalian spatial learning. Learn Mem 6:478–490
31. Shalin SC, Zirrgiebel U, Honsa KJ et al (2004) Neuronal MEK is important for normal fear conditioning in mice. J Neurosci Res 75:760–770
32. Dhaka A, Costa RM, Hu H et al (2003) The RAS effector RIN1 modulates the formation of aversive memories. J Neurosci 23:748–757
33. Poirier R, Jacquot S, Vaillend C et al (2007) Deletion of the Coffin-Lowry syndrome gene Rsk2 in mice is associated with impaired spatial learning and reduced control of exploratory behavior. Behav Genet 37:31–50
34. Lee SE, Simons SB, Heldt SA et al (2010) RGS14 is a natural suppressor of both synaptic plasticity in CA2 neurons and hippocampal-based learning and memory. Proc Natl Acad Sci U S A 107:16994–16998
35. Mazzucchelli C, Vantaggiato C, Ciamei A et al (2002) Knockout of ERK1 MAP kinase enhances synaptic plasticity in the striatum and facilitates striatal-mediated learning and memory. Neuron 34:807–820
36. Ferguson SM, Fasano S, Yang P et al (2006) Knockout of ERK1 enhances cocaine-evoked immediate early gene expression and behavioral plasticity. Neuropsychopharmacology 31:2660–2668
37. Silingardi D, Angelucci A, De Pasquale R et al (2011) ERK pathway activation bidirectionally affects visual recognition memory and synaptic plasticity in the perirhinal cortex. Front Behav Neurosci 5:84

38. Silva AJ, Frankland PW, Marowitz Z et al (1997) A mouse model for the learning and memory deficits associated with neurofibromatosis type I. Nat Genet 15:281–284
39. Costa RM, Federov NB, Kogan JH et al (2002) Mechanism for the learning deficits in a mouse model of neurofibromatosis type 1. Nature 415:526–530
40. Costa RM, Silva AJ (2003) Mouse models of neurofibromatosis type I: bridging the GAP. Trends Mol Med 9:19–23
41. Cui Y, Costa RM, Murphy GG et al (2008) Neurofibromin regulation of ERK signaling modulates GABA release and learning. Cell 135:549–560
42. Mowrer OH (1960) Learning theory and the symbolic processes. Wiley, New York
43. Liss P (1968) Avoidance and freezing behavior following damage to the hippocampus or fornix. J Comp Physiol Psychol 66:193–197
44. Olton DS (1973) Shock-motivated avoidance and the analysis of behavior. Psychol Bull 79:243–251
45. Netto CA, Izquierdo I (1985) On how passive is inhibitory avoidance. Behav Neural Biol 43:327–330
46. Jarvik ME, And Essman WB (1960) A simple one-trial situation for mice. Psychol Rep 6:290
47. Essman WB, Alpern HP (1964) Single trial conditioning: methodology and results with mice. Psychol Rep 14:731–740
48. Kopp R, Bohdanecky Z, Jarvik ME (1966) Long temporal gradient of retrograde amnesia for a well-discriminated stimulus. Science 153:1547–1549
49. Jarvik ME, Kopp R (1967) An improved one-trial passive avoidance learning situation. Psychol Rep 21:221–224
50. Abt JP, Essman WB, Jarvik ME (1961) Ether-induced retrograde amnesia for one-trial conditioning in mice. Science 133:1477–1478
51. Madsen MC, Mc GJ (1961) The effect of ECS on one-trial avoidance learning. J Comp Physiol Psychol 54:522–523
52. Mcgaugh JL, Alpern HP (1966) Effects of electroshock on memory: amnesia without convulsions. Science 152:665–666
53. Mcgaugh JL (1966) Time-dependent processes in memory storage. Science 153:1351–1358
54. Luttges MW, Mcgaugh JL (1967) Permanence of retrograde amnesia produced by electroconvulsive shock. Science 156:408–410
55. Fishbein W, Mcgaugh JL, Swarz JR (1971) Retrograde amnesia: electroconvulsive shock effects after termination of rapid eye movement sleep deprivation. Science 172:80–82
56. Landfield PW, Mcgaugh JL, Tusa RJ (1972) Theta rhythm: a temporal correlate of memory storage processes in the rat. Science 175:87–89
57. Gold PE, Macri J, Mcgaugh JL (1973) Retrograde amnesia gradients: effects of direct cortical stimulation. Science 179:1343–1345
58. Glick SD, Greenstein S (1973) Comparative learning and memory deficits following hippocampal and caudate lesions in mice. J Comp Physiol Psychol 82:188–194
59. Glick SD, Marsanico RG, Greenstein S (1974) Differential recovery of function following caudate, hippocampal, and septal lesions in mice. J Comp Physiol Psychol 86:787–792
60. Boast CA, Zornetzer SF, Hamrick MR (1975) Electrolytic lesions of various hippocampal subfields in the mouse: differential effects on short- and long-term memory. Behav Biol 14:85–94
61. Nagel JA, Kemble ED (1976) Effects of amygdaloid lesions on the performance of rats in four passive avoidance tasks. Physiol Behav 17:245–250
62. Swartzwelder HS (1981) Deficits in passive avoidance and fear behavior following bilateral and unilateral amygdala lesions in mice. Physiol Behav 26:323–326
63. Liang KC, Mcgaugh JL, Martinez JL Jr et al (1982) Post-training amygdaloid lesions impair retention of an inhibitory avoidance response. Behav Brain Res 4:237–249
64. Dunn LT, Everitt BJ (1988) Double dissociations of the effects of amygdala and insular cortex lesions on conditioned taste aversion, passive avoidance, and neophobia in the rat using the excitotoxin ibotenic acid. Behav Neurosci 102:3–23
65. Prado-Alcala RA, Grinberg ZJ, Arditti ZL et al (1975) Learning deficits produced by chronic and reversible lesions of the corpus striatum in rats. Physiol Behav 15:283–287
66. Prado-Alcala RA, Cruz-Morales SE, Lopez-Miro FA (1980) Differential effects of cholinergic blockade of anterior and posterior caudate nucleus on avoidance behaviors. Neuroscience Lett 18:339–345
67. Prado-Alcala RA, Ruiloba MI, Rubio L et al (2003) Regional infusions of serotonin into the striatum and memory consolidation. Synapse 47:169–175
68. Sandberg K, Sanberg PR, Hanin I et al (1984) Cholinergic lesion of the striatum impairs acquisition and retention of a passive avoidance response. Behav Neurosci 98:162–165
69. Boguski MS, Mccormick F (1993) Proteins regulating Ras and its relatives. Nature 366:643–654
70. Fernandez-Medarde A, Santos E (2011) The RasGrf family of mammalian guanine nucleotide exchange factors. Biochim Biophys Acta 1815:170–188
71. Clapcott SJ, Peters J, Orban PC et al (2003) Two ENU-induced mutations in Rasgrf1 and

early mouse growth retardation. Mamm Genome 14:495–505
72. D'isa R, Clapcote SJ, Voikar V et al (2011) Mice lacking Ras-GRF1 show contextual fear conditioning but not spatial memory impairments: convergent evidence from two independently generated mouse mutant lines. Front Behav Neurosci 5:78
73. Yoon B, Herman H, Hu B et al (2005) Rasgrf1 imprinting is regulated by a CTCF-dependent methylation-sensitive enhancer blocker. Mol Cell Biol 25:11184–11190
74. Fasano S, D'antoni A, Orban PC et al (2009) Ras-guanine nucleotide-releasing factor 1 (Ras-GRF1) controls activation of extracellular signal-regulated kinase (ERK) signaling in the striatum and long-term behavioral responses to cocaine. Biol Psychiatr 66:758–768
75. Barco A, Pittenger C, Kandel ER (2003) CREB, memory enhancement and the treatment of memory disorders: promises, pitfalls and prospects. Expert Opin Ther Targets 7:101–114
76. Carlezon WA Jr, Duman RS, Nestler EJ (2005) The many faces of CREB. Trends Neurosci 28:436–445
77. Benito E, Barco A (2010) CREB's control of intrinsic and synaptic plasticity: implications for CREB-dependent memory models. Trends Neurosci 33:230–240
78. Barco A, Marie H (2011) Genetic approaches to investigate the role of CREB in neuronal plasticity and memory. Mol Neurobiol 44:330–349
79. Kandel ER (2012) The molecular biology of memory: cAMP, PKA, CRE, CREB-1, CREB-2, and CPEB. Mol Brain 5:14
80. Pittenger C, Huang YY, Paletzki RF et al (2002) Reversible inhibition of CREB/ATF transcription factors in region CA1 of the dorsal hippocampus disrupts hippocampus-dependent spatial memory. Neuron 34:447–462
81. Pittenger C, Fasano S, Mazzocchi-Jones D et al (2006) Impaired bidirectional synaptic plasticity and procedural memory formation in striatum-specific cAMP response element-binding protein-deficient mice. J Neurosci 26:2808–2813
82. Ennaceur A, Delacour J (1988) A new one-trial test for neurobiological studies of memory in rats. 1: behavioral data. Behav Brain Res 31:47–59
83. Ennaceur A (2010) One-trial object recognition in rats and mice: methodological and theoretical issues. Behav Brain Res 215:244–254
84. Ennaceur A, Neave N, Aggleton JP (1996) Neurotoxic lesions of the perirhinal cortex do not mimic the behavioural effects of fornix transection in the rat. Behav Brain Res 80:9–25
85. Aggleton JP, Keen S, Warburton EC et al (1997) Extensive cytotoxic lesions involving both the rhinal cortices and area TE impair recognition but spare spatial alternation in the rat. Brain Res Bull 43:279–287
86. Bussey TJ, Muir JL, Aggleton JP (1999) Functionally dissociating aspects of event memory: the effects of combined perirhinal and postrhinal cortex lesions on object and place memory in the rat. J Neurosci 19:495–502
87. Mumby DG, Glenn MJ, Nesbitt C et al (2002) Dissociation in retrograde memory for object discriminations and object recognition in rats with perirhinal cortex damage. Behav Brain Res 132:215–226
88. Winters BD, Forwood SE, Cowell RA et al (2004) Double dissociation between the effects of peri-postrhinal cortex and hippocampal lesions on tests of object recognition and spatial memory: heterogeneity of function within the temporal lobe. J Neurosci 24:5901–5908
89. Norman G, Eacott MJ (2004) Impaired object recognition with increasing levels of feature ambiguity in rats with perirhinal cortex lesions. Behav Brain Res 148:79–91
90. Hannesson DK, Howland JG, Pollock M et al (2005) Anterior perirhinal cortex kindling produces long-lasting effects on anxiety and object recognition memory. EJ Neurosci 21:1081–1090
91. Norman G, Eacott MJ (2005) Dissociable effects of lesions to the perirhinal cortex and the postrhinal cortex on memory for context and objects in rats. Behav Neurosci 119:557–566
92. Clark RE, Zola SM, Squire LR (2000) Impaired recognition memory in rats after damage to the hippocampus. J Neurosci 20:8853–8860
93. Hammond RS, Tull LE, Stackman RW (2004) On the delay-dependent involvement of the hippocampus in object recognition memory. Neurobiol Learn Memory 82:26–34
94. Ainge JA, Heron-Maxwell C, Theofilas P et al (2006) The role of the hippocampus in object recognition in rats: examination of the influence of task parameters and lesion size. Behav Brain Res 167:183–195
95. Landi MS, Kreider JW, Lang CM et al (1982) Effects of shipping on the immune function in mice. Am J Vet Res 43:1654–1657
96. Tuli JS, Smith JA, Morton DB (1995) Stress measurements in mice after transportation. Lab Anim 29:132–138
97. Hayssen V (1998) Effect of transatlantic transport on reproduction of agouti and nonagouti deer mice, *Peromyscus maniculatus*. Lab Anim 32:55–64
98. Tabata H, Kitamura T, Nagamatsu N (1998) Comparison of effects of restraint, cage transportation, anaesthesia and repeated bleeding

on plasma glucose levels between mice and rats. Lab Anim 32:143–148
99. Obernier JA, Baldwin RL (2006) Establishing an appropriate period of acclimatization following transportation of laboratory animals. ILAR J 47:364–369
100. Arts JWM, Ohl F, Kramer K (2008) Transportation as major life-event in rats: effects on welfare and limits of adaptation. In: AJ Spink, MR Ballintijn, ND Bogers, F Grieco, LWS Loijens, LPJJ Noldus, G Smit, PH Zimmerman (eds) Proceedings of the measuring behaviour 2008, Maastricht, The Netherlands, 26–29 August 2008, pp 356–357
101. Arts JWM, Kramer K, Arndt SS, Ohl F (2012) The impact of transportation on physiological and behavioral parameters in wistar rats: implications for acclimatization periods. ILAR J 53:82–98
102. Olfe J, Domanska G, Schuett C et al (2010) Different stress-related phenotypes of BALB/c mice from in-house or vendor: alterations of the sympathetic and HPA axis responsiveness. BMC Physiol 10:2
103. Guidotti A, Dong E, Matsumoto K et al (2001) The socially-isolated mouse: a model to study the putative role of allopregnanolone and 5alpha-dihydroprogesterone in psychiatric disorders. Brain Res Brain Res Rev 37:110–115
104. Lapiz MD, Fulford A, Muchimapura S et al (2003) Influence of postweaning social isolation in the rat on brain development, conditioned behavior, and neurotransmission. Neurosci Behav Physiol 33:13–29
105. Voikar V, Polus A, Vasar E et al (2005) Long-term individual housing in C57BL/6J and DBA/2 mice: assessment of behavioral consequences. Genes Brain Behav 4:240–252
106. Fone KC, Porkess MV (2008) Behavioural and neurochemical effects of post-weaning social isolation in rodents-relevance to developmental neuropsychiatric disorders. Neurosci Biobehav Rev 32:1087–1102
107. Laber K, Veatch LM, Lopez MF et al (2008) Effects of housing density on weight gain, immune function, behavior, and plasma corticosterone concentrations in BALB/c and C57BL/6 mice. J Am Assoc Lab Anim Sci 47:16–23
108. Deacon RM (2006) Housing, husbandry and handling of rodents for behavioral experiments. Nat Protoc 1:936–946
109. Ayars GH, Altman LC, Frazier CE et al (1989) The toxicity of constituents of cedar and pine woods to pulmonary epithelium. J Allergy Clin Immun 83:610–618
110. Vesell ES (1967) Induction of drug-metabolizing enzymes in liver microsomes of mice and rats by softwood bedding. Science 157:1057–1058
111. Torronen R, Pelkonen K, Karenlampi S (1989) Enzyme-inducing and cytotoxic effects of wood-based materials used as bedding for laboratory animals. Comparison by a cell culture study. Life Sci 45:559–565
112. Odynets A, Simonova O, Kozhuhov A, Zaitsev T, Verreva A, Gnilomedova L, Rudzish R (1991) Beddings for laboratory animals: criteria of biological evaluation. Lab Zyhvotnye 1:70–76
113. Cunliffe-Beamer TL, Freeman LC, Myers DD (1981) Barbiturate sleeptime in mice exposed to autoclaved or unautoclaved wood beddings. Lab Anim Sci 31:672–675
114. Grailhe R, Waeber C, Dulawa SC et al (1999) Increased exploratory activity and altered response to LSD in mice lacking the 5-HT(5A) receptor. Neuron 22:581–591
115. Zhuang X, Oosting RS, Jones SR et al (2001) Hyperactivity and impaired response habituation in hyperdopaminergic mice. Proc Natl Acad Sci U S A 98:1982–1987
116. Hoeffer CA, Tang W, Wong H et al (2008) Removal of FKBP12 enhances mTOR-Raptor interactions, LTP, memory, and perseverative/repetitive behavior. Neuron 60:832–845
117. Trinh MA, Kaphzan H, Wek RC et al (2012) Brain-specific disruption of the eIF2alpha kinase PERK decreases ATF4 expression and impairs behavioral flexibility. Cell Rep 1:676–688
118. Heyser CJ, Chemero A (2012) Novel object exploration in mice: not all objects are created equal. Behav Processes 89:232–238
119. Calhoun JB (1949) A method for self-control of population growth among mammals living in the wild. Science 109:333–335
120. Calhoun JB (1962) Population density and social pathology. Sci Am 206:139–148
121. Ramsden E (2009) The urban animal: population density and social pathology in rodents and humans. Bull World Health Organ 87:82
122. Ramsden E, Adams J (2009) Escaping the laboratory: the rodent experiments of John B. Calhoun & their cultural influence. J Soc Hist 42:761–792
123. Hock R (2004) Forty studies that changed psychology: explorations in the history of psychological thought. Prentice Hall, New York

# Chapter 10

# Functional Phosphoproteomics of Oncogenic KRAS Signaling

**Putty-Reddy Sudhir and Jeou-Yuan Chen**

## Abstract

Identification of oncogene-mediated phosphorylation events is essential to understanding the molecular determinants responsible for cancer development and progression. Here, we identify KRAS-regulated phosphorylation events using label-free quantitation-based comparative phosphoproteomics analyses of immortalized human bronchial epithelial cells that express oncogenic KRAS as well as cells that do not. Further, we demonstrate integration of the identified phosphorylation events with the Pathway Interaction Database to infer KRAS-regulated pathways, which may have implications in KRAS-associated lung adenocarcinoma development. Taken together, our study provides an overview of the functional phosphoproteomics approach involving cell culture, preparation of whole cell lysates, trypsin digestion, phosphopeptide enrichment, mass spectrometry analyses, label-free quantitative analyses, and signaling pathway analyses to study KRAS-targeted events.

**Key words** Phosphoproteomics, Mass spectrometry, KRAS, Phosphopeptide enrichment, Label-free quantitation, IDEAL-Q, Pathway analysis

## 1 Introduction

The RAS family proteins, including HRAS, KRAS, and NRAS, are small G proteins that function as molecular switches in the regulation of various cellular processes, including proliferation, differentiation, apoptosis, adhesion, and migration. Activation of upstream receptors, such as receptor tyrosine kinases, cytokine receptors, and G protein-coupled receptors, promotes conversion of GDP-bound RAS to the GTP-bound protein. GTP-bound RAS is in the active configuration to interact with various effectors with high affinity, and transduces signals to several downstream pathways [1].

Activation of *RAS* family genes is observed in approximately 30 % of human neoplasms, including cancers of lung, pancreas, and colon. The *KRAS* proto-oncogene is frequently activated by point mutations in codons 12, 13, and 61, giving rise to constitutively active KRAS protein. Active KRAS protein promotes activation of

Lorenza Trabalzini and Saverio Francesco Retta (eds.), *Ras Signaling: Methods and Protocols*,
Methods in Molecular Biology, vol. 1120, DOI 10.1007/978-1-62703-791-4_10, © Springer Science+Business Media, LLC 2014

downstream PI3K/AKT and RAF/MEK/MAPK signaling pathways, leading to cell transformation by deregulating multiple cellular functions and processes such as cell growth, cell cycle progression, cell survival, angiogenesis, and cell migration [1, 2]. However, the roles of activated KRAS in various stages of tumor development are poorly understood. Global identification of potential downstream targets is one approach for examining the oncogenic role of KRAS signaling. Monitoring the entire KRAS signaling network using omics-based approaches at the levels of the transcriptome [3, 4], proteome [5], and posttranslational modifications [5–7] would allow identification of previously unknown targets of KRAS and expand knowledge regarding KRAS-mediated oncogenic signaling.

Posttranslational modifications, including phosphorylation, glycosylation, acetylation, and lipid modification, are known to play fundamental roles in oncogenic signal transduction. Protein phosphorylation, which is reversibly controlled by kinases and phosphatases, is one of the most widely studied posttranslational modifications in cancer biology because of its ubiquitous nature and significance in driving oncogenic signals. Protein phosphorylation defines protein structure, activity, subcellular localization, protein–protein interactions, protein degradation, and other processes. The protein phosphorylation is frequently assessed using site-specific phospho-antibodies; however, they do not allow evaluation of novel phosphorylation events. In contrast, recently developed mass spectrometry (MS)-based phosphoproteomics approaches provide an efficient way to evaluate the known and novel phosphorylation events. It is noted that relatively low abundance phosphoproteins or phosphopeptides need to be enriched prior to mass spectrometry analysis. Two techniques have been adopted by the researchers for the specific enrichment of phosphopeptides [8]: immobilized metal ion (e.g., $Fe^{3+}$) affinity chromatography (IMAC) and metal oxide (e.g., $TiO_2$) affinity chromatography (MOAC). The $Fe^{3+}$- and $TiO_2$-based techniques have been coupled with various fractionation methods (e.g., cation exchange SCX column) to improve the identification of phosphoproteins and phosphosites [8]. Distinct bioinformatics tools, including Ingenuity Pathway Analysis and MetaCore, have been applied to extract the biological significance from the phosphoproteomics data.

Here we use the functional phosphoproteomics approach to study KRAS downstream targets. An outline of the approach is shown in Fig. 1. Briefly, whole cell lysates of Cdk4/hTERT-immortalized human bronchial epithelial cells (HBEC3-KT) and $KRAS^{V12}$-transformed HBEC3-KT (3KTR) cells were prepared, digested with trypsin, and subjected to liquid chromatography (LC)-MS/MS analysis. Samples were analyzed in triplicate and the data obtained from MS were processed using the IDEAL-Q algorithm. IDEAL-Q is an automated tool that performs peptide identification, normalization,

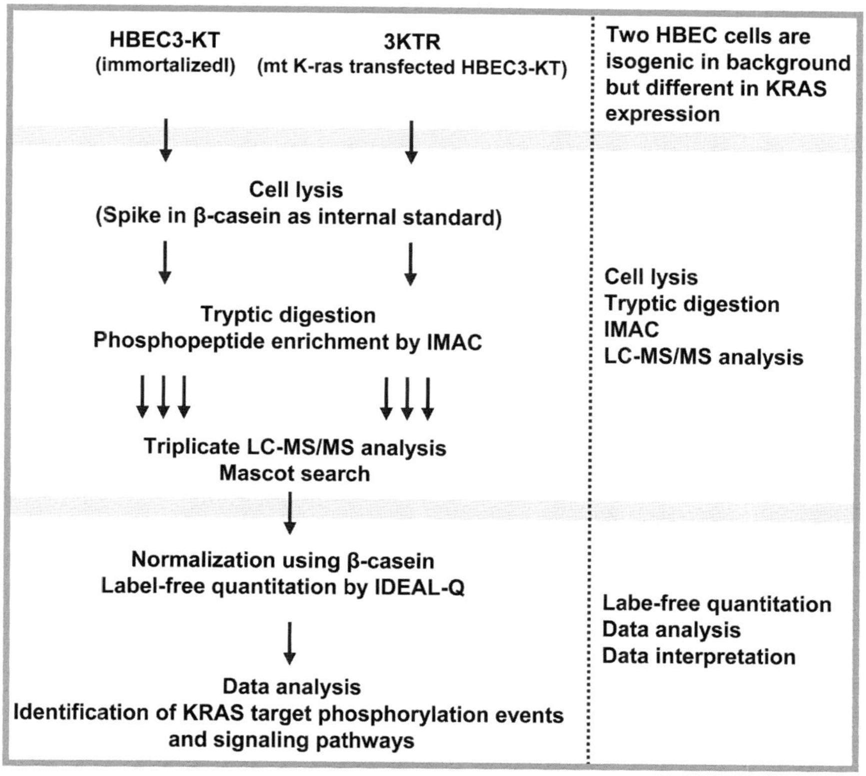


**Fig. 1** Experimental workflow to study oncogenic KRAS signaling by phosphoproteomics approach

and quantitation of raw data obtained from MS. IDEAL-Q considers elution time, signal-to-noise ratio, charge state, and isotopic patterns of identified peptide signals to quantify phosphorylation changes between two samples (i.e., 3-KT and 3KTR cells). The identified phosphorylation events were integrated with the Pathway Interaction Database to evaluate KRAS-targeted signaling pathways.

## 2 Materials

### *2.1 Cell Culture*

1. Cdk4 (cyclin-dependent kinase 4)/hTERT (human telomerase reverse transcriptase)-immortalized human bronchial epithelial cells (HBEC3-KT) and $KRAS^{V12}$-transformed HBEC3-KT cells (3KTR) were used in the procedure described here.
2. Culture medium: K-SFM medium containing 50 μg/mL bovine pituitary extract, 5 ng/mL EGF, and 1 % Penicillin-Streptomycin-Amphotericin B solution.

### *2.2 Cell Lysis and Protein Concentration Determination*

1. Modified RIPA buffer: 1 % NP40 (Igepal CA-630), 300 mM NaCl, 1 mM EDTA, 0.5 mM DTT, 1 mM $NaVO_3$, 10 mM NaF, and cocktail of protease inhibitors.

2. Bovine serum albumin (BSA) as a standard in protein quantitation assay.
3. Bio-Rad Protein Assay reagent (other protein assay reagents may also be used).

### 2.3 Tube-Gel Digestion

1. Eppendorf tube.
2. Gel matrix solution: 5 μL of 40 % (29:1) acrylamide, 0.7 μL of 10 % ammonium persulfate, 0.3 μL of *N,N,N,N′*-tetramethylethylenediamine (TEMED).
3. Bovine β-casein.
4. Trypsin (sequencing grade).
5. Peptide washing solution: 25 mM triethylammonium bicarbonate (TEABC), 50 % HPLC-grade acetonitrile (ACN).
6. Peptide extraction solution: 5 % formic acid (FA), 50 % ACN.

### 2.4 Enrichment of Phosphopeptides

1. 0.1 % acetic acid.
2. 0.2 M ammonium dihydrogen phosphate ($NH_4H_2PO_4$), pH 4.4.
3. Buffer A: 6 % acetic acid.
4. Buffer B: Buffer A containing 50 mM EDTA, 1 M NaCl.
5. Buffer C: Buffer A containing 0.2 M $FeCl_3$.
6. Buffer D: Buffer A containing ACN (Buffer A to ACN ratio of 3:1, v/v).
7. Ni-NTA spin column.
8. 5 cm-long, 500 μm i.d. PEEK microcolumn.
9. Autosampler and Solvent delivery system.

### 2.5 Desalting

1. C18 ZipTip.
2. 20 μL micropipette.
3. Solution A: 0.1 % trifluoroacetic acid (TFA).
4. Solution B: 50 % ACN in 0.1 % TFA.
5. Solution C: 90 % ACN in 0.1 % TFA.

### 2.6 LC-MS/MS Analysis

1. Sample loading buffer: 0.1 % FA in $H_2O$.
2. Elution solution:

   0.1 % FA in $H_2O$.

   0.1 % FA in ACN.
3. LC system: A C18 capillary trap column (2 cm × 180 μm) linked to a C18 capillary separation column (25 cm × 75 μm). Maintain a linear gradient of 0–80 % 0.1 % FA in ACN and 0.1 % FA in $H_2O$ for 120 min to elute the peptides.

4. Q-TOF Mass spectrometer (Waters Q-TOF Premier from Waters Corp, Milford, MA).
5. IDEAL-Q (*Id*entity-based *e*lution time prediction by fr*ag*menta*l* regression-based *q*uantitation), an automated label-free quantitation tool.

# 3 Methods

## 3.1 Culturing Cells

1. Maintain HBEC3-KT and 3KTR cells in K-SFM medium containing 50 μg/mL bovine pituitary extract (BPE) and 5 ng/mL EGF under tissue culture condition in 5 % $CO_2$ at 37 °C [9, 10].
2. Grow the cells to 80 % confluence.

## 3.2 Cell Lysis and Tube-Gel Digestion (See Note 1)

1. Lyse the cells ($1 \times 10^7$) with 0.5 mL modified RIPA buffer, vortex for 1 min, and rotate for 30 min in cold room.
2. Centrifuge at 15,000 × *g* for 5 min, collect supernatant and estimate protein concentration using protein assay reagent.
3. Mix 14 μL of cell lysate (400 μg) with the gel matrix solution in an eppendorf tube (*see* **Note 2**).
4. Cut the solidified gel matrix into small pieces and wash with 25 mM TEABC containing 50 % ACN for 30 min (*see* **Note 3**). Repeat the washing process for 2 more times to remove the detergents (*see* **Note 4**).
5. Wash the gel pieces with pure ACN for 3 min and dry the gel pieces completely in a SpeedVac.
6. Add 25 mM TEABC buffer (twofold to the initial gel matrix volume) and 0.5 μg of β-casein as an internal standard into the eppendorf tube containing gel pieces (*see* **Note 5**).
7. Add trypsin (protein to trypsin ratio of 40:1, w/w) and incubate overnight at 37 °C.
8. Collect peptide solution into an eppendorf tube. Further extract the peptides from the gel pieces with 5 % FA containing 50 % ACN (twofold to the initial gel matrix volume) for 30 min with constant vortexing. Recover the extracted solution and combine into the peptide solution. Repeat the extraction 2 more times (*see* **Note 6**).
9. Dry the peptide solution in a SpeedVac.

## 3.3 Phosphopeptide Enrichment

1. Extract Ni-NTA resin from spin column using 1 mL of 0.1 % acetic acid. Pack 150 μL of Ni-NTA resin into a 5-cm microcolumn.
2. Wash microcolumn with Buffer B for 15 min, and equilibrate with Buffer C for 30 min.

3. Dissolve peptide sample in Buffer A and adjust the pH to 3.0 and final volume to 100 μL (*see* **Note 7**).
4. Load peptide sample into microcolumn. Wash it with Buffer D for 15 min and then with Buffer A for 20 min.
5. Elute peptides using 0.2 M $NH_4H_2PO_4$, pH 4.4 for 15 min. Perform the **steps** from **2** to **5** by means of autosampler with a flow rate of 13 μL/min [7, 11].
6. Collect the eluent and dry it in SpeedVac.

### 3.4 Desalting

1. Redissolve the phosphopeptide sample in 10 μL of Solution A.
2. Attach the Zip-Tip to a 20 μL-pipetman and wash the Zip-Tip by pipeting in and out 10 μL of solution B for 10 times (*see* **Note 8**).
3. Repeat washing the Zip-Tip with Solution A for 10 times.
4. Load phosphopeptides onto Zip-Tip by pipeting in the phosphopeptide solution. Repeat pipeting in and out the solution for 15 times for better binding.
5. Repeat **step 3** to wash the peptide-loaded Zip-Tip with Solution A for 10 times to remove salt from the peptide sample.
6. Elute phosphopeptides by withdraw Solution C into Zip-Tip and pipet it out into a new eppendorf tube (Eluent 1). Repeat this process to elute the sample for 10 times.
7. Recover the phosphopeptides as much as possible by subjecting the solution from **step 4** to another Zip-Tip, which has been conditioned as described in **steps 2** and **3**, and repeat the **steps 4–6** (*see* **Note 9**), and collect the eluent as Eluent 2.
8. Combine Eluents 1 and 2 and dry completely using SpeedVac.

### 3.5 LC-MS/MS Analysis

1. Redissolve the phosphopeptide samples in 0.1 % FA solution for LC-MS/MS analysis [7, 11].
2. Inject the samples into the capillary trap column and separate by the capillary separation column using LCTM system.
3. Elute peptides using the linear gradient described in Subheading 2.6, **item 3**.
4. Operate MS in ESI positive V mode with a resolving power of 10,000.
5. Use the NanoLockSpray source for accurate mass measurement.
6. Acquire data via data directed analysis (DDA). The method included a full sequential MS scan (*m/z* 400–1,600, 0.6 s) and 3 MS/MS (*m/z* 100–1,990, 1.2 s/scan) on the three most intense ions present in the full scan mass spectrum.
7. Analyze each sample for 3 times.

### 3.6 Label-Free Quantitation by IDEAL-Q

1. Convert the raw data files into mzXML format by the program massWolf [5, 7, 12].
2. Export the MASCOT search results in eXtensible Markup Language data (XML) format [5, 7, 12].
3. Upload the data files into Identity-based elution time prediction by fragmental regression-based quantitation (IDEAL-Q) and process the data considering the parameters such as signal-to-noise ratio, charge state, and isotopic patterns [12].
4. Normalize the observed phosphopeptide abundances by using the area of β-casein, the internal standard (*see* **Notes 10** and **11**).
5. Save the data in Excel format or any other readable formats.

### 3.7 Pathway Analysis

1. Infer the activities of individual signaling pathways by integrating the levels of related phosphoprotein signatures determined by IDEAL-Q analysis [7].
2. Sum the representative values of phosphoproteins to define mean $\left(\bar{X}_i\right)$ and standard deviation ($\rho_i$) for each class (e.g., HBEC3-KT and 3KTR).

$$\bar{X}_i = \sqrt{\sum_{j=1}^{k} x_{i,j}^2}$$

$x_{i,j}$ is mean of the values of the phosphoprotein $j$ in class $i$.
$k$ is number of proteins in a given pathway.

$$\rho_i^2 = \frac{\sum_{j=1}^{k} \rho_{i,j}^2}{k}$$

$\rho_{i,j}$ is standard deviation of the values of the phosphoprotein $j$ in class $i$.

3. Calculate the pathway activity ($a_{\mathrm{p}}$) using $t$-test statistics to discriminate between two classes.

$$a_{\mathrm{p}} = \frac{\left|\bar{X}_{c_1} - \bar{X}_{c_2}\right|}{\sqrt{\frac{\rho_{c_1}^2}{n_{c_1}} + \frac{\rho_{c_2}^2}{n_{c_2}}}}$$

$a_{\mathrm{p}}$ is the activity score for pathway p between classes $c_1$ and $c_2$.
$n_{c_1}$ and $n_{c_2}$ are the numbers of replicate experiments for the respective classes.
In our case, $n_{c_1}$ and $n_{c_2}$ are each 3 (Subheading 3.5, **step 7**).

The KRAS-targeted pathways identified with $t$-score >10 are shown in Table 1.

**Table 1**
**Signaling pathways regulated by oncogenic KRAS, identified by functional phosphoproteomics approach (*t*-score >10 are shown)**

| Pathway description | Source | *t*-score (up) |
|---|---|---|
| Agrin in postsynaptic differentiation | biocarta | 36.2 |
| FGF signaling pathway | nci | 36.2 |
| Signaling events mediated by VEGFR1 and VEGFR2 | nci | 31.6 |
| mCalpain and friends in cell motility | biocarta | 31.6 |
| MAPK signaling pathway | biocarta | 31.6 |
| Signaling events activated by c-Met | nci | 31.6 |
| Fc-epsilon receptor I signaling in mast cells | nci | 31.6 |
| Integrins in angioPhosphoproteinsis | nci | 31.6 |
| Keratinocyte differentiation | biocarta | 31.6 |
| Alpha-synuclein signaling | nci | 31.6 |
| BCR signaling pathway | nci | 31.6 |
| Sprouty regulation of tyrosine kinase signals | biocarta | 31.6 |
| Erk and PI-3 kinase necessary for collagen binding in c.e. | biocarta | 31.6 |
| Transcription factor CREB and its extracellular signals | biocarta | 31.6 |
| Regulation of elf4e and p70 S6 kinase | biocarta | 31.6 |
| Trefoil factors-initiated mucosal healing | biocarta | 31.6 |
| Regulation of telomerase | nci | 21.5 |
| HIV-1 Nef: negative effector of Fas and TNF | biocarta | 15.3 |
| Caspase cascade in apoptosis | nci | 15.3 |
| Aurora B signaling | nci | 13.7 |

## 4 Notes

1. Wear the gloves while performing experiments to reduce the keratin contamination.
2. IMAC column loading capacity is equal to the peptides obtained from 400 μg total cell lysate. Use bigger column in case of loading more than 400 μg.
3. Cut the gel matrix into small pieces inside the eppendorf tube to avoid the keratin contamination.
4. The washing step will remove the salts present in the gel pieces. One can see reduced bubbles in the sample after washing the gel pieces for 2 times, indicating reduced detergents used to lyse the cells and residual SDS presents in the gel

pieces. Wash the samples for more times in case the intensity of bubbles not decreased.

5. Adding more than twofold volume of TEABC buffer to the initial gel matrix would reduce the efficiency of tryptic digestion.
6. Extracting the peptides from gel pieces for 3 times would greatly improve the recovery of phosphopeptides.
7. Adjusting the peptide sample solution pH to 3.0 would significantly enhance the phosphopeptide binding to the IMAC column.
8. When withdrawing and pipeting out the solutions and samples, care must be taken not to introduce any air into the Zip-Tip.
9. Zip-Tip desalting of the phosphopeptide sample obtained from IMAC column for 2 times would improve the recovery of the phosphopeptides to a greater extent.
10. Examine the average and standard deviation of the β-casein abundances in HBEC3-KT and 3KTR cells to check the quality of the tube-gel digestion, phosphopeptide enrichment and mass spectrometry performance.
11. To check the activation status of KRAS signaling, examine the MAPK phosphorylation levels by western blot analysis and relative abundance of MAPK-derived phosphopeptides in IDEAL-Q quantitation data output.

## Acknowledgment

This work was supported by the grant to JYC from National Research Program on Genomic Medicine, Taiwan and the postdoctoral fellowship to PRS from the Academia Sinica, Taiwan.

### References

1. Malumbres M, Barbacid M (2003) RAS oncogenes: the first 30 years. Nat Rev Cancer 3:459–465
2. Riely GJ, Marks J, Pao W (2009) KRAS mutations in non-small cell lung cancer. Proc Am Thorac Soc 6:201–205
3. Qian J, Niu J, Li M et al (2005) In vitro modeling of human pancreatic duct epithelial cell transformation defines gene expression changes induced by K-ras oncogenic activation in pancreatic carcinogenesis. Cancer Res 65:5045–5053
4. Tchernitsa OI, Sers C, Zuber J et al (2004) Transcriptional basis of KRAS oncogene-mediated cellular transformation in ovarian epithelial cells. Oncogene 23:4536–4555
5. Sudhir PR, Chen CH, Pavana Kumari M et al (2012) Label-free quantitative proteomics and N-glycoproteomics analysis of KRAS-activated human bronchial epithelial cells. Mol Cell Proteomics 11:901–915
6. Guha U, Chaerkady R, Marimuthu A et al (2008) Comparisons of tyrosine phosphorylated proteins in cells expressing lung cancer-specific alleles of EGFR and KRAS. Proc Natl Acad Sci U S A 105:14112–14117

7. Sudhir P-R, Hsu C-L, Wang M-J et al (2011) Phosphoproteomics identifies oncogenic ras signaling targets and their involvement in lung adenocarcinomas. PLoS One 6:e20199
8. Beltran L, Cutillas PR (2012) Advances in phosphopeptide enrichment techniques for phosphoproteomics. Amino Acids 43:1009–1024
9. Ramirez RD, Sheridan S, Girard L et al (2004) Immortalization of human bronchial epithelial cells in the absence of viral oncoproteins. Cancer Res 64:9027–9034
10. Sato M, Vaughan MB, Girard L et al (2006) Multiple oncogenic changes (K-RAS(V12), p53 knockdown, mutant EGFRs, p16 bypass, telomerase) are not sufficient to confer a full malignant phenotype on human bronchial epithelial cells. Cancer Res 66:2116–2128
11. Tsai CF, Wang YT, Chen YR et al (2008) Immobilized metal affinity chromatography revisited: pH/acid control toward high selectivity in phosphoproteomics. J Proteome Res 7:4058–4069
12. Tsou CC, Tsai CF, Tsui YH et al (2010) IDEAL-Q, an automated tool for label-free quantitation analysis using an efficient peptide alignment approach and spectral data validation. Mol Cell Proteomics 9:131–144

# Chapter 11

# Pull-Down Assay for Analysis of Integrin-Mediated Activation of Rap Proteins in Adherent Platelets

Gianni Francesco Guidetti and Mauro Torti

## Abstract

Rap1 GTPases operate as molecular switches by cycling between a GDP-bound inactive state and a GTP-bound active state and regulate several cellular pathways in response to different stimuli. Circulating blood platelets express high levels of Rap1 proteins, mainly Rap1b, which plays a critical role in platelet adhesion and activation. Rap1 is a key element in the inside-out signaling pathway leading to the conversion of integrins into the high-affinity state for their ligands. In platelets, Rap1b regulates inside-out activation of both integrin $\alpha_{IIb}\beta_3$ and $\alpha_2\beta_1$. In addition, Rap1b is also involved in integrin outside-in signaling. Integrin-mediated platelet adhesion leads to accumulation of GTP-bound Rap1b, which promotes integrin-mediated processes such as spreading and clot retraction. Rap1b is thus a bidirectional regulator of platelet integrin function. Here we describe a method to analyze Rap1b activation induced by platelet adhesion via integrin $\alpha_2\beta_1$.

**Key words** Platelets, Platelet adhesion, Integrin receptors, Rap1b, Collagen, Integrin $\alpha_2\beta_1$, Fibrinogen, Integrin $\alpha_{IIb}\beta_3$

## 1 Introduction

Circulating blood platelets are recruited and activated upon vessel wall injury and promote clot formation. Upon endothelial damage, platelets rapidly adhere to the exposed subendothelial matrix components through different specific membrane receptors, including integrin $\alpha_2\beta_1$ and GPVI, which bind collagen, integrin $\alpha_5\beta_1$, and integrin $\alpha_6\beta_1$, which interact with fibronectin and laminin, respectively, and the GP Ib-IX-V complex, which mediates shear-dependent adhesion to von Willebrand factor (VWF) [1]. Adhesion to the extracellular matrix induces the activation of platelets, which are stimulated to release several soluble agonists, such as ADP and thromboxane $A_2$, which reinforce platelet stimulation by an autocrine/paracrine mechanism. Activated platelets eventually undergo aggregation, a process of cell–cell interaction mediated by the binding of integrin $\alpha_{IIb}\beta_3$ to soluble fibrinogen, which

Lorenza Trabalzini and Saverio Francesco Retta (eds.), *Ras Signaling: Methods and Protocols*,
Methods in Molecular Biology, vol. 1120, DOI 10.1007/978-1-62703-791-4_11, 

acts as a bridge connecting adjacent platelets [2]. Unstimulated platelets are unable to bind soluble fibrinogen, as integrin $\alpha_{IIb}\beta_3$ is locked into an inactive, low-affinity state. Platelet activation culminates in the conversion of integrin $\alpha_{IIb}\beta_3$ into an active conformation with a high affinity for fibrinogen. This process, called inside-out activation, involves GTP-bound Rap1b, which organizes a signaling pathway with cytoskeletal proteins, such as RIAM, kindlin, and talin [3]. On the other hand, Rap1b is a critical player for integrin outside-in signaling, as well, and it is activated upon platelet adhesion mediated by integrins $\alpha_2\beta_1$ and $\alpha_{IIb}\beta_3$ [4–6]. During platelet adhesion, active, GTP-bound Rap1b controls major responses including integrins cross talk, platelet spreading, thrombus formation, and clot retraction [4, 7, 8].

Here we report a detailed method used to evaluate the activation of Rap1b induced by integrin-mediated platelet adhesion. The assay is a modification of a widely used pull-down assay originally developed by Franke et al. [9] to evaluate Rap1b activation in a suspension of blood platelets stimulated with soluble agonists. This assay exploits the ability of an engineered fusion protein, named GST-RalGDS-RBD, to selectively interact with the GTP-bound form of Rap proteins, allowing its precipitation from a cell lysate. It is herein described how to set up a suitable adhesion assay in which platelets are allowed to interact with immobilized, nonfibrillar collagen through integrin $\alpha_2\beta_1$. Upon cell lysis and determination of protein content, active Rap1b is selectively precipitated using the GST-RalGDS-RBD fusion protein coupled to a GSH resin. The presence of Rap1b in the precipitates is then detected by immunoblotting with an anti-Rap1 antibody.

## 2 Materials

Prepare all solutions using MilliQ water and analytical grade reagents.

### 2.1 Dishes Coating

1. Collagen coating solution: 0.1 mg/mL collagen (acid soluble collagen type I purified from bovine tendon) in 0.1 M acetic acid (*see* **Note 1**).
2. Dishes washing buffer: PBS (phosphate buffer saline), 8 mM $Na_2HPO_4$, 2 mM $NaH_2PO_4$, 140 mM NaCl, pH 7.4. Weigh 1.14 g of $Na_2HPO_4$, 0.27 g of $NaH_2PO_4$ monohydrate, and 8.18 g of NaCl in 1 L plastic or glass beaker. Add about 800 mL of water and dissolve the salts. Adjust pH to 7.4 with 1 M HCl solution, transfer to 1 L graduated cylinder, and make up to 1 L with water. Store at 4 °C.

3. Dishes blocking solution: 1 % bovine serum albumin (BSA) in PBS. Slowly add 1 g of BSA powder to 100 mL of PBS under constant stirring.

### 2.2 Platelet Isolation

1. ACD: 130 mM citric acid, 152 mM trisodium citrate, and 112 mM glucose. Weigh 2.73 g of citric acid, 4.47 g of trisodium citrate dihydrate, and 2.02 g of glucose in a beaker; add about 70 mL of water; and dissolve under stirring. Transfer to a graduated cylinder and make up to 100 mL with water.
2. PIPES buffer: 137 mM NaCl, 20 mM PIPES, pH 6.5. Weigh 4.00 g of NaCl and 3.02 g of PIPES (*see* **Note 2**) and add water up to about 400 mL. Put under constant stirring and facilitate dissolution by addition of few drops of 5 M NaOH until the solution is completely clear. Adjust pH to 6.5, transfer to a cylinder, and make the volume up to 500 mL with water.
3. HEPES buffer: 10 mM HEPES, 137 mM NaCl, 2.9 mM KCl, 12 mM $NaHCO_3$, pH 7.4. Weigh in a 500 mL beaker 1.19 g of HEPES (*see* **Note 2**), 4.00 g of NaCl, 0.11 g of KCl, and 0.50 g of $NaHCO_3$. Add 400 mL of water and dissolve the salts under stirring. Adjust the pH to 7.4, transfer to a graduated cylinder, and make up to 500 mL with water.

### 2.3 Platelet Lysis and Rap1 Pull-Down Assay

1. Glucose stock solution: 550 mM glucose. Dissolve 2.48 g of glucose in 25 mL of HEPES buffer.
2. BSA stock solution: 5 % BSA in HEPES buffer. Slowly dissolve 5 g of BSA in 100 mL of HEPES buffer under constant stirring. Prepare suitable aliquots and store at −20 °C.
3. $MgCl_2$ stock solution: 500 mM $MgCl_2$. Dissolve 10.17 g of $MgCl_2$ hexahydrate in 100 mL of water.
4. RGDS: 50 mM RGDS in PBS. Dissolve 10 mg of RGDS peptide (molecular weight 433.4 g/mol) in 0.46 mL of PBS.
5. Lysis buffer 2×: 100 mM Tris–HCl, 400 mM NaCl, 5 mM $MgCl_2$, 2 % Nonidet P-40, and 20 % glycerol. Weigh 1.21 g of Tris–HCl and 2.34 g of NaCl, add 50 mL of water, and dissolve under stirring. As the solution becomes clear, add 1 mL of $MgCl_2$ stock solution, 2 mL of Nonidet P-40, and 20 mL of glycerol (*see* **Note 3**). Adjust pH to 7.4 with 1 M HCl and make up to 100 mL with water. Store at 4 °C. Just before use, add 2 mM PMSF, 10 μg/mL leupeptin, and 10 μg/mL aprotinin from appropriate stock solutions (*see* **Note 4**). Dilute with water to have lysis buffer 1×.
6. GSH resin–GST-RalGDS-RBD probe: mix GSH resin with purified GST-RalGDS-RBD fusion protein and incubate for 30 min under gentle mixing at 4 °C to favor complex formation (*see* **Note 5**).

#### 2.4 SDS-PAGE, Western Blotting, and Immunoblotting

1. Tris-glycine running buffer: 25 mM Tris–HCl, 192 mM glycine, 0.1 % SDS, pH 8.3. Weigh 3.03 g of Tris–HCl, 14.41 g of glycine, and 1 g of SDS. Add 900 mL of water, dissolve under stirring, adjust the pH to 8.3, and make up to 1 L with water.
2. SDS sample buffer 3×: 6 % SDS, 1.5 % dithiothreitol (DTT), 30 % glycerol, and 0.03 % bromophenol blue. Weigh 3 g of SDS, 0.75 g of DTT, and 0.015 g of bromophenol blue in a 50 mL glass beaker. Add 30 mL of Tris-glycine running buffer and dissolve under stirring. Slowly add 15 mL of glycerol, transfer to a cylinder, and make up to 50 mL with Tris-glycine running buffer. Aliquot and store at −20 °C. Dilution to 2× should be done using Tris-glycine running buffer.
3. Membrane blocking solution: 20 mM Tris–HCl, 500 mM NaCl, 5 % BSA, pH 7.5. Weigh 0.48 g of Tris–HCl and 5.84 g of NaCl in a 200 mL beaker. Add approximately 150 mL of water and dissolve the salts under magnetic stirring. Weigh 10 g of BSA and slowly add it to the solution under constant stirring to facilitate dissolution. Adjust the pH to 7.5, transfer to a cylinder, and make up to 200 mL with water. Aliquot and store at −20 °C.
4. Membrane washing buffer (PBST): 0.05 % Tween-20 in PBS. Measure 1 L of PBS in a graduated cylinder, add 0.5 mL of Tween-20, and mix vigorously. Store at 4 °C.
5. Primary antibody solution: anti-Rap1, 1:1,000 dilution in PBST (*see* **Note 6**).
6. Secondary antibody solution: anti-rabbit IgG peroxidase-conjugate, 1:3,000 dilution in PBST.

## 3 Methods

Unless otherwise stated, all the procedures are performed at room temperature.

#### 3.1 Dishes Coating (To Be Performed the Day Before the Assay) (See Note 7)

1. Add 1 mL of collagen coating solution to 60 mm-diameter plastic culture dishes and incubate overnight.
2. The following day, aspirate the coating solution and wash the dishes 3× with 5 mL of PBS, removing the buffer between each wash by aspiration.
3. Incubate the dishes with 2 mL of dishes blocking solution for 2 h to block the nonspecific binding sites. Then, remove the blocking solution by washing the dishes 3×, as described above.

### 3.2 Platelet Isolation (All Centrifugations Must Be Performed on a Swinging-Rotor Bench Centrifuge)

1. Blood should be withdrawn from healthy donors that have not assumed drugs in the previous 2 weeks by venipuncture with a 21G needle, using one-tenth of the final volume of ACD as anticoagulant. Split blood in 5 mL aliquots into 10 mL polypropylene tubes, and centrifuge at 120 × *g* for 10 min.
2. Carefully remove the upper phase (platelet rich plasma, PRP) using plastic, or silanized glass pipettes, and pool the collected aliquots into clean conical 15 mL polypropylene tubes to a maximum final volume of 5 mL per tube (*see* **Note 8**).
3. Estimate the PRP volume and add 0.2 U/mL apyrase and 1 μM $PGE_1$ (*see* **Note 9**). Centrifuge the samples at 720 × *g* for 15 min and remove the supernatant (platelet poor plasma, PPP).
4. Resuspend the platelet pellet in 0.5 mL of PIPES buffer, by pipetting gently until pellet residues are no longer visible, and then add PIPES buffer to a final volume of 5 mL to dilute plasma contaminants.
5. Centrifuge the samples at 720 × *g* for 15 min to collect platelets, remove the supernatant, and resuspend each pellet in 0.5 mL of HEPES buffer.
6. Determine the platelet count and adjust the concentration to $4 \times 10^8$ cells/mL with HEPES buffer (*see* **Note 10**).

### 3.3 Platelet Lysis

1. Remove the buffer from washed dishes prepared according to Subheading 3.1; add to the platelet suspension 5.5 mM glucose, 1 mg/mL BSA, 0.5 mM RGDS (*see* **Note 11**), and 1 mM $MgCl_2$ (*see* **Note 12**) (by addition of a proper dilution from stock solutions); and transfer 0.5 mL of platelet suspension (corresponding to $2 \times 10^8$ cells) to each dish.
2. Incubation is typically protracted for 10–120 min without stirring, then remove non-adherent platelets, and wash the dishes 3× with 5 mL of PBS.
3. Remove the last wash, place the dishes on ice, and add 0.5 mL of ice-cold lysis buffer. Recover the lysates with a cell scraper and transfer to 1.5 mL tubes.
4. Meanwhile, add 0.25 mL of ice-cold lysis buffer 2× to an equal volume of $4 \times 10^8$ cells/mL platelet suspension (corresponding to $1 \times 10^8$ cells) (*see* **Note 13**) to prepare the basal platelet lysate as negative control for Rap activation.
5. Lysis of adherent platelets and basal samples should be protracted under constant mixing for 15–30 min at 4 °C, then clarify the samples by centrifugation at 18,000 × *g* for 10 min at 4 °C.

6. Recover the supernatants and quantify the protein content by the bicinchoninic acid assay. Dilute all the samples to the same protein concentration with lysis buffer (*see* **Note 14**).
7. Transfer 0.1 mL of each sample into a new tube, add 0.05 mL of SDS sample buffer 3×, and dissociate by heating at 95 °C for 3 min for total lysates Western blotting analysis.

### 3.4 Rap1 Pull-Down Assay

1. Add to 0.3 mL of each sample 0.05 mL of the beads mixture coupled to the GST-RalGDS-RBD probe and incubate for 1 h at 4 °C under constant mixing (*see* **Note 5**).
2. Spin down the beads by centrifugation at 18,000 × *g* for 30 s and eliminate the supernatant (*see* **Note 15**).
3. Add 1 mL of ice-cold lysis buffer, gently mix the samples, spin the tubes at 18,000 × *g* for 30 s, and remove the supernatant. Repeat this washing procedure 3× and then accurately eliminate the residual supernatant with an insulin syringe.
4. Dissociate the pellets by addition of 25 μL of SDS sample buffer 2× and heating at 95 °C for 3 min under constant mixing.

### 3.5 SDS-PAGE, Western Blotting, and Immunoblotting

1. Centrifuge the samples at 18,000 × *g* for 1 min to completely precipitate the beads. Load the whole supernatant of the samples from Rap pull-down assay on a 10–20 % SDS-polyacrylamide gradient minigel (*see* **Note 16**). Load 0.03 mL of the corresponding dissociated total lysates for equal loading analysis.
2. Separate pull-down and total lysates samples by SDS-PAGE using the Tris-glycine running buffer under constant voltage.
3. Upon SDS-PAGE, transfer the proteins onto a PVDF membrane by Western blotting (*see* **Note 17**), and then incubate it for 1 h with blocking solution under agitation.
4. Remove the blocking solution, quickly rinse the membrane with PBST and then incubate it with the anti-Rap1 primary antibody solution overnight at 4 °C under mild agitation (*see* **Note 18**).
5. Wash the membrane 3× by incubation with PBST for 5 min under agitation.
6. Add the secondary antibody solution, incubate for 45 min at room temperature and wash 3× with PBST.
7. Add 0.5 mL of enhanced chemiluminescence (ECL) solution and acquire the chemiluminescence with a molecular imager or by film exposure.

## 4 Notes

1. In our experience, integrin $\alpha_2\beta_1$-mediated Rap1b activation has been analyzed in platelets adherent to dishes coated with non-fibrillar collagen, with collagen-derived peptides obtained by CNBr digestion, or with synthetic peptides containing integrin $\alpha_2\beta_1$-binding sequence GFOGER [4, 5, 10]. Non-fibrillar collagen type I mediates platelet adhesion through integrin $\alpha_2\beta_1$ and does not stimulate other platelet receptors for collagen such as GPVI. Indeed, platelets from integrin $\alpha_2$-knockout mice do not bind immobilized non-fibrillar collagen under static conditions [11], and adhesion to this molecule is unable to induce the phosphorylation of FcR γ-chain, a marker of GPVI stimulation [4]. Non-fibrillar collagen is a much cheaper and easily available reagent than other integrin $\alpha_2\beta_1$ ligands. The non-fibrillar conformation of collagen can be preserved by constantly keeping acid soluble collagen or pepsin-digested collagen in acetic acid solution. Typically, collagen dissolution is rather low and should be performed under constant stirring at 4 °C and prolonged for several days.

   The activation of Rap1b induced by platelet adhesion mediated by integrins other than integrin $\alpha_2\beta_1$ can be analyzed by introducing minor modifications. Although unable to bind soluble fibrinogen, unstimulated platelets adhere to immobilized fibrinogen. Therefore, integrin $\alpha_{IIb}\beta_3$-mediated adhesion can be investigated using dishes coated overnight with 100 μg/mL fibrinogen in PBS [5, 6]. Integrin $\alpha_6\beta_1$-mediated Rap1b activation can be analyzed also by setting up an adhesion assay on dishes coated with 50 μg/mL laminin, as previously reported [12], although this procure has not been tested for measuring accumulation of GTP-bound Rap1b. Integrin $\alpha_5\beta_1$ mediates adhesion to fibronectin, that is, a ligand for integrins $\alpha_{IIb}\beta_3$ and $\alpha_V\beta_3$, as well. To our knowledge no information on integrin $\alpha_5\beta_1$-specific ligands used in platelet adhesion assays is available.

2. Here it is considered to use PIPES and HEPES free acid with a molecular weight of 302.4 g/mol and 238.3 g/mol, respectively. It should be noted that for both compounds the sodium salt variant, with a different molecular weight, is commercially available.

3. Anhydrous glycerol must be used to prepare the lysis buffer, as other commercially available versions of glycerol strongly interfere with the BCA assay.

4. Inhibitors stock solutions: 200 mM PMSF dissolved in 2-propanol, 10 mg/mL aprotinin, and 10 mg/mL leupeptin in water. A different mixture of protease inhibitors could be used without significantly affecting the outcome of the assay.

5. GST-RalGDS-RBD fusion protein can be either purchased from international companies or prepared in the laboratory by purification from *Escherichia coli* transformed with the pGEX vector containing the sequence codifying for the probe, using affinity chromatography [9]. Few companies offer detection kits that contain the fusion protein, the GSH resin (either as a mixture or separate), cell lysis buffer, and SDS sample buffer. However, the relative amounts of GST-RalGDS-RBD and GSH resin for the mixture and the volume that should be added to platelet samples should be determined by accurate preliminary experiments using lysate from basal and stimulated platelets. In particular, the use of an excess of probe will nonspecifically precipitate GDP-bound Rap1 as well, thus compromising the evaluation of differences in Rap1b activation. We typically prepare a GSH-Sepharose 50 % slurry in PBS containing the GST-RalGDS-RBD fusion protein in a range of concentration from 0.1 to 0.5 mg/mL.
6. We use an anti-Rap1 antibody from Santa Cruz. Primary antibody solution can be recovered and reused several times.
7. BSA-coated control dishes can be prepared by overnight incubation of the dishes with 1 mL of 1 % BSA in PBS. Adherent platelets, recovered and lysed as described, could be used to determine the level of Rap1b activation induced by nonspecific adhesion to BSA.
8. Do not aspirate the buffy coat to avoid contamination by white blood cells.
9. $PGE_1$ (prostaglandin $E_1$) stimulates Gs-coupled receptors on platelet membrane, inducing the increase of intracellular cAMP, which is a mediator of platelet inhibition. Apyrase metabolizes extracellular ADP, which is a platelet agonist that can be accidentally released by blood cells during manipulation. We purchase $PGE_1$ and Apyrase Grade I from Sigma-Aldrich.
10. We typically count a 1:200 dilution of the platelet preparation using a Burker chamber; however, the use of an automatic cell counter or a flow cytometer would be recommended. This platelet purification procedure typically gives an expected yield of about $2–3 \times 10^9$ platelets from 50 mL of anticoagulated blood.
11. In order to exclude any contribution to Rap1b activation by integrin $\alpha_{IIb}\beta_3$-mediated platelet–platelet interaction induced by adhesion to collagen, and to focus exclusively on integrin $\alpha_2\beta_1$ signaling, experiments should be performed in the presence of RGDS peptide or other selective integrin $\alpha_{IIb}\beta_3$ antagonists. RGDS can be purchased from Sigma-Aldrich.

12. When performing adhesion assays to immobilized fibrinogen, 1 mM $CaCl_2$ instead of $MgCl_2$ should be added to the platelet suspension, since integrin $\alpha_{IIb}\beta_3$ operates in a $Ca^{2+}$-dependent fashion.
13. Platelets in suspension used to set up the basal sample should not contain BSA, that would interfere with the quantitation of platelet proteins.
14. The bicinchoninic acid assay is performed following manufacture's instruction. The analysis of 25 μL of each sample gives a sufficiently reliable measure of absorbance to allow the determination of the protein concentration. Typically, in our experiments, the protein concentration of adherent platelets lysates ranges from 0.1 to 0.5 mg/mL, depending on times of incubation to immobilized ligand.
15. The centrifugation between each washing step can be performed at room temperature or in 4 °C refrigerated centrifuge without introducing significant differences in the outcome of the assay. It may be convenient to add a small amount of uncoupled beads before sample centrifugation to facilitate the visualization of the pellets and the washing procedure. To this purpose we typically add 0.03 mL of Sepharose Cl2B cross-linked 50 % slurry in PBS.
16. We load the samples using a precision 50 μL syringe (Hamilton), which prevents accidental beads aspiration. The use of a 10–20 % polyacrylamide gradient gel for SDS-PAGE is recommended, but a 12 % polyacrylamide gel is suitable, as well.
17. We typically perform Western blotting using a wet blotting apparatus from Bio-Rad; however, different methods and devices can be used. In our conditions, the blotting procedure is performed with a constant current of 200 mA for 2 h.
18. In platelets, Rap1b is the most abundant GTPase, but these cells also contain Rap2b, which, however, is about 10× less abundant than Rap1b. Very low levels of Rap1a and Rap2a have been found in platelets, and Rap2c is not detectable [13–15]. The described method is typically referred to as a Rap1b activation assay for the difference in the expression levels among Rap1 GTPases in platelets; however, the use of an anti-Rap1 antibody will not discriminate between the Rap1a and Rap1b. By using an anti-Rap2 antibody, this assay can be used to determine the extent of Rap2b activation in platelets.

## References

1. Yip J, Shen Y, Berndt MC et al (2005) Primary platelet adhesion receptors. IUBMB Life 57: 103–108
2. Broos K, Feys HB, De Meyer SF et al (2011) Platelets at work in primary hemostasis. Blood Rev 25:155–167
3. Guidetti GF, Torti M (2012) The small GTPase Rap1b: a bidirectional regulator of platelet adhesion receptors. J Signal Transduct 2012:412089
4. Bernardi B, Guidetti GF, Campus F et al (2006) The small GTPase Rap1b regulates the cross talk between platelet integrin alpha-2beta1 and integrin alphaIIbbeta3. Blood 107:2728–2735
5. Guidetti GF, Bernardi B, Consonni A et al (2009) Integrin alpha2beta1 induces phosphorylation-dependent and phosphorylation-independent activation of phospholipase Cgamma2 in platelets: role of Src kinase and Rac GTPase. J Thromb Haemost 7:1200–1206
6. Cipolla L, Consonni A, Guidetti G et al (2013) The proline-rich tyrosine kinase pyk2 regulates platelet integrin alphaIIbbeta3 outside-in signaling. J Thromb Haemost 11:345–356
7. Chrzanowska-Wodnicka M, Smyth SS, Schoenwaelder SM et al (2005) Rap1b is required for normal platelet function and hemostasis in mice. J Clin Invest 115:680–687
8. Zhang G, Xiang B, Ye S et al (2011) Distinct roles for Rap1b protein in platelet secretion and integrin alphaIIbbeta3 outside-in signaling. J Biol Chem 286:39466–39477
9. Franke B, Akkerman JW, Bos JL (1997) Rapid $Ca^{2+}$-mediated activation of Rap1 in human platelets. EMBO J 16:252–259
10. Consonni A, Cipolla L, Guidetti G et al (2012) Role and regulation of phosphatidylinositol 3-kinase beta in platelet integrin alpha2beta1 signaling. Blood 119:847–856
11. Holtkotter O, Nieswandt B, Smyth N et al (2002) Integrin alpha 2-deficient mice develop normally, are fertile, but display partially defective platelet interaction with collagen. J Biol Chem 277:10789–10794
12. Inoue O, Suzuki-Inoue K, McCarty OJ et al (2006) Laminin stimulates spreading of platelets through integrin alpha6beta1-dependent activation of GPVI. Blood 107:1405–1412
13. Torti M, Lapetina EG (1994) Structure and function of rap proteins in human platelets. Thromb Haemost 71:533–543
14. Paganini S, Guidetti GF, Catricala S et al (2006) Identification and biochemical characterization of Rap2C, a new member of the Rap family of small GTP-binding proteins. Biochimie 88:285–295
15. Klinz FJ, Seifert R, Schwaner I et al (1992) Generation of specific antibodies against the rap1A, rap1B and rap2 small GTP-binding proteins. Analysis of rap and ras proteins in membranes from mammalian cells. Eur J Biochem 207:207–213

# Chapter 12

# Combined Pulldown and Time-Lapse Microscopy Studies for Determining the Role of Rap1 in the Crosstalk Between Integrins and Cadherins

**Luca Goitre and Saverio Francesco Retta**

## Abstract

The coordinate modulation of the cellular functions of cadherins and integrins plays an essential role in fundamental physiological and pathological processes, including morphogenesis, tissue differentiation and renewal, wound healing, immune surveillance, inflammatory response, tumor progression, and metastasis. Recent findings state the molecular mechanisms underlying the fine-balanced relationship between cadherins and integrins. In particular, some of the novel results recently obtained raise the possibility of a pivotal role for the small GTPase Rap1 in the functional crosstalk between cadherins and integrins. Considering the importance of the molecular signalling triggered by Rap1, here we provide protocols to study this small GTPase in signalling pathways involving cadherins and integrins.

**Key words** Integrins, Cadherins, Rap1, Small GTPase, Molecular crosstalk

## 1 Introduction

Cadherins and integrins are the major cell–cell and cell-extracellular matrix (ECM) adhesion receptors, respectively, and represent critical determinants of tissue architecture and function both in developing and adult organisms [1, 2].

Cadherins are single-pass transmembrane glycoproteins that support calcium-dependent, homophilic cell–cell adhesions. Together with their cytoplasmic domain interactors, such as β-catenin and p120$^{ctn}$, cadherins constitute the core components of adherens junctions (AJs). These specialized calcium-dependent adhesive structures link the cadherin homophilic adhesion to the actin cytoskeleton and are required for formation and maintenance of stable cell–cell adhesion and differentiated phenotype in all solid tissues [2, 3].

Integrins are heterodimeric transmembrane glycoproteins composed of noncovalently linked α and β subunits, which are

Lorenza Trabalzini and Saverio Francesco Retta (eds.), *Ras Signaling: Methods and Protocols*,
Methods in Molecular Biology, vol. 1120, DOI 10.1007/978-1-62703-791-4_12, © Springer Science+Business Media, LLC 2014

endowed with both structural and regulatory functions. They link the ECM to several distinct cytoplasmic proteins and the actin cytoskeleton at focal adhesion sites. There they provide both outside–in and inside–out transmissions of signals across the plasma membrane that control a number of critical cellular processes, including adhesion, migration, proliferation, differentiation, apoptosis, and gene expression [1].

The postulate that the activity of these two fundamental cell–cell and cell–matrix adhesion receptors must be temporally and spatially coordinated for the proper development and maintenance of tissue architecture, as well as the evidence that coordinate disruption of cadherin-dependent intercellular junctions and induction of integrin-dependent cell motility take place during the epithelial–mesenchymal transition of most malignant tumors [4, 5], suggests that a fine-tuned molecular crosstalk must exist between integrins and cadherins. Indeed, a growing body of evidence now exists to support the crosstalk between members of these two adhesive receptor families [6–17]. Among the molecules that might act as a "traffic light" at the crossroad between cadherin and integrin signalling pathways, Rap1, a member of the Ras subfamily of small GTPases, is emerging as a major candidate. Like other small GTPases, Rap1 cycles between an inactive GDP-bound and an active GTP-bound conformation that allows interactions with effectors, thus triggering cellular responses. This cycle is tightly regulated, both spatially and temporally, by several and markedly distinct guanine nucleotide exchange factors (GEFs) and GTPase-activating proteins (GAPs), which show unique expression profiles and subcellular localization in different cells [18, 19]. It was shown that Rap1 controls both AJs formation [20] and inside–out activation of integrins [18, 21–24]; conversely, it was demonstrated that defective Rap1 activation is related to both loss of AJs [25–27] and malignancy involving integrins and cadherins [27–29]. Indeed, Rap1 was suggested to be one of the key molecules that link cadherins to integrins [30–34]. In particular, it was demonstrated that:

1. Disruption/reduction of E-cadherin cell–cell junctions is associated with an increment in GTP-loaded Rap1.
2. E-cadherin endocytosis is necessary for Rap1 activation promoted by the disassembly of AJs.
3. Activation of Rap1 triggered by E-cadherin endocytosis is required for the acquisition of enhanced integrin-dependent cell–matrix contacts.

Here we describe some methodological approaches to study the involvement of Rap1 in the crosstalk between cadherins and integrins. In particular, we will discuss how the E-cadherin endocytosis-dependent activation of Rap1 and the consequent enhancement of focal adhesion assembly can be studied.

*You will find detailed protocols describing:*

1. How to disrupt or reduce cell–cell junctions
2. How to measure Rap1 activity
3. How to evaluate integrin-dependent cell–matrix contacts

## 2 Materials

### 2.1 Disruption of E-Cadherin Cell–Cell Junctions

1. Cellular lines: human MCF10A cells (mammary epithelial cells), HeLa cells (epithelial cell lines), COS cells (fibroblast-like cell line derived from monkey kidney tissue), NIH 3T3 cells (standard fibroblast cell line), s-End1 cells (endothelial cell line), and FRT cells (Fisher rat thyroid cells).
   Complete media: Dulbecco's modified Eagle's, Coon's F-12 and mammary epithelial cell complete medium, or other suitable media (*see* **Note 1**).
2. Trypsin–EDTA solution (commercially available).
3. Phosphate-buffered saline (PBS): 137 mM NaCl, 2.7 mM KCl, 10 mM $Na_2HPO_4$, 2 mM $KH_2PO_4$, pH 7.4. Dissolve 8 g of NaCl, 0.2 g of KCl, 1.44 g of $Na_2HPO_4$, and 0.24 g of $KH_2PO_4$ in 800 ml of distilled $H_2O$. Adjust the pH to 7.4 with HCl. Add $H_2O$ to 1 l. Sterilize by autoclaving on liquid cycle. Store at room temperature.
4. Cell lysis buffer with protease inhibitors: 50 mM Tris–HCl, pH 7.4, 1 % Igepal CA-630, 150 mM NaCl, 50 mM EDTA, 10 mM sodium pyrophosphate, 1 mM sodium orthovanadate, 10 mM sodium fluoride, 10 μg/ml leupeptin, 4 μg/ml pepstatin, and 10 μg/ml aprotinin (*see* **Note 2**).
5. Glutathione-coupled Sepharose 4B beads (commercially available).
6. Ethylene glycol tetraacetic acid (EGTA). The working solution is 4 mM EGTA in PBS.

### 2.2 Rap1 Activity Assay

#### 2.2.1 Production of RalGDS-RBD/GST Glutathione-Coupled Sepharose

1. Competent TOPF10 *Escherichia coli* cells aliquots ready to be transformed.
2. Expression vector pGEX-RalGDS-RBD DNA.
3. LB broth: stir to suspend 20 g of LB powder (*see* **Note 3**) in 1 l of water. Autoclave for 15 min at 121 °C to sterilize. Store at room temperature or at 4 °C after the bottle has been opened.
4. LB/ampicillin plates: add 40 g of LB–Agar powder (*see* **Note 3**) in 1 l of water. Heat to boiling while stirring to dissolve powder. Autoclave for 15 min at 121 °C to sterilize. Allow to cool slightly (60 °C). Add ampicillin stock solution to 100 μg/ml.

Pour into Petri dishes (20 ml in 60 $cm^2$ dishes). Allow to solidify and store upside down at 4 °C.

5. Ampicillin, stock solution: 50 mg/ml in $H_2O$. Sterilize by filtration through a 0.22-μm filter.
6. 1 M Isopropyl-1-thio-β-D-galactopyranoside (IPTG) stock solution: dissolve 2.38 g of IPTG in 8 ml of distilled $H_2O$. Adjust the volume of the solution to 10 ml with distilled $H_2O$. Sterilize by filtration through a 0.22-μm disposable filter. Aliquot and store at −20 °C.
7. 2× Sample buffer: 125 mM Tris–HCl, pH 6.8, 20 % glycerol, 4 % SDS, 10 % β-mercaptoethanol, and 0.02 % bromophenol blue (*see* **Note 4**).
8. Bacterial lysis buffer with protease inhibitors: 50 mM Tris–HCl, pH 7.4, 0.1 % Triton X-100, 100 mM NaCl, 5 mM $MgCl_2$, 10 % glycerol, 1 mM EDTA, 1 mM DTT, 10 μg/ml leupeptin, 4 μg/ml pepstatin, 0.1 TIU/ml aprotinin, and 1 mM PMSF.
9. Acrylamide/bis mini gel solutions (*see* **Note 5**):
   - Running gel 10 %: 2 ml of 40 % acrylamide, 3.86 ml of $H_2O$, 2 ml of 1.5 M Tris–HCl, pH 8, 80 μl of 10 % SDS, 10 μl of TEMED, and 50 μl of APS (ammonium persulfate).
   - Stacking gel: 0.260 ml of 40 % acrylamide, 1.72 ml of $H_2O$, 0.480 ml of 0.625 M Tris–HCl, pH 6.8, 25 μl of 10 % SDS, 5 μl of TEMED, and 50 μl of APS (ammonium persulfate).
10. Coomassie blue staining solution: dissolve 0.25 g of Coomassie Brilliant Blue (R250) in 90 ml of methanol: $H_2O$ (1:1 v/v) and 10 ml of glacial acetic acid.
11. SDS-PAGE apparatus and solutions.

#### 2.2.2 Rap1 Pull-Down Assay

1. Running buffer: 25 mM Tris base, 192 mM glycine, and 0.1 % SDS (*see* **Note 6**).
2. Dunn buffer: 10 mM sodium bicarbonate, 3 mM sodium carbonate, 20 % methanol, pH 9.9 (*see* **Note 7**).
3. Ponceau S: 0.2 % Ponceau S and 6 % trichloroacetic acid.
4. Tris-buffered saline (TBS): 0.15 M NaCl, 10 mM Tris–HCl, pH 7.4 (*see* **Note 8**).
5. Blocking solution: 5 % w/v bovine serum albumin (BSA) or 5 % w/v nonfat dried milk in TBS.
6. TTBS: 0.3 % Tween 20 in TBS.
7. Substrate for enhanced chemiluminescence (ECL) reaction.
8. Autoradiographic films.

9. Processing chemicals for autoradiography films: developer and fixer solutions.
10. Anti-Rap1 primary antibody at the appropriate dilution in 1 % w/v BSA in TBS.
11. Horseradish peroxidase conjugated secondary antibody. Choose the secondary antibody appropriate to the primary antibody.

### 2.3 Integrin-Dependent Cell–Matrix Contacts Evaluation

#### 2.3.1 Fixed Cells

1. Aseptic glass 1.5 coverslip.
2. 10 μg/ml and 25 μg/ml fibronectin.
3. pEGFP-zyxin vector and suitable transfection reagent.
4. CaCl powder. Use at final concentration of 1.8 mM.
5. 4 % Paraformaldehyde (PFA): dissolve the appropriate amount of paraformaldehyde in PBS in a glass beaker with a stir bar. Heat in a hot plate in the hood at 60 °C. Cool down to 20 °C and adjust pH to 7.4. Store at 4 °C or at −20 °C for longer time (*see* **Note 9**).
6. Mounting reagent. Some mounting reagents are supplied with DAPI and are ready to use.
7. 4′,6-Diamidino-2-phenylindole (DAPI) or Hoechst for nuclear staining (if not supplied with mounting agent).
8. Acid solution: make up 300 ml of two parts of nitric acid to one part of fuming hydrochloric acid in a glass beaker in the hood.

## 3 Methods

### 3.1 Disruption of E-Cadherin Cell–Cell Junctions

In this section, we detail some protocols to disrupt AJs between cells. You can reduce cellular AJs splitting your cells at lower density; alternatively you can disrupt AJs using chemical agents. We provide protocol 3.1.1 to obtain total protein extract from adherent cells and protocol 3.1.2 to collect total protein extract from suspended cells. Moreover, we provide protocol 3.1.3 to evaluate AJs at different time points in adhesion experiments.

#### 3.1.1 Cell Density Experiments: Adherent Cells (Ad)

1. Plate your cell lines. All cell lines indicated in the Materials section must be cultured at 37 °C and 5 % $CO_2$ atmosphere in Dulbecco's modified Eagle's complete medium, except for FRT and MCF10A cells, which require Coon's F-12 complete medium [35] and mammary epithelial cell complete medium, respectively.
2. Harvest cell monolayers by trypsinization and replate them at various densities in serum-free medium; allow cells to recover overnight at 37 °C before measurement of Rap1 activity (*see* **Note 10**).

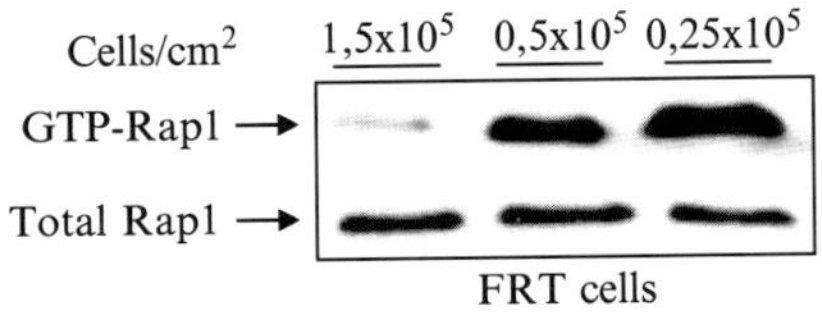


**Fig. 1** The inhibition of Rap1 activity during cell adhesion is significantly influenced by cell density. Cells were harvested by trypsinization, extensively pipetted to generate a maximal dispersed cell suspension, replated at various densities ($0.25 \times 10^5$ to $1.5 \times 10^5$ cells/cm$^2$) in serum-free medium, and allowed to recover overnight at 37 °C before measurement of Rap1 activity. Notice that the down-regulation of Rap1 activity during cell adhesion was clearly dependent on cell density. From ref. [30], with permission

3. Wash twice the cells with ice-cold PBS.
4. Incubate cells with ice-cold cell lysis buffer for 15 min on ice (you can use 0.5 ml of lysis buffer for a 60 cm$^2$ cell dish) (*see* **Note 11**).
5. Harvest the lysate by scraping the plate. Work on ice.
6. Centrifuge lysates at 15,700 × *g* for 10 min at 4 °C.
7. Collect the supernatants and determine total protein concentration. Work on ice.
8. Make preclearing of supernatants containing equal amounts of total proteins (0.5–1 mg) by incubation with glutathione-coupled Sepharose 4B beads for 1–2 h at 4 °C with gentle rotation (*see* **Note 12**).
9. Remove glutathione-coupled Sepharose 4B beads by centrifugation at 15,700 × *g* for 10 min at 4 °C. Transfer the supernatants in fresh microfuge tubes (*see* Fig. 1).

#### *3.1.2 Calcium Switch and Tyrosine Phosphatase Inhibition Procedures: Suspension Cells (Su)*

Use the following procedure if you want to compare adherent and suspension cell (*see* Fig. 2).

1. Plate the cells in complete medium and let them grow.
2. Wash cells twice with PBS.
3. Incubate cell monolayers in serum-free medium supplemented with 4 mM EGTA for 30 min at 37 °C (*see* **Note 13**).
4. Pipette and harvest cells. These are suspension cells (Su).
5. Wash cells once with PBS: resuspend in 5–10 ml of PBS and centrifuge at 400 × *g* for 4 min at room temperature, discharge supernatant.
6. Resuspend the cells (pellet) in ice-cold lysis buffer. If you start from a confluent 60 cm$^2$ dish, you can resuspend the pellet in 0.5 ml of lysis buffer.

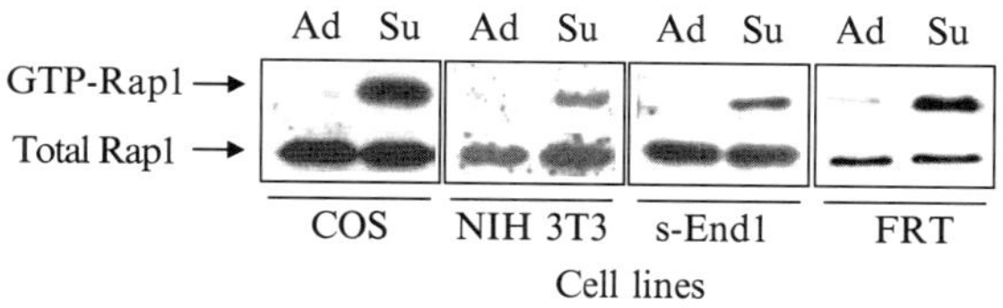


**Fig. 2** Rap1 activity is different in adherent and suspension cells. Confluent cells kept overnight in serum-free medium were either lysed directly onto tissue culture dishes (*Ad*, adherent cells) or harvested by treatment with 5 mM EDTA in PBS and resuspended in serum-free medium before lysis (*Su*, suspension cells). Active, GTP-bound Rap1 (GTP-Rap1) was detected by a pull-down assay using a GST-tagged RalGDS-RBD protein prebound to glutathione–Sepharose beads (*see* Subheadings 2 and 3), whereas Western blotting of whole lysates ensured that relative equal amounts of total Rap1 proteins (Total Rap1) were analyzed. Notice that cell detachment from the substrate promoted a strong activation of Rap1 in all cell line tested. From ref. [30], with permission

7. Incubate the cells with lysis buffer on ice for 15 min. Shake the incubation mix sometimes or put samples at 4 °C with gentle rotation.
8. Centrifuge lysates in a microfuge at 15,700 ×*g* for 10 min at 4 °C. Collect the supernatants and determine total protein concentration. Work on ice.
9. Make preclearing of supernatants containing equal amounts of total proteins (0.5–1 mg) by incubating supernatants with glutathione-coupled Sepharose 4B beads for 1–2 h at 4 °C with gentle rotation.
10. Remove glutathione-coupled Sepharose 4B beads by centrifugation at 15,700 ×*g* for 10 min at 4 °C. Transfer the supernatants in fresh microfuge tubes.

*3.1.3 Cell-Substrate Adhesion Studies*

1. Plate the cells in complete medium and let them grow.
2. Wash cells twice with PBS.
3. Incubate cell monolayers in serum-free medium supplemented with 4 mM EGTA for 30 min at 37 °C.
4. Pipette and harvest cells.
5. Resuspend cells in serum-free medium and allow them to adhere for the indicated time to tissue culture dishes coated with fibronectin.
6. Wash twice the cells with ice-cold PBS.
7. Incubate cells with ice-cold cell lysis buffer for 15 min on ice (use 0.5 ml of lysis buffer in a 60 $cm^2$ dish).
8. Harvest the lysate by scraping the plate. Work on ice.
9. Centrifuge lysates in a microfuge at 15,700 ×*g* for 10 min at 4 °C.

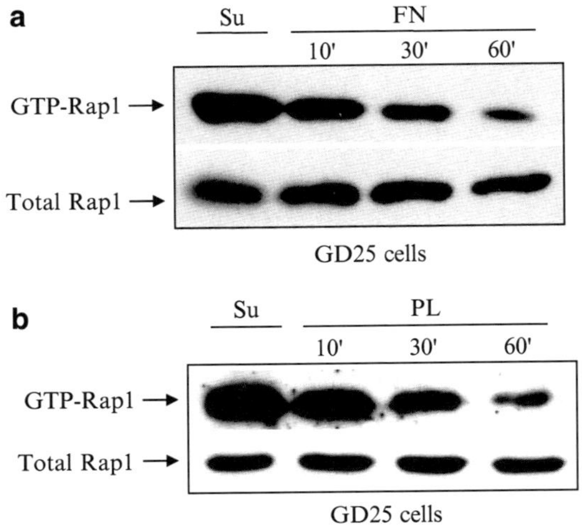


**Fig. 3** Rap1 activity is gradually downregulated during cell adhesion. Confluent cells kept overnight in serum-free medium were harvested by treatment with 5 mM EDTA in PBS and resuspended in serum-free medium. (**a** and **b**) Resuspended cells were either lysed (*Su*, suspension cells) or replated at high density and allowed to adhere for the indicated time frames to tissue culture dishes coated with 10 μg/ml substrate proteins, including fibronectin (*FN*) (**a**), poly-L-Lysine (*PL*) (**b**), before measurement of active Rap1. Notice that cell adhesion induced a progressive downregulation of Rap1 activity, which was independent of the adhesive substrate used and reached the maximal level at about 1 h. From ref. [30], with permission

10. Collect the supernatants and determine total protein concentration. Work on ice.
11. Make preclearing of supernatants containing equal amounts of total proteins (0.5–1 mg) by incubating with glutathione-coupled Sepharose 4B beads for 1–2 h at 4 °C with gentle rotation.
12. Remove glutathione-coupled Sepharose 4B beads by centrifugation at 15,700 ×*g* for 10 min at 4 °C. Transfer the supernatants in fresh microfuge tubes (*see* Fig. 3a, b).

### 3.2 Rap1 Activity Assay

The activation of Rap1 is detected using an established pull-down method based on the specific binding of a GST fusion protein containing the Rap-binding domain of RalGDS (RalGDS-RBD/GST) to the active, GTP-bound form of Rap1 [36]. Cell lysates used in this assay are produced as described in Subheading 3.1.

#### 3.2.1 Production of RalGDS-RBD/GST Glutathione-Coupled Sepharose

Perform the experiment close to a Bunsen burner:

1. Take an aliquot of competent TOPF10 *E. coli* from −80 °C freezer; maintain the Eppendorf on ice.
2. Thaw the bacteria by gently shaking the Eppendorf with your fingers.

3. Add 10 μl (10–100 ng of DNA) of the expression vector pGEX-RalGDS-RBD DNA to the bacteria.
4. Incubate 20 min on ice.
5. Incubate 2 min at 42 °C or 5 min at 37 °C.
6. Add 1 ml of pre-warmed LB broth (without antibiotic).
7. Incubate 30 min at 37 °C.
8. Spin bacteria and discard supernatant. Suspend bacteria in 50–200 μl of remaining supernatant.
9. Spread suspended transformed bacteria on a pre-warmed LB/ampicillin plate.
10. Recover plates at 37 °C overnight. Put the plates upside down (*see* **Note 14**).
11. Optional. Make mini prep of plasmidic DNA to control the presence of vector pGEX-RalGDS-RBD.
12. Optional. Pick one bacterial colony and inoculate it in a tube containing 5 ml of LB with antibiotic.
13. Optional. Grow bacteria at 37 °C with gentle rotation or shaking for some hours (2–6 h).
14. Use the 5 ml of culture at **steps 12** and **13** or pick one bacterial colony to inoculate 50 ml LB/ampicillin. Grow with shaking at 37 °C.
15. The day after you can save 600 μl of bacterial culture added with 400 μl of sterile glycerol. Store it at −80 °C.
16. Inoculate the overnight bacterial culture in 500 ml of LB/ampicillin.
17. Grow with shaking at 37 °C until an $OD_{600}$ of approximately 0.6 is reached (2–3 h).
18. When $OD_{600}$ = 0.5–0.6, save on ice 1 ml of bacterial culture broth (***not induced sample***) and induce the RalGDS-RBD expression by adding isopropyl-1-thio-β-D-galactopyranoside (IPTG) to 0.1 mM final concentration.
19. Continue to grow the culture for an additional 3 h. Save 1 ml of bacterial culture broth on ice (***induced sample***).
20. Harvest the bacteria by centrifugation at 800–1,000 ×*g* for 35 min at 4 °C.
21. Spin induced and not induced samples (from **steps 18** and **19**) and resuspend the pellet in 100 μl of sample buffer 2×. You can store them at −20 °C.
22. Remove medium obtained at **step 20** and then resuspend the pellet in bacterial lysis buffer containing protease inhibitors (15 ml). Lyse the cells on ice by mild sonication. You can freeze the samples at −20 °C for some hours or overnight before sonication, the freezing and thawing improve bacterial

lysis. If desired, at this stage the pelleted bacteria can be stored indefinitely at −80 °C.

23. Centrifuge the sample at 800–1,000 × *g* for 35 min at 4 °C. Transfer the supernatant to a fresh tube and store the pellet at −20 °C (*see* **Note 15**).
24. Incubate RalGDS-RBD/GST fusion protein-containing supernatant with 0.1 ml of glutathione-coupled Sepharose 4B beads for 2 h at 4 °C with gentle rotation (*see* **Note 12**).
25. Centrifuge 10 min at 15,700 × *g* at 4 °C; wash pellet three times in lysis buffer. (Each time resuspend in large amount, 1 ml, of lysis buffer and centrifuge for 10 min at 15,700 × *g* at 4 °C.)
26. Resuspend the RalGDS-RBD/GST glutathione-coupled Sepharose beads in 0.5 ml of lysis buffer containing protease inhibitors.
27. Estimate the amount of fusion proteins by SDS-PAGE and Coomassie blue staining. Analyze 20 μl of washed beads and 20 μl of induced and not induced samples (*see* **Notes 16** and **17**).

#### 3.2.2 Rap1 Pull-Down Assay

1. Rate the resuspended beads (50 μl for each sample, you have enough beads for ten samples) (Subheading 3.2.1, **step 26**) into the precleared samples of cellular lysates (*see* **Note 18**).
2. Incubate for 2 h at 4 °C with gentle rotation.
3. Harvest the complexes Rap1-RDB/GST-glutathione-coupled Sepharose beads by centrifugation at 15,700 × *g* for 5 min at 4 °C in a microfuge.
4. Wash two to three times with lysis buffer (you can reduce protease inhibitor concentration).
5. Dry the pellet with a Hamilton syringe.
6. Add 25 μl of 2× sample buffer.
7. Denature the proteins by heat treatment at 100 °C for 3–5 min.
8. Centrifuge the samples at 15,700 × *g* for 20 s at room temperature in a microfuge.
9. Load samples into an 8–10 % SDS-polyacrylamide gel electrophoresis. You must compare Rap1 pull-down samples with total extracts from the same samples (50 μg of total extraction stored; *see* **Notes 18** and **19**).
10. Transfer proteins from SDS-polyacrylamide gel to a solid support (nitrocellulose membrane or other membranes) (*see* **Note 20**).
11. Stain the nitrocellulose membrane with Ponceau S (*see* **Note 21**).
12. Block binding sites for immunoglobulins on the nitrocellulose membrane with 5 % w/v BSA or 5 % w/v nonfat dried milk for 1 h at room temperature with agitation.

13. Rinse the membrane with TBS.
14. Bind the primary anti-Rap1 antibody to the target protein. Incubate the blot with a diluted anti-Rap1 primary antibody solution (1 % w/v BSA in TBS) overnight at 4 °C with agitation. The day after, remove solution.
15. Rinse with TTBS at room temperature. Rinse once rapidly, then wash three to four times for 10 min with agitation.
16. Incubate the nitrocellulose filter with the secondary antibody in TTBS at recommended dilution (usually 1:5,000) for 1 h at room temperature with agitation. Remove solution.
17. Rinse as in **step 15**.
18. Perform detection protocol. Rinse with TBS and apply ECL solution to completely cover the surface of the membrane for the recommended time (usually 1 min). Some ECL reagents must be used in the dark! Drain off reagent and wrap the membrane in plastic wrap. Expose the membrane to film as required (from 1 s to 30 min; you have to try!) and develop film (*see* **Note 22**).
19. Normalize total protein extracts (start from **step 13**) with a suitable antibody (tubulin, vinculin, glyceraldehyde-3-phosphate dehydrogenase).

### 3.3 Integrin-Dependent Cell–Matrix Contacts Evaluation

The activation of Rap1—triggered by E-cadherin endocytosis—is followed by the acquisition of enhanced integrin-dependent cell–matrix contacts [30, 34]. Here we describe a method to evaluate cellular focal adhesions formation following the disruption/reduction of E-cadherin cell–cell junctions and consequent increment in GTP-loaded Rap1. We suggest using EGFP-zyxin as marker for the detection of focal adhesion assembly [30, 37–39].

#### 3.3.1 Fixed Cells

1. Clean coverslips (*see* **Note 23**):
   - Place coverslips (1.5 coverslips are of appropriate thickness) a few at a time in the acid solution. Allow them to sit for 2 h with occasional swirling.
   - Discard the acid into adequate waste.
   - Wash the coverslips thoroughly in running tap water until the pH of the washing is back to 5.5–6.0.
2. Sterilize coverslips in a hot air oven (200 °C for 3 h), or by autoclaving them, flaming them under laminar flow hood, or other suitable methods. You can move sterile glass coverslips with a sterile Pasteur pipette linked to a vacuum pump.
3. Put aseptic glass coverslips into cellular dishes (*see* **Note 24**).

4. Coat glass coverslips:
   - Cover the floor of cell dishes with 10 μg/ml fibronectin.
   - Put the dishes at 4 °C overnight or at 37 °C for 2 h.
   - Remove excess fibronectin solution by aspiration.
5. Plate the cells (you may use FRT cells).
6. Transfect EGFP-zyxin into cells with the appropriate method (*see* **Note 25**).
7. Wait 12–48 h until cells express EGFP-zyxin. Verify EGFP-zyxin expression by fluorescence microscopy.
8. In one well, disrupt cell–cell junctions by EGTA treatment. Wash cells twice with PBS and then incubate cell monolayer in serum-free medium supplemented with 4 mM EGTA for 30 min at 37 °C.
9. In another well, allow adherens junctions to reform after EGTA treatment by further incubation with fresh medium containing 1.8 mM $Ca^{2+}$ for 30–60 min at 37 °C. In order to get cells from **steps 8** and **9** ready at the same time, we suggest to start by treating cells at **step 9** with EGTA for 30 min, then add $Ca^{2+}$ for 30 min and immediately start to treat with EGTA for 30 min cells at **step 8**.
10. Remember that you need the untreated control. You may use the third well of the 6-well plate for the untreated control. Change the medium and add fresh serum-free medium.
11. Wash cells twice with PBS.
12. Fix for 10 min with 4 % PFA (*see* **Note 26**).
13. Rinse the cells briefly for three times with PBS.
14. Rinse coverslips briefly in water to remove salts.
15. Dry by carefully touching the edges of the coverslips with a piece of paper.
16. Mount coverslips on a microscope slide with cells down in accordance with your mounting agent manufacturer's instructions (*see* **Note 27**). You can mount two coverslips on a microscopic slide, and you should mount them very closely.
17. Allow to dry overnight (at room temperature and in the dark) to obtain a hard mount.
18. Make your analysis under a fluorescent microscope equipped with a camera.

#### *3.3.2 Time-Lapse Video Microscopy*

Alternatively, you may perform a time-lapse video microscopy analysis. You need a suitable equipment with a chamber maintaining constant temperature and $CO_2$ atmosphere.

1. Plate cells (FRT cells) into sterile, optical quality, glass bottom dishes, coated by overnight treatment with 25 μg/ml fibronectin, at 4 °C or 2 h incubation at 37 °C.

2. Transfect EGFP-zyxin into cells with the appropriate method.
3. Wait 12–48 h until cells express EGFP-zyxin.
4. Wash cell twice with PBS and then incubate cell monolayers in serum-free medium.
5. Start time-lapse video microscopy analysis.
6. To induce AJ disassembly, treat cells with 4 mM EGTA for 30 min at 37 °C.
7. Wash cells with PBS.
8. To restore AJs, incubate cells with fresh medium containing 1.8 mM $Ca^{2+}$ for 30–60 min at 37 °C.
9. Pause/resume time-lapse video microscopy analysis whenever required (*see* Fig. 4).

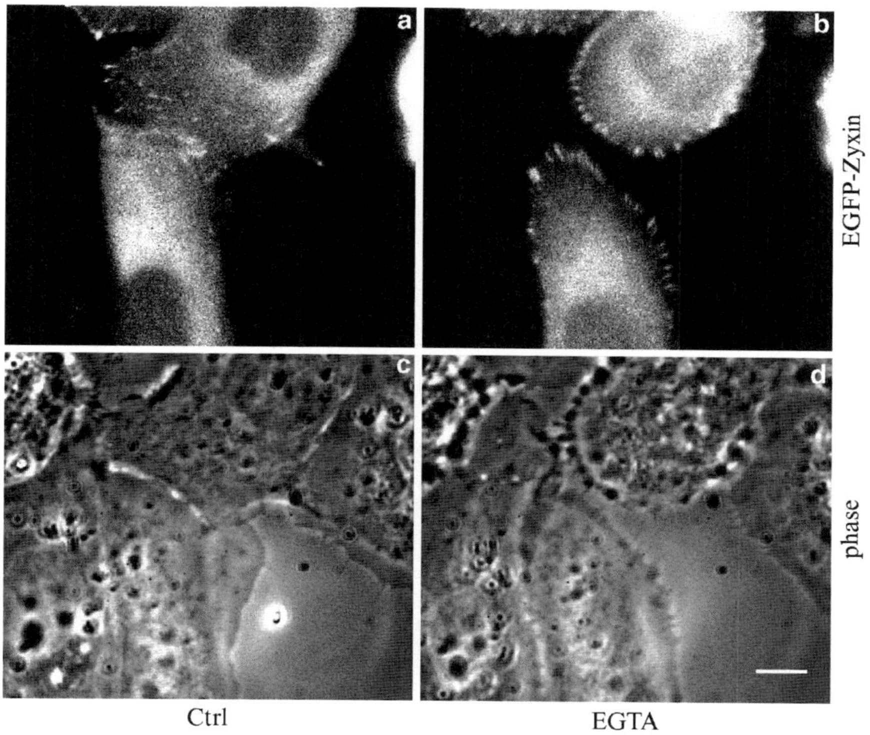


**Fig. 4** AJs disassembly induces a Rap1 activity-dependent assembly of focal adhesions. FRT cells transiently transfected with EGFP-zyxin, as a marker for focal adhesions, were plated on FN-coated coverslips and grown to confluence for 12 h. Living cells were then analyzed by time-lapse microscopy during disruption of cell–cell junctions by EGTA treatment in serum-free medium. The two selected frames of the video sequence show the subcellular distribution of EGFP-zyxin before cell treatments (*Ctrl*) and after 30 min of EGTA treatment (*EGTA*). Notice the assembly of EGFP-zyxin-containing focal adhesions upon EGTA treatment. From ref. [30], with permission

## 4 Notes

1. Complete medium is a medium supplemented with 10 % fetal calf serum (FCS) or fetal bovine serum (FBS), 2 mM glutamine, and 100 U/ml penicillin/streptomycin. Some media containing a stabilized form of L-glutamine, L-alanyl-L-glutamine, do not need glutamine supplementation.
2. You may substitute leupeptin, pepstatin, aprotinin, and PMSF with a premix of proteases inhibitors diluted according to the manufacturer's instruction.
3. LB (Luria-Bertani) Broth powder (commercially available): 10 g/L Tryptone, 5 g/L yeast extract, and 5 g/L NaCl.

   LB–Agar powder (commercially available): 15 g/L agar, 10 g/L Tryptone, 5 g/L yeast extract, and 5 g/L NaCl.
4. If your samples have a very low protein concentration, you can use 5× Laemmli sample buffer: 312 mM Tris–HCl, pH 6.8, 50 % glycerol, 10 % SDS, 25 % β-mercaptoethanol, and 0.05 % bromophenol blue. Bromophenol blue can be added by dipping a tip into the powder and then into the Laemmli solution. Take care to have a dark blue solution. Store at −20 °C.
5. Make the solutions in tubes under chemical hood.

   TEMED or APS must be added just prior to transfer solutions into gel cassette.
6. Stock 10× running buffer: 30 g of Tris base, 144 g of glycine, and 10 g of SDS. Store at 4 °C.
7. Stock 5× Dunn buffer: 4.2 g of sodium bicarbonate, 1.59 g of sodium carbonate in 1 l of deionized water. Make 1× working solution by mixing 200 ml of 5× Dunn buffer, 200 ml of methanol, and 600 ml of deionized water. Store at room temperature.
8. Stock 10× TBS (Tris-buffered saline) solution: 300 ml of 5 M NaCl, 200 ml of Tris–HCl pH 7.4, 500 ml of $H_2O$.
9. When you use PFA, you must work under chemical hood.
10. Split cells as follows:
    (a) Aspirate medium.
    (b) Wash twice (or more) with pre-warmed PBS.
    (c) Add 1× trypsin–EDTA working solution (0.5 ml for a 60 $cm^2$ Petri culture dish).
    (d) Rock dish four to five times to let trypsin coat cells.
    (e) Place dish in $CO_2$ incubator at 37 °C for 2–5 min.
    (f) Check for cell detachment. If no cells are dislodged, put the dish back at 37 °C for a few minutes.

(g) Resuspend cells in the medium at the appropriate density.

You should plate your cells calculating that at least one plate should be very confluent the day you will perform the lysis.

11. Prepare a sufficient volume of cell lysis buffer for all the procedures you will perform. Take into account that you need lysis buffer:

(a) To resuspend glutathione-coupled Sepharose 4B beads.

(b) To bring to volume cellular extracts to be loaded on SDS-PAGE gel.

(c) To perform washes during Rap1 pull-down assay (in this case you can reduce the amount of proteases inhibitors).

You should use fresh lysis buffer when you lyse cells. Maintain lysis buffer on ice.

12. Wash glutathione-coupled Sepharose 4B beads three times with ice-cold PBS and resuspend in lysis buffer (100 μl for samples). Use 25–30 μl of Sepharose beads for sample.

Do not freeze nor heat more than 40 °C the glutathione-coupled Sepharose 4B beads.

13. As an alternative method to EGTA treatment, in some experiments cadherin-mediated intercellular adhesions were disrupted by treating confluent epithelial cells for different time frames with serum-free medium containing 5 μM phenylarsine oxide (PAO), a selective tyrosine phosphatase inhibitor [40]. Alternatively, you can use 5 mM EDTA in PBS, or DECMA-1, a neutralizing antibody against E-cadherin [30].

14. If you observe very few colonies, reduce the ampicillin concentration (both in LB broth and in LB-agar plates) till 20 μg/ml.

15. If you need extra RalGDS-RBD/GST fusion protein, you can restart by sonicating bacteria.

16. To perform SDS-PAGE analysis, assemble the slab gel cassette (glass plates and spacers) in the casting apparatus. Choose 0.75 mm spacers for gel to be stained with Coomassie blue (Subheading 3.2.1, **step 27**, *Production of RalGDS-RBD/GST glutathione-coupled Sepharose*) and 1.5 spacer for Western blotting procedure (Subheading 3.2.2, **step 9**, *Rap1 pull-down assay*).

(a) Transfer running gel from the tube to the gel cassette; overlay the gel gently with water.

(b) Let the gel polymerize for 30–60 min (under chemical hood).

(c) Remove water from the surface of running gel.

(d) Pipette the stacking gel into the gel cassette.

(e) Immediately insert the proper comb.

(f) Let it polymerize.

(g) Remove the comb and rinse the wells with running buffer solution (we suggest to use the running buffer described in Subheading 2).

(h) Insert the gel cassette into the electrophoresis apparatus and add running buffer solution.

(i) Load the protein samples into the wells using a Hamilton syringe. Samples must be added with 2× loading sample buffer, boiled for 3–5 min and then centrifuged. Alternatively, you can use a pipettor to load samples. When you use samples with beads, take care to escape the beads. Hamilton syringe does not aspire beads.

(j) Attach the electrode wires. Apply 100 V (constant voltage) until the bromophenol blue dye is out from the gel. You may run the gel at higher voltage (110 V), but you should put the case into ice to avoid overheating.

17. Coomassie Brilliant Blue staining:

(a) Soak the gel in at least 5 volumes of Coomassie Brilliant Blue staining solution and place it on a slowly rotating platform for a minimum of 4 h at room temperature. Alternatively you can warm it up in a microwave oven for a few seconds. Do not exceed to avoid evaporation of your solution.

(b) Remove the staining solution and save it for future use. Wash the gel with water. Destain the gel by soaking it in methanol/acetic acid solution on a slowly rocking platform for 4–8 h, change the destaining solution three or four times. Alternatively you can soak the gel in a large amount of water and boil it in a microwave oven for some minutes.

18. Remember to store at least 50 μg of total cell protein extracts from Subheading 3.1.1, **step 9**; Subheading 3.1.2, **step 10**; or Subheading 3.1.3, **step 12** before adding the RalGDS-RBD/GST glutathione-coupled Sepharose beads.

19. Remember loading the molecular weight marker.

20. To transfer proteins from gel to nitrocellulose membrane, we generally use a semi-dry transfer apparatus. Move the nitrocellulose membrane with tweezers (do not touch it with your hands and gloves).

(a) Use a spatula to take out the running gel from the plates (remove the stacking gel).

(b) Soak the gel for 5 min in Dunn buffer.

(c) Soak the nitrocellulose membrane in Dunn buffer for 30 min (you sholud perform this step while the gel is still running).

(d) Place six pieces of Whatman 3 MM filter paper soaked in Dunn buffer on the electric plate, followed by the nitrocellulose membrane, the gel, and six pieces of filter paper soaked in Dunn buffer. At each step, ensure that no air bubbles are present between layers, using a pipette to roll out any bubbles. Place the other electrode plate on top of the sandwich according to the manufacturer's instructions.

(e) Transfer for 30–60 min at 15 V (constant voltage).

21. Staining of the nitrocellulose membrane with Ponceau S:

    (a) Incubate the membrane 1 min in Ponceau S solution with gentle agitation.

    (b) Discard Ponceau S (you can reuse it) and wash the filter in deionized water at room temperature. Do not pour water directly on the membrane.

    (c) You can save an image of your Ponceau S.

    (d) Sign the molecular weight bands on the membrane with a pencil.

    (e) Wash with 0.3 % Tween in TBS (TTBS).

22. The development of the autoradiographic film may be performed either with a developing machine or manually. To do it manually, soak the film in developing solution for 1 min, rinse it under water, soak it in fixing solution for 1 min, and rinse under water. After developing the film, remember to report on it the markers signed on the membrane.

23. Alternatively:

    (a) Incubate coverslips for 2 h in 2N NaOH.

    (b) Rinse extensively in deionized water.

24. We suggest to use 6-well plates and 3 coverslips/well.

25. The transfection method to be used depends on the cellular line you use. For example, FRT should be transfected by a nonliposomal lipid reagent.

26. Fixing with 4 % PFA may be done at room temperature for 10 min. However, it is better to fix your samples 10 min on ice. PFA should cover the cells. Work under chemical hood.

27. Some mounting reagents are ready to use and are supplied with DAPI staining; in this case:

    (a) Place a drop (20 μl) of mounting medium on a clean microscope slide.

    (b) Gently put the coverslip onto the mounting medium, cell-side down, so that no air bubbles are trapped (help yourself with tweezers and a syringe needle).

(c) Blot away the excess mounting medium with filter paper.

If your mounting reagent is not supplied with DAPI or Hoechst, you have to perform the nuclear staining in accordance with the manufacturer's instruction. Generally you have to dilute DAPI or Hoechst in PBS, stain the cells with this solution, and then, after the appropriate time, wash the cell with PBS.

## References


1. Hynes RO (2002) Integrins: bidirectional, allosteric signaling machines. Cell 110: 673–687
2. Wheelock MJ, Johnson KR (2003) Cadherins as modulators of cellular phenotype. Annu Rev Cell Dev Biol 19:207–235
3. Peifer M, Yap AS (2003) Traffic control: P120-catenin acts as a gatekeeper to control the fate of classical cadherins in mammalian cells. J Cell Biol 163:437–440
4. Frame MC (2002) Src in cancer: deregulation and consequences for cell behaviour. Biochim Biophys Acta 1602:114–130
5. Thiery JP (2002) Epithelial-mesenchymal transitions in tumour progression. Nat Rev Cancer 2:442–454
6. Avizienyte E, Wyke AW, Jones RJ et al (2002) Src-induced de-regulation of e-cadherin in colon cancer cells requires integrin signalling. Nat Cell Biol 4:632–638
7. Gimond C, van Der Flier A, van Delft S et al (1999) Induction of cell scattering by expression of beta1 integrins in beta1-deficient epithelial cells requires activation of members of the rho family of gtpases and downregulation of cadherin and catenin function. J Cell Biol 147:1325–1340
8. Hintermann E, Yang N, O'Sullivan D et al (2005) Integrin alpha6beta4–erbb2 complex inhibits haptotaxis by up-regulating e-cadherin cell–cell junctions in keratinocytes. J Biol Chem 280:8004–8015
9. Hodivala KJ, Watt FM (2004) Evidence that cadherins play a role in the downregulation of integrin expression that occurs during keratinocyte terminal differentiation. J Cell Biol 124: 589–600
10. Huttenlocher A, Lakonishok M, Kinder M et al (1998) Integrin and cadherin synergy regulates contact inhibition of migration and motile activity. J Cell Biol 141:515–526
11. Lu Q, Paredes M, Zhang J et al (1998) Basal extracellular signal-regulated kinase activity modulates cell–cell and cell–matrix interactions. Mol Cell Biol 18:3257–3265
12. Monier-Gavelle F, Duband JL (1997) Cross talk between adhesion molecules: control of n-cadherin activity by intracellular signals elicited by beta1 and beta3 integrins in migrating neural crest cells. J Cell Biol 137: 1663–1681
13. Nelson CM, Pirone DM, Tan JL et al (2004) Vascular endothelial-cadherin regulates cytoskeletal tension, cell spreading, and focal adhesions by stimulating rhoa. Mol Biol Cell 15: 2943–2953
14. Retta SF, Cassara G, D'Amato M et al (2001) Cross talk between beta(1) and alpha(v) integrins: beta(1) affects beta(3) mrna stability. Mol Biol Cell 12:3126–3138
15. Schreider C, Peignon G, Thenet S et al (2002) Integrin-mediated functional polarization of caco-2 cells through e-cadherin—actin complexes. J Cell Sci 115:543–552
16. von Schlippe M, Marshall JF, Perry P et al (2000) Functional interaction between e-cadherin and alphav-containing integrins in carcinoma cells. J Cell Sci 113:425–437
17. Weber GF, Bjerke MA, DeSimone DW (2011) Integrins and cadherins join forces to form adhesive networks. J Cell Sci 124:1183–1193
18. Bos JL, de Rooij J, Reedquist KA (2001) Rap1 signalling: adhering to new models. Nat Rev Mol Cell Biol 2:369–377
19. Ohba Y, Kurokawa K, Matsuda M (2003) Mechanism of the spatio-temporal regulation of ras and rap1. EMBO J 22:859–869
20. Knox AL, Brown NH (2002) Rap1 gtpase regulation of adherens junction positioning and cell adhesion. Science 295:1285–1288
21. Bivona TG, Wiener HH, Ahearn IM et al (2004) Rap1 up-regulation and activation on plasma membrane regulates t cell adhesion. J Cell Biol 164:461–470
22. Caron E (2003) Cellular functions of the rap1 gtp-binding protein: a pattern emerges. J Cell Sci 116:435–440
23. Dustin ML, Bivona TG, Philips MR (2004) Membranes as messengers in t cell adhesion signaling. Nat Immunol 5:363–372

24. Katagiri K, Maeda A, Shimonaka M et al (2003) Rapl, a rap1-binding molecule that mediates rap1-induced adhesion through spatial regulation of lfa-1. Nat Immunol 4:741–748
25. Hogan C, Serpente N, Cogram P et al (2004) Rap1 regulates the formation of e-cadherin-based cell–cell contacts. Mol Cell Biol 24: 6690–6700
26. Price LS, Hajdo-Milasinovic A, Zhao J (2004) Rap1 regulates e-cadherin-mediated cell–cell adhesion. J Biol Chem 279:35127–35132
27. Yajnik V, Paulding C, Sordella R et al (2003) Dock4, a gtpase activator, is disrupted during tumorigenesis. Cell 112:673–684
28. Hattori M, Minato N (2003) Rap1 gtpase: functions, regulation, and malignancy. J Biochem 134:479–484
29. Ricono JM, Huang M, Barnes LA et al (2009) Specific cross-talk between epidermal growth factor receptor and integrin alphavbeta5 promotes carcinoma cell invasion and metastasis. Cancer Res 69:1383–1391
30. Balzac F, Avolio M, Degani S et al (2005) E-cadherin endocytosis regulates the activity of rap1: a traffic light gtpase at the crossroads between cadherin and integrin function. J Cell Sci 118:4765–4783
31. Chrzanowska-Wodnicka M (2010) Regulation of angiogenesis by a small gtpase rap1. Vascul Pharmacol 53:1–10
32. Ferro E, Goitre L, Retta SF et al (2012) The interplay between ros and ras gtpases: physiological and pathological implications. J Signal Transduct 2012:365769. doi:10.1155/2012/365769
33. Goitre L, Pergolizzi B, Ferro E et al (2012) Molecular crosstalk between integrins and cadherins: do reactive oxygen species set the talk? J Signal Transduct 2012:807682. doi:10.1155/2012/807682
34. Retta SF, Balzac F, Avolio M (2006) Rap1: a turnabout for the crosstalk between cadherins and integrins. Eur J Cell Biol 85:283–293
35. Ambesi-Impiombato FS, Parks LA, Coon HG (1980) Culture of hormone-dependent functional epithelial cells from rat thyroids. Proc Natl Acad Sci U S A 77:3455–3459
36. Franke B, van Triest M, de Bruijn KM et al (2000) Sequential regulation of the small gtpase rap1 in human platelets. Mol Cell Biol 20:779–785
37. Rottner K, Krause M, Gimona M et al (2001) Zyxin is not colocalized with vasodilator-stimulated phosphoprotein (vasp) at lamellipodial tips and exhibits different dynamics to vinculin, paxillin, and vasp in focal adhesions. Mol Biol Cell 12:3103–3113
38. Zaidel-Bar R, Ballestrem C, Kam Z et al (2003) Early molecular events in the assembly of matrix adhesions at the leading edge of migrating cells. J Cell Sci 116:4605–4613
39. Zaidel-Bar R, Cohen M, Addadi L et al (2004) Hierarchical assembly of cell-matrix adhesion complexes. Biochem Soc Trans 32: 416–420
40. Retta SF, Barry ST, Critchley DR et al (1996) Focal adhesion and stress fiber formation is regulated by tyrosine phosphatase activity. Exp Cell Res 229:307–317

# Chapter 13

# Fluorescence Microscopy Study of Rap1 Subcellular Localization

**Luca Goitre, Valentina Cutano, and Saverio Francesco Retta**

## Abstract

The Ras-related GTPase Rap has been implicated in multiple cellular functions. In particular, Rap1 is a crucial regulator of both inside-out integrin activation and outside-in E-cadherin-mediated signaling. Thus, Rap1 was proposed as a fundamental regulator of the cross talk between cadherins and integrins. We provide microscopic techniques to study subcellular localization of Rap1 protein in the crosstalk between integrins and cadherins.

**Key words** Rap1, Small GTPase, Integrins, Cadherins, Molecular crosstalk, Immunofluorescence

## 1 Introduction

Rap1 is a member of the Ras GTPase superfamily. It was described in 1989 by Kitayama [1] as Krev-1 with anti-oncogenic activity and subsequently identified by Bourne [2] as a Ras-related protein [3]. Like others small G-proteins, Rap1 cycles between a GTP-bound (active) and a GDP-bound (inactive) state, thereby acting as a molecular on/off switch. This switch mechanism is spatially and temporally regulated by a plethora of guanine nucleotide exchange factors (GEFs) and GTPase-activating proteins (GAPs), which are multi-domain proteins that undergo intricate cellular control [3–8]. A number of recent studies have firmly established that Rap1 regulates the inside-out activation of most integrins by mediating their polarized spatial redistribution and stabilization in an active conformation [5, 9–12], suggesting that Rap1 activation is required for the induction and maintenance of integrin-mediated cell adhesion. In particular, it was shown that Rap1 mediates integrin activation by forming a complex containing talin in combination with a Rap1 effector, Rap1-GTP-interacting adaptor molecule (RIAM), which targets talin to integrins [13, 14].

Lorenza Trabalzini and Saverio Francesco Retta (eds.), *Ras Signaling: Methods and Protocols*,
Methods in Molecular Biology, vol. 1120, DOI 10.1007/978-1-62703-791-4_13, © Springer Science+Business Media, LLC 2014

On the other hand, Rap1 is a target for E-cadherin-mediated outside-in signaling [8, 15]. Rap1 is required for cell–cell junction formation, for instance after their destabilization by EGTA-mediated chelation of extracellular calcium, but it is not required for the maintenance of existing junctions [16, 17]. This suggests that Rap signaling at cell–cell contacts occurs specifically during contact formation and that Rap1 may respond to junction destabilization in order to maintain junctional integrity. Indeed, junction destabilization by hepatocyte growth factor (HGF) or calcium chelation elevates Rap1GTP levels in both epithelial and endothelial cells [6, 16, 17]. In particular, it was shown that in epithelial cells a strong activation of Rap1 occurs upon adherens junction disassembly and is triggered by E-cadherin internalization and endocytic trafficking to the perinuclear Rab11-positive recycling endosome compartment [8, 15]. Furthermore, it was demonstrated that the E-cadherin endocytosis-dependent activation of Rap1 is associated with and controlled by an increased Src kinase activity and is paralleled by the association of Rap1 with a subset of E-cadherin–catenin complexes that does not contain p120$^{ctn}$ [15]. Conversely, Rap1 activity is suppressed by the formation of E-cadherin-dependent cell–cell junctions as well as by agents that inhibit either Src activity or E-cadherin internalization and intracellular trafficking [15].

The finding that E-cadherin plays a major role in the outside-in regulation of Rap1 activity during cell adhesion, combined with the well-established pivotal role of Rap1 in the inside-out regulation of integrin adhesive functions, has raised the possibility that Rap1 could be placed at the crossroad between both signaling pathways. Indeed, using GFP-zyxin as a marker for the transition of integrin-mediated adhesions from weak focal complexes to stable focal adhesions [18, 19], and time-lapse video microscopy analyses, it was demonstrated that in epithelial cells the activation of Rap1 triggered by disruption of cell–cell junctions and E-cadherin endocytosis is associated with and required for the formation of integrin-mediated adhesive structures corresponding to mature focal adhesions. This implies changes in local adhesive strength of integrins and suggests the existence of an inverse relationship between the assembly/disassembly of cadherin- and integrin-dependent adhesive structures [8, 15]. Importantly, these findings have also suggested a major role for Rap1 in transmitting information from cadherin-based to integrin-based adhesive structures to couple cadherin inhibition to integrin activation during the remodelling of epithelial tissues [8, 15].

In this chapter we describe a fluorescence microscopic approach to study subcellular localization of Rap1 protein in the crosstalk between integrins and cadherins.

## 2 Materials

### 2.1 Cleaning Coverslips

1. Acid solution: Two parts of nitric acid, one part of chloridric acid. Make up the solution in a glass beaker. Operate under chemical hood.
2. No. 1.5 coverslips.
3. 6-Well plates.

### 2.2 Coating Coverslips

1. PBS buffer: 137 mM NaCl, 2.7 mM KCl, 10 mM $Na_2HPO_4$, 2 mM $KH_2PO_4$, pH 7.4. Dissolve 8 g of NaCl, 0.2 g of KCl, 1.44 g of $Na_2HPO_4$, and 0.24 g of $KH_2PO_4$ in 800 ml of distilled $H_2O$. Adjust the pH to 7.4 with HCl. Add $H_2O$ to 1 l. Sterilize by autoclaving on liquid cycle. Store at room temperature.
2. Fibronectin (lyophilized and sterile powder): Prepare a 25 μg/ml solution in sterile PBS.

### 2.3 Cell Growth and Treatments

1. Cell lines of your interest (e.g., FRT cells).
2. Suitable cell medium.
3. Complete medium: Medium supplemented generally with 10 % fetal calf serum (FCS) or fetal bovine serum (FBS), 2 mM glutamine, 100 U/ml penicillin/streptomycin. Some media containing a stabilized form of L-glutamine, L-alanyl-L-glutamine, do not need glutamine supplementation.
4. Cytochalasin D.
5. Ethylene glycol tetraacetic acid (EGTA): Working solution: 4 mM EGTA in PBS.

### 2.4 Fixation and Immunofluorescence

1. $NH_4Cl$ Stock solution: 1 M in PBS; working solution: 50 mM in PBS.
2. 4 % paraformaldehyde (PFA): Dissolve the appropriate amount of PFA in PBS in a glass beaker with a stir bar. Heat in a hot plate in the hood at 60 °C. Cool down to 20 °C, and adjust pH to 7.4. Store at 4 °C or at −20 °C for longer time (*see* **Note 1**).
3. PBS–0.5 % Triton-X 100.
4. PBS–1 % bovine serum albumin (BSA).
5. Anti-Rap1 and -E-cadherin primary antibodies.
6. Secondary fluorescent antibodies: choose the secondary antibody appropriate for primary antibody.
7. Mounting reagent: Some mounting reagents are supplied with DAPI and antifade reagents and are ready to use.
8. 4′,6-Diamidino-2-phenylindole (DAPI) or Hoechst for nuclear staining: Not necessary if you use a mounting reagent supplied with DAPI.

## 3 Methods

### 3.1 Cleaning Coverslips (See Note 2)

1. Place coverslips (a few at a time) into the acid solution (300 ml), and allow them to sit for 2 h with occasional swirling.
2. Decant the acid carefully into an adequate waste receptacle.
3. Wash the coverslips extensively in running tap water until the pH of the wash water is back to 5.5–6.0.
4. Store the coverslips in a covered container suitable to be sterilized in a hot air oven (*see* **Note 3**).
5. Sterilize the coverslips in a hot air oven at 180 °C for 3 h. You can use a sterilization indicator to check sterility (*see* **Note 4**).
6. Put coverslips into cell culture dishes. You can use a Pasteur pipette connected with a vacuum pump to grab coverslips (*see* **Note 5**).

### 3.2 Coating Coverslips

To analyze the effect of Rap1 on cell adhesion to extracellular matrix proteins, coverslips must be coated with a solution containing the extracellular matrix protein of interest. In our experience, we used fibronectin (*see* **Note 6**).

#### 3.2.1 Fibronectin Coating

1. Put coverslips in your cell culture dish (you can use 6 well/dish with 3 cover slips for each well).
2. Coat coverslips with fibronectin.
3. Incubate for 2 h at 37 °C (you can use the cell incubator) or overnight at 4 °C.
4. Remove the excess of fibronectin by aspiration.
5. Wash once with PBS.

### 3.3 Cell Growth and Treatments

1. Plate your cell lines into a 6-well plate containing two or three cleaned coverslips for well (coated, if necessary). You should use at least three different wells (two to be treated and one for control).
2. Grow cell lines until confluence. Many cell lines must be cultured at 37 °C and 5 % $CO_2$ atmosphere in Dulbecco's modified Eagle's complete medium; some cell lines require other complete media (e.g., FRT require Coon's F-12 complete medium [20]).
3. Treat one well with 1 μM cytochalasin D for 30 min to prevent E-cadherin internalization and then with 4 mM EGTA for 30 min to disrupt cell–cell contacts (*CytD+ EGTA sample*). Use serum-free medium.

4. Treat another well with 4 mM EGTA for 30 min to disrupt cell–cell contacts (*EGTA sample*). Use serum-free medium.
5. Leave one well untreated (*Ctrl sample*). Use serum-free medium.

### 3.4 Fixation and Immunofluorescence

1. Rinse cells twice with PBS (*see* **Note 7**).
2. Fix with 4 % PFA for 10 min (*see* **Notes 8** and **9**).
3. Rinse 3× with PBS.
4. Quench with 50 mM $NH_4Cl$ for 20 min. This step is crucial in reducing nonspecific background.
5. Rinse 3× with PBS.
6. Permeabilize with PBS–0.5 % Triton-X 100 for 1 min.
7. Rinse 3× with PBS.
8. Block with PBS–1 % BSA for 30–60 min at room temperature. This passage blocks nonspecific protein-binding sites.
9. Rinse 3× with PBS.
10. Incubate with primary antibodies to E-cadherin and Rap1 for 1 h at room temperature. Primary antibodies should be diluted (according to the manufacturer's guideline) in PBS–0.1 % BSA (*see* **Note 10**).
11. Rinse 3× with PBS.
12. Incubate with appropriate secondary antibody for 1 h at room temperature. Make appropriate dilution (according to the manufacturer's guideline) of secondary antibodies in PBS–0.1 % BSA.
13. Rinse 3× with PBS.
14. Soak coverslips briefly in $H_2O$ to remove salts.
15. Wick excess water from coverslip by carefully touching the edges of the cover slip with a piece of paper.
16. Mount coverslip with cells down on a drop (20 μl) of mounting medium supplied with nuclear staining on a clean microscope slide. Gently lower the coverslips onto the mounting medium, cell side down, so that no air bubbles become trapped (*see* **Note 11**).
17. Gently wick off excess liquid and allow to dry overnight at room temperature in the dark to obtain a hard mount.
18. Observe samples at confocal immunofluorescence microscopy, store at 4 °C, and preserve them from light (*see* Fig. 1).

For alternative methods, *see* **Note 12**.

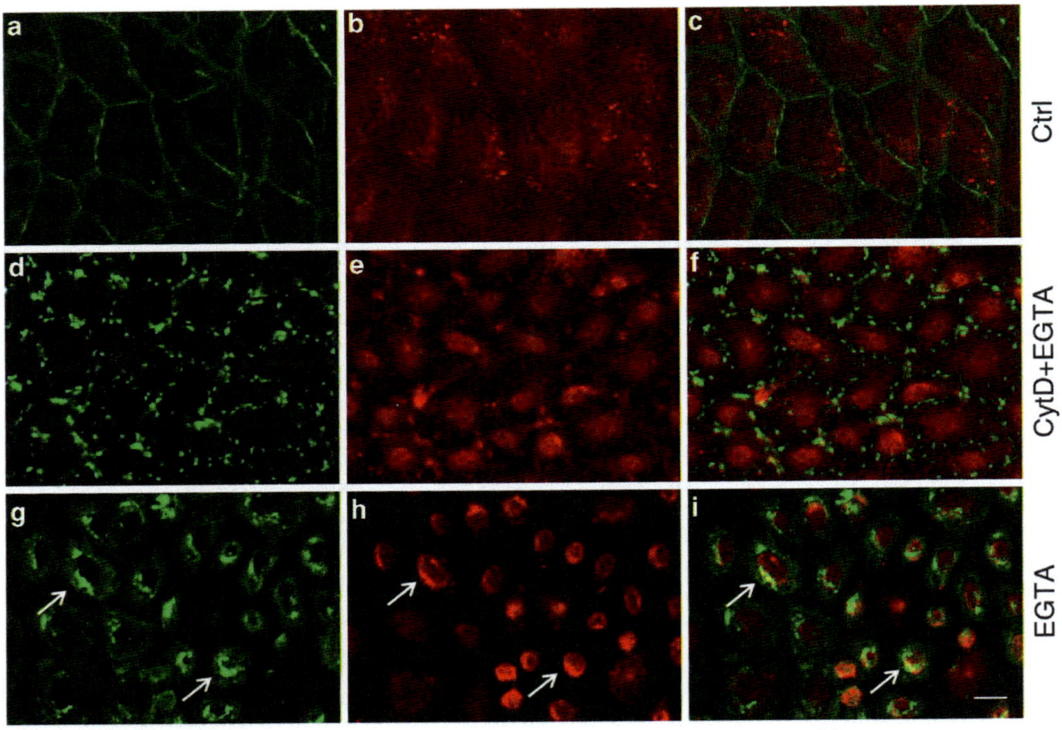


**Fig. 1** Rap1 redistribution promoted by the disassembly of AJs requires E-cadherin internalization and endocytic trafficking. Confluent FRT cells were left untreated (*Ctrl*), treated with EGTA (4 mM) for 30 min to disrupt cell–cell adhesion (*EGTA*), or pretreated with CytD for 30 min before EGTA treatment (*CytD+ EGTA*) to prevent E-cadherin internalization. Cells were then fixed, permeabilized, and double stained with antibodies to E-cadherin (*E-cadh*) and Rap1 (*Rap1*) followed by FITC- and RITC-conjugated secondary antibodies, respectively, and analyzed by confocal immunofluorescence microscopy. E-cadherin and Rap1 distribution was compared by merging images (*Merge*). Notice that both E-cadherin and Rap1 were redistributed to and partially colocalized at the perinuclear cytoplasm upon depletion of extracellular $Ca^{2+}$ by EGTA (panels **g**, **h**, **i**). In contrast, in cells pretreated with CytD prior to exposure to EGTA, a majority of E-cadherin remained at the cell surface despite the loss of cell–cell adhesion, whereas Rap1 mostly remained diffuse into the cytoplasm (panels **d**, **e**, **f**) as it was in untreated cells (panels **a**, **b**, **c**). From ref. 15, with permission

## 4 Notes

1. When you use PFA you must work under chemical hood.
2. It is very important that coverslips used in fluorescence microscopy are very clean. Although they look clean, they may have a thin film of grease that will not allow tissue culture cells to adhere well.
3. As an alternative, you may clean coverslips by using a basic solution:
   (a) Incubate coverslips for 2 h in 2 N NaOH.
   (b) Rinse extensively in $dH_2O$.

4. Alternative methods to sterilize coverslips:
   (a) Autoclave coverslips, and then let them dry before use.
   (b) Flame coverslips with Bunsen burner under laminar flow hood. This is the fastest way, but the flame can break coverslips.
5. You can move coverslips with sterile tweezers.
   We suggest to use 6-well plates and 3 coverslips/well.
6. As an alternative, coverslips may be coated with poly-L-lysine:
   (a) Prepare an adequate amount of 0.5 mg/ml poly-L-lysine in $H_2O$.
   (b) Put coverslips in your cell culture dish.
   (c) Coat coverslips with the poly-L-lysine solution.
   (d) Incubate at room temperature for 10 min.
   (e) Wash three times with sterile $H_2O$ or PBS.
   (f) Allow to air-dry for at least 45 min (under laminar flow hood).
7. In this section you can substitute PBS with Tris-buffered saline (TBS): 0.15 M NaCl, 10 mM Tris–HCl, pH 7.4.
8. Fixation with 4 % PFA may be done at room temperature for 10 min. However, it is better to fix your samples for 10 min on ice. PFA should cover the cells. Work under chemical hood.
9. You may fix your cells with alternative methods including glutaraldehyde, cold methanol, and cold acetone (work under chemical hood).
10. To incubate with the primary antibodies, you can take the coverslips from the well with a needle and tweezers, wick excess buffer from coverslip by carefully touching the edges of the coverslips with a piece of paper, and turn it on a drop (30 μl) of diluted antibody (cell side down). You can lay down the drops on a parafilm piece.
    We suggest to put the parafilm piece in a box containing a humidified 3 M paper layer. This procedure avoids consuming a lot of antibody. Remember to leave a coverslip without primary antibody (but only with PBS–0.1 % BSA) as a negative control.
11. If your mounting reagent is not supplied with DAPI or Hoechst, before mounting you have to perform the nuclear staining according to the manufacturer's instruction. Generally you have to dilute DAPI or Hoechst in PBS, stain the cells with this solution, and then, after the appropriate time, wash the cell with PBS.
12. Alternative methods:
    You can use fluorescent-tagged proteins rather than primary and secondary antibodies. You can perform the same analysis by transiently cotransfecting FRT cells—or other cell lines—with EGFP-tagged E-cadherin and mRFP-tagged Rap1 or vice versa.

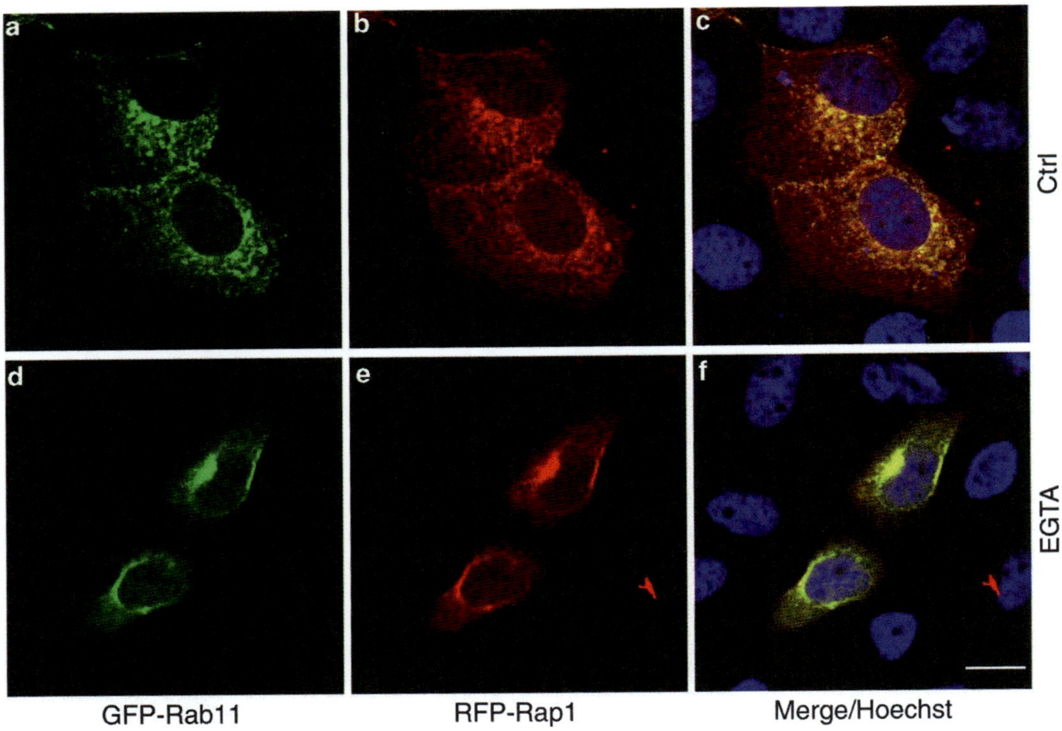


**Fig. 2** Perinuclear localization of Rap1 in Rab11 recycling endosome upon cell–cell junction disruption by EGTA. FRT cells were transiently cotransfected with EGFP-tagged Rab11, a marker for perinuclear recycling endosomes, and mRFP-Rap1 and grown to confluence for 12 h on glass coverslips. Cells were then either left untreated (*Ctrl*) or treated with EGTA (4 mM) for 30 min to disrupt cell–cell adhesion (*EGTA*) and analyzed by confocal immunofluorescence microscopy. Notice that the perinuclear pool of mRFP-Rap1 significantly colocalized with EGFP-Rab11. From ref. 15, with permission

FRT cells can be transiently transfected with 1.5 μg of cDNA constructs using the FuGENE 6 Transfection Reagent. After transfection, replate cells for 12–24 h in serum-free medium onto glass coverslips coated with 25 μg/ml fibronectin.

You may also cotransfect cells in cell culture dishes containing coverslips.

Moreover, you can transiently cotransfect cells with fluorescent-tagged Rap1 and fluorescent-tagged Rab11, a marker for perinuclear recycling endosome, to observe the perinuclear pool of Rap1 colocalized with Rab11 after disruption of cell–cell junctions upon EGTA treatment (*see* Fig. 2).

## References

1. Kitayama H, Sugimoto Y, Matsuzaki T et al (1989) A ras-related gene with transformation suppressor activity. Cell 56:77–84
2. Bourne HR, Sanders DA, McCormick F (1990) The GTPase superfamily: a conserved switch for diverse cell functions. Nature 348: 125–132
3. Jeyaraj SC, Unger NT, Chotani MA (2011) Rap1 GTPases: an emerging role in the cardiovasculature. Life Sci 88:645–652

4. Bos JL, de Rooij J, Reedquist KA (2001) Rap1 signalling: adhering to new models. Nat Rev Mol Cell Biol 2:369–377
5. Caron E (2003) Cellular functions of the rap1 GTP-binding protein: a pattern emerges. J Cell Sci 116:435–440
6. Gloerich M, Bos JL (2011) Regulating rap small g-proteins in time and space. Trends Cell Biol 21:615–623
7. Ohba Y, Kurokawa K, Matsuda M (2003) Mechanism of the spatio-temporal regulation of ras and rap1. EMBO J 22:859–869
8. Retta SF, Balzac F, Avolio M (2006) Rap1: a turnabout for the crosstalk between cadherins and integrins. Eur J Cell Biol 85:283–293
9. Bivona TG, Wiener HH, Ahearn IM et al (2004) Rap1 up-regulation and activation on plasma membrane regulates t cell adhesion. J Cell Biol 164:461–470
10. Bos JL, de Bruyn K, Enserink J et al (2003) The role of rap1 in integrin-mediated cell adhesion. Biochem Soc Trans 31:83–86
11. Dustin ML, Bivona TG, Philips MR (2004) Membranes as messengers in t cell adhesion signaling. Nat Immunol 5:363–372
12. Katagiri K, Maeda A, Shimonaka M et al (2003) Rapl, a rap1-binding molecule that mediates rap1-induced adhesion through spatial regulation of lfa-1. Nat Immunol 4:741–748
13. Han J, Lim CJ, Watanabe N et al (2006) Reconstructing and deconstructing agonist-induced activation of integrin alphaiibbeta3. Curr Biol 16:1796–1806
14. Kim C, Ye F, Ginsberg MH (2011) Regulation of integrin activation. Annu Rev Cell Dev Biol 27:321–345
15. Balzac F, Avolio M, Degani S et al (2005) E-cadherin endocytosis regulates the activity of rap1: a traffic light gtpase at the crossroads between cadherin and integrin function. J Cell Sci 118:4765–4783
16. Asuri S, Yan J, Paranavitana NC et al (2008) E-cadherin dis-engagement activates the rap1 gtpase. J Cell Biochem 105:1027–1037
17. Hogan C, Serpente N, Cogram P et al (2004) Rap1 regulates the formation of e-cadherin-based cell–cell contacts. Mol Cell Biol 24: 6690–6700
18. Rottner K, Krause M, Gimona M et al (2001) Zyxin is not colocalized with vasodilator-stimulated phosphoprotein (vasp) at lamellipodial tips and exhibits different dynamics to vinculin, paxillin, and vasp in focal adhesions. Mol Biol Cell 12:3103–3113
19. Zaidel-Bar R, Cohen M, Addadi L et al (2004) Hierarchical assembly of cell–matrix adhesion complexes. Biochem Soc Trans 32: 416–420
20. Ambesi-Impiombato FS, Parks LA, Coon HG (1980) Culture of hormone-dependent functional epithelial cells from rat thyroids. Proc Natl Acad Sci U S A 77:3455–3459

# Chapter 14

# An In Vitro System to Evaluate the Scaffold Function of the RalA Effector Protein RalBP1

David F. Kashatus

## Abstract

The in vitro kinase assay is an invaluable method to identify and characterize kinase substrates and to determine specific site(s) of phosphorylation. However, the use of recombinant proteins does not adequately account for the contribution of accessory proteins, such as scaffolds, that can contribute to the phosphorylation reaction in vivo. We describe here a method to evaluate the contribution of specific scaffold proteins to kinase reactions using a modified in vitro kinase assay.

**Key words** RalBP1, Drp1, Kinase assay, RalA

## 1 Introduction

Protein kinases are a huge and diverse family of proteins that play an important regulatory role in nearly every biological process [1, 2]. These important enzymes catalyze the transfer of a phosphate group from ATP to specific amino acid residues on their substrates. The specificity of these reactions is regulated in large part by the specificity of the direct interaction between the kinase and the substrate protein. However, there are a number of ways that additional proteins can influence the reaction, by changing the activity of the kinase, masking or revealing the site of phosphorylation or binding, or acting as a scaffold and directly bringing the kinase and substrate together.

The radioactive in vitro kinase assay has been used for several decades to study phosphorylation [3], and despite the myriad new ways that have been developed to study this important regulatory modification [4], it remains an important tool. Kinases are a group of proteins that have proven accessible to small-molecule inhibition [5], and they are central to the signaling pathways that mediate several human diseases [6–9]. For these reasons, research efforts from academic, government, and industry labs have generated a

Lorenza Trabalzini and Saverio Francesco Retta (eds.), *Ras Signaling: Methods and Protocols*, Methods in Molecular Biology, vol. 1120, DOI 10.1007/978-1-62703-791-4_14, © Springer Science+Business Media, LLC 2014 wealth of medium- and high-throughput data on kinase substrates, novel roles for kinases, and new regulatory proteins [10]. With all of this new data, the in vitro assay remains a critical tool to confirm the data and to fill in key molecular details.

In this chapter we describe a modification of the classic in vitro kinase assay [11]. By keeping the concentration of kinase and substrate low, and by titrating a putative accessory protein into the reaction, we are able to evaluate the contribution of the accessory protein to the kinase reaction. In this case, we are evaluating the role of the RalA effector protein RalBP1 as a scaffold between the kinase complex Cdk1/cyclin B and one of its substrates, Drp1. We have described below the generation of the recombinant proteins used in the assay, the setup of the kinase reaction, and the electrophoresis and analysis of the reaction products.

## 2 Materials

Prepare all reagents using deionized $H_2O$ and analytical grade reagents. Diligently follow all safety regulations when working with hazardous chemicals and radiation. Properly dispose of all hazardous and radioactive waste according to the regulations of your institution.

### 2.1 Generation of GST-RalBP1 and GST-Drp1

1. pGEX-5X-2 RalBP1 and pGEX-5X-2 $Drp1^{518–736}$ were described previously [11] and are available from Addgene.
2. BL21 (DE3) competent cells: Thaw on ice, and aliquot 50 μL/tube in 1.5 mL Eppendorf tubes. Store at −80 °C.
3. S.O.C. medium: To 900 mL of $H_2O$ add 20 g of bacto tryptone, 5 g of bacto yeast extract, 2 mL of 5 M NaCl, 2.5 mL of 1 M KCl, 10 mL of 1 M $MgCl_2$, 10 mL of 1 M $MgSO_4$, 20 mL 1 M glucose. Transfer to a 1 L graduated cylinder and bring volume to 1 L with $H_2O$. Autoclave or sterile filter, and store at 4 °C in 50 mL aliquots.
4. LB-AMP plates: Weigh out 12.5 g of Luria-Bertani agar and add to a 500 mL bottle with 300 mL of $H_2O$. Autoclave for 30 min. Allow bottle to cool slowly in a 65 °C water bath. Add 300 μL of 100 mg/mL ampicillin and slowly pour 15 mL of liquid agar into twenty 100 cm petri dishes (*see* **Note 1**). Allow agar to solidify on a flat surface at room temperature and then store at 4 °C.
5. LB-AMP medium: Weigh out 7.5 g of Luria-Bertani media and add to a 500 mL bottle with 300 mL of $H_2O$. Autoclave for 30 min. Allow bottle to cool to room temperature. Add 300 μL of 100 mg/mL ampicillin.
6. Isopropyl β-D-1-thiogalactopyranoside (IPTG): To make a 500 mM stock solution, dissolve 1 g of IPTG powder in 8.39 mL of $H_2O$.

7. Phosphate-buffered saline (PBS, 10×): Add ~500 mL of $H_2O$ to a 1 L glass beaker. Put the beaker on a stir plate and add a magnetic stir bar. While stirring, weigh out and transfer to the beaker 80 g of NaCl, 2 g of KCl, 14.4 g of $Na_2HPO_4 \cdot 2H_2O$, and 2.4 g of $KH_2PO_4$. pH to 7.4 by slowly adding 10 N HCl. Transfer to a 1 L graduated cylinder, and bring volume to 1 L with $H_2O$.
8. Lysis buffer—PBS–Triton X-100 (PBS-T): In a 500 mL glass beaker, add ~300 mL of $H_2O$, 50 mL of 10× PBS, and 5 mL of Triton X-100 (*see* **Note 2**). Stir with a magnetic stir bar until Triton X-100 is fully dissolved. Transfer to 500 mL graduated cylinder, and bring volume to 500 mL with $H_2O$. Store at 4 °C for up to 1 week. When ready to use, add protease inhibitor stock solutions up to 1 h prior to use at the following final concentrations (*see* below): 2 μg/mL aprotinin, 2 μg/mL leupeptin, 2 μg/mL pepstatin A, and 1 mM PMSF.
9. Protease inhibitors, stock solutions:

   5 mg/mL Aprotinin.

   50 mg/mL Leupeptin.

   1 mg/mL Pepstatin.

   1.74 mg/mL (100 mM) Phenylmethanesulfonyl fluoride (PMSF), dissolve in anhydrous ethanol.
10. Elution/dialysis buffer: 100 mM Tris, pH 8.0, 120 mM NaCl. Add 200 mL of 1 M Tris, pH 8.0 and 48 mL of 5 M NaCl to a 2 L graduated cylinder. Bring to 2 L with $H_2O$. Remove 50 mL (in 50 mL conical) for washes and elution, and add the remaining 1,950 mL to a 2 L beaker for dialysis. Store at 4 °C. For elution from glutathione beads, add 0.1 g of reduced glutathione powder to 21.69 mL of elution buffer, and rock gently to dissolve.
11. Glutathione sepharose 4B: To prewash beads, remove 1 mL of slurry (*see* **Note 3**) and add to 15 mL conical tube, add 10 mL of lysis buffer, invert tube two to three times, and pellet beads by spinning at 1,000 × *g* for 1 min.
12. Slide-A-Lyzer Dialysis Cassettes, 10 K MWCO (Pierce): Equilibrate the cassette by submerging in elution buffer (without glutathione) for 2 h at 4 °C.
13. Amicon Ultra-0.5 Centrifugal Filter Device, 10,000 NMWL (Millipore).
14. Bio-Rad Protein Assay Dye Reagent Concentrate.
15. Bovine serum albumin (BSA)—80 mg/mL protein standard. Dilute in $H_2O$ to make 1, 3, 5, 7, and 9 μg/μL protein standards.

### 2.2 In Vitro Kinase Reaction Components

1. Kinase buffer: 20 mM HEPES–KOH, pH 7.4, 15 mM EGTA, 20 mM $MgCl_2$. To make 10 mL, add 200 μL of 1 M HEPES–KOH pH 7.4, 300 μL of 0.5 M ethylene glycol tetraacetic acid (EGTA), and 200 μL of 1 M $MgCl_2$ to a 15 mL conical tube and bring volume to 10 mL with $H_2O$. Store at 4 °C.
2. Dithiothreitol (DTT): To make 1 M DTT, dissolve 1 g of DTT powder in 6.48 mL of $H_2O$.
3. 100 mM ATP (adenosine 5′-triphosphate disodium salt hydrate): Dissolve 1 g of ATP-disodium salt in 18.15 mL of $H_2O$. Make 500 μL aliquots, and store them at −80 °C to limit spontaneous ATP hydrolysis. Limit the number of freeze/thaw cycles.
4. ATP [γ-$^{32}$P]: 3,000 Ci/mmol, 5 mCi/mL.
5. Recombinant Active Cdk1/CyclinB1, GST-Tagged (Sigma). Supplied at 0.1 μg/μL.

### 2.3 Polyacrylamide Gel Electrophoresis Components

1. 30 % Acrylamide/Bis solution (29:1).
2. 20 % SDS: In a 500 mL beaker, dissolve 20 g of SDS powder in ~80 mL $H_2O$ (*see* **Note 4**). Transfer to 100 mL graduate cylinder, and bring volume to 100 mL with $H_2O$.
3. Upper gel buffer: 0.5 M Tris–HCl, pH 6.8, 0.4 % SDS. Weigh 60.6 g of Tris and add to a glass beaker with ~700 mL of $H_2O$. While stirring, adjust pH to 6.8 using 10 N HCl. Add 20 mL of 20 % SDS. Transfer to 1 L graduated cylinder, and bring to 1 L with $H_2O$.
4. Lower gel buffer: 1.5 M Tris–HCl, pH 8.8, 0.4 % SDS. Weigh 181.7 g of Tris and add to a glass beaker with ~700 mL of $H_2O$. While stirring, adjust pH to 8.8 using 10 N HCl. Add 20 mL of 20 % SDS. Transfer to 1 L graduated cylinder, and bring to 1 L with $H_2O$.
5. *N, N, N, N′* -tetramethyl-ethylenediamine (TEMED): Store at 4 °C.
6. 10 % ammonium persulfate (APS): Dissolve 1 g of APS powder in 10 mL of $H_2O$. Store at 4 °C.
7. 5× SDS running buffer: Add ~700 mL of $H_2O$ to a 1 L beaker. Add 72 g of glycine, 15 g of Tris, and 5 g of SDS (*see* **Note 4**). Transfer to graduated cylinder, and bring volume to 1 L with $H_2O$.
8. 2× SDS sample buffer: In a 50 mL conical tube, add 5 mL of Tris–HCl, pH 6.8, 5 mL of glycerol, 5 mL of 20 % SDS, 500 μL of 2-mercaptoethanol (2-ME), and 250 μg of bromophenol blue. Bring volume to 25 mL with $H_2O$. Make 5 mL aliquots in 15 mL conical tubes and store at −20 °C.
9. Protein molecular weight standard: SeeBlue Plus2 (Invitrogen).

10. Coomassie staining solution: In a 1 L graduated cylinder, add 400 mL of $H_2O$, 500 mL of methanol, and 100 mL of acetic acid (glacial). Transfer to 1 L bottle, add 2 g of Coomassie brilliant blue R250, and mix by inversion.
11. Coomassie destain: In a 1 L graduated cylinder, add 830 mL of $H_2O$, 100 mL of methanol, and 70 mL acetic acid (glacial).
12. Phosphorimager.

## 3 Methods

### 3.1 Generation of GST-RalBP1 and GST-Drp1

1. Thaw a 50 μL vial of competent BL-21 cells on ice.
2. Add 25 ng of pGEX-5X-2 RalBP1 or pGEX-5X-2 Drp1$^{518-736}$ to the vial of cells, tap gently to mix, and incubate on ice for 30 min. As the process for generating these proteins is identical, the procedure for one will be described.
3. Heat shock bacteria/DNA mix by placing the tube in a 42 °C heat block for 45 s and then returning it to ice for 2 min.
4. Add 500 μL of SOC media to the mixture, transfer total contents to a 15 mL snap-cap tube, and incubate in a 37 °C rocking incubator for 1 h.
5. Remove 100 μL of SOC/bacteria/DNA mix and pipette onto a pre-warmed LB-AMP plate. Spread mixture on the plate using a sterilized glass bar. Incubate the plate upside down overnight in a 37 °C incubator.
6. Remove the plate from the incubator, and pick a single colony by touching the tip of a sterile toothpick or pipette tip to the colony and subsequently submerging it in a 14.5 mL snap-cap tube pre-loaded with 5 mL of LB-Amp media. Incubate in a 37 °C rocking incubator for 4–6 h.
7. Transfer the entire 5 mL culture to a 250 mL Erlenmeyer flask containing 50 mL of LB-Amp media. Incubate in a 37 °C rocking incubator overnight.
8. Transfer the entire 55 mL culture to a 1 L Erlenmeyer flask containing 445 mL of LB-Amp media. Incubate in a 37 °C rocking incubator for up to 3 h or until the optical density (measured on a spectrophotometer at 600 nm wavelength; $OD_{600}$) reaches between 0.8 and 1.0.
9. Add 400 μL of 0.5 M IPTG to the culture, and return it to the rocking 37 °C incubator for 3 h (*see* **Note 5**).
10. Transfer the bacterial culture to a 500 mL centrifuge bottle and centrifuge at 5,000 × *g* for 15 min (*see* **Note 6**).
11. Resuspend pellet in 40 mL of lysis buffer (+ protease inhibitors) by vigorously pipetting up and down. Transfer resuspension to a 50 mL conical tube and place on ice.

12. Sonicate the bacterial suspension for 1 min at maximum output (0.5 s on–0.5 s off). Return to ice for 1 min, and then repeat sonication (*see* **Note 7**).
13. Transfer suspension to a 50 mL polypropylene centrifuge and centrifuge for 15 min at 10,000 × *g*.
14. While the bacterial suspension is spinning, add 2 mL of prewashed glutathione sepharose 4B beads to a 50 mL conical tube on ice.
15. Transfer supernatant from **step 13** (discard the pellet) to the 50 mL conical tube containing the glutathione beads, seal tightly, and rock the tube at 4 °C for 3 h (or overnight).
16. Pellet the beads in a tabletop centrifuge at no more than 500 × *g* for 1 min.
17. Wash beads three times with 15 mL of elution buffer (without glutathione). Transfer to a 15 mL conical tube during last wash.
18. To elute the GST-RalBP1 (or GST-$Drp1^{518-736}$) from the beads, add 1 mL of elution buffer + 15 mM reduced glutathione to the beads and resuspend by gently tapping the side of the tube. Rotate at 4 °C for 30 min.
19. Pellet the beads in a tabletop centrifuge at no more than 500 × *g* for 1 min, and remove supernatant to a clean Eppendorf tube. Repeat previous elution step two times with 500 μL of elution buffer + glutathione.
20. Carefully transfer elution to a Slide-A-Lyzer 10 K Dialysis Cassette using a 3 mL syringe with an 18 gauge, 1 in. beveled needle (*see* **Note 8**). Float the cassette in 2 L of elution buffer (without glutathione) in a large (>2 L) beaker with a stir bar at 4 °C overnight. Gently stir elution buffer with cassette on a stir plate at the lowest speed.
21. Carefully remove elution from the Slide-A-Lyzer 10 K Dialysis Cassette using a 3 mL syringe with an 18 gauge, 1 in. beveled needle (*see* **Note 9**). Transfer to a 2 mL Eppendorf tube.
22. Transfer elution to four Amicon Ultra-0.5 Centrifugal Filter Devices (add 500 μL/device; use a fifth if the total volume is greater than 2 mL). Centrifuge at 14,000 × *g* in a microcentrifuge for 15 min at room temperature. Remove insert, invert, and place in a new collection tube. Pulse in a microcentrifuge to collect concentrated sample (*see* **Note 10**).
23. Determine protein concentration using Bio-Rad Protein Assay (*see* **Note 11**).
24. Aliquot recombinant protein in 0.5 mL Eppendorf tubes (10 μL/tube) and store at −80 °C for up to 1 year.

### 3.2 In Vitro Kinase Reaction

1. Thaw 1 aliquot each of GST-RalBP1 and GST-Drp1$^{518-736}$, in addition to recombinant GST-Cdk1/cyclin B1 on ice.
2. Dilute GST-Drp1$^{518-736}$ to 200 ng/μL with sterile $H_2O$ in a 1.5 mL Eppendorf tube.
3. Dilute GST-RalBP1 to 1 μg/μL with sterile $H_2O$ in a 1.5 mL Eppendorf tube.
4. From the 1 μg/μL stock, make a 0.1, 0.2, and 0.5 μg/μL dilution of GST-RalBP1 with sterile $H_2O$.
5. Label five 1.5 mL Eppendorf tubes 1–5, and set up substrate mix as follows: add 1 μL of GST-Drp1$^{518-736}$ to each tube, plus 1 μL of $H_2O$ (tube 1) or 1 μL of GST-RalBP1 at 0.1 μg/mL (tube 2), 0.2 μg/mL (tube 3), 0.5 μg/mL (tube 4), or 1 μg/mL (tube 5).
6. Set up kinase reaction master mix as follows (*see* **Note 12**): 112.75 μL of kinase buffer, 1 μL of 0.1 M DTT, 5 μL of 100 mM ATP, 5 μL of ATP [$\gamma$-$^{32}$P], 1.25 μL of GST-Cdk1/cyclin B1.
7. Add 25 μL of kinase reaction master mix to each substrate tube. Incubate in a 30 °C heat block for 30 min.
8. Add 25 μL of 2× SDS sample buffer to each tube to stop the kinase reaction.

### 3.3 Polyacrylamide Gel Electrophoresis

1. To prepare the resolving gel, mix 5.2 mL of $H_2O$, 3.125 mL of 4× lower gel buffer, 4.2 mL of 30 % acrylamide/bis solution, 65 μL of APS, and 25 μL TEMED in a 15 mL conical tube on ice (*see* **Note 13**), and cast a gel within a 8.3 cm × 7.3 cm × 1.5 mm cassette (use approximately 8 mL to leave room for the stacking gel). Gently overlay with 1 mL of isopropyl alcohol.
2. Following polymerization of lower gel, wash away the isopropyl alcohol with $H_2O$ prior to pouring the stacking gel.
3. To prepare the stacking gel, mix 3.1 mL of $H_2O$, 1.25 mL of 4× upper gel buffer, 0.65 mL of 30 % acrylamide/bis solution, 15 μL of APS, and 15 μL of TEMED in a 15 mL conical tube on ice. Cast the gel on top of the resolving gel, and immediately insert a 10-well comb, taking care to avoid bubbles.
4. Following polymerization, remove the comb carefully and gently wash out the wells with $H_2O$.
5. From the 5× SDS running buffer, make up 500 mL of 1× SDS running buffer with $H_2O$. Set up the electrophoresis apparatus with the gel and the running buffer. Flush the wells with running buffer using a pipette or a syringe.
6. Heat the kinase reactions to 95 °C for 5 min. Spin down the reactions in a microcentrifuge for 30 s at 3,000 × *g*. Load 10 μL

of protein molecular weight standard followed in consecutive lanes by the 5× 50 μL kinase reactions.

7. Electrophorese at 100 V until the samples have entered the gel, and then increase voltage to 180 V. Continue until the dye front is 1–2 mm from the bottom of the gel (*see* **Note 14**).
8. Pry open the plates with a spatula, and carefully transfer the gel to a shallow dish (*see* **Note 15**). Carefully wash the gel twice with $H_2O$, and then add approximately 50 mL of Coomassie staining solution. Gently rock the gel for 15 min at room temperature.
9. Carefully discard the staining solution as liquid radioactive waste, wash the gel one to two times with $H_2O$, and then add 50 mL of Coomassie destain. Add several crumpled kimwipes to the destain to collect the excess dye.
10. Destain for several hours, replacing the kimwipes as they turn blue, until the gel becomes clear and distinct protein bands become visible.
11. Carefully seal the gel in plastic wrap (*see* **Note 16**) and place in a cassette with a phospho-screen. Expose the gel to the screen for 24 h (*see* **Note 17**).
12. Develop the screen using a phosphorimager.

## 4 Notes

1. To avoid bubbles, do not shake or mix agar too vigorously following autoclave step. If there are bubbles on the plates, they can be removed by carefully exposing them to a flame with a handheld torch.
2. Triton X-100 is very viscous, and thus it is very hard to measure using a serological pipette. Another way to measure 5 mL is to slowly pour the Triton X-100 into a 15 mL conical tube up to the 5 mL line, add 5 mL 1× PBS, and then rock on a nutator at room temperature for 10–15 min. The resulting 50 % Triton X-100 can easily be transferred without leaving a residue.
3. When pipetting beads, either use a wide-bore pipette tip or cut the end of a standard pipette tip with a clean razor blade to avoid the beads clogging the tip and to avoid pipetting inaccurately.
4. Wear a mask and use a fume hood when working with SDS.
5. It is often beneficial to perform a small-scale version of this induction to ensure that your protein is properly expressed. To perform this preliminary experiment, add IPTG to a 5 mL culture (at $OD_{600}$ 0.8–1.0) and remove 1 mL every hour for 3 h.

Pellet the bacteria, resuspend in 100 μL of 2× SDS loading buffer, and run 10 μL of each time point (0, 1, 2, and 3 h) on a 10 % acrylamide gel. Coomassie stain the gel, and observe whether a prominent band appears at the correct molecular weight in the IPTG-induced samples. If it does, you may proceed with the larger scale induction.

6. The bacterial pellet can be frozen at –80 °C indefinitely at this point. When ready to continue the protocol, remove the centrifuge bottle with the pellet from the freezer and thaw it on ice.
7. All sonicators are slightly different. This sonication scheme has worked on several different manufacturers but may need to be adjusted slightly if you get incomplete lysis.
8. Alternate forms of dialysis, such as dialysis tubing, can be used for this step, but we prefer the Slide-A-Lyzer system because of the ease. Closely follow the manufacturer's instructions on adding the eluate to the cassette, taking special care not to pierce the membrane with the needle.
9. As detailed in the manufacturer's instructions, remove the dialyzed solution from a different corner of the cassette than the one in which it was added.
10. Certain proteins will become insoluble when they are concentrated and precipitate out of solution. In this case, you may need to experiment with different buffer conditions for the dialysis step to find the optimal buffer for your protein. Start by adjusting the pH of the buffer or altering the concentration of NaCl. If there is still precipitation, add a reducing agent such as DTT or 2-ME.
11. We use the Bio-Rad protein assay system according to the manufacturer's instructions, but any commercially available protein assay system will suffice. The final concentration of the protein is dependent on several factors and may end up outside of the linear range of the protein assay. If so, several dilutions may need to be tested before an accurate protein concentration can be achieved.
12. Use a master mix to minimize differences due to pipetting inaccuracies.
13. Keeping the mix on ice will help to prevent premature polymerization. Conversely, addition of higher volumes of APS and TEMED and/or placing the cast gel at 37 °C can speed up the process.
14. Each reaction will have large amounts of free γ-$^{32}$P-ATP, which will run through the gel at about the level of the dye front. If the dye front runs off the gel, the buffer will be contaminated with $^{32}$P and the cleanup will become much more involved.

15. Before transferring the gel, use a razor blade to trim off excess acrylamide, disposing the trimmed portions as radioactive waste. Cutting the gel just above the dye front will remove the free γ-$^{32}$P-ATP, which will reduce personal exposure to $^{32}$P. In addition, the presence of a large signal at the bottom of the gel can often mask the appearance of weaker signals elsewhere on the gel, making data interpretation more difficult.
16. The gel can be dried using a gel dryer prior to exposure, but with the exception noted below, this is not necessary and it risks cracking the gel. The gel will stay wet for several days if it is properly sealed in plastic wrap.
17. If a phospho-screen or phosphorimager is not available, film can be used to develop the kinase assay. In this case, the gel should be dried on a gel dryer, and the film and the gel should be placed in a cassette between two intensifying screens and then stored at −80 °C until being developed. When using film, a longer exposure time may be required to achieve the same signal intensity as with the phospho-screen.

## References


1. Manning G, Whyte DB, Martinez R et al (2002) The protein kinase complement of the human genome. Science 298:1912
2. Hanks SK (2003) Genomic analysis of the eukaryotic protein kinase superfamily: a perspective. Genome Biol 4:111
3. Burnett G, Kennedy EP (1954) The enzymatic phosphorylation of proteins. J Biol Chem 211:969–980
4. Jia Y, Quinn CM, Kwak S et al (2008) Current in vitro kinase assay technologies: the quest for a universal format. Curr Drug Discov Technol 5:59–69
5. Montagut C, Settleman J (2009) Targeting the RAF-MEK-ERK pathway in cancer therapy. Cancer Lett 283:125–134
6. Cuny G (2009) Kinase inhibitors as potential therapeutics for acute and chronic neurodegenerative conditions. Curr Pharm Des 15:3919–3939
7. Dancey J, Sausville EA (2003) Issues and progress with protein kinase inhibitors for cancer treatment. Nat Rev Drug Discov 2:296–313
8. Ždychová J, Komers R (2005) Emerging role of Akt kinase/protein kinase B signaling in pathophysiology of diabetes and its complications. Physiol Res 54:1–16
9. Gaestel M, Kotlyarov A, Kracht M (2009) Targeting innate immunity protein kinase signalling in inflammation. Nat Rev Drug Discov 8:480–499
10. Goldstein DM, Gray NS, Zarrinkar PP (2008) High-throughput kinase profiling as a platform for drug discovery. Nat Rev Drug Discov 7:391–397
11. Kashatus DF, Lim K-H, Brady DC et al (2011) RALA and RALBP1 regulate mitochondrial fission at mitosis. Nat Cell Biol 13:1–10

# Chapter 15

# Analysis of the Rit Subfamily GTPase-Mediated Signaling and Neuronal Differentiation and Survival

**Geng-Xian Shi and Catherine N. Kaminski**

## Abstract

The Rit subfamily of GTPases is a founding branch within the Ras family of small G-proteins and preserves unique sequences in the G2 effector loop domain and the *C*-terminus. Rit proteins regulate a diversity of signal transduction pathways, some of which are similar to and others of which differ from the pathways that are regulated by other Ras family GTPases. Rit proteins have been demonstrated to be essential regulators in neuronal differentiation and survival. Here, we describe the materials and methods utilized to characterize cellular signaling for the Rit subfamily of G-proteins in neuronal differentiation and survival.

**Key words** Rit GTPase, Ras GTPase, Signal transduction, Neuronal differentiation, Survival

## 1 Introduction

Rit and Rin, together with *Drosophila* orphan *dRic*, comprise a novel Rit subfamily of the Ras-related GTPases [1]. Like other Ras-related GTPases, Rit proteins act as intracellular switches by cycling between an active GTP-bound form and an inactive GDP-bound form [1]. Rit proteins share a higher homology in sequence with each other and possess a unique G2 effector-loop domain and a conserved sequence in the *C*-terminus that contains a cluster of polybasic amino acids (aa) but lacks the classic CAAX motif, making them differ from other Ras family GTPases [2]. These unique features suggest that Rit proteins may couple to distinct effectors and/or cellular compartments in cellular responses [1, 2]. Indeed, Rit proteins were found to stimulate a variety of signaling cascades including the mitogen-activated protein kinases (MAPKs) and to regulate diverse cellular functions including neuronal differentiation, survival, and neurogenesis, in ways that are either similar to or distinct from that regulated by other Ras proteins [3–12]. This chapter describes the methods utilized to elucidate the role for Rit proteins on neurite outgrowth and cell survival as well as to

Lorenza Trabalzini and Saverio Francesco Retta (eds.), *Ras Signaling: Methods and Protocols*,
Methods in Molecular Biology, vol. 1120, DOI 10.1007/978-1-62703-791-4_15, 

characterize the signaling networks critical for Rit proteins in these cellular processes in pheochromocytoma PC6 cells.

A variety of methods have been adapted to this chapter.

### 1.1 PC6 Cell Culture and Transfection

PC6 cells are a subline of pheochromocytoma PC12 cells. PC6 cells produce neurites in response to neurotrophic factors and grow as well-isolated cells rather than in clumps, making them an excellent cell model for determining Rit protein-mediated neuronal differentiation and survival [3–9, 11, 12].

PC6 cells appear to resist most commercial transfection reagents. Fortunately, we routinely obtain transfection efficiencies in the range of 50–90 % using either effectene or transgin at the optimized ratio of lipid to DNA.

### 1.2 Cell Lysate Preparation

We used kinase lysis buffer (KLB) to prepare total cell lysates for all of the experiments described in this chapter. The purpose is to avoid dephosphorylation of the effectors that could lead to improper interpretations of Rit protein-mediated cellular signaling if protein phosphorylation is critical for Rit protein-dependent cellular responses in these processes.

### 1.3 SDS-PAGE Gel Electrophoresis and Immunoblotting

We used SDS-PAGE electrophoresis to fractionate lysate proteins and immunoblotting to detect levels of either phosphorylated or total proteins in the lysates.

### 1.4 Immunoprecipitation and GST Pulldown Assays

We utilized either immunoprecipitation (i.p.) or GST pulldown (p.d.) to assess the association of Rit proteins with their effectors, kinase activity, or activation of proteins lacking sensitive immunoblotting reagents.

### 1.5 Nonradioactive Assay for the Determination of Rit Activation

The nonradioactive assay for Rit activation is established upon the enormous difference in the association of a specific binding domain within a downstream effector protein with the GTP-bound versus GDP-bound forms of Rit proteins in vitro. The GST-RGL3-RBD precipitation method has been demonstrated as a reliable and reproducible probe for detecting GTP loading to Rit in various cellular systems and in response to diverse cellular signals [4, 6, 7, 9–12]. While the GTP/GDP dissociation rates for both Rit and Rin are rapid and a high percentage of GTP-bound Rit/Rin is maintained in the cells under normal conditions, the assay should be carefully manipulated and optimized to ensure lower basal levels of Rit-GTP and higher specificity of GTP loading to Rit proteins. The procedures described in this chapter have been optimized for the assay in PC6 cells and can be adapted to other cell types by slight modification [4, 6, 12].

### 1.6 Small Hairpin RNA-Mediated Gene Silencing

Small hairpin RNA (shRNA)-mediated gene silencing is a useful tool for determining the critical role for a gene in cellular responses. By using the method described here, we have successfully identified a variety of effectors and signaling cascades attributable to Rit-mediated cellular signaling and function.

### 1.7 Signaling Analysis

We have used transient expression of constitutively active (CA) Rit proteins in PC6 cells to identify signaling pathways that are provoked by Rit proteins or expression of dominant negative (DN) mutants or shRNA constructs to validate the signaling pathways that are dependent upon Rit signaling. These methods have revealed several signaling cascades attributable to Rit protein-mediated functions [3–5, 7–9, 11].

### 1.8 In Vitro Kinase Assay

The in vitro kinase assay is a valuable tool for characterizing the kinase activity activated during cellular processes and/or by signal transduction cascades. We have used this assay to assess the ability of activated Rit proteins to elevate Raf and p38 MAPK kinase activities and to evaluate the contribution of Rit proteins to the Raf and the p38 signaling pathways in response to diverse extracellular stimuli [4, 7–9].

### 1.9 Analysis of the Association of Rit Proteins with Their Effectors

To assess the association of Rit proteins with their effectors, we utilized either i.p. or p.d. approaches.

### 1.10 Neurite Outgrowth Analysis

Neurite outgrowth is the first step in neuronal differentiation and plays an important role in the development of the central nervous system. This process is regulated by complex mechanisms including Ras-related GTPase-mediated activation of various signaling networks. We have demonstrated the ability of Rit subfamily proteins to induce neurite outgrowth and the contribution of Rit proteins to neurotrophic factor-mediated neuronal differentiation [7–9, 11] using the methods described in this chapter.

### 1.11 Cell Survival/Death Analysis

Rit GTPase has been shown to be an essential element in cellular resistance to stresses through its ability to control the evolutionarily conserved p38-dependent survival signaling [3, 4]. The protocol reported here describes the methods used to characterize Rit-mediated protection.

## 2 Materials

Prepare all solutions using distilled deionized water (DDW), and store all reagents at 4 °C unless indicated otherwise. All waste disposal should follow the NIH waste disposal guide and applicable federal, state, and local laws and regulations and executive orders.

### 2.1 PC6 Cell Culture and Transfection

1. *Dulbecco's modified Eagle's medium complete medium* (DMEM-CM): Use DMEM containing high glucose (4,500 mg/L) and L-glutamine for PC6 cell cultures. Prepare DMEM-CM by adding 10 % fetal bovine serum (FBS), 5 % horse donor serum, 100 IU/ml penicillin, and 100 μg/ml streptomycin.
2. *PBS*: Prepare 10× PBS stock solution by dissolving 80 g NaCl, 2.4 g $KH_2PO_4$, 14.4 g $Na_2HPO_4$, and 2 g KCl into DDW to a final volume of 1 L. Store the stock solution at room temperature (RT). Prepare 1× PBS by 1:10 dilution with DDW and autoclave before use.
3. *Trypsin–EDTA solution*: Use commercially available 0.05 % (w/v) trypsin solution containing 0.481 mM EDTA.
4. *Cell stock solution*: Prepare cell stock solution containing 50 % DMEM, 40 % FBS, and 10 % dimethylsulfoxide (DMSO).
5. *Opti-Mem I*: Use serum-free alternative medium Opti-Mem I (Invitrogen/GIBCO) for PC6 cell transfection.
6. *Transfection reagents*: Lipid-based cell transfection reagents, Effectene (Qiagen) or Transgin (APS-Bio, Gaithersburg, MD), may be used.

### 2.2 Cell Lysate Preparation

1. *1× PBS*: Prepare as described in Subheading 2.1.
2. *KLB*: Prepare and store the stock solution (500 ml) containing 10 mM HEPES-KOH (pH 7.6), 150 mM NaCl, 50 mM KF, 50 mM β-glycerol phosphate, 2 mM EGTA (pH 8.0), 10 % glycerol, and 1 % Triton X-100. Right before use, add sodium orthovanadate ($Na_3VO_4$, final concentration 2 mM) and 1× protease inhibitor cocktail to appropriate volume (depends on the usage of each experiment) of the stock solution to form the buffer KLB.
3. *4× Laemmli sample loading buffer*: Prepare by mixing 20 ml glycerol, 25 ml 20 % SDS, 6.25 ml 2 M Tris–HCl (pH 6.8), 5 ml β-mercaptoethanol, and 0.01 % (w/v) bromophenol blue and then store at RT. Add to cell lysates to a final concentration of 1×.
4. *Rubber scrapers*: Commercially available scrapers can be used.
5. *Protein assay reagents*: Commercially available protein assay reagents can be used.

### 2.3 SDS-PAGE Gel Electrophoresis and Immunoblotting

1. *30% (w/v) acrylamide solution*: Dissolve 290 g acrylamide and 8 g bis-acrylamide in DDW to a final volume of 1 L. Filter with filter paper, and store at RT.
2. *1.5 M Tris–HCl (pH 8.8)*: Dissolve 90.9 g Tris base in 400 ml DDW, adjust the pH with HCl using a pH meter to pH 8.8, and then add DDW to a final volume of 500 ml. Filter with filter paper, and store at RT.

3. *1 M Tris–HCl (pH 6.8)*: Dissolve 60.6 g Tris base in 400 ml DDW, adjust the pH with HCl using a pH meter to pH 6.8, and add DDW to a final volume of 500 ml. Filter with filter paper, and store at RT.
4. *10% (w/v) SDS*: Dissolve 50 g SDS in DDW to a final volume of 500 ml and store at RT.
5. *10% ammonium persulfate (APS) solution*: Dissolve 1 g of APS in 10 ml DDW and store at 4 °C.
6. *TEMED*: Commercial products can be used.
7. *Pre-stained protein ladder marker*: Commercially available protein ladder can be used.
8. *SDS-PAGE running buffer*: Prepare the 10× stock solution by dissolving 30 g Tris base, 144 g glycine, and 10 g SDS in DDW to a final volume of 1 L. Dilute 1:10 with DDW to 1× before use. Store at RT.
9. *Western transfer buffer*: Prepare Western transfer buffer by dissolving 5.8 g Tris base, 2.9 g glycine, and 0.37 g SDS in DDW to a volume of 800 ml and then add 200 ml methanol. Store at RT.
10. *PBS–Tween-20 solution (PBST)*: Prepare PBST by adding Tween-20 into 1× PBS to a final concentration of 0.1 % (w/v). Store at RT.
11. *Western blocking buffer*: Dissolve casein (Sigma) into PBST solution to 1 % by heating up to 40–50 °C while stirring the solution until completely dissolved. Aliquot and store at –20 °C. Once thawed, return to 4 °C for storage after each use.
12. *Nitrocellulose membrane and chemiluminescence (ECL) solutions*: Commercial reagents can be used.
13. *Antibodies*: Commercially available primary antibodies and HRP-conjugated species-specific secondary antibodies can be purchased.

### 2.4 Immunoprecipitation and GST Pulldown Assays

1. *KLB*: Prepare as described in Subheading 2.2.
2. *Protein-G-sepharose and glutathione-sepharose*: Commercial products can be used. Before use, wash the beads (30 μl/sample) sequentially with 10 volumes of 1× PBS and KLB at 4 °C, 500 ×*g*, for 5 min. Suspend the pelleted beads with KLB and keep on ice for use.
3. *Antibodies*: Anti-Flag and anti-Myc antibodies can be purchased.
4. *5 M NaCl*: Prepare 5 M NaCl with DDW and store at RT.
5. *1× Laemmli sample loading buffer*: Dilute 4× Laemmli sample loading buffer (*see* Subheading 2.2) with KLB to 1× before use.

### 2.5 Nonradioactive Assay for the Determination of Rit Activation

1. *Plasmids*: pGEX-KG-RGL3-RBD and p3×Flag-CMV-10-hRit$^{WT}$ vectors can be obtained from Dr. Andres, Department of Biochemistry, University of Kentucky College of Medicine, 741 S. Limestone St., Lexington, KY 40536. Empty Flag vector is commercially available.
2. *Transfection reagents*: Use the reagents described in Subheading 2.1.
3. *KLB*: Prepare as described in Subheading 2.2.
4. *1× Laemmli sample loading buffer*: Prepare as described in Subheading 2.4.
5. *GST-RGL3-RBD agarose*: Prepare GST-RGL3-RBD agarose beads as follows:
   (a) Prepare the following reagents first: (1) 5 ml of 10 mM isopropyl β-D-thiogalactoside (IPTG) in DDW; (2) 10 ml of TES buffer: 50 mM Tris–HCl (pH 7.5), 40 mM EDTA, and 25 % sucrose; (3) 2 ml of 20 mg/ml lysozyme in DDW; (4) 10 ml of buffer A: 10 mM Tris–HCl (pH 7.5), 1 mM dithiothreitol (DTT), and 1 mM phenylmethylsulfonyl fluoride (PMSF); (5) 1 L of buffer B: 20 mM HEPES-KOH (pH 7.6), 100 mM KCl, 1 mM DTT, and 1 mM PMSF; (6) 5 ml of 10 % (v/v) Triton X-100; and (7) 10 ml of 1 M $CaCl_2$ in DDW.
   (b) Incubate pGEX-KG-RGL3-RBD transformed *Escherichia coli* of BL21DE3 in 500 ml LB medium at 37 °C by vigorous shaking to an $OD_{600}$ of 0.5–0.6.
   (c) Add 5 ml of 10 mM IPTG solution into the culture to a final concentration of 0.1 mM.
   (d) Induce protein expression at RT for 3 h, and then harvest the cells by centrifugation at 4 °C, 5,000 ×*g*, for 10 min.
   (e) Discard the supernatant, and resuspend the bacterial pellets with 10 ml of TES buffer.
   (f) Add 2 ml of 20 mg/ml lysozyme, mix well, and then incubate on ice for 1 h.
   (g) Add 10 ml of buffer A and mix well.
   (h) Add 25 ml of buffer B and mix well.
   (i) Add 1/10 volume of 10 % Triton X-100 and mix well.
   (j) Spin cell extract at 4 °C, 100,000 ×*g*, for 30 min, and collect the supernatant.
   (k) Incubate the supernatant with 5 ml glutathione-agarose with end-to-end rotation at 4 °C for 3 h.
   (l) Pellet the GST-RGL3-RBD-bound agarose beads by brief centrifugation at 4 °C, 500 ×*g*, for 5 min.

(m) Wash the beads sequentially with 50 ml (10 volumes) of buffer B three times, buffer B + 1/5 volume of 5 M NaCl once, buffer B + 1/50 volume of 5 M NaCl + 1/50 volume of 1 M $CaCl_2$ once, buffer B + 1/5 volume of 5 M NaCl once, buffer B + 1/50 volume of 5 M NaCl once, and then buffer B twice.

(n) Resuspend the agarose pellets with 25 ml buffer B. Aliquot 500 μl/tube, and store the beads at −80 °C until use.

(o) Take a small aliquot and boil in 1× Laemmli sample loading buffer. Spin and run supernatant on 10 % SDS-PAGE gel, and estimate the concentration of bound GST-RGL3-RBD proteins following staining with Coomassie blue. Normally, a 500 μl resin suspension contains about 100 μg of recombinant GST-RGL3-RBD proteins and can be used for ten pulldown assays.

6. *Anti-Flag antibody*: Commercially available.

### 2.6 Small Hairpin RNA-Mediated Gene Silencing

1. *shRNA constructs*: We generally use the pSUPER-neo/GFP vector (OligoEngine). This vector contains a polymerase-H1-RNA gene promoter that allows the synthesis of shRNA transcripts and a PGK promoter that directs the co-expression of GFP used for detecting transfected cells and of a neomycin resistance gene used for G418 selection. Rat Rit-specific shRNA construct shRit208 and a control scramble shRNA construct shCTR can be provided by Dr. Andres.
2. *G418 selection medium*: Dissolve G418 (Geneticin, GIBCO) in DMEM-CM to 400 μg/ml and filter with a 0.22 μm filter to sterilize.
3. *KLB*: Prepare as described in Subheading 2.2.
4. *SDS-PAGE and immunoblotting reagents*: Prepare as described in Subheading 2.3.
5. *Antibodies*: Purchase the anti-Rit polyclonal antibody from Abcam or the anti-Rit monoclonal antibody from Sigma. Purchase the following antibodies: actin (Sigma), HSP27, MK2 (Cell Signaling), SOS1, SOS2, and C3G (Santa Cruz).
6. *RT-PCR reagents*: RNA isolation kit, reverse transcription kit, Oligo$(dT)_{18}$ primers, Taq DNA polymerase, and dNTPs can be purchased. Design gene-specific primers using Vector NTI software, and order the primers.
7. *DNA electrophoresis reagents and device*: Agarose, 10 mg/ml ethidium bromide (EB), 10× TAE buffer, and 1 kb DNA ladder marker can be purchased. DNA agarose gel running device can be purchased.

### 2.7 Signaling Analysis

1. *Plasmids*: Request the following plasmids from Dr. Andres: p3×Flag-CMV-10-Rit$^{WT}$, -Rit$^{Q79L}$ (constitutively active form), -Rit$^{S35N}$ (dominant negative form), -Rin$^{WT}$, -Rin$^{Q78L}$ (CA), -Rin$^{S34N}$ (DN), -H-Ras$^{WT}$, -H-Ras$^{Q61L}$ (CA), or -H-Ras$^{S17N}$ (DN) and shRNA constructs of shCTR, shRit208, and shRin99; GST, Flag, or myc-tagged WT, CA, or DN forms of B-Raf, C-Raf, p38 MAPK, MK2, HSP27, and AKT.
2. *G418 selection medium*: Prepare as described in Subheading 2.6.
3. *KLB*: Prepare as described in Subheading 2.2.
4. *4× Laemmli sample loading buffer*: Prepare as described in Subheading 2.2.
5. *Antibodies*: Commercially available phospho-specific and total protein-specific antibodies, as well as anti-Flag and anti-Myc antibodies, can be used.

### 2.8 In Vitro Kinase Assay

1. *Plasmids*: As described in Subheading 2.7.
2. *G418 selection medium*: Prepare as described in Subheading 2.6.
3. *KLB*: Prepare as described in Subheading 2.2.
4. *Kinase assay buffer (KAB)*: Prepare the stock solution (100 ml) containing 25 mM Tris–HCl, pH 7.5, 5 mM β-glycerol phosphate, 2 mM DTT, 0.1 mM $Na_3VO_4$, and 10 mM $MgCl_2$. Store the solution at −20 °C before use. For kinase assay reactions, add ATP to a final concentration of 200 μM. For radioactive analysis, add an additional 10 μCi of [γ-$^{32}$P]-ATP (Perkin Elmer).
5. *Substrates*: Purchase myelin basic protein (MBP), recombinant inactive MEK1, ERK2 (GE Healthcare), and GST-ATF2 (Cell Signaling).
6. *Scintillation reagents*: Purchase Whatman P81 paper, scintillation liquid and vials.
7. *7.5% phosphoric acid*: Dilute phosphoric acid with DDW to final concentration of 7.5 % (v/v).

### 2.9 Analysis of the Association of Rit Proteins with Their Effectors

1. *Plasmids*: As described in Subheading 2.7.
2. *KLB*: Prepare as described in Subheading 2.2.
3. *1× Laemmli sample loading buffer*: Prepare as described in Subheading 2.4.
4. *Biotinylated antibodies*: Generate biotinylated anti-Flag and -Myc antibodies using an EZ-link sulfo-NHS-biotin and biotinylation kit.

### 2.10 Neurite Outgrowth Analysis

1. *Plasmids*: Flag-tagged Rit, Rin, and H-Ras constructs as well as shRNA constructs of shCTR, shRit208, and shRin99 are described in Subheading 2.7.

2. *G418 selection medium*: Prepare as described in Subheading 2.6.
3. *Neurotrophic factors*: NGF (R&D Biosystem) or PACAP38 (Bachem) can be purchased.
4. *4% (w/v) Paraformaldehyde (PFA) solution*: Add 4 g into 90 ml of 1× PBS. Heat the suspension, and adjust the pH with 10 N NaOH (about 40–60 μl) until the solution is completely clear. Add 1× PBS to a final volume of 100 ml. Store at 4 °C before use. The solution can be used within 2 days. Otherwise, fresh solution should be prepared.
5. *Antibodies*: Anti-Flag and FITC-conjugated goat anti-mouse IgG can be purchased.
6. *Inverted microscope system*: An inverted microscope equipped with fluorescent filters and the necessary software for neurite analysis and counting can be used.

#### 2.11 Cell Survival/Death Analysis

1. *Plasmids*: Flag-Rit$^{WT, Q79L}$, H-Ras$^{WT, Q61L}$, shCTR, and shRit208 are described in Subheading 2.7.
2. *KLB*: Prepare as described in Subheading 2.2.
3. *4× Laemmli sample loading buffer*: Prepare as described in Subheading 2.2.
4. *DNA isolation buffer*: 20 mM Tris–HCl, pH 7.5, 50 mM NaCl, 10 mM EDTA, pH 8.0, 1 % (w/v) SDS, 50 μg/ml RNase A, and 100 μg/ml proteinase K.
5. *TE buffer*: Prepare 1× TE buffer containing 20 mM Tris-HCl, pH 8.0, and 1 mM EDTA, pH 8.0.
6. *DNA electrophoresis reagents and device*: Use 100 bp DNA ladder as a marker. Other reagents are the same as described in Subheading 2.6.

## 3 Methods

#### 3.1 PC6 Cell Culture and Transfection

1. Maintain PC6 cells in DMEM-CM in a humidified 37 °C atmosphere containing 5 % $CO_2$.
2. To maintain the cells, initiate the culture at a density of $2 \times 10^4$ cells/cm$^2$ and change the medium every 2 days (*see* **Note 1**). When the cell confluence reaches 70–90 %, passage and reseed the cells.
3. To passage the cells, release the cells from the tissue culture plates by incubating with trypsin–EDTA solution at room temperature briefly (~1 min). After sitting at RT for 5 min, pipette the cells off with DMEM-CM, pellet the cells by centrifugation at 500 ×*g* for 5 min, and resuspend the cells with DMEM-CM. Count cell number with a hemocytometer under a microscope.

4. Seed the cells at different densities for distinct experimental purposes. For example, seed the cells at $1 \times 10^5$ cells/cm$^2$ for transfections and at $2 \times 10^4$ cells/cm$^2$ for cell growth maintenance.
5. To prepare cell stocks, resuspend cells to a density of $5 \times 10^6$ cells/ml with cell stock solution, then aliquot into cryovials, and freeze overnight at −80 °C before storage in a liquid nitrogen tank.
6. To recover cell stocks, thaw the vials rapidly in a 37 °C water bath; transfer the cells immediately to prewarmed DMEM-CM; and then pellet, resuspend, and plate the cells for regular cultures.
7. For cell transfection, plate PC6 cells at $1 \times 10^5$ cells/cm$^2$ in 6-well plates. 12–16 h later, replace the medium with 1 ml of Opti-Mem I medium. If studies involve pharmacological inhibition, apply pretreatment with inhibitors that are specific to different kinases at this time for at least 30 min before the addition of DNA–lipid mixtures.
8. Prepare the transfection mixtures at the optimized ratio of DNA to lipids (*see* **Note 2**) and incubate at RT for 20 min to allow the formation of DNA–lipid complexes. When combination of different plasmids is required, avoid exceeding the total amount of DNA for the transfection over 2 μg/well in 6-well plates.
9. Add the DNA–lipid mixtures dropwise onto the cells with gentle swirling, and incubate the cultures in a 37 °C incubator for 9–16 h to gain maximum transfection efficiency.
10. Re-feed the cells and/or replate with DMEM-CM subsequently (*see* **Note 3**), and allow the cells to recover for the indicated periods of time prior to experimental analysis.

### 3.2 Cell Lysate Preparation

1. Aspirate the medium, and wash the cells with ice-cold 1× PBS briefly.
2. Add appropriate amount of KLB and harvest using a rubber scraper.
3. After incubation on ice for 30 min, collect the detergent-soluble fraction of the lysates after centrifugation at 4 °C, 13,000 × *g*, for 10 min.
4. Quantitate protein concentrations using protein assay reagents.
5. Prepare cell lysates following adjustment of equal protein concentrations.
6. Subject the lysates to distinct experimental analyses. For direct analysis of total cell lysates using Western blot, add 4× Laemmli sample loading buffer in the lysates to a final concentration of 1×, mix well, and then boil cell lysates for 5 min.

### *3.3 SDS-PAGE Gel Electrophoresis and Immunoblotting*

1. Assemble the gel casting system first.
2. Prepare the resolving gel solution according to the percentage of gel to be used. Mix 2.5 ml 1.5 M Tris–HCl (pH 8.8), 100 μl 10 % SDS, 100 μl 10 % APS, and appropriate volume of 30 % acrylamide solution and then add DDW to 10 ml. Add 10 μl TEMED into the resolving gel solution, mix well, and pour into the casting system to a level 1 cm below the teeth of comb. Load about 300 μl DDW on the top of gel and then set at RT for about 15 min to ensure complete gel polymerization. Once the gel has polymerized, wash the top of gel with DDW three times and aspirate DDW completely.
3. Prepare the stacking gel solution by mixing 3.63 ml DDW, 0.67 ml 30 % acrylamide solution, 0.625 ml 1 M Tris–HCl (pH 6.8), 50 μl 10 % SDS, and 50 μl 10 % APS. Add 5 μl TEMED, mix well, and pour onto the top of resolving gel. Insert comb and set at RT for 15 min for complete polymerization.
4. Assemble the gel running system, pull out the comb, and then add 1× SDS-PAGE running buffer to a level higher than the short plate.
5. Load the samples and pre-stained protein marker to individual wells.
6. Run the gel with a constant voltage of 200 V until the dye front begins running off the gel.
7. Assemble the transfer sandwich in a transfer cassette: Sequentially lay three pieces of filter paper, the gel, the nitrocellulose membrane, and then three pieces of filter paper.
8. Insert the cassette into a transfer cell with the membrane located at the positively charged side. Add Western transfer buffer to a level that covers the gel.
9. Perform the transfer with a constant voltage of 30 V overnight.
10. Disassemble the transfer cassette, and save the membrane.
11. Block the membrane with Western blocking solution for 1 h at RT with consistent rocking, and then discard the blocking solution.
12. Incubate the membrane with primary antibody for the appropriate durations (*see* **Note 4**), and then remove the antibody.
13. Wash the membrane with PBST at RT three times for 10 min each.
14. Incubate the membrane with the appropriate HRP-conjugated secondary antibody at the manufacturer's suggested dilution with Western blocking solution at RT for 1 h with consistent rocking, and then discard the antibody.

15. Wash the membrane with PBST at RT three times for 10 min each.
16. Discard PBST, and spray ECL to cover the membrane.
17. Expose the membrane to X-ray film, and develop the film. Multiple exposures may be needed to ensure proper intensity of signals.

### 3.4 Immunoprecipitation and GST Pulldown

1. Unless indicated otherwise, perform immunoprecipitations by incubating 500 μg of total protein lysates with 30 μl of protein-G-sepharose together with the appropriate antibodies (*see* **Note 5**) at 4 °C with end-to-end rotation for the indicated duration. Save the remaining cell lysates for equal protein loading analysis.
2. Pellet the sepharose beads containing the bound proteins by centrifugation at 4 °C, 500 ×*g*, for 5 min, and carefully remove the supernatants by aspiration.
3. Extensively wash the pellets with ice-cold buffers. Washing procedures are varied for distinct experimental purposes and will be described in each individual protocol.
4. After washing, resuspend and subject the resultant pellets containing the bound proteins to analysis.
5. Perform GST pulldown following the same protocol for i.p. except that glutathione-agarose resin is used instead of protein-G-sepharose and antibody.

### 3.5 Nonradioactive Assay for the Determination of Rit Activation

1. Transfect PC6 cells with 0.5 μg of Flag-Rit$^{WT}$ per well in 6-well plates. To minimize the basal GTP levels that can result from high expression of Flag-Rit$^{WT}$, co-transfect 1.5 μg of Flag-empty vector (Flag-EV) together with Flag-Rit$^{WT}$ to reduce the copies of Flag-Rit$^{WT}$ being introduced into the cells.
2. 36 h after transfection, starve PC6 cells with serum-free DMEM for 5 h and subsequently stimulate with extracellular stimuli (such as nerve growth factor (NGF) [9], PACAP38 [7], or stresses (Fig. 1) [4]) for different durations. Harvest the cells, and prepare the total cell lysates.
3. Incubate 200 μg total cell lysates with the GST-agarose beads containing approximately 10 μg of GST-RGL3-RBD at 4 °C with end-to-end rotation for 1 h (*see* **Note 6**). Meanwhile, save the remaining total cell lysates for determining equal loading of the overexpressed Flag-Rit$^{WT}$ proteins to each assay.
4. Recover the GTP-Rit-bound GST-RGL3-RBD resin by brief centrifugation (500 ×*g*, 5 min) and sequentially wash with KLB once, KLB plus 1/10 volume of 5 M NaCl twice, and KLB two more times. Boil the pelleted agarose beads containing the bound GTP-Rit in 1× Laemmli sample loading buffer

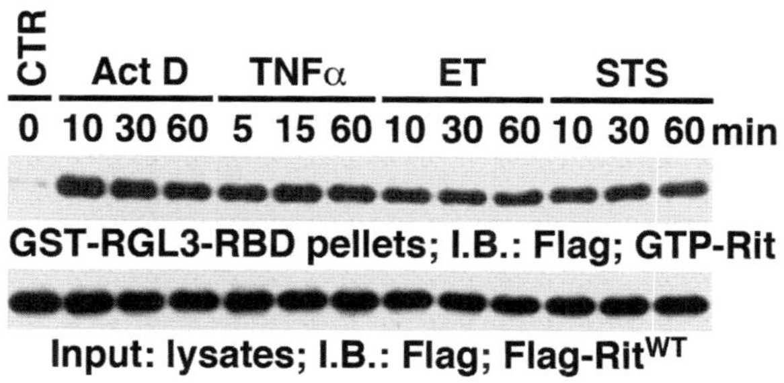


**Fig. 1** Stresses trigger Rit activation. PC6 cells were transiently transfected with a Flag-tagged wild-type Rit (Flag-Rit$^{WT}$) in vector p3×Flag-CMV-10 and starved with serum-free DMEM for 5 h prior to stimulation with actinomycin D (Act D, 2 μM), TNFα (100 ng/ml), etoposide (ET, 40 μM), or staurosporine (STS, 100 nM) for the indicated times. GTP-loaded Rit was recovered from total cell lysates (200 μg) by precipitation using GST-RGL3-RBD as bait and detected by immunoblotting with anti-Flag monoclonal antibody. The equal input of over-expressed Flag-Rit$^{WT}$ was examined by immunoblotting 5 μg of total cell lysates

for 5 min, and resolve the supernatant containing the released GTP-Rit with SDS-PAGE gels. Load 5 μg of total cell lysates for equal loading validation.

5. The GTP-Rit levels are determined by the amount of Flag-Rit precipitated by the GST-RGL3-RBD pulldowns (*see* **Note 7**). Detect the Flag-Rit levels by immunoblotting using anti-Flag monoclonal antibody (1:3,000, RT, 1-h incubation) (Fig. 1).

### *3.6 Small Hairpin RNA-Mediated Gene Silencing*

1. The gene-specific target sequences are usually suggested by OligoEngine, and the sequences with the highest potential in gene silencing are selected. Synthesize the target sense 19 nucleotide sequence, separated by a 9-nucleotide noncomplementary spacer (TCTCTTGAA) from the reverse complement of the same 19-nucleotide sequence, and subsequently subclone into the *Bgl II* and *Hind III* sites of the pSUPER-neo/GFP vector and verify the resultant construct with DNA sequencing using the M13 reverse primer. Generate an shRNA construct containing the sense sequence TTCTCCGAACGTGTCACGT that does not have a predicted target site in the rat genome in the pSUPER-neo/GFP vector as negative control (shCTR).
2. Transfect PC6 cells with 2 μg of either shCTR- or gene-specific shRNA (*see* **Note 8**) and subject to G418 selection for 60 h. Change the G418 selection medium every day to ensure removal of all untransfected cells.
3. Prepare the cell lysates, examine the levels of gene expression by immunoblotting with a gene-specific antibody, and compare the gene expression levels to those observed from the shCTR-expressing cells. Determine the equal loading by immunoblotting with β-actin.

4. When the gene-specific antibody is not commercially available, RT-PCR can be used in place of immunoblotting to validate the shRNA-mediated gene silencing.
   (a) Extract and isolate total RNA following the manufacturer's protocol from the cells transfected with gene-specific shRNA or control shCTR constructs.
   (b) Synthesize the first-strand cDNA using reverse transcription from 2 μg of total RNA as the template and oligo$(dT)_{18}$ as primers following the manufacturer's manual.
   (c) Evaluate gene expression levels by PCR using the first-strand cDNA as a template and gene-specific primers. Perform 30–40 cycles of PCR at the optimal annealing temperature calculated from the primers. Use β-actin amplification as a control.
   (d) Run PCR products in a 1 % agarose gel containing EB prior to detecting the intensity of PCR products under an ultraviolet (UV) light. Compare the intensities of the gene bands observed between testing and control samples.

### *3.7 Signaling Analysis*

1. Transfect PC6 cells with 2 μg of the appropriate plasmids per well in 6-well plates. Depending upon the nature of the experiment, the transfections are varied.

   Perform co-transfection of WT forms of effectors with CA mutants of the Rit proteins when necessary (*see* **Note 9**). Use pretreatment (>30 min) of effector-specific inhibitors such as PD98059 (MEK inhibitor, 10 μM), SB203580 (p38 inhibitor, 10 μM), LY294002 (PI3K inhibitor, 10 μM), and ddA (adenylyl cyclase inhibitor, 50–100 μM), among others [3–12], to determine the role for Rit proteins in the activation of different signaling pathways. Use empty vector and/or vehicle pretreatment (such as DMSO) as controls.

   If an effector-specific inhibitor is not available, co-transfect the DN form (0.5 μg) of an effector or the effector-specific shRNA (1.5 μg) to disrupt the activity of the effector. Use empty vectors and/or shCTR as negative controls. Confirm the specificity of shRNA-mediated signaling analysis using a reconstitution assay that involves introduction of a low expression of the WT gene derived from a different species that cannot be silenced by the shRNA. The signaling deficits caused by the shRNA-mediated gene silencing should be restored by the expression of the exogenous WT gene.

   To determine the requirement of Rit protein signaling for cellular responses to extracellular signals, transfect the DN mutants or the shRNA constructs specific for Rit proteins (Fig. 2) to disrupt Rit protein signaling. In substitution of extracellular stimulation, co-transfect the CA form of an

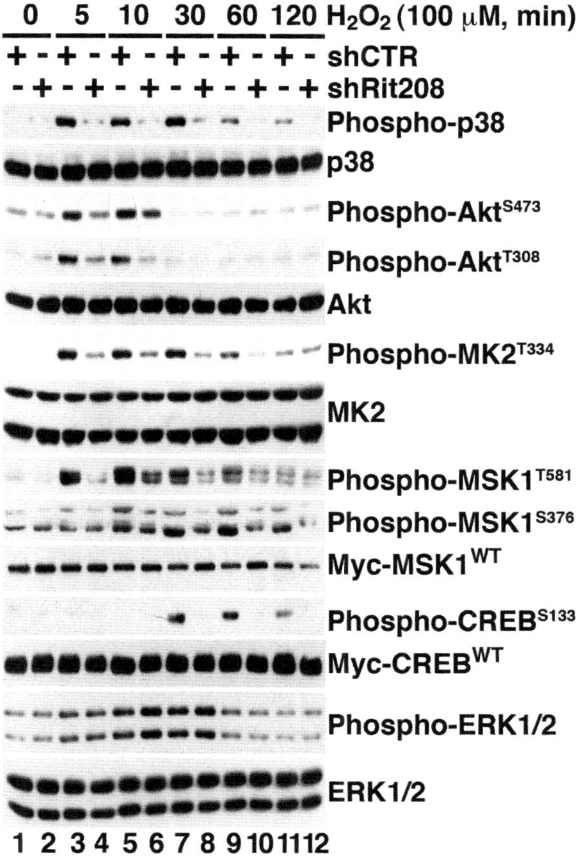


**Fig. 2** Loss of Rit attenuates cellular signaling in response to stress. PC6 cells were transfected with 2 μg of shRit208 or shCTR in the presence of Myc-tagged MSK1$^{WT}$ and CREB$^{WT}$ (50 ng). The transfected cells were enriched by G418 selection (400 μg/ml, 60 h) and then stimulated with hydrogen peroxide ($H_2O_2$, 100 μM) for the indicated duration after 5-h starvation with serum-free DMEM. Total cell lysates were subsequently prepared with kinase lysis buffer (KLB) and subjected to immunoblotting with the indicated phospho-specific antibodies, while the expression levels of individual protein kinases were also detected and shown. *Note* that Rit silencing attenuates $H_2O_2$-induced phosphorylation of p38 MAPK, MK2, Akt, MSK1/2, and CREB but not that of ERK1/2

upstream effector that triggers activation of a specific signaling cascade to define the role of Rit proteins in the activation of the signaling cascade.

2. After transfection, enrich transfected cells by G418 selection prior to starvation with serum-free DMEM for 5 h. When necessary, treat cells with an extracellular stimulus for different durations at a fixed concentration (Fig. 2) or at varied concentrations for a fixed duration to establish the dynamic role for Rit in the activation of signaling cascades being tested.

For example, receptor ligands (such as NGF [8, 9], PACAP38 [5, 7, 11], or EGF [4, 8, 9]), stresses (such as hydrogen peroxide ($H_2O_2$), TNFα, or etoposide [3, 4]), compounds that activate a specific effector (such as 8-Br-cAMP (PKA) and 8-CPT-2′-Me-cAMP (Epac) [5, 7, 11]), or biological toxins (such as cholera toxin, Gsα [5, 7, 11]) were used in our previous studies.

3. Prepare the total cell lysates, and subject the lysates to either direct immunoblotting analysis or i.p./p.d. analysis (*see* **Note 9**). Detect the activation of individual effectors by immunoblotting with effector-specific phospho-antibodies (Fig. 2) (*see* **Note 4**).

### *3.8 In Vitro Kinase Assay*

1. Transfect PC6 cells with different strategies designed for distinct experimental purposes.

   To determine the isoform-specific activation of kinases, transfect Flag-tagged CA-Rit$^{Q79L}$ or –RinQ$^{78L}$ (1.5 μg) along with 0.5 μg of distinct isoforms of WT kinases tagged with a different epitope tag (such as GST-tagged B-Raf or C-Raf [8, 9]; Myc-tagged p38α, γ or δ [7, 8]). Use Flag-tagged CA-H-Ras$^{Q61L}$ and Flag-EV transfection as controls.

   To evaluate the requirement of Rit proteins for extracellular signal-mediated activation of either Raf kinases or p38 MAPKs, transfect PC6 cells with 2 μg of Rit protein-specific shRNA plasmids [4, 8]. Use shCTR transfection as a control.

2. Enrich the transfected PC6 cells by G418 (400 μg/ml) selection for 60 h prior to starvation with serum-free DMEM for 5 h. Leave the cells untreated (for CA mutants) or stimulate the cells with an extracellular stimulus (for shRNA-mediated gene targeting). Subject soluble cell lysates (500 μg) to either immunoprecipitation with an epitope tag-specific antibody (for transfected kinases) or a kinase-specific antibody (such as phospho-p38, B-Raf, or C-Raf antibody, for endogenous kinases) or to GST pulldowns (for GST-fused kinases). Save the remaining cell lysates for equal loading validation and/or additional analyses on the activation of other effectors.

3. Wash the resulting precipitates with KLB twice by brief centrifugation (500 × *g*, 5 min). If needed, divide samples into two parts at this time, half for the kinase assay and the other half for immunoblotting with a phospho-specific antibody. Further wash the samples for phospho-blotting with KLB twice prior to Western blot analysis, whereas wash the samples with KAB twice before the kinase assay. Carefully remove all liquid remaining after the last wash.

4. Incubate the resulting resin with reaction mixtures containing kinase-specific substrates and ATP at 30 °C for 30 min to detect the kinase activities.

   Determine Raf kinase activities by a coupled kinase cascade assay with MBP as the substrate. Mix the GST-Raf kinase-bound resin gently with 38 μl of reaction mixtures containing 0.4 μg of recombinant-inactive MEK1, 1 μg of recombinant-inactive ERK2, 20 μg of MBP, 10 μCi of [$\gamma$-$^{32}$P]-ATP, and 500 μM ATP in KAB buffer and then subject to the reaction.

Evaluate p38 kinase activity using purified GST-ATF-2 as substrate. Mix the resulting p38-bound resin (from i.p. using anti-Myc or anti-phospho-p38 antibodies) gently with 30 μl of reaction mixtures in KAB containing 200 μM ATP and 2 μg of GST-ATF-2 prior to the reaction.

5. Detect the levels of kinase activity by the following methods:

   To analyze Raf kinase activity using scintillation counting, terminate the reaction by adding 25 μl of the reaction mixture onto 2 × 2 cm Whatman P81 paper. Wash the filters three times with 7.5 % phosphoric acid (5 min/wash) followed by one wash with acetone. After air-dried, insert the filters into scintillation vials and add 4 ml of scintillation liquid to each vial. Quantify the radioactivity using a scintillation counter. Perform at least three independent experiments with triplicate reactions for each experiment to ensure a statistically significant difference.

   To detect Raf kinase activity using radiographic image analysis, stop the reaction by boiling in 1× Laemmli sample loading buffer for 5 min and fractionate the samples with SDS-PAGE. Dry the gel and expose the gel to X-ray film at –80 °C or analyze the intensity of radioactivity using phospho-imager.

   p38 kinase activity is determined by the levels of phospho-ATF2 using immunoblotting. Terminate the kinase reaction by boiling in 1× Laemmli sample loading buffer prior to fractionation with SDS-PAGE. Determine the p38 kinase activity by immunoblotting with anti-phospho-ATF2 antibody.

### 3.9 Analysis of the Association of Rit Proteins with Their Effectors

1. Co-transfect PC6 cells with a combination of distinct epitope-tagged constructs (for example, pEBG-Rit or p3×Flag-CMV-10-Rit versus pCMV-Myc-B-Raf$^{WT}$ or -C-Raf$^{WT}$ (1.0 μg) [8, 9]; or p3×Flag-CMV-10-Rit versus pCMV-Myc-p38$\alpha$/$\gamma^{WT}$, -MK2$^{WT}$, -HSP27$^{WT}$, or -Akt$^{WT}$ (Fig. 3) [4]). Use empty pEBG, p3×Flag-CMV-10, and/or pCMV-Myc vectors as controls.
2. Prepare the cell lysates at 60 h after transfection and subject to either immunoprecipitation (with anti-Flag antibody for Rit or anti-Myc antibody for effectors) or GST pulldown. Extensively wash the resulting precipitates with ice-cold KLB once, ice-cold KLB plus 1/10 volume of 5 M NaCl twice, and ice-cold KLB two more times. Release the bound proteins by boiling in 1× Laemmli sample loading buffer for 5 min prior to fractionation with SDS-PAGE.
3. Detect the co-immunoprecipitated proteins subsequently as well as the i.p./p.d. efficiencies and equal protein loading to the assay by immunoblotting (Fig. 3) (*see* **Note 10**).

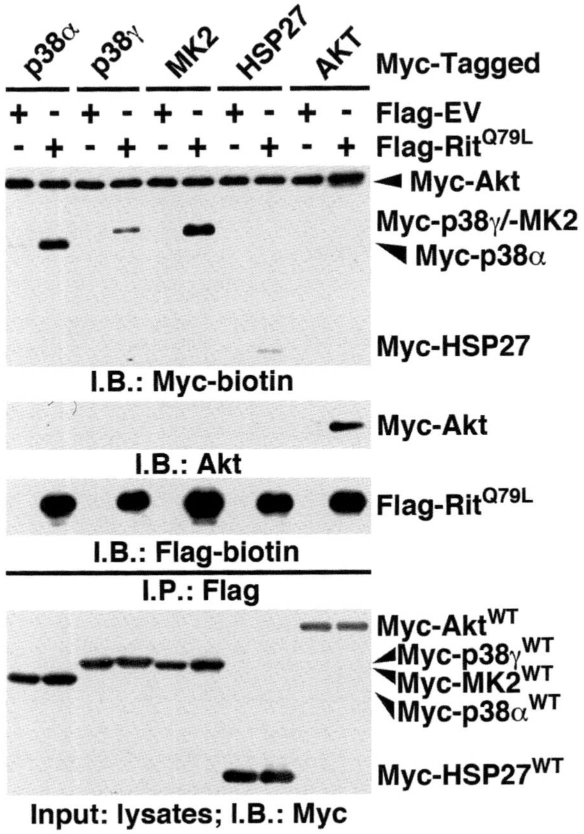


**Fig. 3** Association of Rit with p38, MK2, HSP27, and Akt. PC6 cells were transiently transfected with 1 μg of either empty Flag vector (*Flag-EV*) or Flag-Rit$^{Q79L}$ in the presence of 1 μg of Myc-tagged WT-p38α, -p38γ, -MK2, -HSP27, or -Akt. The protein association was determined by immunoprecipitation using anti-Flag antibody. The co-immunoprecipitated proteins were detected by immunoblotting with biotinylated anti-Myc antibody. A nonspecific band from the anti-Myc-biotin blot was shown closely to that for Myc-Akt and interfered with the visualization of Myc-Akt. This issue was solved by immunoblotting with anti-Akt antibody. Levels of Flag-Rit$^{Q79L}$ in the immunoprecipitates and equal loading of Myc-tagged proteins in the lysates were examined and shown

#### 3.10 Neurite Outgrowth Analysis

1. Transfect PC6 cells in 6-well plates using different strategies as described in Subheading 3.7. Use inhibitors to determine the requirement of specific signaling cascades for Rit protein-mediated neurite outgrowth, whereas co-transfect DN mutants or gene-specific shRNA to examine the contribution of Rit proteins and/or their effectors to neurotrophic factor- and/or Rit protein-stimulated neurite outgrowth [5, 7–9, 11].
2. After transfection, dislodge the cells by gentle pipetting and divide the cells equally onto four 35 mm diameter tissue culture dishes in 1.5 ml of selection medium per dish. To determine the requirement of Rit protein for neurotrophic factor

(NTF)-induced neurite outgrowth, apply NTFs such as NGF (100 ng/ml) or PACAP38 (10 or 20 nM) to the cultures at this point to initiate neurite outgrowth.

3. Replace the medium every day for the first 3 days to ensure the efficient removal of all untransfected cells and then every 2 days until the neurite outgrowth is assessed.
4. Score the remaining cells for neurite outgrowth on days 3 and 7. Analyze the percentage of neurite-bearing cells, the average number of neurites per cell body, the average neurite length, and the average number of branch points per neurite (*see* **Note 11**) under a phase-contrast microscope with a 20× lens objective using imaging software (Fig. 4) (*see* **Note 12**). Count at least 200 cells in 9–12 random fields per experiment. Perform at least three independent experiments with each experiment in triplicate to obtain a statistically significant difference.

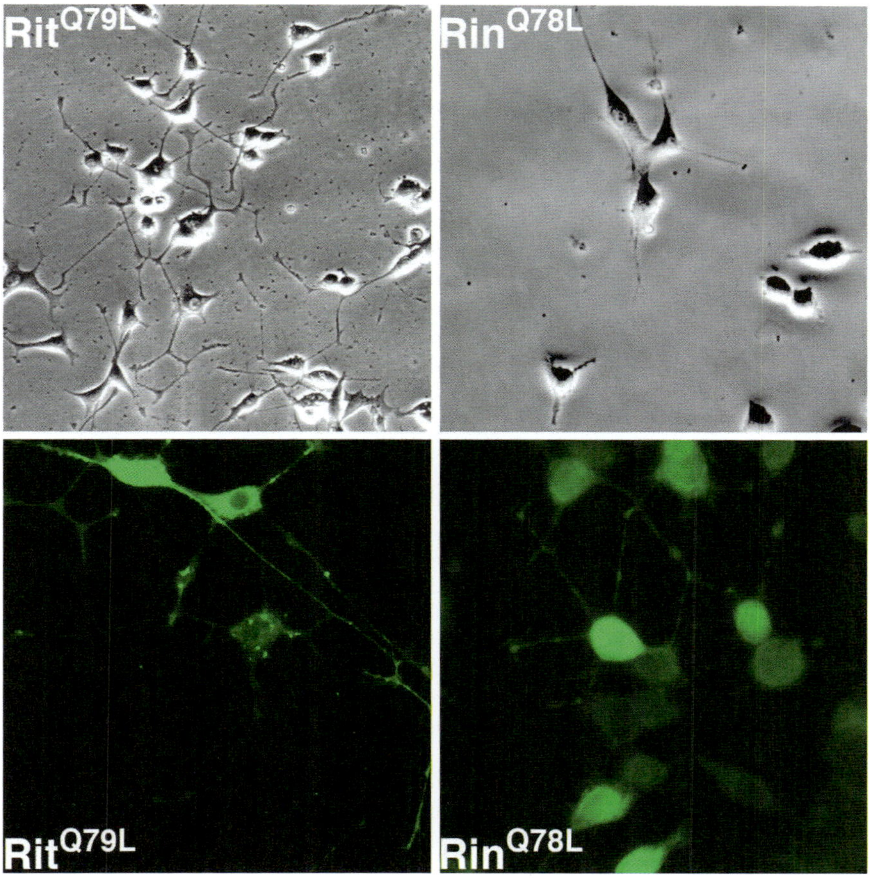


**Fig. 4** Rit and Rin induce neurite outgrowth. PC6 cells transiently expressing Flag-Rit$^{Q79L}$ (*left two panels*) or -Rin$^{Q78L}$ (*right two panels*) were maintained in DMEM complete medium supplemented with G418 (400 μg/ml). At day 7, the ability of constitutively active Rit or Rin to induce neurite outgrowth was analyzed by phase-contrast microscopy (*upper two panels*) or visualized by immunostaining with anti-Flag antibody and FITC-conjugated goat–anti-mouse IgG secondary antibody under a fluorescent microscope (*lower two panels*)

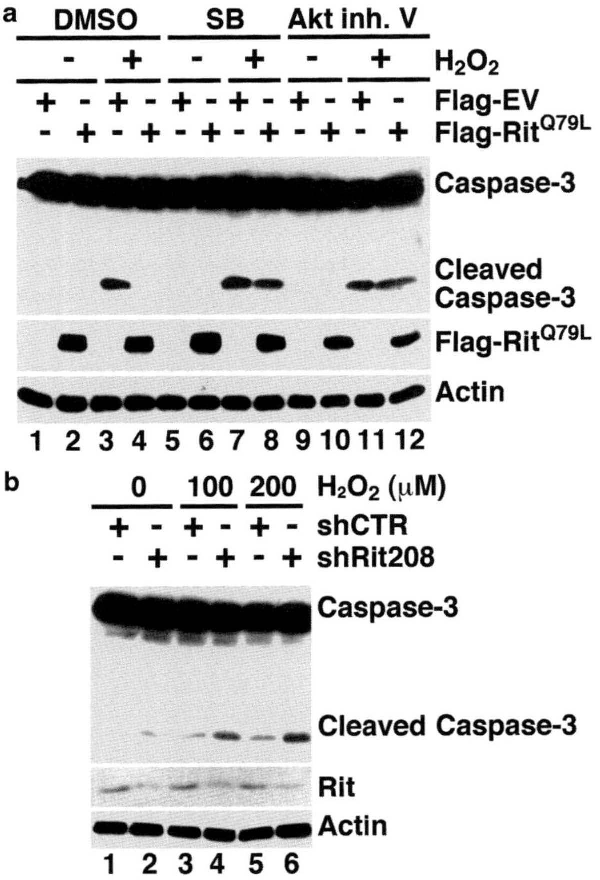


**Fig. 5** Rit confers cellular resistance to stress in PC6 cells. (**a**) Expression of active Rit (Flag-Rit$^{Q79L}$) resulted in a significant reduction in the levels of the activated caspase-3 (cleaved caspase-3) induced by $H_2O_2$ (200 μM, 15 h) (compare *lane 4* with *lane 3*), suggesting a protective role for Rit against reactive oxygen species. The Rit-mediated protection was reversed by the pretreatment of either p38 inhibitor SB203580 (SB, 10 μM) (*lane 8*) or Akt inhibitor V (Akt inh. V, 2 μM) (*lane 12*), indicating the requirements of both p38 and Akt signaling for the Rit-dependent cellular survival. (**b**) Rit loss (*lanes 4* and *6*) increased caspase-3 cleavage (cell death) triggered by exposure to $H_2O_2$ (200 μM, 15 h)

### 3.11 Cell Survival/Death Analysis

1. Transfect PC6 cells in 6-well plates using different strategies as described in Subheading 3.7. Flag-Rit$^{Q79L}$ is used to determine the protective role of Rit signaling against stresses (gain-of-function analysis). In contrast, shRit208-mediated Rit silencing is used for the loss-of-function analysis (Fig. 5). Use pretreatments of effector-specific inhibitors (Fig. 5a) or co-expression of DN mutants or shRNAs that disrupt the activities of effectors to elucidate the requirement of distinct signaling cascades for Rit-mediated protective signaling [3, 4].
2. After transfection, dislodge the PC6 cells by gentle pipetting, then replate the cells at a density of $2 \times 10^4$ cells/cm$^2$, and culture the cells for an additional 60 h to allow maximal

expression of the transfected plasmids and recovery of the cells following transfection.

3. Induce cell death by exposure to stresses at concentrations and durations sufficient to activate caspase-3 (Fig. 5) and/or DNA fragmentation, the hallmarks of cell death. The following concentrations and durations were used in our experiments: actinomycin D, 2 μM, 12 h; etoposide, 40 μM, 6 or 15 h; $H_2O_2$, 200 μM, 6 or 24 h; sorbitol, 0.2 M, 6 h; staurosporine, 100 nM, 12 h; TNFα, 100 ng/ml, 6 or 24 h [3, 4].
4. Collect the cells along with the medium by pipetting. Pellet the cells by centrifugation at 4 °C, 3,600 × *g*, for 10 min, and then wash the cells once with ice-cold PBS.
5. To assess cell death by immunoblotting, prepare total cell lysates with KLB and subject the lysates to fractionation with 14 % SDS-PAGE. Determine cell death by the elevation of the levels of activated (cleaved) caspase-3 (17, 19 kDa) by immunoblotting with anti-caspase-3 antibody (Fig. 5).
6. To examine cell death by DNA fragmentation, harvest the cells (1 well of cells per sample in 6-well plates) as described in **step 4** and lyse the cells with 500 μl of DNA isolation buffer. Incubate the cell lysates at 55 °C overnight, and extract the lysates subsequently with phenol:chloroform:isoamyl alcohol (25:24:1) three times. Precipitate small-molecular-weight DNAs with 0.6 volume of isopropanol, and pellet the DNAs by centrifugation at RT, 7,500 × *g*, for 10 min. Wash the DNAs with 70 % ethanol, air-dry, and then dissolve the DNA pellets in 20 μl of 1× TE buffer. Resolve the DNAs by 2 % agarose gel electrophoresis at low voltage (50 V) with 100 bp DNA ladder as a marker, and visualize the laddering DNAs under UV using a gel imaging system.

## 4 Notes

1. Proper handling of the PC6 cells is critical for the studies described in this chapter. Otherwise, the cells will display spontaneous neuritogenesis, which compromises the cellular responses and results in difficulty in measuring Rit-dependent signaling and function. Once the cells display over 5 % spontaneous neurites or low transfection efficiency, establish a new culture from the cell stocks.
2. The variation in transfection efficiency for PC6 cells can be large. These variations may result from differences in the quality of varied lots of transfection reagents or from varied sensitivity of different cell stocks to the transfection reagents. For these reasons, the transfection should be periodically optimized. We have achieved relatively consistent transfection efficiencies

with lipid–DNA ratios of 1:4:10 (DNA:enhancer:effectene) or 1:10 (DNA:transgin).

3. For the signaling analysis and the protein–protein association assays, drug selection (for example, G418 [Geneticin, 400 μg/ml] for vectors containing a neomycin resistance gene such as p3×Flag-CMV-10 (−14) or pCMV-Myc) may be applied to enrich the transfected cells. For Rit protein activation and cell survival assays, drug selection should be absolutely avoided.
4. The duration of incubation with primary phospho-specific antibodies varies and mostly depends on the affinity of the antibody. Optimization is therefore required. We usually begin with 1:1,000 dilution (with 5 % bovine serum albumin [BSA] in PBST) and 1 h incubation at RT or following the manufacturer's recommendation. Modifications should subsequently be made. It is also important to find highly specific primary antibodies.
5. For i.p., the amount of antibodies and/or total cell lysates and the incubation duration used for each experiment are varied. These parameters depend on both the affinity of the antibody and the strength of the protein–protein association. When using anti-Flag or anti-Myc monoclonal antibodies, we usually incubate 500 μg total lysates with 4 μg antibody and 30 μl protein G-sepharose for 1 h. For other antibodies, perform the initial experiment upon the manufacturer's suggestion and then optimize binding conditions accordingly.
6. Subject the cell lysates to the assay immediately. Freeze–thaw may result in the dissociation of GTP from Rit and Rin and is not suggested for this assay.
7. To assess GTP-Rin levels, express a minimum Ras-interacting domain (RID) of Raf (1–140 aa) fused to GST in the pGEX-KG vector (GST-Raf-RID) in *E. coli* BL21 DE3 and then couple the recombinant proteins to glutathione-agarose beads following the protocol described for GST-RGL3-RBD. Perform the Rin activation assay following the protocol described for the Rit activation assay [5, 8].
8. shRNA constructs that efficiently knock down endogenous gene expression in our studies are (except indicated, all are rat gene specific) shRit208 [9], shRin99 [8], shSOS2-3434, shC3G-128 and -2739 [11], shHSP27-344 [5], shMK2-547 and human Rit gene-specific shRit490 [4] (determined by protein levels), shEpac1-1501 [7], and shPACR1-384 [5, 11] (determined by mRNA levels).
9. When the Rit protein-triggered activation of an effector is weak or the affinity of an effector-specific antibody is not high enough to detect the modification and/or the expression of endogenous protein, the WT form of the effector can be co-transfected [3] and/or i.p. can be performed [8] to increase the sensitivity of the analysis. Similarly, co-transfection may be

necessary to determine the role of Rit proteins in the activation of individual members within a family that possess high sequence homology and/or close cellular signaling.

10. Use biotinylated antibodies when the mobility of a protein is similar to that of either the heavy (IgH) or the light (IgL) chain of the antibody. For example, Rit, Rin, and Ras proteins migrate near IgL, but they are successfully detected from the immunoprecipitates using a biotinylated anti-Flag antibody (Fig. 3) [4, 8, 9].
11. A positive neurite-bearing cell is defined as a cell which bears neurite(s) with length(s) exceeding one diameter of the cell body. The false and/or spontaneous neurite outgrowth should occur at a low frequency (<5 %) in the empty vector transfected cells or the neurotrophic factor-free cells. Otherwise, higher frequency of neurite outgrowth will interfere with the analysis.

    Count the number of neurites in a cell by including all neurites with lengths longer than one cell body. Determine the neurite length by measuring the length of the longest neurite in each cell. Count the number of neurite branch points from the longest neurite on each neurite-bearing cell.
12. Accurate quantification of neurite outgrowth should be performed by counting live cultures under an inverted microscope (Fig. 4, upper two panels). The cells can also be fixed with methanol:acetone (3:1) at 4 °C for 45 min followed by three washes with PBS and storage in PBS at 4 °C before counting. Alternatively, the neurites can also be visualized by immunostaining with anti-Flag antibody (1:1,000, 4 °C overnight) and FITC-conjugated secondary antibody (1:1,000, RT, 3 h) (Fig. 4, lower two panels) after fixation.

## Acknowledgments

We thank Dr. D.A. Andres for his support to the preparation of this manuscript.

### References

1. Rojas AM, Fuentes G, Rausell A et al (2012) The Ras protein superfamily: evolutionary tree and role of conserved amino acids. J Cell Biol 196:189–201
2. Wes PD, Yu M, Montell C (1996) RIC, a calmodulin-binding Ras-like GTPase. EMBO J 15:5839–5848
3. Shi GX, Cai W, Andres DA (2012) Rit-mediated stress resistance involves a p38-mitogen- and stress-activated protein kinase 1 (MSK1)-dependent cAMP Response Element-binding Protein (CREB) activation cascade. J Biol Chem 287:39859–39868
4. Shi GX, Jin L, Andres DA (2011) A Rit GTPase-p38 mitogen-activated protein kinase survival pathway confers resistance to cellular stress. Mol Cell Biol 31:1938–1948
5. Shi GX, Jin L, Andres DA (2008) Pituitary adenylate cyclase-activating polypeptide 38-mediated Rin activation requires Src and contributes to the regulation of HSP27 signaling during neuronal differentiation. Mol Cell Biol 28:4940–4951

6. Lein PJ, Guo X, Shi GX et al (2007) The novel GTPase Rit differentially regulates axonal and dendritic growth. J Neurosci 27:4725–4736
7. Shi GX, Rehmann H, Andres DA (2006) A novel cyclic AMP-dependent Epac-Rit signaling pathway contributes to PACAP38-mediated neuronal differentiation. Mol Cell Biol 26:9136–9147
8. Shi GX, Han J, Andres DA (2005) Rin GTPase couples nerve growth factor signaling to p38 and b-Raf/ERK pathways to promote neuronal differentiation. J Biol Chem 280: 37599–37609
9. Shi GX, Andres DA (2005) Rit contributes to nerve growth factor-induced neuronal differentiation via activation of B-Raf-extracellular signal-regulated kinase and p38 mitogen-activated protein kinase cascades. Mol Cell Biol 25:830–846
10. Shao H, Andres DA (2000) A novel RalGEF-like protein, RGL3, as a candidate effector for Rit and Ras. J Biol Chem 275:26914–26924
11. Shi GX, Jin L, Andres DA (2010) Src-dependent TrkA transactivation is required for pituitary adenylate cyclase-activating polypeptide 38-mediated Rit activation and neuronal differentiation. Mol Biol Cell 21:1597–1608
12. Andres DA, Shi GX, Bruun D et al (2008) Rit signaling contributes to interferon-gamma-induced dendritic retraction via p38 mitogen-activated protein kinase activation. J Neurochem 107:1436–1447

# Chapter 16

# Immunofluorescence Methods in Studies of the GTPase Ran and Its Effectors in Interphase and in Mitotic Cells

Giulia Guarguaglini, Valeria de Turris, and Patrizia Lavia

## Abstract

The GTPase RAN, the regulators of its nucleotide-bound state, and its effectors represent a specialized network in the RAS GTPase superfamily and regulate the localization of macromolecules (RNAs and proteins) in subcellular compartments in interphase cells and at the mitotic apparatus when cells divide. Essential cell cycle processes, e.g., replication, repair, transcription, export of transcribed RNAs out of the nucleus, assembly of the mitotic apparatus, kinetochore function, chromosome segregation, nuclear reorganization, and rebuilding of the nuclear envelope and nuclear pores at mitotic exit, ultimately depend on RAN's ability to orchestrate localization of key target factors in space and time. To achieve this, RAN network members acquire themselves dynamic localization patterns. Biochemical fractionation protocols describe where the bulk of RAN network members localize. Immunofluorescence methods have revealed more subtle and complex patterns, with specific populations of RAN network components associating with cellular structures, or organelles, where they play crucial roles as spatial regulators for a large set of macromolecules. These localization studies are important to understand RAN modes of action and to identify new targets of RAN control. Here we describe methods for the visualization of RAN network members and effectors in mammalian cells.

**Key words** Immunofluorescence, GTPase RAN, Importin beta, Exportin/CRM1, Nucleocytoplasmic transport, Mitosis, Mitotic spindle, Nuclear envelope, Nucleus, Spatial control

## 1 Introduction

RAN network members and effectors—the latter belonging to the family of transport receptors (or karyopherins) and including nuclear import and export vectors—have a dynamic localization and shuttle between the nucleus and the cytoplasm throughout the interphase cell cycle. Most of these shuttling factors also comprise fractions that associate with particular cellular structures or organelles (e.g., the nuclear envelope, nuclear pores, and centrosomes in interphase, and spindle microtubules, spindle poles, and chromosomal kinetochores in mitosis), where they orchestrate spatial regulatory processes of critical biological importance.

Lorenza Trabalzini and Saverio Francesco Retta (eds.), *Ras Signaling: Methods and Protocols*,
Methods in Molecular Biology, vol. 1120, DOI 10.1007/978-1-62703-791-4_16, © Springer Science+Business Media, LLC 2014

For many years, the RAN network and effectors have been studied at the biochemical and structural levels as regulators of nucleocytoplasmic transport. These studies have had seminal value in deducing general paradigms for vectorial transport, but their inherent limit was that they examined a pool of different populations of molecules, "averaging" their behavior, and missed the identification of localized fractions with subtle, finely tuned functions.

The development of immunofluorescence protocols has led to identification of localized populations of RAN network members at their site of action. Immunofluorescence has therefore acquired increasing importance in co-localization studies and in identifying targets of RAN and its effectors. It is worth noting that, for many RAN network members, fluorescently tagged chimeric proteins—which necessarily entail some level of overexpression—fail to reveal specific localization patterns at particular subcellular sites or structures. These may also be difficult to appreciate when using standard immunofluorescence protocols, in which cells are directly fixed prior to being permeabilized to allow antibody entry into the cell ("fix first then permeabilize"). These procedures fix the entire pool of the target protein, including its soluble component, anywhere in the cell, and reveal where, by and large, the bulk of that protein grossly localizes at any given time: for example, RAN and its guanine exchange factor RCC1 are typically depicted in interphase nuclei and the GTP hydrolysis-activating factors RANGAP1 and RANBP1 in the cytoplasm. The "fix first" approach may be useful in quantitative immunofluorescence assays, but it usually covers up fractions of factors with specific localization and function. Regulatory and functional studies may rather take advantage from the use of protocols that can depict localized fractions at their target site. This requires a detergent-based step to solubilize diffuse molecules prior to fixation ("first solubilize then fix" approach).

Subheading 3 illustrates protocols designed to visualize RAN network components associated with interphase and mitotic structures in mammalian (human and murine) cells. These protocols include general common steps to be used for all purposes, including the preparation of cover slips (Subheading 3.1) and basic immunostaining (Subheading 3.9). They also detail permeabilization steps designed to solubilize non-associated proteins prior to fixation and decrease the contribution of diffuse signals to the image; these steps vary in buffer composition and timing, depending on the nature (membranous, proteinaceous, etc.) of the structure or organelle to be visualized, on the abundance of the protein under investigation, and on the nature of its interaction with particular structures. Specifically:

- Subheading 3.2 describes an optimized protocol to detect RAN and RANBP1 centrosomal fractions in interphase and mitotic cells [1, 2]; under standard fixation conditions, these restricted fluorescence signals are "masked" by the strong staining of RAN in nuclei and of RANBP1 in the cytoplasm, respectively.

- Subheading 3.3 describes an optimized procedure to visualize RANGTP using conformation-specific antibodies to the GTP-bound form. Our experience is with anti-RAN serum AR-12 [3], generated in Ian Macara's laboratory against a RAN-specific C-terminal peptide that extrudes when RAN is bound to GTP (*see* comment in ref. 4). Commercial antibodies are also available as indicated in the procedure. Preserving the GTP-bound RAN conformational specificity requires that the antibody be incubated with unfixed permeabilized cell samples. This procedure depicts RANGTP at centrosomes [2] and nuclear envelopes in interphase ([5], ***Supplementary Material***) and at spindle poles, microtubules [5, 6], and kinetochores [7] in mitosis.
- Subheading 3.4 describes a protocol adaptation for importin beta, a major RANGTP effector and vector for nuclear protein import. Importin beta is a shuttling molecule and carries protein cargoes into the nucleus; after nuclear envelope breakdown it acts as a negative regulator of several processes in mitosis. Subheading 3.4 describes how a mild extraction step prior to fixation eliminates the soluble importin beta fraction: in interphase the nuclear pore-associated fraction is depicted as a sharp staining punctating the nuclear envelope; in mitosis, sharp signals can be seen at spindle poles and along microtubules [6].
- Subheadings 3.5 and 3.6 report procedures for RANGAP1 (RAN GTPase-activating protein 1) and for the exportin CRM1 (a RANGTP effector transporting proteins out of the nucleus into the cytoplasm). RANGAP1 and CRM1 localize on opposite sides of the interphase nuclear envelope (to the cytoplasmic and nuclear faces, respectively), and both are recruited to the spindle microtubules and kinetochores in mitosis [8–11].
- Subheadings 3.7 and 3.8 briefly report standard fixation protocols, based on formaldehyde and methanol, respectively. The protocol in Subheading 3.7 is particularly suitable for the purpose of quantitative studies, in which the total amount of a given RAN component has to be assessed. Both of these protocols can also be used to visualize RCC1, the guanine exchange factor for RAN, which interacts stably with chromatin by binding histones H2A and H2B: RCC1 is the only RAN network member for which the total and the localized protein pools coincide. In this case, therefore, the fixation procedure is flexible and may be chosen to accommodate conditions required to co-visualize other protein(s) of interest in the same experiment.

## 2 Materials

### 2.1 Cover Slip Coating

1. Sterile glass cover slips, 0.16–0.18 mm thick.
2. Sterile deionized water.
3. Petri dish.
4. Plastic grid.
5. Parafilm.
6. Forceps.
7. Sterile poly-L-lysine solution stored at 4 °C.

### 2.2 Fixation and Permeabilization

1. Phosphate-buffered saline (PBS; 1×): 137 mM NaCl, 2.7 mM KCl, 10 mM $Na_2HPO_4$, 2 mM $KH_2PO_4$, pH 7.4.
2. Permeabilization solutions: PBS, PHEM, or PEM buffer containing Triton X-100 at the final concentration indicated in the individual protocols (*see* **Note 1**).
3. 0.1 M Glycine in PBS.
4. PHEM stock solution (2×): 90 mM Pipes, 90 mM Hepes, 20 mM EGTA, 10 mM $MgCl_2$. Dilute to 1× just before use, and add phenylmethylsulfonyl fluoride (PMSF) 1 mM final concentration (*see* **Notes 2** and **3**).
5. PEM solution (1×): 20 mM Pipes, 10 mM EGTA, 1 mM $MgCl_2$.
6. Absolute methanol stored at −20 °C.
7. Paraformaldehyde (PFA) solution: Prepare fresh at 3–3.7 % in 1× fixation buffer plus 30 mM sucrose (*see* **Note 4**).
8. Transport buffer (2×): 220 mM KOAc, 4 mM $Mg(OAc)_2$, 1 mM EGTA, 40 mM Hepes. Store at 4 °C. Dilute to 1× with sterile deionized water before use.
9. Transport buffer plus digitonin: Dilute the digitonin stock solution (0.5 % in sterile deionized water; *see* **Notes 5** and **6**) 1:100 in transport buffer to obtain a final digitonin concentration of 0.005 %. Add inhibitors to a final concentration of 2 mM for DL-dithiothreitol (DTT) and 1 mg/ml each for leupeptin, pepstatin A, and aprotinin (*see* **Notes 7–10**).

### 2.3 Staining

1. PBS containing 0.05 % Tween-20 (PBST) (*see* **Note 1**).
2. Blocking and antibody diluent solution (PBST-B): PBS containing 0.05 % Tween-20, 3 % bovine serum albumin (BSA).
3. 4′,6-diamidino-2-phenylindole (DAPI) solution: Store the stock solution (5 mg/ml in sterile deionized water) in small aliquots at −20 °C (*see* **Note 11**). When needed, dilute one aliquot to 0.1 μg/ml in sterile deionized water (working solution) and store at 4 °C for 3–4 weeks (*see* **Note 12**).

4. Vectashield (Vector): Store protected from light at 4 °C.
5. Microscope slides.
6. Nail polish.
7. Parafilm.
8. Forceps.
9. Multiwell plates.

### 2.4 Primary Antibodies

1. RAN: Mouse monoclonal clone 20/Ran (BD Transduction Laboratories), working dilution 0.25 μg/ml; goat polyclonal C-20 (Santa Cruz Biotechnology, Inc.), working dilution 2 μg/ml.
2. RANGTP: Rabbit antiserum AR-12 ([3]; source: courtesy of Dr. Ian Macara, Center for Cell Signalling, Charlottesville, VA, USA), working dilution 1:50. Commercial monoclonal antibodies have recently been made available from NewEast Biosciences, but their use in immunofluorescence assays has not been reported in publications as yet.
3. RCC1: Goat polyclonal C-20 (Santa Cruz Biotechnology, Inc.), working dilution 2 μg/ml.
4. RANGAP1: Rabbit H-180 or goat N-19, both polyclonal (both from Santa Cruz Biotechnology, Inc.), working dilutions 1 and 4 μg/ml, respectively.
5. RANBP1: Goat polyclonal M-19 for murine cells and C-19 for human cells (both from Santa Cruz Biotechnology, Inc.), working dilutions 2 and 8 μg/ml, respectively.
6. Importin beta: Mouse monoclonal clone 23/Karyopherin β (BD Transduction Laboratories), working dilution 5 μg/ml; mouse monoclonal anti-NTF97/importin beta [3E9] (Abcam), ascites, working dilution 1:1,000.
7. CRM1: Rabbit polyclonal H-300, raised against region 772–1071 (Santa Cruz Biotechnology, Inc.), working dilution 4 μg/ml.

## 3 Methods

Specific fixation and/or permeabilization procedures are adapted to visualize the indicated RAN network member; their ideal use is discussed in Subheadings 1 and 4.

Carry out all procedures at room temperature unless otherwise specified.

All solutions are intended 1× unless otherwise specified.

For general advice and tips on immunofluorescence procedures, also *see* [12].

### 3.1 Coating of Cover Slips with Poly-L-Lysine

1. Handle sterile cover slips under a sterile tissue culture hood (*see* **Note 13**); using forceps put them on a plastic grid support.
2. Place about 150 μl of poly-L-lysine onto square 18 × 18 mm cover slips (*see* **Note 14**); let stand for 5 min.
3. Take one cover slip at the time using forceps (*see* **Note 15**), and wash it in a Petri dish filled with sterile deionized water.
4. Put cover slips side up on a clean tissue, and let them dry overnight under UV light.
5. Store them at 4 °C, dry, in a clean Petri dish sealed with parafilm.
6. For immunofluorescence assays, place a UV-sterile poly-L-lysine-coated cover slip in a culture dish and grow cells directly on the cover slip; when ready, treat cell samples as required according to your chosen experimental protocol.

### 3.2 Fixation and Permeabilization for RAN and RANBP1 at Centrosomes and at the Mitotic Apparatus

1. Wash cells in PBS (*see* **Note 16**).
2. Permeabilize cells briefly; suggested conditions are 30 s in PHEM containing 0.5 % Triton X-100 for RANBP1 and 1 min in PHEM containing 0.1 % Triton X-100 for RAN (*see* **Note 17**).
3. Fix cells in 3 % PFA in PHEM plus 30 mM sucrose for 10 min (*see* **Notes 4, 18, 19**, and **20**).
4. Wash once with PBS.
5. Quench in 0.1 M glycine for 10 min (*see* **Note 21**).
6. Wash twice with PBS.
7. Follow the immunostaining procedure described in Subheading 3.9.

### 3.3 Permeabilization for RANGTP-Specific Immunostaining (See Note 22)

1. Wash cells in PBS (*see* **Note 16**).
2. Permeabilize cells in PHEM containing 1 % Triton X-100 for 1 min.
3. Wash once in PHEM.
4. Incubate with AR-12 rabbit antiserum diluted 1:50 in PBST-B (*see* **Note 23**) for 15 min at room temperature.
5. Wash once in PBST (*see* **Note 23**).
6. Wash twice in PBS.
7. Fix cells in 3.7 % PFA in PBS plus 30 mM sucrose for 10 min (*see* **Note 4**).
8. Wash twice in PBS.
9. Incubate in −20 °C cold methanol for 2 min at −20 °C.
10. Wash twice in PBS.
11. Follow the immunostaining procedure described in Subheading 3.9 (*see* **Note 24**).

### 3.4 Fixation and Permeabilization for Importin Beta at Spindle Microtubules and Nuclear Envelope

1. Wash cells in PHEM (*see* **Note 16**).
2. Permeabilize cells in PHEM containing 1 % Triton X-100 for 4 min.
3. Wash for 2 min in PHEM.
4. Fix cells in 3.7 % PFA in PHEM plus 30 mM sucrose for 10 min (*see* **Note 4**).
5. Wash twice in PBS.
6. Incubate in –20 °C cold methanol for 2 min at –20 °C.
7. Wash twice with PBS.
8. Follow the immunostaining procedure described in Subheading 3.9.

### 3.5 Fixation and Permeabilization for RANGAP1 at Kinetochores and Nuclear Envelope

1. Wash cells in transport buffer (*see* **Note 16**).
2. Permeabilize cells in transport buffer containing 0.005 % digitonin (*see* **Note 25**) for 4 min.
3. Wash once in PBS.
4. Fix cells in 3.7 % PFA in PBS plus 30 mM sucrose for 10 min (*see* **Note 4**).
5. Wash once with PBS.
6. Quench in 0.1 M glycine for 10 min (*see* **Note 21**).
7. Wash twice with PBS.
8. Follow the immunostaining procedure described in Subheading 3.9.

### 3.6 Fixation and Permeabilization for CRM1 at Kinetochores and Nuclear Envelope

1. Wash cells in PEM (*see* **Note 16**).
2. Permeabilize cells in PEM containing 0.1 % Triton X-100 for 1 min.
3. Wash once in PEM.
4. Fix cells with 3.7 % PFA in PEM plus 30 mM sucrose for 10 min (*see* **Note 4**).
5. Wash once with PBS.
6. Quench in 0.1 M glycine for 10 min (*see* **Note 21**).
7. Wash twice with PBS.
8. Follow the immunostaining procedure described in Subheading 3.9.

### 3.7 Direct PFA Fixation Procedure (See Note 18)

1. Perform two washes of cells in PBS.
2. Incubate cells in 3.7 % PFA in PBS plus 30 mM sucrose for 10 min at room temperature (*see* **Note 4**).
3. Wash once in PBS.
4. Permeabilize fixed cells in PBS containing 0.1 % Triton X-100 for 5 min.
5. Proceed as described from **step 4** in Subheading 3.2.

### 3.8 Direct Methanol Fixation (See Note 19)

1. Wash cells in PBS.
2. Incubate in −20 °C cold methanol for 6 min at −20 °C.
3. Wash twice with PBS.
4. Proceed for immunofluorescence staining (Subheading 3.9).

### 3.9 Immuno-fluorescence Staining (See Note 26)

1. Blocking: Incubate cover slips treated according to one of the procedures described in Subheadings 3.2–3.8 for 30 min with PBST-B (*see* **Note 27**).
2. Incubation with primary antibody: Dilute antibodies to the indicated concentration in PBST-B and incubate for 1 h (*see* **Notes 27** and **28**).
3. Wash three times in PBST for 5 min; gentle agitation is recommended (*see* **Note 29**).
4. Incubation with secondary antibody (*see* **Note 30**): Dilute antibodies to the indicated final concentration in PBST-B and incubate for 45 min.
5. DNA staining (*see* **Note 31**): Incubate cover slips for 1 min in DAPI solution.
6. Wash three times for 5 min in PBST; gentle agitation is recommended.
7. During the washing step, clean up microscope slides in ethanol, dry, and label them (*see* **Note 32**).
8. Mount cover slips on clean microscope slides in Vectashield (*see* **Note 33**).
9. Seal mounted cover slips with nail polish (*see* **Note 34**).

## 4 Notes

1. Cut off the pipette tip just enough to broaden up the opening end, and very slowly aspire Triton X-100 or Tween-20; be careful to take just exactly the right volume, as the stock solution is very dense and viscous. Note that it sticks to the outside of the plastic pipette; absorb excess on parafilm.
2. Work under a chemical hood as PMSF is highly toxic. To prepare a PMSF stock solution in isopropanol, weight an empty sterile Falcon tube, then add 1 g of PMSF, and calculate the amount of isopropanol required to make it 100 mM (1 g of PMSF in 57.4 ml of isopropanol). Dissolve it by heating (40–50 °C in water bath). Aliquot the solution and keep at −20 °C for long-term storage. PMSF powder has a shelf life of 3 years. The stock solution is stable for at least 9 months at 2–8 °C.
3. Vortex the PHEM solution while adding PMSF to help dissolving it.

4. The addition of sucrose to PFA helps to preserve cytoskeleton structures.
5. Digitonin is highly toxic, so open the tube containing the powder under a chemical hood. Wear gloves, mask, and eye protection.
6. The digitonin solution should always be prepared fresh. If needed, keep at 4 °C for a week.
7. Work with DTT under a chemical hood as it is highly toxic. To make a DTT stock solution, weight an empty sterile Falcon, then add 1 g of DTT, and calculate the required volume of sterile deionized water required to make it 1 M (1 g of DTT in 6.5 ml of water). Mix well, aliquot, and store at −20 °C.
8. Prepare a 10 mg/ml leupeptin stock solution in sterile deionized water: mix well, aliquot, and store at −20 °C.
9. Pepstatin A has limited solubility in water. Acetic acid must be added to dissolve it in methanol or DMSO. To prepare a 1 mg/ml stock solution in 10 % acetic acid in methanol, add part of the final volume directly into the powder container, mix well, transfer to a Falcon tube, and then bring the solution to the final volume; mix well, aliquot, and store at −20 °C. The stock solution is stable for at least 1 week at 4 °C or for months at −20 °C. If the solution turns yellow the reagent is hydrolyzing.
10. The aprotinin stock comes dissolved in 0.9 % NaCl in 0.9 % benzyl alcohol. The concentration depends on the lot. Prepare small aliquots and store at 2–8 °C.
11. DAPI is readily soluble in water and has instead limited solubility in PBS.
12. The DAPI solution should be stored in the dark and shielded from light by wrapping the tube in aluminum foil to prevent photobleaching.
13. Always handle cover slips under a tissue culture hood to preserve sterility. Drain off excess liquid from one edge of the cover slip onto clean tissue at each step.
14. The volume of poly-L-lysine may vary depending on the cover slip surface area. Be careful to add enough poly-L-lysine to cover the entire surface. To prevent cover slips from sticking to plastic surfaces during preparation, use a plastic grid support rather than a flat support.
15. Remember that only one side of the cover slip is coated at this point! Always keep the coated surface side up.
16. Perform all steps under a chemical hood. Bring buffers to room temperature before applying to cells to avoid fragmentation of microtubules. Completely remove medium by washing cells in the same buffer that will be used for fixation (e.g., PBS, PHEM).

Handle the cell preparations gently, taking special care during washes before fixation to preserve mitotic cells (may tend to detach even from poly-L-lysine-coated slides). Use a 1 ml manual pipette. After fixation you may use a vacuum pump to change solutions, but take care to aspire the liquids off from one side of the culture dish, not from the center of the cover slip. It may be helpful to check the state of the cultures under an inverted microscope before and after fixing to make sure that mitotic cells (round and rifrangent) were not lost during the procedure.

17. The indicated conditions work well in our experience, yet the ideal timing and Triton X-100 concentration in the permeabilization step may need further adjustment depending on the cell type and relative abundance of the target protein.
18. PFA fixation is based on protein cross-linking and, when used directly on the cell preparation, preserves the entire pool of the protein under investigation (i.e., both soluble and associated to specific structures). A permeabilization step is needed for the antibody to enter the cell. Because permeabilization will solubilize and remove diffusible molecules, its schedule—before or after fixation—will depend on whether the experimental goal is to visualize the total amount and global distribution of the protein of interest (e.g., in quantitative immunofluorescence studies) or their association with particular structures (e.g., in functional and co-localization studies), as indicated in Subheading 1.
19. Note that not all antibodies will work in PFA, because some antigens may become inaccessible after the cross-linking reaction. Methanol fixation may be used as an alternative procedure. The methanol-based fixation procedure is quick, being based on protein precipitation, and allows for simultaneous fixation and permeabilization.
20. In studies of centrosome function, RANBP1 and RAN can be directly immunostained on purified centrosome preparations rather than on whole cells. Isolated centrosomes can be prepared as described in [13]. Sediment the isolated centrosomes onto cover slips at 20,000 × $g$ for 15 min at 4 °C and then fix using the methanol method as described in Subheading 3.8 (*see* **Note 19**).
21. Autofluorescence can be generated by the reaction of aldehyde-containing fixing solutions with amines and proteins. Quenching by adding bland amino groups (e.g., glycine) is therefore recommended.
22. It is mandatory that cells be unfixed during incubation with the conformational antibody in order to preserve the native GTP-bound conformation of RAN.
23. It is advised to increase the concentration of Tween-20 to 0.1 % (rather than 0.05 % Tween-20, as used in standard PBST and PBST-B).

24. The AR-12 antiserum-based protocol can be used in quantitative immunofluorescence assays to measure the intracellular RANGTP abundance [5] or in qualitative analyses, e.g., to assess RANGTP dynamic relocalization in response to particular conditions [7]. NewEast Biosciences provides a recent commercial source of antibodies for GTP-bound Ran.
25. Incubation may require optimization depending on the cell type: e.g., 4 min for HeLa cells and 2 min for NIH/3T3 cells.
26. For blocking and antibody incubation, carefully place the cover slips with the cell side up on a parafilm sheet using forceps. Remember to label positions. Be sure to cover the entire surface with the solution, e.g., about 80 μl per 18 × 18 cover slips.
27. This step will help eliminating nonspecific antibody interactions. As an alternative blocking agent, fetal bovine serum (FBS, Santa Cruz Biotechnology, Inc.) may be used: block cover slips by incubating in PBS containing 20 % FBS and 0.1 % Tween-20, 30 min at room temperature; dilute primary and secondary antibodies at the desired concentration in PBS containing 5 % FBS and 0.1 % Tween-20, and then proceed as described using 0.1 % Tween-20 in the washing steps.
28. Keeping the cover slips on parafilm, remove liquids from an edge by aspiration with a vacuum pump, taking care not to remove cells, and then apply the primary antibody solution. Depending on the experimental goal, co-labeling may be performed with an independent primary antibody to visualize the associated cellular structure, e.g., alpha-tubulin for spindle microtubules, lamin B1 for the nuclear envelope, pericentrin for centrosomes, and CREST (centromere protein human antiserum, Antibodies Incorporated) for kinetochores.
29. Using forceps carefully remove cover slips from the parafilm, and put them in multiwell plates. Remember to label positions. Carefully change washing solutions using a vacuum pump.
30. Secondary antibodies are labeled with fluorescent dyes: from this step onwards, shield all solution-containing tubes and cover slips from light, covering them with aluminum foil during incubations and washing steps.
31. Shield the DAPI solution from light and cover the cover slips with foil at all steps of the staining procedure. This step can be skipped if DNA staining is not required. In this case move directly to the next washing step.
32. This step will remove dust and grease.
33. It is important to use a mounting medium with antifading properties to prevent both the fading of fluorophores over time and photobleaching during microscope observation. We routinely use Vectashield, but other media are also effective. To mount cover slips on microscope slides, carefully remove all liquid traces

from previous steps by adsorbing on clean tissue. Put a drop of mounting media on a clean microscope slide, and then place the cover slip with cell side down over the drop. Carefully remove any air bubbles, and absorb excess of mounting solution by covering it with tissue for 5–10 min.

34. For long-term storage cover slips can be permanently sealed around their perimeter with transparent nail polish. Mounted slides should be stored at 4 °C or −20 °C depending on the directions indicated for the mounting medium that was used (check in the datasheet).

## Acknowledgements

We are grateful to Dr. Pietro Cirigliano and to staff of Nikon Instruments S.p.A. (Italy) for collaborative support in implementation and development of the imaging platform at the CNR-IBPM Nikon Reference Center (National Research Council of Italy-Institute of Molecular Biology and Pathology). Work in our laboratory is supported by grants from the Italian Association for Cancer Research (IG10164).

### References

1. Di Fiore B, Ciciarello M, Mangiacasale R et al (2003) Mammalian RanBP1 regulates centrosome cohesion during mitosis. J Cell Sci 116: 3399–3411
2. Kéryer G, Di Fiore B, Celati C et al (2003) Part of Ran is associated with AKAP450 at the centrosome: involvement in microtubule-organizing activity. Mol Biol Cell 14: 4260–4271
3. Richards SA, Lounsbury KM, Macara IG (1995) The C terminus of the nuclear RAN/TC4 GTPase stabilizes the GDP-bound state and mediates interactions with RCC1, RAN-GAP, and HTF9A/RANBP1. J Biol Chem 270:14405–14411
4. Macara IG (1999) Nuclear transport: randy couples. Curr Biol 9:R436–R439
5. Tedeschi A, Ciciarello M, Mangiacasale R et al (2007) RANBP1 localizes a subset of mitotic regulatory factors on spindle microtubules and regulates chromosome segregation in human cells. J Cell Sci 120:3748–3761
6. Ciciarello M, Mangiacasale R, Thibier C et al (2004) Importin beta is transported to spindle poles during mitosis and regulates Ran-dependent spindle assembly factors in mammalian cells. J Cell Sci 117:6511–6522
7. Torosantucci L, De Luca M, Guarguaglini G et al (2008) Localized RanGTP accumulation promotes microtubule nucleation at kinetochores in somatic mammalian cells. Mol Biol Cell 19:1873–1882
8. Joseph J, Tan SH, Karpova TS et al (2002) SUMO-1 targets RanGAP1 to kinetochores and mitotic spindles. J Cell Biol 156:595–602
9. Joseph J, Liu ST, Jablonski SA et al (2004) The RanGAP1-RanBP2 complex is essential for microtubule-kinetochore interactions in vivo. Curr Biol 14:611–617
10. Arnaoutov A, Azuma Y, Ribbeck K et al (2005) Crm1 is a mitotic effector of Ran-GTP in somatic cells. Nat Cell Biol 7:626–632
11. Roscioli E, Di Francesco L, Bolognesi A et al (2012) Importin-β negatively regulates multiple aspects of mitosis including RANGAP1 recruitment to kinetochores. J Cell Biol 196:435–450
12. Spector DL, Goldman RD (eds) (2006) Basic methods in microscopy. Protocols and concepts from cells: a laboratory manual. Cold Spring Harbor Laboratory Press, Cold Spring Harbor, NY
13. Moudjou M, Bornens M (1994) Isolation of centrosomes from cultured animal cells. In: Celis JE (ed) Cell biology: a laboratory handbook. Academic Press, pp 595–604

# Chapter 17

# High-Resolution Scanning Electron Microscopy for the Imaging of Nuclear Pore Complexes and Ran-Mediated Transport

**Lihi Shaulov, Boris Fichtman, and Amnon Harel**

## Abstract

High-resolution scanning electron microscopy provides three-dimensional surface images of nuclear pore complexes (NPCs) embedded in the nuclear envelope. Here, we describe a method for exposing the nuclear surface in mammalian tissue culture cells for imaging by scanning electron microscopy. Hypotonic treatment is followed by low-speed centrifugation onto polylysine-coated silicon chips, without the use of detergents. This helps to preserve NPCs close to their native morphology, embedded in undamaged nuclear membranes. This method is particularly advantageous for combining high-resolution imaging of NPCs with mammalian genetic systems.

**Key words** Scanning electron microscopy, Nuclear pore complex, Nuclear envelope, Mammalian cell culture

## 1 Introduction

Scanning electron microscopy (SEM) is a powerful technique for acquiring high-resolution surface images of nuclear envelopes and nuclear pore complexes (NPCs; [1–4]). To do this, one must first gain access to the outer surface of the cell nucleus, a task which has previously been achieved by the use of permeabilizing detergents or dry fracturing techniques [5–7]. A notable exception concerns the huge nuclear envelopes of amphibian oocytes, which can be manually removed and have provided a wealth of structural information in the pioneering work of Terry Allen and colleagues [1, 3, 8].

We have recently developed a new method for exposing the nuclei of mammalian tissue culture cells by repeated hypotonic treatments and without the use of detergents [9]. This method paves the way for combining genetic manipulation in mammalian systems with three-dimensional surface imaging by SEM. Importantly, field emission SEM imaging does not depend on

Lorenza Trabalzini and Saverio Francesco Retta (eds.), *Ras Signaling: Methods and Protocols*,
Methods in Molecular Biology, vol. 1120, DOI 10.1007/978-1-62703-791-4_17, © Springer Science+Business Media, LLC 2014

averaging techniques and reveals subtle differences between individual NPCs. The effects of certain Ran GTPase mutants on NPC morphology have previously been investigated by microinjection of the recombinant mutant proteins into *Xenopus laevis* oocytes [10]. Using the new methodology it now becomes possible to test the effect of point mutants in Ran and its effectors as well as mutations in key nucleoporins and dominant negative mutants that block specific nuclear transport pathways [11–13]. This strategy can be applied to cell culture lines following transfection and also to the analysis of primary cultures derived from transgenic animals. In particular, it is interesting to test the potential effect of specific mutations on the appearance of the "central plug" or the "transporter," which has long been thought to represent large cargo caught in transit through the central channel of the NPC [14–16].

A detailed protocol is described here for COS-1 cells and has been successfully adjusted to three other types of adherent cells (immortalized cell lines and primary culture). Practical considerations for applying the procedure to additional cell types are discussed in Subheading 4.

## 2 Materials

### 2.1 Cell Culture, Buffers, and Growth Media

1. Phosphate-buffered saline (PBS): 2.67 mM KCl, 1.47 mM $KH_2PO_4$, 138 mM NaCl, 8.1 mM $Na_2HPO_4$, pH 7.4. Store at 4 °C.
2. Use a trypsin (0.05 %)–EDTA (0.02 %) solution for cell detachment. Store at −4 °C.
3. Cell culture medium: To 500 ml of Dulbecco's modified Eagle medium (DMEM) with 4.5 g/l D-glucose add 5 ml of 200 mM glutamine, 50 ml of fetal bovine serum (FBS), 5 ml of 100 mM sodium-pyruvate, 5 ml of penicillin–streptomycin solution (10,000 units/ml). Filter the mixture through a 0.22 μm sterile vacuum filtration system. Store at 4 °C (*see* **Note 1**).
4. Cell freezing medium: To 40 ml of cell culture medium add 5 ml of dimethyl sulfoxide (DMSO) and 5 ml of FBS. Store at 4 °C.

### 2.2 Buffers for Hypotonic Treatment

1. Hypotonic buffer: 15 mM Tris–HCl, 10 mM NaCl, 3 mM $MgCl_2$, pH 7.4. Prepare 50 ml of the buffer from separate stock solutions (e.g., 1 M Tris–HCl, 5 M NaCl, 1 M $MgCl_2$) and ultrapure, sterilized water. Filter through a 0.22 μm filter, divide into 5 ml aliquots, and freeze at −20 °C (*see* **Note 2**).
2. Hypotonic buffer + 10 % glycerol (*see* **Notes 2** and **8**).
3. PBS + 10 % glycerol (*see* **Note 2**).

***2.3 SEM Preparation Reagents***

1. Polylysine coating solution: Dissolve poly-L-lysine hydrobromide (*see* **Note 3**) in ultrapure water at 0.1 mg/ml, filter through a 0.22 μm filter, and divide into 120 μl aliquots, each suitable for coating six 5 × 5 mm chips. Store at −20 °C.
2. 4″ Silicon Wafer in 5 × 5 mm chips (Ted Pella).
3. Carbon adhesive tabs.
4. Aluminum specimen mounting stubs.
5. 3 % glutaraldehyde: Dilute 8 % glutaraldehyde (EM grade) in PBS to a final concentration of 3 %, immediately before use (*see* **Note 4**).
6. 0.2 M cacodylate buffer, pH 7.4 (*see* **Note 5**).
7. 1 % osmium tetroxide: Dilute 2 % osmium tetroxide in 0.2 M cacodylate buffer to a final concentration of 1 %, immediately before use (*see* **Notes 4** and **5**).
8. Graded ethanol series: 30, 50, 70, 90, 95, and 100 % ethanol. Prepare from absolute, molecular biology-grade ethanol and ultrapure water, filter through a 0.22 μm filter, and store at −20 °C.

***2.4 Equipment***

1. Refrigerated bench-top centrifuge with fixed-angle and swing-bucket rotors (e.g., 5810R, Eppendorf).
2. Critical point dryer (e.g., CPD030, Bal-Tec) attached to a high-purity liquid $CO_2$ cylinder.
3. Chromium sputter-coating apparatus (e.g., K575X, Emitech).
4. High-resolution field emission scanning electron microscope (e.g., Zeiss ULTRA *plus*). An in-lens detector for secondary electrons is required for surface imaging. If immunogold labeling is performed, a separate EsB detector for backscattered electrons will be required to localize gold particles [9].

## 3 Methods

***3.1 Experimental Setup***

Prepare the following items for SEM sample preparation on the day of the experiment and before treating the cells.

1. Plan on using four 5 × 5 mm silicon chips for one million COS-1 cells. This number may vary for other cell types according to the dimensions of the adherent cells and the efficiency of nuclear exposure and recovery.
2. Spread a drop of 20 μl of 0.1 mg/ml polylysine over the surface of each chip (shiny side up) and incubate for 20 min at room temperature. Wash the chip in filtered ultrapure water and air-dry (*see* **Note 6**).
3. Prepare an improvised low-speed centrifugation device, as shown in Fig. 1, for each silicon chip. Place the chip with the polylysine-coated side facing upwards on the flat bottom of the device (Fig. 1b–d; *see* **Note 7**).

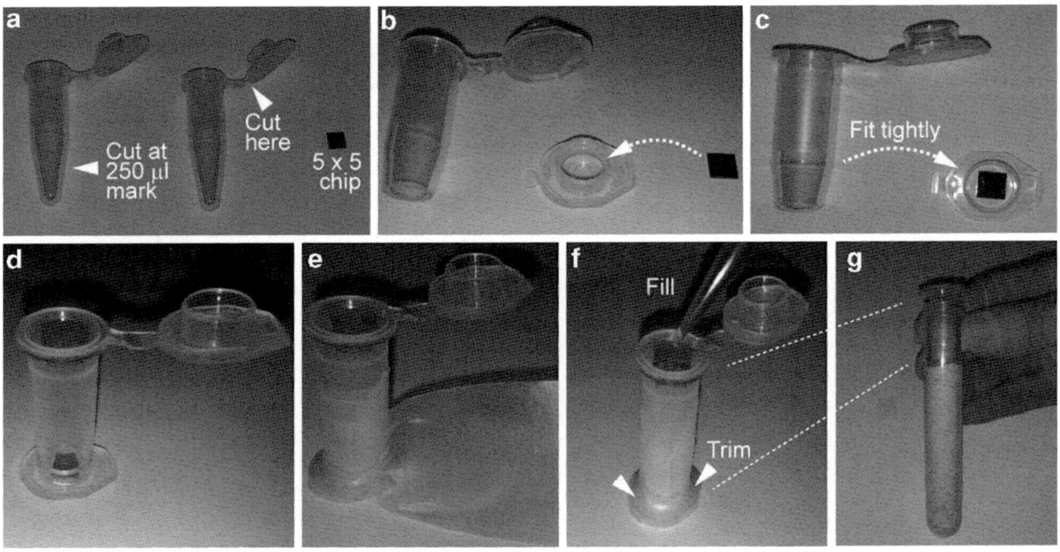


**Fig. 1** Preparation of a low-speed centrifugation device for silicon chips. (**a**) Use a razor blade to cut one 1.5 ml Eppendorf tube at the 250 μl mark and remove the lid from a second tube. (**b**) Place a polylysine-coated 5 × 5 mm silicon chip with the shiny side facing upwards inside the lid. (**c**, **d**) Place the cut bottom of the first tube inside the lid so that it fits tightly, and (**e**) wrap parafilm around the tube to prevent leakage. (**f**) Trim the protruding edges of the lid at the bottom of the device, and fill with 900 μl PBS + 10 % glycerol, followed by the suspension of treated cells. (**g**) Place inside a 14 ml Falcon tube stuffed with paper up to the 10.5 ml mark, and spin in a swing-bucket rotor

4. Right before starting the hypotonic treatment of the cells, check each centrifugation device by filling it with 900 μl PBS + 10 % glycerol (Fig. 1f) to ensure that there are no leaks.

### 3.2 Hypotonic Treatment of Tissue Culture Cells

1. Grow cells under standard conditions to 70–80 % confluence in 100 mm dishes.
2. Detach cells by trypsinization, add growth medium, and pellet the cells by centrifugation for 3 min at 400 x*g* and room temperature.
3. Wash twice in PBS (resuspend and centrifuge each time) in preparation for the hypotonic treatment. Alternatively freeze cells in freezing medium (use 2 ml for one 100 mm dish) for storage or shipment (*see* **Note 8**).
4. Gently resuspend ~250,000 cells for each silicon chip in 200 μl of hypotonic buffer inside a 1.5 ml Eppendorf tube and incubate on ice for 5 min.
5. Pass the cells twice through a syringe connected to a 21 gauge needle and centrifuge for 3 min at 1,500 x*g* and 4 °C in a fixed-angle rotor.
6. Optional: Repeat **steps 4** and **5** with 200 μl of hypotonic buffer + 10 % glycerol (*see* **Note 9**).
7. Resuspend each cell pellet in 100 μl of PBS + 10 % glycerol.

8. Gently add the cell suspension to the 900 µl of buffer inside each improvised centrifugation device from **step 4** in Subheading 3.1. Centrifuge for 10 min at 800 × *g* and 4 °C in a swing-bucket rotor to pellet the cells and exposed nuclei onto the polylysine-coated silicon chip.

### 3.3 Preparation for SEM

1. Carefully transfer the silicon chips, using high-precision tweezers, into a 24-well cell culture plate filled with PBS. Keep track of chip numbering, and take care not to overturn the chips. Up to three chips can be handled in one well. Avoid drying (*see* **Note 10**).
2. Fix in 3 % glutaraldehyde for 30 min at room temperature.
3. Wash 3 × 2 min in 0.1 M cacodylate buffer pH 7.4.
4. Postfix in osmium tetroxide for 10 min at room temperature.
5. Wash 3 × 2 min in ultrapure water.
6. Dehydrate through a graded ethanol series: 30, 50, 70, 90, 95 and 100 %, 2 × 2 min each wash.
7. Transfer the silicon chips to a critical point drying apparatus filled with ethanol as an intermediate fluid. Exchange into high-purity liquid $CO_2$ and subsequently from liquid to gas, according to the manufacturer's instructions. Samples should be stored under vacuum, in a desiccator, from this point on (*see* **Note 11**).
8. Coat the samples with 3–4 nm of chromium (for COS-1 cells). For other cell types it may be advantageous to use a thinner coating layer (*see* **Note 12**).
9. Mount silicon chips on aluminum specimen stubs using double-side carbon adhesive tabs. Attach the stubs to the SEM viewing stage. For best results, perform SEM imaging on the same day and minimize the exposure of the samples to air. Although they can be stored under vacuum for days to weeks, chromium oxidation causes a gradual deterioration of the specimens over time.

### 3.4 High-Resolution SEM Imaging

High-resolution field emission SEM imaging can only be performed after professional training on the appropriate instrument. However, once this condition is met, it is relatively easy to analyze several chips in one session and search large areas over a wide range of magnifications. Successful exposure and preservation of the nuclear envelope and NPCs can only be determined by viewing the samples at high magnification. Although many nuclei can be identified, only some of these are suitable for high-resolution imaging because of the problem of specimen charging (*see* **Note 12**). SEM images of COS-1 cells at low magnification (a field of treated cells) and at higher magnification (NPCs on the surface of an exposed nucleus) are shown in Fig. 2. Note one example (arrowhead in Fig. 2b) of an exceptionally large "central plug," or particle, occupying most of the space of the transport channel of one NPC.

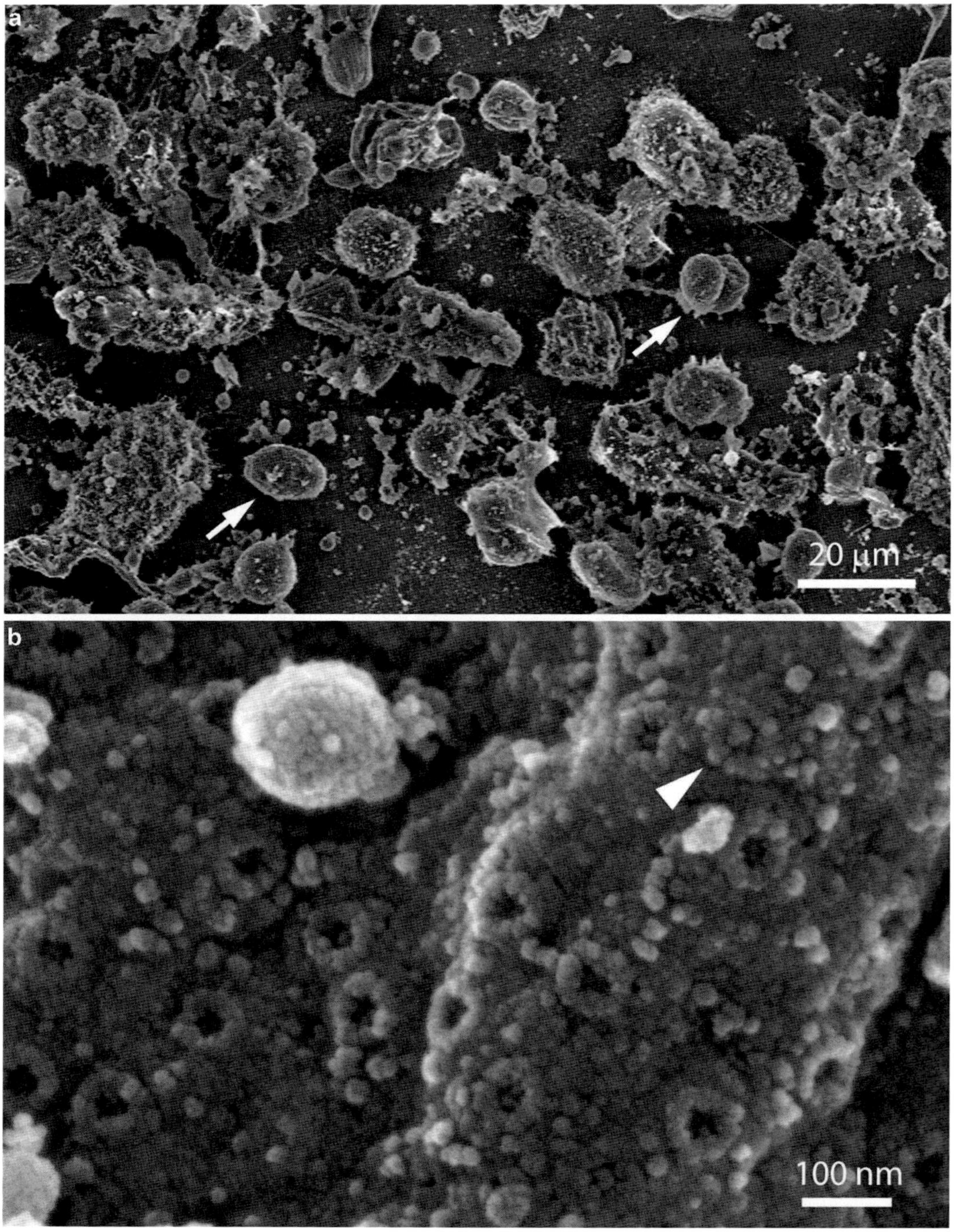


**Fig. 2** SEM imaging of COS-1 cells following hypotonic treatment and the exposure of nuclei. (**a**) A low-magnification image showing a field of cells with *arrows* pointing at two spherical bodies identified as nuclei (this can only be done by zooming in to higher magnifications). (**b**) Selected surface area of an exposed nucleus shown at high magnification. NPCs are embedded in the undamaged membrane surface of the nuclear envelope. *Arrowhead* points to an exceptionally large particle caught in transit through the central transport channel of one NPC

## 4 Notes

1. Cell culture media should be prepared in a laminar flow hood (class II safety cabinet) and stored at 4 °C according to the manufacturer's instructions. Warm media to 37 °C prior to use, and use sterile technique in the laminar flow hood until the hypotonic treatment stage.
2. All the buffers for hypotonic treatment should be filtered through 0.22 μm filters, frozen in aliquots, and thawed before use. It is essential to keep these solutions free from any form of aggregates or particles.
3. Various commercially available polylysine reagents produce markedly different results. We find that poly-L-lysine hydrobromide (Sigma; P1524) ensures reproducible results and recommend the preparation of a large quantity of stock solution to be stored in frozen aliquots.
4. Glutaraldehyde and osmium tetroxide should be purchased in single-use, sealed ampoules. Working solutions for fixatives should be prepared with 0.22 μm filtered buffers, immediately before use. In our hands, the addition of formaldehyde to glutaraldehyde in the fixation solution did not improve the final results.
5. Sodium cacodylate has been the buffer of choice for some electron microscopy applications since the 1960s; however, it is rather sensitive. We recommend the use of commercially available 0.2 M solutions of this buffer (e.g., Electron Microscopy Sciences).
6. Before coating the chips with polylysine, it is advisable to engrave numbers or letters in one corner with a diamond pen. This helps in keeping track of multiple samples and treatments, and the markings can be detected by the naked eye as well as by SEM imaging at low magnifications. The comparison of different treatments is particularly important for applying this method to additional cell types.
7. In our hands, this type of improvised low-speed centrifugation device gives the best and most consistent results, with a high yield of cells and nuclei spread over the entire surface of the silicon chips.
8. No significant differences in the efficiency of the procedure were detected between thawed and directly treated cells for COS-1, HeLa cells, and mouse lung fibroblasts. In practical terms, this means that cell lines can be maintained and frozen in one location and shipped on dry ice for the preparation for SEM in a different location.

9. We have tested this procedure on the COS-1, HeLa, and U2OS cell lines and on a primary culture of mouse lung fibroblasts. The detailed protocol is described for COS-1 cells. For each cell type, the length of incubation and the exact handling procedure need to be optimized empirically. Thus, one incubation for 5 min on ice and gentle handling are sufficient to release COS-1 cell nuclei, while HeLa cell nuclei require 2 × 20-min incubations and more aggressive handling in pipetting and in the passage through a 21 gauge needle. The successful exposure of nuclei and the degree of preservation of the nuclear envelope can only be monitored by SEM imaging at high magnifications. Therefore, it is important to prepare multiple samples and vary the length of treatment when testing a new cell type.
10. It is essential to prevent the chips from drying, even for a few seconds. We find that it is easier to continuously treat chips in the same well than to transfer them with tweezers. To do this, work simultaneously with two pipetmans: one for removing the liquid covering the chips in the well and the other for introducing a new solution. Avoid large bursts of liquid to prevent the chips from overturning.
11. Critical point drying is another delicate stage of the preparation. Simple air drying of specimens can cause the collapse of nuclear structure or severe deformation as the result of surface tensions at the boundary between the evaporating liquid and air. When a critical point drying apparatus is operated correctly, ethanol is replaced with liquid $CO_2$ which is changed to vapor at 35 °C and the correct pressure, without a change in density. This eliminates surface tension effects and is essential for producing high-quality SEM images.
12. COS-1 cell nuclei are particularly susceptible to specimen charging and therefore require a thicker chromium coat. The other cell types we have tested can be imaged with a 1–2 nm coating layer, which is more suitable for immunogold labeling and the detection of gold particles by an EsB detector.

## Acknowledgments

The authors thank Eugenia Klein and Michael Elbaum for creative brainstorming and invaluable advice on SEM analysis. This work was supported by a research grant from the Israel Science Foundation (1072/10) to A.H.

## References


1. Goldberg MW, Allen TD (1992) High resolution scanning electron microscopy of the nuclear envelope: demonstration of a new, regular, fibrous lattice attached to the baskets of the nucleoplasmic face of the nuclear pores. J Cell Biol 119:1429–1440
2. Goldberg MW, Allen TD (1993) The nuclear pore complex: three-dimensional surface structure revealed by field emission, in-lens scanning electron microscopy, with underlying structure uncovered by proteolysis. J Cell Sci 106:261–274
3. Goldberg MW, Allen TD (1995) Structural and functional organization of the nuclear envelope. Curr Opin Cell Biol 7:301–309
4. Harel A, Orjalo AV, Vincent T et al (2003) Removal of a single pore subcomplex results in vertebrate nuclei devoid of nuclear pores. Mol Cell 11:853–864
5. Allen TD, Rutherford SA, Murray S et al (2007) Visualization of the nucleus and nuclear envelope in situ by SEM in tissue culture cells. Nat Protoc 2:1180–1184
6. Drummond SP, Rutherford SA, Sanderson HS et al (2006) High resolution analysis of mammalian nuclear structure throughout the cell cycle: implications for nuclear pore complex assembly during interphase and mitosis. Can J Physiol Pharmacol 84: 423–430
7. Walther TC, Alves A, Pickersgill H et al (2003) The conserved Nup107-160 complex is critical for nuclear pore complex assembly. Cell 113:195–206
8. Allen TD, Rutherford SA, Murray S et al (2007) A protocol for isolating Xenopus oocyte nuclear envelope for visualization and characterization by scanning electron microscopy (SEM) or transmission electron microscopy (TEM). Nat Protoc 2:1166–1172
9. Shaulov L, Harel A (2012) Improved visualization of vertebrate nuclear pore complexes by field emission scanning electron microscopy. Structure 20:407–413
10. Goldberg MW, Rutherford SA, Hughes M et al (2000) Ran alters nuclear pore complex conformation. J Mol Biol 300:519–529
11. Gorlich D (1997) Nuclear protein import. Curr Opin Cell Biol 9:412–419
12. Hodge CA, Tran EJ, Noble KN et al (2011) The Dbp5 cycle at the nuclear pore complex during mRNA export I: dbp5 mutants with defects in RNA binding and ATP hydrolysis define key steps for Nup159 and Gle1. Genes Dev 25:1052–1064
13. Terry LJ, Shows EB, Wente SR (2007) Crossing the nuclear envelope: hierarchical regulation of nucleocytoplasmic transport. Science 318:1412–1416
14. Jarnik M, Aebi U (1991) Toward a more complete 3-D structure of the nuclear pore complex. J Struct Biol 107:291–308
15. Beck M, Forster F, Ecke M et al (2004) Nuclear pore complex structure and dynamics revealed by cryoelectron tomography. Science 306:1387–1390
16. Hoelz A, Debler EW, Blobel G (2011) The structure of the nuclear pore complex. Annu Rev Biochem 80:613–643

# Chapter 18

# Effector Recruitment Method to Study Spatially Regulated Activation of Ras and Rho GTPases

**Lauren P. Huff, Molly J. DeCristo, and Adrienne D. Cox**


## Abstract

Ras and Rho family GTPases control a wide variety of cellular processes, and the signaling downstream of these GTPases is influenced by their subcellular localization when activated. Since only a minority of total cellular GTPases is active, observation of the total subcellular distribution of GTPases does not reveal where active GTPases are localized. In this chapter, we describe the use of effector recruitment assays to monitor the subcellular localization of active Ras and Rho family GTPases. The recruitment assay relies on preferential binding of downstream effectors to active GTPases versus inactive GTPases. Tagging the *G*TPase-*b*inding-*d*omain (GBD) of a downstream effector with a fluorescent protein produces a probe that is recruited to compartments where GTPases are active. We describe an example of a recruitment assay using the GBD of PAK1 to monitor Rac1 activity and explain how the assay can be expanded to determine the subcellular localization of activation of other GTPases.


**Key words** Active GTPases, Subcellular localization, Recruitment, Fluorescence, Binding domain

## 1 Introduction

The Ras and Rho families of small GTPases comprise 55 family members that act as cellular signaling nodes [1]. Numerous extracellular signals converge on these GTPases, which then regulate diverse downstream signaling cascades. The subcellular localization of GTPase activation strongly influences its specific outcomes. For example, when K-Ras4B is active at the plasma membrane, it signals for cell survival, but when it is active at internal membranes such as endoplasmic reticulum and mitochondria, it signals for cell death [2]. However, the predominant subcellular localization of a GTPase does not necessarily correlate with where that GTPase is active [3] (*see* Table 1). A given stimulus may lead to activation of less than 10 % or as much as 40 % of the available cellular Ras [4] or Rho [5, 6]. Therefore, typical observation of GTPase distribution (via immunofluorescence or fractionation) primarily detects the inactive pools of that GTPase, and alternative techniques must

Lorenza Trabalzini and Saverio Francesco Retta (eds.), *Ras Signaling: Methods and Protocols*,
Methods in Molecular Biology, vol. 1120, DOI 10.1007/978-1-62703-791-4_18, © Springer Science+Business Media, LLC 2014

**Table 1**
**GTPase-binding domains (GBDs) of Ras and Rho family downstream effectors can be used to determine the subcellular locations of active, GTP-bound GTPases, which are frequently distinct from the global locations of total GTPase pools**

| GTPase | Subcellular localization: Active GTPase pools | Total GTPase pools |
|---|---|---|
| *RAS family* | | |
| Di-Ras1 | PM [32][a], vacuolar membranes [32][a] | PM [32], vacuolar membranes [32] |
| Di-Ras2 | PM [33][b], [32][a], vacuolar membranes [32][a] | PM [32], vacuolar membranes [32] |
| Di-Ras3 (ARH1, NOEY2) | PM [33][b] | |
| E-Ras | PM [34][a] | PM [34] |
| Gem (Kir) | PM [33][b] | Cytosol [35], nucleus [35] |
| H-Ras | PM [25, 33, 36][b] (non-lipid rafts [37, 38]), Golgi [25, 37], endoplasmic reticulum (ER) [25] | PM [39, 40] (lipid rafts [37, 38]), Golgi [40] |
| K-Ras (4A) | PM [33][b] | PM [41][a] |
| K-Ras (4B) | PM [25], [33][b] (non-lipid rafts [38]), mitochondria [2][a] | PM [40, 41] (non-lipid rafts [38]) |
| N-Ras | PM [25], [33][b] (lipid rafts [37]), Golgi [25, 37] | PM [41] (non-lipid rafts [37]) |
| NKIRas1 | Cytosol [33][b], nucleus [33][b] | |
| NKIRas2 | Cytosol [33][b], nucleus [33][b] | |
| R-Ras | PM [33][b], focal adhesions [42], endosomes [43] | Perinuclear region [42], Golgi [42], early and recycling endosomes [43] > PM [42, 44] |
| R-Ras2 (TC21) | PM [33][b] (lipid rafts and non-lipid rafts [45]), ER [45] | PM [44] (multiple microdomains [45]) > focal adhesions [45], perinuclear region (ER and cis-Golgi [45]), recycling endosomes [45] |
| R-Ras3 (M-Ras) | PM [33][b] (non-lipid rafts [46]) | PM [44] (non-lipid rafts [46]) |
| Rad | PM [33][b] | Cytosol [47], nucleus [47] |
| RalA | PM [33, 48, 49][b], perinuclear region (recycling endosomes [48]), mitochondria [50] | Punctate cytoplasmic vesicles [48] |
| RalB | PM [33, 48][b] | Punctate cytoplasmic vesicles [48], endomembranes [51] |
| Rap1A | PM [17], perinuclear region [36] | Perinuclear endosomes [8], late endosomes/lysosomes [52] |
| Rap1B | PM [53][a], cytosol [53][a] | Late endosomes/lysosomes [52] |
| Rap2A | PM [33][b], vesicles [33][b] | Golgi [52] |
| Rap2B | PM [33][b], vesicles [33][b] | |
| Rap2C | | PM [54], cytosol [55], recycling endosomes [56] |
| RasD1 | Cytosol [33][b], nucleus [33][b] | |
| RasD2 | PM [33][b], vesicles [33][b] | |
| RasL 10A (RRP22) | Nucleolus [57] | Nucleus [57] |
| RasL 10B (RRP17) | | Perinuclear region [58] |
| RasL 11A | | Nucleolus [59] |

| | | |
|---|---|---|
| RasL 11B | Cytosol [33][b], nucleus [33][b] | Cytosol [60] |
| RasL 12 | Nucleus [33][b] > cytosol [33][b] | |
| Rem1 (Ges) | PM [33][b] | Cytosol [47], nucleus [47] |
| Rem2 | | Cytosol [61], nucleus [61] |
| Rerg | Nucleus [33][b] > cytosol [33][b] | Cytosol [62] > nucleus [62] |
| Rheb1 | Endomembranes [34][a] | Endomembranes [34] |
| Rheb2 | Endomembranes [34][a] | Endomembranes [34] |
| Rit1 | PM [33][b] | PM [63] |
| Rit2 (Rin) | Nucleus [33][b] > PM [33][b] | PM [63] |
| *RHO family* | | |
| Cdc42 | Trans-Golgi [64] > PM [18, 19, 40], cell periphery [64] | PM [11, 40], Golgi [11, 40], ER [40], nuclear envelope [40] |
| Chp (RhoV) | PM [65], focal adhesions [66][b] | PM [11], endomembranes [11], Golgi [65] |
| Rac1 | PM [33, 40][b], lamellipodia, lamellae [3, 40, 67] | Cytosol [31] > PM [40], nuclear envelope [3] |
| Rac2 | Membranes [68], PM [33][b] (vs. not PM [40])[a] | Golgi [40], ER [40], cytosol [11], endomembranes [69], nuclear envelope [40] > PM [40] |
| Rac3 | Membranes [31], PM [31][a], [33][b] | Membranes [31], PM [31][a], endomembranes [11] |
| RhoA | PM [19, 40], [66][b], cell periphery [15], nucleus [70] | Cytosol [40, 71] > perinuclear region [15] |
| RhoB | PM [33][b], endosomes [66][b], [72][a] | PM [40], Golgi [40], early endosomes [71], pre-lysosomes [71] |
| RhoBTB1 | Vesicles (not lysosomes or Golgi) [66][a] | Vesicles (not lysosomes or Golgi) [66] |
| RhoBTB2 | Vesicles (not lysosomes or Golgi) [66][a] | Vesicles (not lysosomes or Golgi) [66] |
| RhoC | PM [33][b], [66][b], perinuclear region [72][b] | PM [73], cytosol [71] |
| RhoD | PM [33][b], vesicles [33][b], [66][b] | PM [74], endosomes [74] |
| RhoF (Rif) | PM [33][b], perinuclear region [66][b] | PM [11] |
| RhoG | PM [33][b], [75][a], vesicles [33][b], mitochondria [66][b] | PM [11], endosomes [11], mitochondria [66] |
| RhoH (TTF) | PM [33][b], cytosol [76][a] | Cytosol [76] |
| Rnd1 | PM [33][b], [77][a], adherens junctions [77][a] | PM [77], adherens junctions [77] |
| Rnd2 (Rho7) | Endosomes [78][a] > cytosol [78][a] | Endosomes [11] > cytosol [11] |
| Rnd3 (RhoE) | PM [33][b], [79][a], internal membranes [79][a], Golgi [79][a], cytosol [79][a] | PM [78], internal membranes [78], Golgi [11], cytosol [78] |
| TC10 | Perinuclear region [80], vesicles [33][b], [80] > PM [33][b] (lipid rafts [81]), focal adhesions [66][b] | PM [40, 80] (lipid rafts) [82], actin filaments [11], perinuclear endosomes [40, 80], recycling endosomes [80], exocytosing vesicles [80] |
| TCL | PM [33][b], vesicles [33][b], [66][b] | PM [11], endosomes [11] |
| Wrch-1 (RhoU) | PM [22] | PM [22], endomembranes [22] |

All currently known Ras and Rho family members are listed. *Column 2* lists subcellular locations where the GTPase has been detected in an active conformation, whereas *column 3* lists locations where the GTPase has been described in significant abundance

*PM* plasma membrane

[a]Predicted location, based on information in the associated reference

[b]Subcellular localization determined in the associated reference by ectopic expression of a constitutively active GTPase

be utilized to determine where a GTPase is active. In this chapter we describe the use of effector recruitment assays to monitor the subcellular localization of active Ras and Rho family small GTPases.

Ras and Rho family GTPases are active when GTP bound and inactive when GDP bound. Guanine nucleotide exchange factors (GEFs) catalyze exchange of GDP for GTP to activate GTPases, and GTPase-activating proteins (GAPs) catalyze GTP hydrolysis to inactivate them. Effector molecules that transmit GTPase signals downstream preferentially bind to active, GTP-bound GTPases. For example, Raf-1, a key downstream effector of Ras, has a 1,000–10,000-fold higher affinity for Ras-GTP than for Ras-GDP [7, 8]. Each GTPase signals to an array of disparate downstream effectors, each of which contains a *G*TPase-*b*inding-*d*omain (GBD) that is responsible for this preferential interaction. For instance, Ras downstream effectors contain *R*as-*b*inding *d*omains (RBD) or *R*as *a*ssociation (RA) domains, and downstream effectors of Rac, a Rho family GTPase, contain either *p*21-*b*inding *d*omains (PBD) or *C*dc42 and *R*ac *i*nteractive *b*inding (CRIB) domains. Many of these domains share little sequence identity, but they have a common structure and can be identified via databases such as SMART (the online *S*imple *M*odular *A*rchitecture *R*esearch *T*ool [http://smart.embl-heidelberg.de/]). Other downstream effectors (and GBDs) have only been identified empirically. For example, residues 292–362 of POSH have been shown to preferentially bind to Rac-GTP, so this is considered its PBD even though it shows little similarity to that of other downstream Rac effectors [9]. Notably, each downstream effector shows varying degrees of specificity among GTPases (Table 2). For example, POSH is thought to be specific to Rac1 and does not bind other Rho GTPases tested [9], whereas mDia2 is a downstream effector for many Rho GTPases (RhoA, RhoB, RhoC, RhoD, RhoF, Rac1, Cdc42) [10–12].

In an effector recruitment assay, the GBD of a downstream effector is tagged with a fluorescent molecule such as GFP. This fluorescent probe is recruited to locations where there are active GTPases that the GBD binds. Thus, one can monitor changes in the subcellular localization of active GTPases due to different stimuli and also determine if specific subcellular pools of a GTPase are active or inactive. This method is similar to the commonly used biochemical pulldowns for active GTPases, in which GBD sequences are expressed in bacteria, bound to solid support such as Sepharose or agarose beads, and used to retrieve active GTPases from cellular lysates [13, 14]. Both assays rely on the preferential binding of GBD to active versus inactive GTPases. Table 2 compares the use of probes in each assay. In pulldown assays, GBD specificity for the GTPase of interest is less important, because the pulldown is followed by immunoblotting with an antibody specific for the desired GTPase. However, in localization recruitment assays, identification of the GTPase which is responsible for recruitment of the GBD is much more heavily dependent on probe choice and experimental design (*see* **Notes 2, 4, 5**).

Other assays that have been created to observe changes in the subcellular localization of active GTPases include FRET-based biosensors [3, 15] and fractionation followed by GBD pulldowns [16]. Each assay has advantages and disadvantages. Although FRET-based biosensors can be highly sensitive and are GTPase specific, their use requires amenable cell conditions, sophisticated microscopy, and time-intensive imaging and processing. These considerations limit the number of cells that can be observed in a single assay and thereby limit throughput of an overall population. In addition, FRET-based biosensors generally require exogenous expression of the GTPase, whereas recruitment assays can often be used to monitor changes in endogenous GTPase activation [17–19]. Fractionation pulldown assays can measure changes in endogenous GTPase activation yet do not require as specific a probe as needed for recruitment assays, but they are technically challenging. Additionally, hydrolysis of GTP into GDP during the hours it takes to fractionate the cells can decrease the signal-to-noise ratio. Finally, recruitment assays yield better spatiotemporal resolution than do fractionation pulldowns, as recruitment assays can be used to observe more transient GTPase activation and to observe active pools of GTPases in microdomains within the cell that may be difficult to isolate by a biochemical fractionation protocol. Recruitment assays also have disadvantages, however. Negative results can be related to a variety of factors (*see* **Notes 5** and **6**) and can therefore be hard to interpret, and certain GTPases are easier to monitor with available probes than others (*see* **Note 5** and Table 2). Additionally, quantification of this type of assay can be challenging (*see* **Note 17**). However, many of these issues can be attenuated with proper experimental setup, as described in Subheading 4.

In Subheading 3 of this chapter, we describe a method whereby a fluorescent GBD probe is used to reveal the subcellular localization of an active Ras or Rho small GTPase. We illustrate this method with an example in which a GFP-PAK1-PBD probe shows the location of active Rac1 (Fig. 1). In this example, an exogenously expressed Rac1 mutant contains two mutations, one to make it constitutively active (Q61L) and an additional mutation (C189S) at a residue that has been previously described to cause nuclear accumulation of Rac1 [20]. PAK1 is a key effector of Rac1 [21]. In resting cells, GFP-PAK1-PBD is not localized to specific subcellular compartments but is distributed diffusely throughout the nucleus and cytoplasm (Fig. 1). In contrast, in cells expressing the constitutively active, nuclearly localized Rac1 mutant, GFP-PAK1-PBD highlights the nucleus. Images of these cells are shown as a striking example of GBD recruitment by the presence of an active GTPase, in this case, to the cell nucleus. In Subheading 4 and Table 2 we describe variations of this technique that enable it

**Table 2**
**GTPase specificity of GBD probes**

| GBD probe | Amino acid residues (species) | Fluorophore | Pulldown assay | | Recruitment assay | | Notes |
|---|---|---|---|---|---|---|---|
| | | | GTPase binds | GTPase does NOT bind | GTPase recruits | GTPase does NOT recruit | |
| Raf-1 RBD | 51–131 (human) [23–25], 1–149 [14, 83] | GFP [24, 25], RFP [37] | N-Ras [23], K-Ras [14, 23], H-Ras [14, 23, 83], R-Ras [84, 85] | Rap1A [23], Rit1 [63], Rit2 [63] | N-Ras [23, 25, 37], K-Ras [23, 25], H-Ras [23–25, 37] | Rap1A [25] | Rap1A binding affinity 50× less than Ras [86] |
| RalBP1 RBD | 397–518 [87] | | RalA [87], RalB [88] | | | | |
| RalGDS RBD | 786–883 (human) [8] | GFP [8] | Rap1 [84], Rap2 [84], Ras [8] | | Rap1 [8, 17] | Ras [8] | Ras binding affinity 50× less than Rap [86] |
| N-WASP CRIB | 230–288 [89], 215–295 [18], 142–276 [90], 201–321 (human) [66] | GFP [18, 19, 89], RFP [19] | Cdc42 [18], Rac1 [18], Rac2 [66], Rac3 [66], TCL [66], TC10 [66], Chp [65] | RhoA [18, 66], RhoB [66], RhoC [66], RhoD [66], RhoG [66], RhoH [66], Ras [18], Chp [66], Wrch-1 [66], Rnd1 [66], Rnd2 [66], Rnd3 [66], RhoBTB1 [66], RhoBTB2 [66], Rif [66] | Cdc42 [18, 19, 89] | RhoA [89], Rac1 [89] | Rac1/2/3 and TC10 binding affinity lower than Cdc42 [18, 90]; binds N-terminally truncated but not full-length Chp [65] |

| | | | | | | | |
|---|---|---|---|---|---|---|---|
| PAK1 PBD | 67–150 (human) [6], 65–150 (human) [67], 69–150 [31], 75–118 (human) [22], 56–267 (human) [66], 1–252 (rat) [10], 135–227 (yeast) [6] | YFP [67], GFP [22] | Rac1 [6, 66], Rac2 [6, 66, 91], Rac3 [31, 66], Cdc42 [6, 10, 66], TC10 [66, 81, 82], Wrch1 [22, 66], TCL [66], Rnd1 [66], Rnd2 (slightly) [66], RhoBTB2 [66], Rif [66] | RhoA [10, 66], RhoB [66], RhoC [66], RhoD [66], RhoG [66], RhoH [66], Chp [66], Rnd3 [66], RhoBTB1 [66], | Rac1 [67, 85], Wrch1 [22] | Cdc42 [67] | |
| POSH GBD | 292–362 (mouse) [9], 292–396 [80] | | Rac1 [9] | RhoA [9], Cdc42 [9] | | | Has been used as a FRET probe for TC10 [80] |
| Rhotekin RBD | 7–89 (mouse) [5], 1–90 [92], 1–89 [66], 40–107 (mouse)[a] | GFP | RhoA [5, 66], RhoB [66, 91], RhoC [66, 91, 92], TC10 [66], (slightly TCL, RhoH, RhoD, Rif, Rnd1 [66]) | Rac1 [5, 66], Rac2 [66], Rac3 [66], Cdc42 [5, 66], RhoG [66], Wrch-1 [66], Chp [66], Rnd2 [66], Rnd3 [66], RhoBTB1 [66], RhoBTB2 [66] | RhoA [19, 29, 85] | | |

GBD probes were initially used to determine the levels of active GTPases (by using pulldown assays) and since then have been adapted to determine the localization of GTPase activity (by using recruitment assays). The demonstrated specificity is similar, but not identical, in each assay. Common GBD probes are described based on the amino acid residues and species (when specified) of the downstream effector from which they were derived. Fluorophores that have been attached to the GBD are also listed. The tested specificity for GTPases in each assay is given

[a]SMART database alignment for GBDs

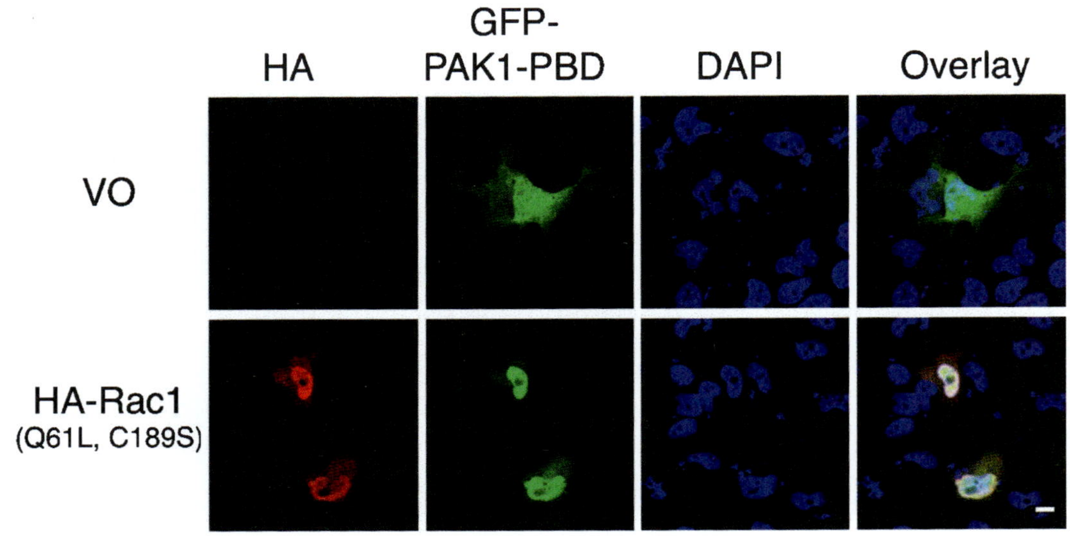


**Fig. 1** Nuclearly localized, constitutively active Rac1 (Rac1-Q61L-C189S) recruits GFP-PAK1-PBD to the nucleus. An example of conditions described Subheading 3 is shown. GFP-PAK1-PBD (*green*) is localized diffusely throughout the cell in cells co-expressing only empty pCGN vector. In contrast, when HA-tagged Rac1-61L-C189S (*red*) is ectopically expressed and localizes to the nucleus, GFP-PAK1-PBD is then observed highlighting the nucleus. This acts as easily visualized proof of principle that GFP-PAK1-PBD is recruited to areas of active, GTP-bound Rac1. DAPI serves as a nuclear marker. Images were taken using confocal microscopy at a magnification of ×40. Scale bar represents 10 μm

to be used to observe the location of active forms of other Ras and Rho family small GTPases, both exogenous and endogenous. We have previously utilized this method to examine the subcellular localization of active Wrch-1 [22] and Ras [23–25] and recently used it to determine spatially regulated activation of Rho family GTPases by the RhoGEF Ect2 (Huff, DeCristo, Cox et al., Genes & Cancer, in press).

## 2 Materials

### *2.1 Plating and Transfecting Cells*

1. Cover slips (microscope cover glass, 12 mm circles, 0.13–0.17 mm thick).
2. 70 % ethanol.
3. Tissue culture hood, equipped with a vacuum flask and attached aspirator.
4. Pasteur pipets, autoclaved for sterility.
5. Dulbecco's phosphate-buffered saline (DPBS, 1× with calcium chloride and magnesium chloride).
6. 10 cm tissue culture plates and 12-well plates (sterile polystyrene dishes).

7. Fine-tip tweezers.
8. Fibronectin: Reconstitute in DPBS to a concentration of 2.5 μg/ml, and filter-sterilize using a 0.2 μm filter. This solution can be stored at 4 °C for many months.
9. RPMI-1640 cell culture medium (RPMI-1640 1× with L-glutamine) or other growth medium appropriate for the cells of interest.
10. Fetal bovine serum (FBS): Mix 1 volume of FBS with 9 volumes of RPMI-1640 to produce complete RPMI growth medium.
11. Tissue culture incubator: Set to 37 °C and 5 % $CO_2$, or appropriate conditions for cells of interest.
12. Trypsin (0.05 % trypsin with EDTA).
13. Expression plasmids for proteins of interest, including the fluorescent GBD probe (here we used pEGFP-C1-PAK1-PBD, a.a. 75–118 [*see* Table 2]; pCGN-HA empty vector control; and pCGN-HA-Rac1-Q61L-C189S).
14. Polystyrene 5 ml round-bottom tubes.
15. Lipid-mediated transfection reagent optimal for the cells to be transfected.

### 2.2 Fixing and Staining Cells

1. 10× phosphate-buffered saline (PBS) solution: 1,370 mM NaCl, 100 mM $Na_2HPO_4$, 18 mM $KH_2PO_4$, 27 mM KCl in $ddH_2O$, pH to 7.4; mix 9 volumes of distilled water with 1 volume of 10× PBS for 1× working solution.
2. Paraformaldehyde (PFA): Dilute to working solution of 4 % PFA using 1 volume of 10× PBS, 2 volumes of 20 % PFA, 7 volumes of $ddH_2O$.
3. Triton X-100.
4. Parafilm M or similar wrap.
5. Anti-HA antibody (Covance, HA.11 monoclonal antibody clone 16B12 mouse; this is the best validated antibody available for this purpose, and we highly encourage its use); other secondary antibodies may be appropriate depending on the epitope tag or the primary antibody used to detect the GTPase of interest.
6. Alexa Fluor® 594 or similar fluorescent anti-mouse antibody that works well in contrast to green GFP color.
7. Lab wipes.
8. Mounting medium with DNA stain such as 4′,6-diamidino-2-phenylindole, DAPI.

#### *2.3 Imaging and Data Analysis*

1. Fluorescent confocal microscope, equipped with appropriate filters for the secondary antibodies used and with a camera (*see* **Note 19**).
2. Push-button manual cell counter.

## 3 Methods

#### *3.1 Plating and Transfecting Cells*

*Perform these steps in a tissue culture hood, using sterile technique.*

1. Place cover slips in a 10 cm plate.
2. Sterilize the cover slips by covering them completely with 70 % ethanol.
3. Aspirate the ethanol off the cover slips, and rinse them twice in sterile DPBS, leaving any excess DPBS remaining after the second rinse for ease of retrieving single cover slips.
4. Place a single cover slip in each well of a 12-well plate.
5. Cover each cover slip with 1 ml of 2.5 μg/ml sterile fibronectin (*see* **Note 1**) and incubate at room temperature for 20 min.
6. Remove the fibronectin from each well in a sterile manner, and store it at 4 °C for future use.
7. Aspirate residual fibronectin from the well and from under the cover slip by prodding the edges of the cover slip with a Pasteur pipet attached to an aspirator, until the cover slip sticks to the bottom of the well and does not move when prodded further with the aspirator.
8. Add 1 ml complete growth medium to each well and place the 12-well plate in the incubator to equilibrate.
9. Remove a 75 % confluent 10 cm plate of cells (numerous cell types are appropriate; *see* **Note 2**) from the incubator.
10. Aspirate the growth medium, and rinse the cells 1× in DPBS to remove excess medium.
11. Add 1 ml trypsin and return the plate to the incubator to speed cell detachment.
12. Once the cells have detached, resuspend them in 7 ml of complete growth medium and add 250 μl of the cell suspension to each well of the prepared 12-well plate containing fibronectin-coated cover slips (*see* **Note 3**).
13. Allow cells to adhere overnight.
14. Transfect 1 well of cells with both 1 μg empty vector (e.g., pCGN; *see* **Note 4**) and 1 μg of the expression plasmid for the probe (e.g., pEGFP-C1-PAK1-PBD; *see* **Note 5**), and transfect another well of cells with 1 μg each of the GTPase of

interest (e.g., pCGN Rac1-Q61L-C189S) and the probe (e.g., pEGFP-C1-PAK1-PBD; *see* **Note 6**).

(a) Mix the DNA for each well with 100 μl serum-free growth medium in a polystyrene tube.

(b) In a separate tube, combine 12 μl of transfection reagent and 200 μl serum-free growth medium.

(c) Incubate both mixtures for 15 min at room temperature.

(d) Add 100 μl of transfection reagent mixture to each tube of DNA.

(e) Incubate for 45 min at room temperature.

(f) Add each DNA/transfection reagent mixture to the appropriate well of cells, dropwise.

### 3.2 Fixing and Staining Cells

1. 24 h after transfection, remove the growth medium from the cells by aspiration, and rinse twice with PBS (*see* **Note 7**). *After this step, minimize exposure of cells to light, to prevent photobleaching (see* **Note 8***).*
2. Remove the PBS by aspiration, and then fix the cells by adding 4 % PFA to each well and incubating for 30 min at room temperature (*see* **Note 9**). *[Pause point: Cells can remain in PFA overnight at 4 °C, shielded from light.*]
3. Rinse the cells 5× with PBS.
4. Permeabilize the cells by incubating them in 0.2 % Triton X-100 (diluted in PBS) for 5 min (*see* **Note 10**).
5. Rinse the cells 5× with PBS to remove Triton X-100 (*see* **Note 11**).
6. Incubate the cover slips in anti-HA antibody at a dilution of 1:500 for 1 h at room temperature (*see* **Note 12**).
7. Rinse the cells 5× in PBS to remove excess primary antibody.
8. Incubate the cover slips in Alexa Fluor 594 anti-mouse antibody at a dilution of 1:500 for 2 h at room temperature (*see* **Note 13**).
9. Rinse the cells 5× in PBS to remove excess secondary antibody.
10. Soak the cover slips in dd$H_2O$ for 5 min (*see* **Note 14**).
11. Remove excess water from the cover slip by holding it perpendicular to a lab wipe, touching it gently to the wipe, and dabbing off excess water. *Do not touch the cell-containing surface.*
12. Mount each cover slip on a slide by applying 10 μl of Vectashield with DAPI to a slide and placing the cover slip cell side down on the Vectashield (*see* **Notes 15** and **16**). *[Pause point: Mounted cover slips can be stored for weeks to months at 4 °C, shielded from light.]*

### 3.3 Imaging and Data Analysis

1. Quantify localization of GFP-PAK1-PBD under each condition (*see* **Note 17**):
   (a) On an appropriately equipped fluorescent microscope, use a red filter to visualize the Alexa Fluor® 594 (excitation maximum, 594 nm/emission maximum, 618 nm) and thereby identify cells expressing the HA-tagged Rac1 or other GTPase protein of interest.
   (b) Switch to the green filter to visualize GFP (excitation maximum, 488 nm/emission maximum, 509 nm) and thereby observe where GFP-PAK1-PBD is localized.
   (c) Expression patterns can be divided into categories, such as diffuse nuclear and cytoplasmic localization (probe is evenly distributed between the two compartments), nuclear highlighted (probe is greater than 2× brighter in the nucleus than the cytoplasm), nuclear excluded (probe in the nucleus is greater than 2× dimmer than probe in cytoplasm), and others (plasma membrane highlighted/endomembrane highlighted, etc.). Classify each red (HA-Rac1-expressing) cell into the appropriate category based on localization of GFP-PAK1-PBD, and count it towards the given category using a push-button cell counter.
   (d) Repeat **steps (a)**–**(c)** for 100 cells per condition. Since vector-only cells do not visibly express the HA tag, skip **step (a)** for this condition.
   (e) The percentage of cells with the GFP-PAK1-PBD probe in each category can be compared between conditions.
2. Repeat the assay a minimum of three times (*see* **Note 18**).
3. Record images of representative cells (this may require the use of a different, camera-equipped microscope; *see* **Note 19**). Recruitment of the probe can be visualized by using overlaid images of the probe and GTPase (*see* **Note 20**). Overlay with DAPI can confirm where the recruitment is relative to the nucleus.

## 4 Notes

1. It is easier to visualize subcellular compartments in cells that are well spread. Rho family GTPases regulate the actin cytoskeleton, cell adhesion, and cell spreading [26]. Therefore, overexpression of GBDs downstream of Rho GTPases can result in undesired cell rounding and decreased cell adhesion. This can be minimized by coating the cover slips with fibronectin and also by thorough assay optimization (*see* **Notes 6** and 7). If fibronectin causes unintended signaling consequences, alternatives to increase cell adhesion include coating cover slips with 0.1 % fish gelatin or treating them with poly-L-lysine.

2. There are many factors to consider when choosing the appropriate cells for an effector recruitment assay.
   (a) Spread, adherent cells with easily visible organelles will allow for better identification of subcellular compartments than rounded or suspended cells.
   (b) Cell choice can be used to help maximize probe specificity while monitoring endogenous GTPase activation (*see* **Note 5**). For example, POSH-PBD is a good probe for endogenous Rac1 activation because it does not bind to other Rho GTPases such as RhoA and Cdc42. While it has not been determined if POSH is a downstream effector for Rac2 in addition to Rac1, it likely is, given that the effector-binding loops of Rac1 and Rac2 are identical [27]. Therefore, changes in POSH-PBD localization could be due to activation of either Rac1 or Rac2. However, Rac2 is expressed largely in hematopoietic cells and some tumor cells. If it is desirable to maximize probe specificity for Rac1, POSH-PBD can be used in cell types that do not express Rac2.
   (c) To increase the signal-to-noise ratio, it is best to use cells with low basal levels of GTPase activation when observing the effect of overexpressed GTPases.
   (d) There are numerous examples of cell type differences in GTPase activation [28]; thus, cells of relevance to the topic of study must be used. Effector recruitment assays have also been successfully utilized to monitor GTPase activation in model organisms, such as *Xenopus* [19, 29].
3. Given an approximately 24-h doubling time, it is optimal that cells are ~40 % confluent on the day after plating, so they will still be in logarithmic growth phase when they are fixed. However, this can vary with cell lines and doubling times. Additionally, low density on the day of transfection can lead to low transfection efficiency. If this occurs, multiple cover slips may be needed for each condition in order to acquire enough cells to count.
4. In the example illustrated here, vector-only was used as a negative control and the Rac1 mutant served as a proof of principle that PAK1-PBD could be recruited to the nucleus by expression of active, nuclearly localized Rac1. Other comparisons could include constitutively active Rac1 that recruits PAK1-PBD or POSH-PBD to the plasma membrane; dominant negative mutants of Rac1 that fail to bind any effectors; and effector domain mutants of Rac1 or non-Rac HA-tagged small GTPases, both of which fail to bind PAK or POSH. Assays for endogenous GTPase activation require more rigorous controls. For example, to confirm that recruitment in response to

a given stimulus is due to activation of a specific GTPase, knockdown of the GTPase should impair recruitment. To examine whether expression of a GEF causes probe recruitment due to GTPase activation, a GEF mutant with impaired GEF activity should be used to determine if recruitment is then reduced. Additionally, controls for potential contributions of other structural or functional elements of the regulator, such as scaffolding functions, should also be included.

5. Probe selection greatly impacts the performance of recruitment assays and the conclusions regarding specificity that can be drawn from them. As mentioned previously, some GBDs are very specific, while others are more promiscuous among GTPases. The specificity of the GBD probe aids in the determination of which GTPases are activated by a given stimulus in the observed location. A less specific probe may be desirable at first to explore the activation of multiple GTPases at once. However, this can also be complicated and lead to false-negative results if two GTPases have opposing effects on the same probe, and eventually a more specific probe will likely be the most useful. Table 2 lists previously described probes and their GTPase specificity; however, none of the probes have been tested for specificity among all GTPases. Therefore, other methods (*see* **Notes 2** and **4**) to confirm specificity should also be used. Additionally, as new discoveries about which residues define the GTPase specificity of a GBD are made, designer probes can be made with a desired specificity. For example, based on current knowledge, PAK1-PBD can be mutated to interact specifically with Cdc42 instead of Rac [30].

   Additionally, the ability of a probe to localize to a given location must be considered. For example, if a probe is too large (GFP adds an additional 27 kDa) then it may not be able to enter the nucleus, even if there are active GTPases present. Ideally, under basal conditions, effective probes will be expressed diffusely throughout the cytoplasm and nucleus prior to GTPase manipulation. If this is not the case, then performing proof-of-principle experiments with constitutively activated GTPases targeted to specific subcellular regions (such as the one described here with Rac in the nucleus) can be used to determine if a given probe is capable of interacting with active GTPases in a specific location.

6. The amounts of GBD probe and GTPase ectopically expressed will likely require optimization based on the probe and cell type. As mentioned above, Rho GTPases regulate the cytoskeleton, and overexpression of GBD probes derived from their downstream effectors (or the GTPases themselves) can have undesired consequences [26, 31]. Therefore, using the lowest amount of probe that can be easily observed is advantageous.

We recommend testing multiple concentrations of each probe and GTPase to determine the best concentration for cell health, transfection efficiency, and low probe expression.

Stoichiometry is very important in these assays. We caution that a lack of observed recruitment may be due to stoichiometry problems causing a false-negative result. If an expected effect is not seen with endogenous GTPases, one should also attempt the experiment with overexpression of the GTPase of interest. This caution is exemplified with Raf1-RBD, which can detect changes in exogenous but not endogenous Ras localization, presumably due to relatively low levels of endogenous Ras [8, 25].

7. As with the amount of probe added, timing must also be optimized. Upon expression of the GBD for long periods of time, some cells will change in morphology and may require fixation prior to 24 h to preserve cell health. Alternatively, some cells are slower to transcribe the probe and will require more time before optimal probe expression is observed. Since the probes are fluorescently labeled and suitable for live cell imaging (e.g., GFP), the preliminary assay used to optimize DNA concentrations for transfection (*see* **Note 6**) can be observed at multiple time points (e.g., 12, 24, 36, 48, 60, 72 h) to determine optimal timing.
8. To minimize photobleaching, cells expressing the fluorescent probe should be kept in the dark. A cardboard box can be wrapped with aluminum foil and then used to cover the cover slips between steps during the fixation and staining process. After mounting the cover slips on slides, the slides should be kept in a dark box at 4 °C prior to counting and imaging.
9. The assay can also be performed with a fluorescently labeled GTPase (or regulator) or in response to stimulation. In this case, one can perform the assay with live cells, and fixation is not necessary. However, fixation is useful to avoid phototoxicity and to establish more clearly defined time points.
10. All other time points related to fixing and staining can be estimated. However, it is important that incubation with Triton X-100 is monitored more closely. Leaving detergent on the cells for too long can cause cellular degradation.
11. Some primary antibodies require blocking at this step to prevent nonspecific binding. Consult the immunofluorescence staining instructions provided by the manufacturer.
12. To conserve antibody, tape a piece of Parafilm to a lab bench so that it is taut and free of wrinkles. Use a pipet to dispense 40 μl of diluted antibody onto the Parafilm for each cover slip. Place the cover slips with the cell side facing down onto the antibody droplets for incubation. For later rinsing steps, return cover slips to the 12-well plate with the cell side of the cover slip facing up.

13. In the described assay, the PAK1-PBD probe was tagged with EGFP (excitation maximum, 488 nm/emission maximum, 509 nm), which pairs well with Alexa Fluor® 594 (excitation maximum, 594 nm/emission maximum, 618 nm) because their excitation and emission spectra do not overlap. Other fluorescent probes can be paired, but it is very important that their spectra do not overlap. Otherwise, the observed "recruitment" might be an artifact of bleed-through of the fluorescence of one probe into the wavelength of the other probe. To rule out bleed-through, it is important to use both channels to image cells expressing only the probe and cells staining only for the GTPase, and confirm that the other channel is blank in each scenario.
14. Rinsing with water helps remove residual salt (in the PBS) from the cover slip. Areas on the cover slip with salt crystals cannot be viewed usefully under the microscope.
15. In this assay, mounting medium containing DAPI was used. DAPI stains nucleic acids and acts as a nuclear marker. The assay can be expanded to look at recruitment to various other subcellular locations by utilizing different markers and stains. For example galactosyl transferase (GalT) marks the Golgi [8]; transferrin marks endosomes [22]; and MitoTracker® Red stains mitochondria [2]. Many antibody suppliers provide useful lists of their antibodies that are specific to distinct subcellular compartments.
16. It is important to avoid bubbles in the mounting medium, since this will obstruct viewing of the cells. Be sure that bubbles are not released while pipetting the mounting medium onto the slide. Bubbles can also be generated from mounting the cover slip too quickly. While holding the cover slip at a 45° angle to the slide, place an edge of the cover slip onto the mounting medium and slowly lower the cover slip towards the slide.
17. In the described assay, it is impossible to count the cells blindly with respect to the presence of the exogenously expressed GTPase (i.e., vector-only cells are not red as are Rac1-expressing cells); however, whenever possible, images should be counted blind.

    Quantitation can also be done by analysis using image-processing applications such as ImageJ (available free as public domain software from the US National Institutes of Health [http://rsbweb.nih.gov/ij/] and other download sites). Each cell that expresses both Rac and the probe can be imaged, and ImageJ can then be used to quantify the pixel intensity in the area of interest (e.g., the nucleus) and another area in the cell (e.g., the cytoplasm). An arbitrary threshold can be set (e.g., for nuclear recruitment, a ratio of greater than 2:1 for nuclear:cytoplasmic intensity), and cells can be assigned to categories using this quantitative approach. When using this method, it is very important that the images are taken in the linear range of the microscope so that none of the pixel intensities are saturated (this can be controlled by exposure time and gain on the microscope). Also, it is important to compare the ratios and not the raw pixel intensities in a given area. The raw pixel intensity is a measure of how much GBD is being expressed, whereas the ratio signifies the distribution of the GBD among different locations. A change in the distribution (ratio) of the GBD among conditions is what is of interest in recruitment assays. The laborious nature of this quantitative method of analysis may limit the number of cells that can be analyzed, so care must be taken to produce an accurate representation of the population as a whole. The more cells counted (using any method) the better, because typically cell populations are heterogeneous, with a range of distributions of the GBD probe due to differences in the activation state of a variety of pathways within a given cell. Recruitment assays by definition compare at least two conditions, where the location of the probe is expected to change upon conditions favoring recruitment. The key task of data analysis is to accurately determine the representation of specific subpopulations among a heterogeneous overall population of cells under a given set of experimental conditions.

18. Recruitment assay results are regulated by a variety of factors that cannot always be adequately controlled, including serum freshness, cell density, and cell passage number. This can cause results to vary significantly upon repetition. Therefore, it is not uncommon to require repetition of the assay 5–8 times before obtaining statistically significant and biologically meaningful results with a low standard deviation.
19. Imaging cells on a confocal microscope allows imaging of a narrow slice in the *z*-plane. Overlap observed with a confocal microscope therefore represents co-localization in *x*-, *y*-, and *z*-planes within a cell, whereas overlap observed with an epifluorescent microscope represents co-localization only in the *x*- and *y*-planes. If a confocal microscope is unavailable, an epifluorescent microscope with the capability of visualizing the chosen fluorophores can still be used for less precise observations of recruitment.
20. Recruited GBD probes should overlay well with activated GTPases. But, using recruitment assays to visualize endogenous activation of a GTPase by a regulator will likely result in an imperfect overlay between the probe and the regulator, since the regulator may activate the GTPase in this location and then be released from its close proximity.

## References


1. Cox AD, Der CJ (2010) Ras history: the saga continues. Small GTPases 1:2–27
2. Bivona TG, Quatela SE, Bodemann BO et al (2006) PKC regulates a farnesyl-electrostatic switch on K-Ras that promotes its association with Bcl-XL on mitochondria and induces apoptosis. Mol Cell 21:481–493
3. Kraynov VS, Chamberlain C, Bokoch GM et al (2000) Localized Rac activation dynamics visualized in living cells. Science 290:333–337
4. Pertz O (2010) Spatio-temporal Rho GTPase signaling—where are we now? J Cell Sci 123:1841–1850
5. Ren XD, Kiosses WB, Schwartz MA (1999) Regulation of the small GTP-binding protein Rho by cell adhesion and the cytoskeleton. EMBO J 18:578–585
6. Benard V, Bohl BP, Bokoch GM (1999) Characterization of rac and cdc42 activation in chemoattractant-stimulated human neutrophils using a novel assay for active GTPases. J Biol Chem 274:13198–13204
7. Herrmann C, Martin GA, Wittinghofer A (1995) Quantitative analysis of the complex between p21ras and the Ras-binding domain of the human Raf-1 protein kinase. J Biol Chem 270:2901–2905
8. Bivona TG, Philips MR (2005) Analysis of Ras and Rap activation in living cells using fluorescent Ras binding domains. Methods 37:138–145
9. Tapon N, Nagata K, Lamarche N et al (1998) A new rac target POSH is an SH3-containing scaffold protein involved in the JNK and NF-kappaB signalling pathways. EMBO J 17: 1395–1404
10. Alberts AS, Bouquin N, Johnston LH et al (1998) Analysis of RhoA-binding proteins reveals an interaction domain conserved in heterotrimeric G protein beta subunits and the yeast response regulator protein Skn7. J Biol Chem 273:8616–8622
11. Vega FM, Ridley AJ (2007) SnapShot: Rho family GTPases. Cell 129:1430
12. Lammers M, Meyer S, Kuhlmann D et al (2008) Specificity of interactions between mDia isoforms and Rho proteins. J Biol Chem 283:35236–35246
13. de Rooij J, Bos JL (1997) Minimal Ras-binding domain of Raf1 can be used as an activation-specific probe for Ras. Oncogene 14:623–625
14. Taylor SJ, Shalloway D (1996) Cell cycle-dependent activation of Ras. Curr Biol 6: 1621–1627
15. Pertz O, Hodgson L, Klemke RL et al (2006) Spatiotemporal dynamics of RhoA activity in migrating cells. Nature 440:1069–1072
16. Guilluy C, Dubash AD, Garcia-Mata R (2011) Analysis of RhoA and Rho GEF activity in whole cells and the cell nucleus. Nat Protoc 6:2050–2060
17. Bivona TG, Wiener HH, Ahearn IM et al (2004) Rap1 up-regulation and activation on plasma membrane regulates T cell adhesion. J Cell Biol 164:461–470
18. Kim SH, Li Z, Sacks DB (2000) E-cadherin-mediated cell-cell attachment activates Cdc42. J Biol Chem 275:36999–37005
19. Benink HA, Bement WM (2005) Concentric zones of active RhoA and Cdc42 around single cell wounds. J Cell Biol 168:429–439
20. Michaelson D, Abidi W, Guardavaccaro D et al (2008) Rac1 accumulates in the nucleus during the G2 phase of the cell cycle and promotes cell division. J Cell Biol 181:485–496
21. Manser E, Leung T, Salihuddin H et al (1994) A brain serine/threonine protein kinase activated by Cdc42 and Rac1. Nature 367:40–46
22. Alan JK, Berzat AC, Dewar BJ et al (2010) Regulation of the Rho family small GTPase Wrch-1/RhoU by C-terminal tyrosine phosphorylation requires Src. Mol Cell Biol 30:4324–4338
23. Sherman LS, Atit R, Rosenbaum T et al (2000) Single cell Ras-GTP analysis reveals altered Ras activity in a subpopulation of neurofibroma Schwann cells but not fibroblasts. J Biol Chem 275:30740–30745
24. Bivona TG, Perez De Castro I, Ahearn IM et al (2003) Phospholipase C gamma activates Ras on the Golgi apparatus by means of RasGRP1. Nature 424:694–698
25. Chiu VK, Bivona T, Hach A et al (2002) Ras signalling on the endoplasmic reticulum and the Golgi. Nat Cell Biol 4:343–350
26. Huveneers S, Danen EH (2009) Adhesion signaling—crosstalk between integrins, Src and Rho. J Cell Sci 122:1059–1069
27. ten Klooster JP, Hordijk PL (2007) Targeting and localized signalling by small GTPases. Biol Cell 99:1–12
28. Wang L, Zheng Y (2007) Cell type-specific functions of Rho GTPases revealed by gene targeting in mice. Trends Cell Biol 17: 58–64
29. Park TJ, Mitchell BJ, Abitua PB et al (2008) Dishevelled controls apical docking and planar polarization of basal bodies in ciliated epithelial cells. Nat Genet 40:871–879
30. Reeder MK, Serebriiskii IG, Golemis EA et al (2001) Analysis of small GTPase signaling pathways using p21-activated kinase mutants that selectively couple to Cdc42. J Biol Chem 276:40606–40613

31. Mira JP, Benard V, Groffen J et al (2000) Endogenous, hyperactive Rac3 controls proliferation of breast cancer cells by a p21-activated kinase-dependent pathway. Proc Natl Acad Sci U S A 97:185–189
32. Kontani K, Tada M, Ogawa T et al (2002) Di-Ras, a distinct subgroup of ras family GTPases with unique biochemical properties. J Biol Chem 277:41070–41078
33. Heo WD, Inoue T, Park WS et al (2006) PI(3,4,5)P3 and PI(4,5)P2 lipids target proteins with polybasic clusters to the plasma membrane. Science 314:1458–1461
34. Takahashi K, Nakagawa M, Young SG et al (2005) Differential membrane localization of ERas and Rheb, two Ras-related proteins involved in the phosphatidylinositol 3-kinase/mTOR pathway. J Biol Chem 280: 32768–32774
35. Beguin P, Nagashima K, Gonoi T et al (2001) Regulation of Ca2+ channel expression at the cell surface by the small G-protein kir/Gem. Nature 411:701–706
36. Mochizuki N, Yamashita S, Kurokawa K et al (2001) Spatio-temporal images of growth-factor-induced activation of Ras and Rap1. Nature 411:1065–1068
37. Roy S, Plowman S, Rotblat B et al (2005) Individual palmitoyl residues serve distinct roles in H-ras trafficking, microlocalization, and signaling. Mol Cell Biol 25:6722–6733
38. Prior IA, Harding A, Yan J et al (2001) GTP-dependent segregation of H-ras from lipid rafts is required for biological activity. Nat Cell Biol 3:368–375
39. Willingham MC, Pastan I, Shih TY et al (1980) Localization of the src gene product of the Harvey strain of MSV to plasma membrane of transformed cells by electron microscopic immunocytochemistry. Cell 19:1005–1014
40. Michaelson D, Silletti J, Murphy G et al (2001) Differential localization of Rho GTPases in live cells: regulation by hypervariable regions and RhoGDI binding. J Cell Biol 152:111–126
41. Hancock JF, Paterson H, Marshall CJ (1990) A polybasic domain or palmitoylation is required in addition to the CAAX motif to localize p21ras to the plasma membrane. Cell 63:133–139
42. Furuhjelm J, Peranen J (2003) The C-terminal end of R-Ras contains a focal adhesion targeting signal. J Cell Sci 116:3729–3738
43. Takaya A, Kamio T, Masuda M et al (2007) R-Ras regulates exocytosis by Rgl2/Rlf-mediated activation of RalA on endosomes. Mol Biol Cell 18:1850–1860
44. Ohba Y, Mochizuki N, Yamashita S et al (2000) Regulatory proteins of R-Ras, TC21/R-Ras2, and M-Ras/R-Ras3. J Biol Chem 275:20020–20026
45. Calvo F, Crespo P (2009) Structural and spatial determinants regulating TC21 activation by RasGRF family nucleotide exchange factors. Mol Biol Cell 20:4289–4302
46. Ehrhardt A, David MD, Ehrhardt GR et al (2004) Distinct mechanisms determine the patterns of differential activation of H-Ras, N-Ras, K-Ras 4B, and M-Ras by receptors for growth factors or antigen. Mol Cell Biol 24:6311–6323
47. Beguin P, Mahalakshmi RN, Nagashima K et al (2006) Nuclear sequestration of beta-subunits by Rad and Rem is controlled by 14-3-3 and calmodulin and reveals a novel mechanism for Ca2+ channel regulation. J Mol Biol 355:34–46
48. Shipitsin M, Feig LA (2004) RalA but not RalB enhances polarized delivery of membrane proteins to the basolateral surface of epithelial cells. Mol Cell Biol 24:5746–5756
49. Takaya A, Ohba Y, Kurokawa K et al (2004) RalA activation at nascent lamellipodia of epidermal growth factor-stimulated Cos7 cells and migrating Madin-Darby canine kidney cells. Mol Biol Cell 15:2549–2557
50. Kashatus DF, Lim KH, Brady DC et al (2011) RALA and RALBP1 regulate mitochondrial fission at mitosis. Nat Cell Biol 13: 1108–1115
51. Martin TD, Mitin N, Cox AD et al (2012) Phosphorylation by protein kinase C alpha regulates RalB small GTPase protein activation, subcellular localization, and effector utilization. J Biol Chem 287:14827–14836
52. Pizon V, Desjardins M, Bucci C et al (1994) Association of Rap1a and Rap1b proteins with late endocytic/phagocytic compartments and Rap2a with the Golgi complex. J Cell Sci 107: 1661–1670
53. Hodges-Loaiza HB, Parker LE, Cox AD (2011) Prenylation and phosphorylation of Ras superfamily small GTPases. In: Hrycyna CA, Bergo MO, Tamanoi F (eds) The enzymes. Academic, Burlington, pp 43–69
54. Paganini S, Guidetti GF, Catricala S et al (2006) Identification and biochemical characterization of Rap2C, a new member of the Rap family of small GTP-binding proteins. Biochimie 88:285–295
55. Guo Z, Yuan J, Tang W et al (2007) Cloning and characterization of the human gene RAP2C, a novel member of Ras family, which activates transcriptional activities of SRE. Mol Biol Rep 34:137–144
56. Uechi Y, Bayarjargal M, Umikawa M et al (2009) Rap2 function requires palmitoylation and recycling endosome localization. Biochem Biophys Res Commun 378:732–737
57. Elam C, Hesson L, Vos MD et al (2005) RRP22 is a farnesylated, nucleolar, Ras-related

protein with tumor suppressor potential. Cancer Res 65:3117–3125
58. Rybkin II, Kim MS, Bezprozvannaya S et al (2007) Regulation of atrial natriuretic peptide secretion by a novel Ras-like protein. J Cell Biol 179:527–537
59. Pistoni M, Verrecchia A, Doni M et al (2010) Chromatin association and regulation of rDNA transcription by the Ras-family protein RasL11a. EMBO J 29:1215–1224
60. Pezeron G, Lambert G, Dickmeis T et al (2008) Rasl11b knock down in zebrafish suppresses one-eyed-pinhead mutant phenotype. PLoS One 3:e1434
61. Beguin P, Mahalakshmi RN, Nagashima K et al (2005) Roles of 14-3-3 and calmodulin binding in subcellular localization and function of the small G-protein Rem2. Biochem J 390:67–75
62. Finlin BS, Gau CL, Murphy GA et al (2001) RERG is a novel ras-related, estrogen-regulated and growth-inhibitory gene in breast cancer. J Biol Chem 276:42259–42267
63. Lee CH, Della NG, Chew CE et al (1996) Rin, a neuron-specific and calmodulin-binding small G-protein, and Rit define a novel subfamily of ras proteins. J Neurosci 16:6784–6794
64. Nalbant P, Hodgson L, Kraynov V et al (2004) Activation of endogenous Cdc42 visualized in living cells. Science 305:1615–1619
65. Aronheim A, Broder YC, Cohen A et al (1998) Chp, a homologue of the GTPase Cdc42Hs, activates the JNK pathway and is implicated in reorganizing the actin cytoskeleton. Curr Biol 8:1125–1128
66. Aspenstrom P, Fransson A, Saras J (2004) Rho GTPases have diverse effects on the organization of the actin filament system. Biochem J 377:327–337
67. Srinivasan S, Wang F, Glavas S et al (2003) Rac and Cdc42 play distinct roles in regulating PI(3,4,5)P3 and polarity during neutrophil chemotaxis. J Cell Biol 160:375–385
68. Philips MR, Pillinger MH, Staud R et al (1993) Carboxyl methylation of Ras-related proteins during signal transduction in neutrophils. Science 259:977–980
69. Wennerberg K, Der CJ (2004) Rho-family GTPases: it's not only Rac and Rho (and I like it). J Cell Sci 117:1301–1312
70. Dubash AD, Guilluy C, Srougi MC et al (2011) The small GTPase RhoA localizes to the nucleus and is activated by Net1 and DNA damage signals. PLoS One 6:e17380
71. Adamson P, Paterson HF, Hall A (1992) Intracellular localization of the P21rho proteins. J Cell Biol 119:617–627
72. Wang L, Yang L, Luo Y et al (2003) A novel strategy for specifically down-regulating individual Rho GTPase activity in tumor cells. J Biol Chem 278:44617–44625
73. Dietrich KA, Schwarz R, Liska M et al (2009) Specific induction of migration and invasion of pancreatic carcinoma cells by RhoC, which differs from RhoA in its localisation and activity. Biol Chem 390:1063–1077
74. Murphy C, Saffrich R, Grummt M et al (1996) Endosome dynamics regulated by a Rho protein. Nature 384:427–432
75. Samson T, Welch C, Monaghan-Benson E et al (2010) Endogenous RhoG is rapidly activated after epidermal growth factor stimulation through multiple guanine-nucleotide exchange factors. Mol Biol Cell 21:1629–1642
76. Li X, Bu X, Lu B et al (2002) The hematopoiesis-specific GTP-binding protein RhoH is GTPase deficient and modulates activities of other Rho GTPases by an inhibitory function. Mol Cell Biol 22:1158–1171
77. Nobes CD, Lauritzen I, Mattei MG et al (1998) A new member of the Rho family, Rnd1, promotes disassembly of actin filament structures and loss of cell adhesion. J Cell Biol 141:187–197
78. Madigan JP, Bodemann BO, Brady DC et al (2009) Regulation of Rnd3 localization and function by protein kinase C alpha-mediated phosphorylation. Biochem J 424:153–161
79. Foster R, Hu KQ, Lu Y et al (1996) Identification of a novel human Rho protein with unusual properties: GTPase deficiency and in vivo farnesylation. Mol Cell Biol 16:2689–2699
80. Kawase K, Nakamura T, Takaya A et al (2006) GTP hydrolysis by the Rho family GTPase TC10 promotes exocytic vesicle fusion. Dev Cell 11:411–421
81. Okada S, Yamada E, Saito T et al (2008) CDK5-dependent phosphorylation of the Rho family GTPase TC10(alpha) regulates insulin-stimulated GLUT4 translocation. J Biol Chem 283:35455–35463
82. Chiang SH, Hou JC, Hwang J et al (2002) Cloning and functional characterization of related TC10 isoforms, a subfamily of Rho proteins involved in insulin-stimulated glucose transport. J Biol Chem 277:13067–13073
83. Baschieri F, Farhan H (2012) Crosstalk of small GTPases at the Golgi apparatus. Small GTPases 3:80–90
84. Ohba Y, Ikuta K, Ogura A et al (2001) Requirement for C3G-dependent Rap1 activation for cell adhesion and embryogenesis. EMBO J 20:3333–3341
85. Wozniak MA, Kwong L, Chodniewicz D et al (2005) R-Ras controls membrane protrusion and cell migration through the spatial regulation of Rac and Rho. Mol Biol Cell 16:84–96

86. Herrmann C, Horn G, Spaargaren M et al (1996) Differential interaction of the ras family GTP-binding proteins H-Ras, Rap1A, and R-Ras with the putative effector molecules Raf kinase and Ral-guanine nucleotide exchange factor. J Biol Chem 271:6794–6800
87. Rosario M, Paterson HF, Marshall CJ (2001) Activation of the Ral and phosphatidylinositol 3′ kinase signaling pathways by the ras-related protein TC21. Mol Cell Biol 21:3750–3762
88. Hofer F, Berdeaux R, Martin GS (1998) Ras-independent activation of Ral by a Ca(2+)-dependent pathway. Curr Biol 8:839–842
89. Cannon JL, Labno CM, Bosco G et al (2001) Wasp recruitment to the T cell:APC contact site occurs independently of Cdc42 activation. Immunity 15:249–259
90. Hemsath L, Dvorsky R, Fiegen D et al (2005) An electrostatic steering mechanism of Cdc42 recognition by Wiskott-Aldrich syndrome proteins. Mol Cell 20:313–324
91. Saito S, Liu XF, Kamijo K et al (2004) Deregulation and mislocalization of the cytokinesis regulator ECT2 activate the Rho signaling pathways leading to malignant transformation. J Biol Chem 279: 7169–7179
92. Lucey M, Unger H, van Golen KL (2010) RhoC GTPase activation assay. J Vis Exp 22, pii: 2083

# Chapter 19

# Real-Time Visualization and Quantification of Native Ras Activation in Single Living Cells


## Christoph Biskup and Ignacio Rubio



## Abstract

Members of the Ras family of small guanosine triphosphate phosphohydrolases are GDP/GTP-binding proteins that function as pivotal molecular switches in multiple cell biological processes. The prototypical Ras family members K-Ras, N-Ras, and H-Ras, in particular, have been the focus of intense research for the last 30 years owing to their critical function as signalling nodes in the control of cell growth and proliferation and as drivers of oncogenic transformation. One aspect that has attracted much attention in recent times is the spatial control of Ras activity, which is dictated largely by a series of posttranslational modifications that do effectively govern the subcellular distribution and trafficking of Ras. Accordingly, strong emphasis has been placed on developing methodological microscopy-based approaches for the visualization of active Ras-GTP complexes at subcellular resolution. Here we describe the use of a collection of fluorescent affinity probes for the real-time visualization of Ras-GTP in live cells. These probes are multivalent and thus feature high avidity/affinity to Ras-GTP, which obviates the over-expression of Ras and enables one to image endogenous Ras-GTP formation. In addition, this chapter details the use of automated segmentation strategies for the unbiased quantification of probe-derived fluorescence at individual subcellular sites like the plasma membrane and endomembranes.



**Key words** Ras, Imaging, Affinity probe, Live cell, Quantification


## 1 Introduction

The ability to image biochemical and biophysical reactions including molecular interactions, dynamics, or changes in conformational and/or activity status at subcellular resolution has had a tremendous impact in modern cell biology and signal transduction research. Two major lines of research and technological development have been essential to pave the way for this massive and ever-continuing surge of imaging applications: Firstly, the cloning and characterization of fluorescent proteins (FP), along with the generation of a vast collection of spectrally distinct FP variants, have enabled the creation of genetically encoded fluorescent probes. Secondly, the development and improvement of fluorescence microscopy-based methodologies including confocal laser scanning microscopes with improved


Lorenza Trabalzini and Saverio Francesco Retta (eds.), *Ras Signaling: Methods and Protocols*,
Methods in Molecular Biology, vol. 1120, DOI 10.1007/978-1-62703-791-4_19, © Springer Science+Business Media, LLC 2014

optics, a variety of lasers for fluorescence excitation, and sensitive detectors, which allow detection of faint fluorescence signals, have increased considerably the spectrum of biological imaging applications. In addition to these two technological aspects, the ability to image biochemical or biophysical processes critically depends on the availability of appropriate fluorescent biosensors or reporter probes for the accurate, specific, and sensitive detection of the particular event under investigation. A family of important signal transduction mediators that has been particularly amenable to imaging approaches is the Ras superfamily of small guanosine triphosphate phosphohydrolases (GTPases). Ras, as the prototype GTPase, and other small GTPases or G-proteins can adopt two distinct conformations depending on whether they are bound to GDP or GTP. Ras-GTP represents the biologically active variant, and its formation and accumulation are tightly regulated by enzymatic systems in cells. In its GTP-bound conformation Ras features markedly higher affinity than Ras-GDP for Ras effector proteins, a group that includes PI3K class I catalytic subunits and the 3 Ser/Thr kinase Raf isoforms (collectively Raf) [1]. Raf interacts with Ras-GTP via an evolutionary conserved Ras-binding domain (RBD) of about 150 amino acids in size located at its N-terminus [2–4]. Since RBD discriminates between the inactive and active form of Ras this domain has been successfully exploited to generate fluorescent reporters of Ras activation in live cells. In 2001 Matsuda and co-workers described the first application of RBD for the visualization of active Ras in living cells via FRET [5, 6]. In that study a pentameric fusion protein termed RAICHU, composed of Ras, RBD, a pair of FRET donor and acceptor fluorescent proteins, and a membrane anchoring sequence, reported activation of Ras at the cell surface of EGF-stimulated fibroblasts. Strictly speaking, however, RAICHU probes do not visualize Ras activation but rather report on the local overall activity of Ras-activating (guanine nucleotide exchange factors or GEFs) and Ras-inactivating activities (GAPs or GTP phosphohydrolase-activating proteins), a feature that may not always be fully superimposable on the actual pattern of Ras-GTP formation. In 2002 the group of Mark Philips reported the use of EGFP–RBD as a recruitment-based probe to decorate Ras-GTP in living cells [7]. In a series of studies EGFP–RBD was used to image the formation of Ras-GTP in numerous cellular backgrounds, but EGFP–RBD was never sensitive enough to visualize endogenous levels of Ras-GTP, and thus all studies relied on the simultaneous over-expression of Ras proteins [8–12]. As in the case of the RAICHU fusion proteins, the requirement for Ras over-expression raises the concern of artefactual effects resulting from the mislocalization or the misregulation of Ras. Our laboratory has generated and characterized a new generation of RBD recruitment-based probes that enable the visualization of endogenous Ras-GTP levels. These probes, which we termed EGFPx3–RBDx3 or E3-R3 (Fig. 1), are composed of three EGFP (or Cherry)

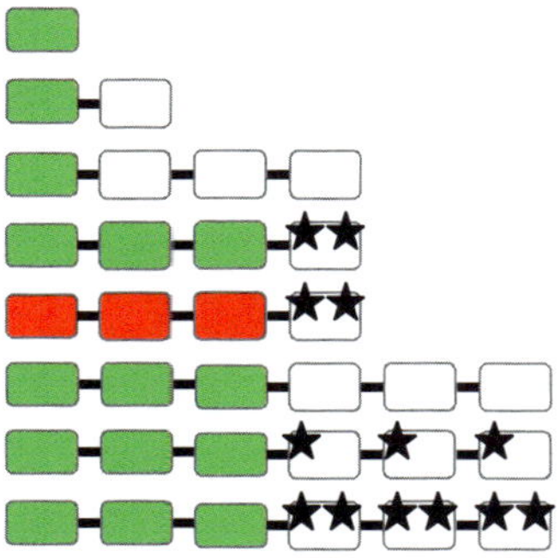

| *Modular composition* | *Short name* | *Mw* |
|---|---|---|
| EGFP | E1 | 28 |
| EGFP-RBD | E1-R1 | 37 |
| EGFP-RBDx3 | E1-R3 | 56 |
| EGFPx3-RBD(R59A/N64D)x1 | E3-R1(A/D) | 93 |
| Cherryx3-RBD(R59A/N64D)x1 | C3-R1(A/D) | 93 |
| EGFPx3-RBDx3 | E3-R3 | 111 |
| EGFPx3-RBD(R59A)x3 | E3-R3(A) | 111 |
| EGFPx3-RBD(R59A/N64D)x3 | E3-R3(A/D) | 111 |

**Fig. 1** Schematic representation of fluorescent reporter probes for Ras-GTP visualization described herein. *Asterisks* symbolize R59A and N64D point mutations. *Mw* molecular weight in kDa. Figure reprinted with modifications from ref. 16. Copyright 2010, The American Association of Immunologists, Inc

and three RBD modules and differ from the original EGFP–RBD biosensor in two major features: (1) Owing to the oligomerization of three RBD modules the probes have markedly enhanced affinity/avidity for Ras-GTP (as determined via GAP protection assays and isothermal calorimetry using recombinant purified proteins [13]). (2) The E3–R3 probes exhibit threefold brighter fluorescence as a result of EGFP trimerization [14]. The latter property, in particular, enables one to curb the expression levels of the probe without suffering a loss in fluorescence, thus increasing the chances of observing robust probe translocations upon cellular Ras-GTP formation. In addition, in a second round of optimization we introduced well-characterized point mutations into each of the RBD modules in order to generate a collection of probes with graded affinity for Ras-GTP [15] (*see* Fig. 1). The creation of these attenuated reporter probes turned out to be particularly important for studies with cells that rely heavily on Ras signalling for growth and survival. In such cell types, which in our hands included the COS and Jurkat cell lines, expression of probes containing triple wild-type RBD (R3) leads to widespread cell death [13, 16], but we were able to image Ras-GTP in these cells by using the attenuated probes E3–R3(A) and E3–R3(A/D). Using any of these probes we and other laboratories have meanwhile imaged live cell endogenous Ras-GTP formation in a variety of cellular systems [12, 16–19] (*see* also Chapter 24 in this book).

## 2 Materials

### *2.1 Equipment and Instrumentation*

1. Laminar flow hood for cell culture and other basic cell culture instrumentation.
2. Inverted laser scanning confocal microscope equipped minimally with argon and helium/neon lasers for EGFP and Cherry/RFP/bodipy excitation with Ar 488 nm and HeNe 543 nm laser

lines, respectively. LSM unit should contain 505–550 (or similar) band-pass filter for EGFP emission ("green channel") and 560 "long-pass" filter for RFP/Cherry emission ("red channel"). Individual configurations may differ depending on the fluorophores used.

3. Microscopy heating stage and chamber (optional: $CO_2$ atmosphere controlled). The microscope stage should accommodate 35 mm cell culture dishes.
4. Glass-bottomed cell culture dishes. These are available from various commercial suppliers (e.g., Mattek, Ibidi) or can be self-made as follows: Have your institution's mechanics drill holes with an appropriate diameter (e.g., 1 cm) into common 35 mm cell culture dishes. Use an inert adhesive (we use RTV615kit from KVD, Bad Wimpfen, Germany) to glue a glass slide from beneath to the bottom of the dish. Optionally use glass slides with grid (*see* **Note 8**). Sterilize dishes either by plasma sterilization (e.g., at a hospital's sterilization facilities) or by UV and/or short-time soaking in ethanol. After use dishes can be washed with a mild soap using a soft mop, rinsed in deionized water, dried, sterilized, and reused several times.
5. Optionally, depending on the cell type under investigation: Transfection/electroporation device (*see* **Note 6**).
6. General molecular biology reagents for restriction and PCR-based cloning procedures.

### 2.2 Transfection/Transduction of E3–R3 Probes into Cells

1. Collagen I, 0.5 mg/ml stock in water, frozen and aliquoted at –80 °C, or other matrix proteins, as convenient.
2. Glass-bottomed 35 mm cell culture dishes (*see* **step 4** in Subheading 2.1).
3. Sterile PBS: 100 mM NaCl, 20 mM $NaH_2PO_4$, 80 mM $Na_2HPO_4$, pH 7.4.
4. Transfection reagent and/or device.
5. Plasmids:
   (a) E3–R3 or attenuated version E3–R3(A/D) or E3–R3(A) (*see* Fig. 1).
   (b) Optional: Internal control: C3–R1(A/D) (*see* Fig. 1 and **Note 7**).
   (c) Optional: Additional genetically encoded fluorescent probes, e.g., Golgi or other organelle markers.
   (d) Optional: Expression plasmids for N-Ras, H-Ras, or K-Ras (*see* **Note 2**).
   (e) Alternative: Lentiviral particles coding for the same constructs as above.
6. Cell culture medium with serum as appropriate for the cell type under investigation (*see* **Note 1**).

7. Antibiotics for cell culture, e.g., ready-to-use 100× penicillin/streptomycin stock from Life Technologies or other suppliers.
8. Diacylglycerol (DAG) for increased probe expression levels in Jurkat T-cells and other leukocytes.

### 2.3 Golgi Labelling with Vital Golgi Marker

1. BODIPY-TR-C5-ceramide (Molecular Probes/Life Technologies). We use ready-to-dissolve ceramide complexed to BSA. Prepare a 500 μM solution of BODIPY-TR-C5-ceramide in deionized water exactly as described by the manufacturer. Aliquot in small portions, and snap freeze at −20 °C. Dilute stock 100-fold in serum-free medium or any buffered saline to make a 5 μM staining solution.
2. Serum-free medium as appropriate for the cell type under investigation. Alternatively any buffered and balanced saline will do.

### 2.4 Imaging Ras-GTP Formation in Living Adherent Cells

1. Starvation medium: Cell culture medium without serum as suited for the cell type under investigation.
2. 1 M HEPES, pH 7.4, stock solution, sterile filtered.
3. Medium for imaging: Starvation medium supplemented with HEPES. Buffering is not required if microscopy equipment includes on-stage $CO_2$ atmosphere control.
4. Appropriate stimulus for Ras activation, e.g., EGF for Ras stimulation in most mesenchymal and epithelial cells (*see* **Note 1**). Prepare stimulus solution in the imaging medium described in **item 2** at a concentration fivefold higher than the final working concentration.
5. Timer.
6. Software for image analysis and Ras-GTP quantification (*see* Appendix and **Note 10**).

### 2.5 Imaging Ras-GTP Formation in Living Suspension Cells

1. Glass-bottomed cell culture dishes (*see* **step 4** in Subheading 2.1).
2. Poly-L-lysine (PLL): Prepare a 1 mg/ml stock in deionized water, aliquot, and store at −80 °C.
3. Sterile PBS.
4. Fatty acid-free/endotoxin-low BSA: Dissolve 10 % (w/v) in PBS, aliquot, and store at −20 °C.
5. Starvation medium: Cell culture medium without serum as suited for the cell type under investigation supplemented with 0.2 % of fatty acid-free/endotoxin-low BSA.
6. Medium for imaging: Appropriate medium buffered with HEPES (e.g., RPMI, $HCO_3$-free, supplemented with sterile 50 mM HEPES pH 7.4) containing 0.2 % of fatty acid-free/endotoxin-low BSA. Buffering with HEPES is not required if microscopy equipment includes on-stage $CO_2$ atmosphere control.

7. Appropriate stimulus for Ras activation, e.g., anti-CD3 and anti-CD28 antibodies for TCR stimulation in T-cells or phorbol ester for Ras activation in most leukocytes. Prepare stimulus solution in the imaging medium at a concentration fivefold higher than the final working concentration.
8. Timer.
9. Software for image analysis and Ras-GTP quantification (*see* Appendix and **Note 10**).

## 3 Methods

### 3.1 Construction of Expression Plasmids for E3–E3 Ras-GTP Reporter Probes

1. Plasmids coding for trimeric RBD-3 in fusion with EGFP are generated by the sequential PCR-based appending of RBD cassettes to EGFP in the background of the pEGFP-C2 vector (Clontech). Figure 2 depicts the architecture of the E1–R1, R2, and R3 constructs. Primers for PCR were designed such that adjacent RBD domains were separated by 5-amino acid linkers [13].
2. In a second step a dimeric EGFP sequence was appended to the 5′end of the E1–R3 constructs. We refer to a detailed cloning description (including primer sequence information) in the supplementary material to ref. 13 and in the experimental section to ref. 16. All expression plasmids are available upon request from the authors.

### 3.2 Transfection of Adherent Cells with E3–R3 Reporters

1. Prepare a 100 μg/ml solution of collagen I in water by diluting 1:4 the stock solution.
2. Cover the glass bottom of the dishes with a drop of the collagen I solution and let stand for at least 2 h at RT.
3. Wash dishes 1× with sterile PBS and 2× with sterile deionized water. Let dishes dry under the laminar flow box. Coated dishes can be stored in a sealed container in the cold for several days.
4. 2 days before imaging seed out cells on collagen-coated dishes in standard culture medium containing serum. Let grow overnight, ideally to 70–80 % confluency.
5. Next day transfect cells with the E3–R3 reporter plasmid of choice (*see* **Note 5**) using your favorite transfection approach. We found that the use of as different approaches as lipofection, electroporation, or microporation did not compromise the performance of the probes in a number of cell types (*see* **Note 6**).
6. Optionally include antibiotics in the medium at this step to avoid contamination.
7. Alternatively use retroviral or lentiviral transduction to express the E3–R3 reporters in your target cells. In this case optimal reporter protein expression may be achieved at 3–4 days after infection.

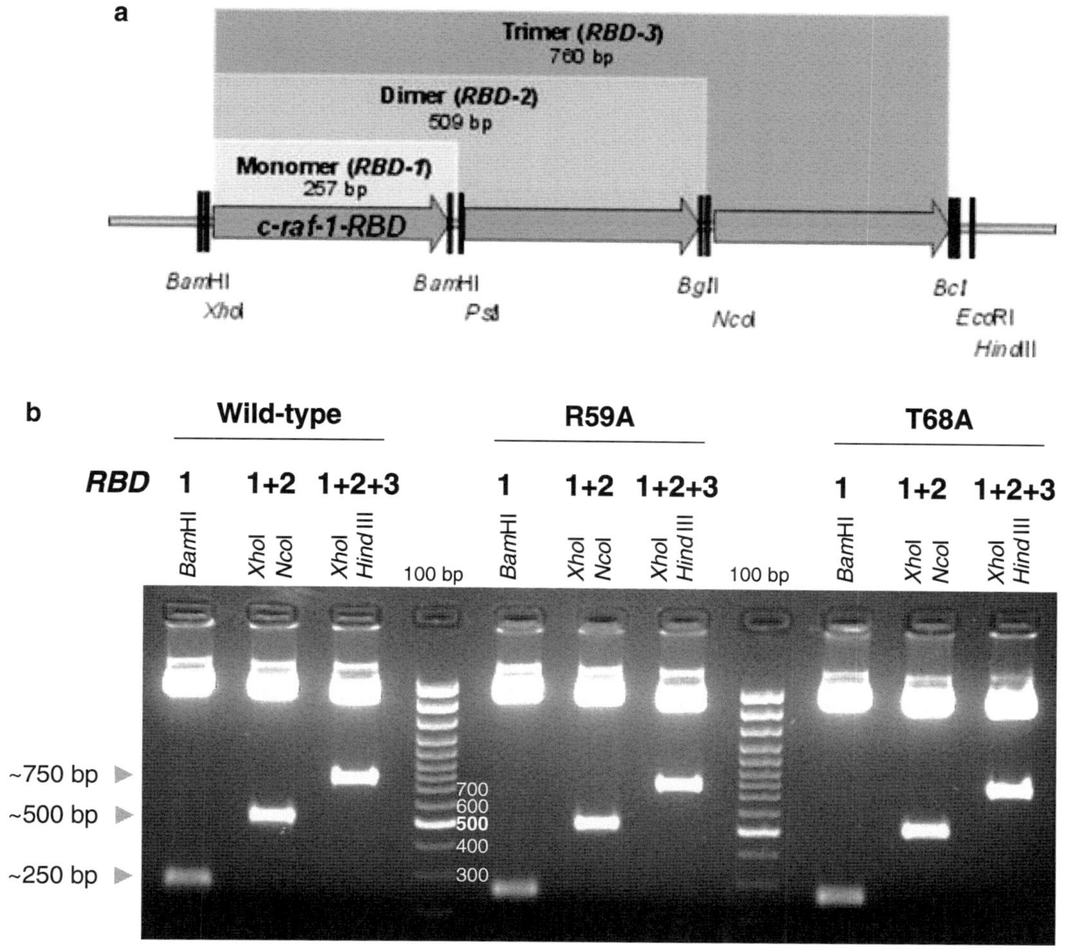


**Fig. 2** General cloning strategy and restriction analysis of the RBDx3 cassette. (**a**) Architecture of the RBDx3 modules. (**b**) Example of restriction analysis for three RBD modules (wild type (wt), R59A, and T68A). Note that, by choosing the right restriction nucleases, RBDx1, RBDx2, and RBDx3 cassettes can be individually excised from the parental R3 plasmid. Refer to refs. 13, 16 for further details

### 3.3 Transfection of Suspension Cells with E3–R3 Reporters

1. One day prior to transfection split leukocytes 1:2. Grow at a cell density suited for the particular cell type.
2. Transfect cells with the E3–R3 reporter plasmid of choice using your favorite transfection approach. As before, the use of transfection approaches such as lipofection, electroporation, or microporation did not compromise the performance of the probes in lymphocyte cell types (*see* **Note 6**).
3. Optionally include antibiotics in the medium at this step to avoid contamination.
4. Alternatively use retroviral or lentiviral transduction to express the E3–R3 reporters in your target cells. In this case optimal reporter protein expression may be achieved at 3–4 days after infection.

5. Optional: In some lymphocyte cell lines the CMV-driven expression of the E3–R3 reporter polypeptide can be very poor. To boost expression we sometimes supplement the medium with 10 μM DAG 3–4 h after transfection.

### 3.4 Labelling of Golgi Apparatus with Vital Dye

Co-labelling of subcellular structures or organelles can be useful to better observe and identify sites of Ras-GTP accumulation. We routinely label the Golgi apparatus using a vital fluorescent dye as follows:

1. Remove medium from cell dish (adherent cells) or spin down cells grown in suspension and aspirate off medium.
2. Incubate cells with BODIPY-TR-C5-ceramide staining solution: For adherent cells apply about 250 μl of staining solution to the cells on the glass bottom; for suspension cells add 100 μl staining solution/$10^6$ cells, and resuspend well the cell pellet.
3. Incubate in the dark for 30 min at 4 °C. For temperature-sensitive cells 10 min at room temperature may also work.
4. Wash at least twice with ice-cold or tempered medium, respectively.
5. Add fresh warm medium and place cells in incubator for 10–20 min prior to imaging.

### 3.5 Confocal Imaging of Ras-GTP in Adherent Cells

1. One day after transfection or 3–5 days after viral transduction (Subheading 3.2) prepare cells for imaging. Deprive cells of serum if required for optimal Ras activation: incubate your cells in starvation medium. If necessary, serum deprivation can be extended to overnight incubation.
2. Optional: If required, before moving to the microscope, label subcellular organelles with vital dyes (Subheading 3.4).
3. Treat your cells with inhibitors or other cues as required for your experiment.
4. Replace medium with 1.2 ml of warm imaging medium.
5. Place the dish on the microscope stage and keep thermostated.
6. Check the configuration of your LSM (*see* **Note 4**). Laser intensity and detector gain should be adjusted such that almost the entire bit range of the image is exploited. To avoid photobleaching it is strongly recommended to increase the detector gain to higher voltages before increasing the laser intensity. Since dark noise increases especially at high voltages, it might be advisable not to increase the detector gain to the maximum. In any case, laser intensity and detector gain should be adjusted such that recorded intensities do not exceed the maximum value in any pixel (i.e., 255 in case of an 8-bit image or 65,535 in case of a 16-bit image). Otherwise higher intensities would be clipped to the maximum value and a quantitative image

analysis would not be possible. Likewise no values should be clipped at zero by adjusting the offset to the background.

In addition, all images should be obtained with the same settings. Since the recorded intensities do not increase in a linear fashion with all parameters, it is difficult, if not almost impossible, to correct for different settings.

Sticking to these rules is essential for recording a set of images that can be analyzed quantitatively. This might not yield images with a perfect contrast and brightness. But, these parameters can be optimized later, when the images are processed (*see* **step 2** in Subheading 3.7).

7. Focus the cell layer in the dish using transmission light. We recommend using objective lenses with 63× or higher magnification.
8. Switch to the track configured to record E3–R3, and select a cell for stimulation and imaging (*see* **Note 4**).
9. Remove the lid of on-stage chamber.
10. Focus the cell as accurately as possible and switch to the scanning mode.
11. Acquire one or more images. These are the control images before stimulation.
12. Stimulate the cells by adding 300 μl of the fivefold concentrated stimulus solution dropwise to the dish. Be careful not to touch the dish or the optics!
13. Start timer.
14. Replace the lid of the on-stage thermostated chamber.
15. Check immediately for possible cell or focus drift as a consequence of these manipulations. In case of drift, restore the focus as quickly as possible (*see* **Note 8**).
16. Acquire more images at the required time points.

### 3.6 Confocal Imaging of Ras-GTP in Suspension Cells

1. Prepare in advance PLL-coated glass-bottomed dishes as follows. Cover the glass bottom with 150 μg/ml PLL solution, and incubate for 2 h to overnight under the sterile cabinet. Remove PLL solution (we recover and reuse the solution 2–3 times) and wash 1× with sterile PBS and 2× with sterile water.

   Let the dishes dry under the laminar flow. Dishes can be stored in a sealed container for several days at 4 °C.
2. One day after transfection or 3–5 days after viral transduction prepare for imaging: deprive your cells of serum if required for optimal Ras activation. Incubate your cells in starvation medium. If necessary, serum deprivation can be extended to overnight incubation.
3. Optional: If required, before moving to the microscope, label subcellular organelles with vital dyes (*see* Subheading 3.4).

4. Treat your cells with inhibitors or other cues as required for your experiment.
5. Add a drop of your cell suspension (about 250 µl containing optimally between 100,000 and 400,000 cells) to the center of the glass bottom of a pre-warmed PLL-coated dish.
6. Let stand in incubator for at least 10 min. Most cells should have settled and stuck to the bottom by then.
7. Replace medium with 1.2 ml of warm imaging medium.
8. Place the dish on the microscope stage, and keep cells tempered.
9. Continue with **step 6** in Subheading 3.5.

Figure 3 shows a representative image series for a Jurkat T-cell expressing E3–R3(A/D) and C3–R1(A/D) challenged with anti-CD3 and anti-CD28 Abs (*see* also **Note 7**).

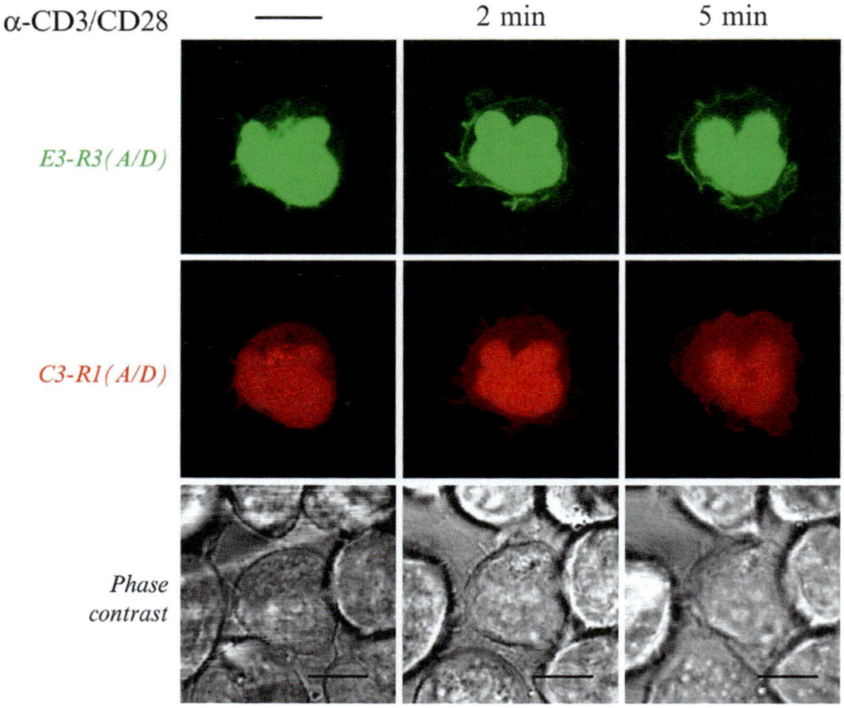


**Fig. 3** Ras-GTP visualization in Jurkat T-cell. Time series of confocal images of a Jurkat T-cell co-expressing E3–R3(A/D) and C3–R1(A/D) challenged with antibodies to CD3 and CD28. This panel illustrates essentially the outcome of an experiment as described in Subheading 3.6. Note that whilst E3–R3(A/D) redistributes to the PM following T-cell receptor triggering and concomitant Ras-GTP formation, the negative-control reporter probe C3–R1(A/D) does not undergo ostensible relocalization in the same cell (*see* **Note 7**). Scale bars: 10 µm. Original magnification ×63

### *3.7 Data Processing and Quantification of E3–R3 Redistribution*

1. Loading the image (step 1 of the program listed in the Appendix): Usually the format of images, in which the microscope images are saved, is documented by the manufacturer. Often along with the image a variety of meta-information such as the configuration is saved. Knowing the structure of these files these data can, in principle, be retrieved directly from the original files.

   However, instead of writing interfaces the image data can be exported to common file formats such as tif and then imported into the image analysis program. In MATLAB, images can be loaded with the function *imread*.
2. Adjusting contrast and brightness (step 2) (*see* **Note 11**): When images are acquired the parameters should be adjusted such that the bit range of an image is optimally used while none of the recorded intensities in any pixel of the image series should exceed the bit range (*see* **step 6** in Subheading 3.5). By obeying to this rule images might not be optimally adjusted for contrast and brightness. This flaw can now be corrected for by mapping the intensity values to new values such that the entire bit range of an image is exploited. Brightness and contrast can be further optimized when a low percentage of pixel values is allowed to assume maximal values after the mapping procedure. In *MATLAB* adjustment of the image intensity values can be done with the function *imadjust* of the *Image Processing Toolbox*.
3. Filtering the image (step 3): To reduce noise the images have to be filtered. In principle, several algorithms can be used to filter an image. We decided to use a simple median filter, which effectively reduces "salt and pepper" noise. Alternatively a Gaussian filter can be used. However, median filtering preserves the edges better.
4. General strategy for image segmentation: To quantify the distribution of the E3–R3 probe within a cell the compartments of interests must be identified and delineated first. This process is called segmentation. In general, segmentation algorithms subdivide an image into its constituent regions or objects. The segmentation process is not trivial. The accuracy of the segmentation determines the success or the failure of the image analysis. Here, in steps 5–10 several strategies to segment an image are discussed. With the masks created in these steps the average concentration of the Ras-GTP probe in the marked compartments is calculated in step 11.

   To illustrate the algorithms a short program written in *MATLAB* is listed in the Appendix. In the *MATLAB* routines the segmentation algorithm yields masks, which are binary images. Pixels, which are inside the mask, have the value 1; pixels, which are outside the mask, have the value 0. By multiplying the image containing the fluorescence intensities with the binary mask all pixels outside the mask are set to zero, while pixels being inside the mask retain their value. These data can be then easily processed to calculate the average fluorescence intensity in the compartment of interest (step 11).

5. Segmentation by manual definition of regions of interest: In principle, segmentation of an image can be done manually by the user. Most programs distributed with fluorescence microscopes allow for defining regions of interests (ROIs) by drawing circles, squares, or arbitrary shapes around the structures, in which the fluorescence signal has to be measured. The software then determines the average fluorescence inside the region of the individual image or in a series of images taken subsequently in a time series. However, since living cells move and change their shape, ROIs have in many cases to be defined individually for each image of a time series, which makes this approach cumbersome. Analyzing many experiments in this way not only is time consuming but will also be prone to errors introduced by the user. Especially, when fluorescence signals are weak, it is difficult to delineate the structures of interest. One example is shown in Fig. 4. Here, the nucleus can be easily delineated by the user thanks to a bright fluorescence signal stemming from the nuclear accumulation of the E3–R3 probes (*see* **Note 9**). But, due to the weak fluorescence signal it is difficult to delineate the plasma membrane (*see* **Note 3**). In the case of flat culture cells that do not have a round shape like the lymphocytes investigated in this sample the procedure becomes even more difficult: By drawing a tight ROI around the membrane a high average value will be obtained; by including pixels from the background into the calculation of the average a lower fluorescence intensity value will be obtained. Thus, the result strongly depends on the diligence of the user. Here, a reliable, objective quantification of fluorescence intensities is only possible when images are segmented by automated software routines. Some strategies are explained in the following steps.

   In *MATLAB* a ROI can be created by the user via the function *roipoly*. This function allows the user to specify a polygonal region of interest and returns a binary image that can be used as a mask. In the example shown in the Appendix, this function is only used to select one cell if many cells are recorded in an image (step 4).

6. Definition of regions of interest by thresholding (step 9): If a marker is used to stain a specific compartment, often a threshold can be used to identify the compartment. In the example shown here, the Golgi apparatus was stained with BODIPY-TR-C5-ceramide. Fluorescence originating from the marker was collected in the "red channel," so that the red image can be used to delineate the Golgi apparatus. A mask for the Golgi apparatus can be simply created by selecting pixels exceeding a

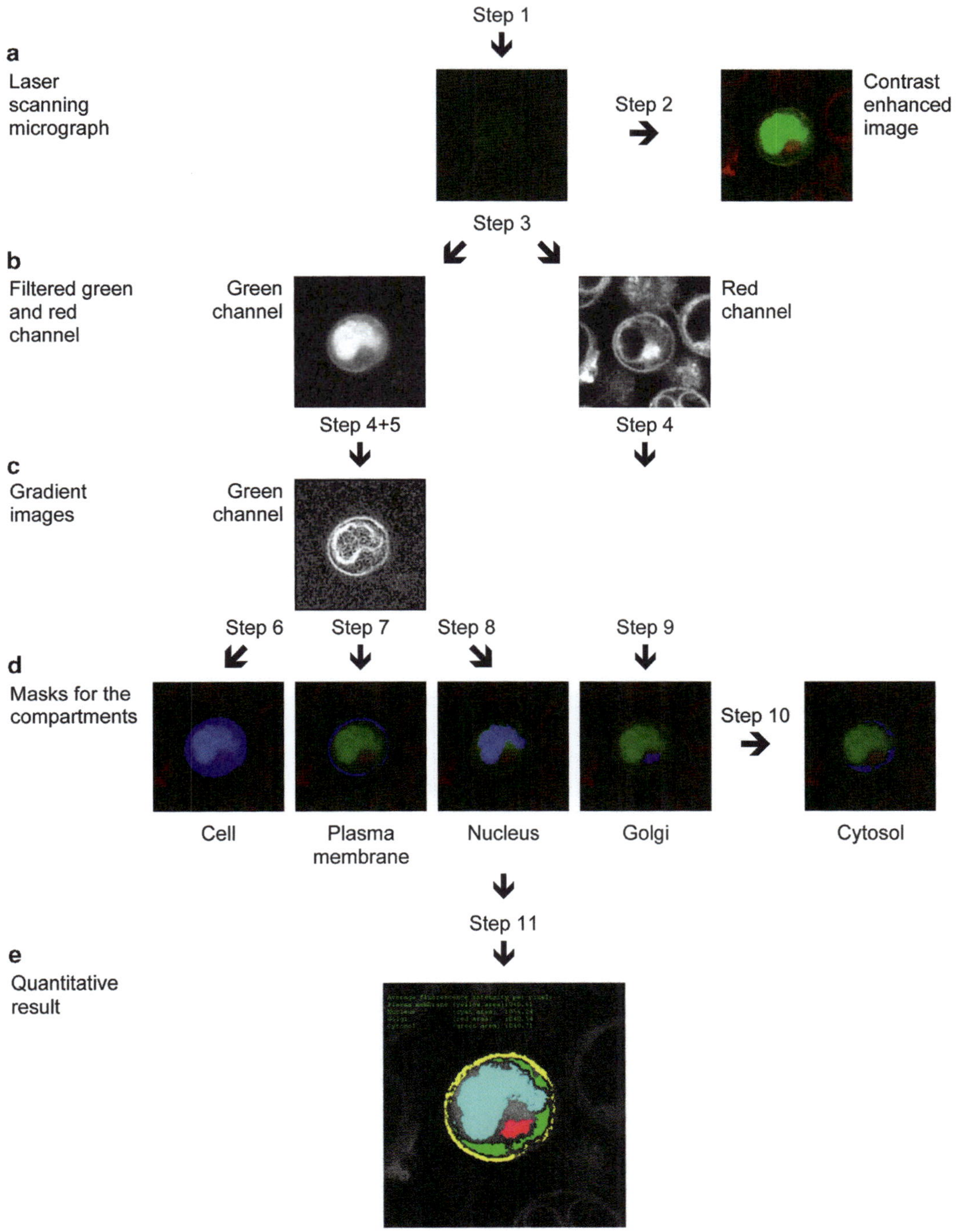


**Fig. 4** Major steps of the automated segmentation process. The indicated steps refer to the program listed in the Appendix. (**a**) Confocal laser scanning micrograph of a Jurkat cell. *Left*: original image, *right*: contrast-enhanced image. The *green channel* was used to record EGFP fluorescence, whereas BODIPY-TR-C5-ceramide fluorescence was recorded in the red channel. (**b**) Filtered image. (**c**) Gradient image of the *green channel*. (**d**) Masks for the cell, plasma membrane, nucleus, Golgi apparatus, and cytosol (*see* Subheading 3.7 for a detailed description of the segmentation algorithms). (**e**) Segmentation result. Pixel intensities in the marked regions were averaged

set threshold. In *MATLAB* this can be done by the operation ">." The command

$$Mask = Image > Threshold;$$

yields a binary image which is stored in the variable "Mask." Here, "Image" is a matrix containing all pixel intensities; the variable "Threshold" contains the threshold value. Holes within the mask can be filled with the function *imfill.*

Identifying cellular compartments in this way is straightforward and easy. Unfortunately, the use of some organelle markers is limited. Also, not all markers can be simultaneously used, since it might not be possible to separate them by their spectrum from each other or from the fluorescent probe. Especially when the fluorescence from the probe is weak, cross talk with the organelle markers should be avoided. In such a situation the noise from a brightly fluorescent organelle marker can be higher than the signal from a weak probe, so that these signals cannot be disentangled by spectral unmixing algorithms. In these cases the signal from the probe has to be used to delineate the subcellular structure of interest (*see* next paragraph).

7. Definition of ROIs by identifying gradients (steps 5–8): The fluorescence of the E3–R3 probe localized in the plasma membrane is weak. It can hardly be detected in the green channel of the original image and is only visible in the contrast-enhanced image. Average fluorescence intensities in the images differ only by a few units or even fractions of a unit from the background, so that applying a threshold like in step 6 is not feasible. What however distinguishes the image from its background is a steep gradient with respect to neighboring pixels.

   The MATLAB routine, which can be found in the Appendix, identifies the steepest gradient of each pixel (step 5). Then, in a second step, pixels with the steepest gradient can be selected. In this way the cell can be distinguished from the background (step 6) and the plasma membrane (step 7) and the nucleus (step 8) can be identified.

8. Exclusion of regions with overlapping compartments (step 10): Although thin cross sections with a size of less than 1 μm can be recorded with a confocal laser scanning microscope, organelles might overlap within one slice, so that the fluorescence signal recorded in the corresponding pixels cannot be attributed to one compartment. In the example shown here, the Golgi apparatus overlaps with the nucleus. To allow for a reliable quantification of the fluorescence originating from the respective compartments, these regions have to be excluded from the analysis.

9. Creating a mask for the cytosol (step 10): The cytosol is the area within the mask of the cell, which does not overlap with

the masks created for all other compartments. Again, for an unambiguous quantification only regions that do not show any overlap are chosen.

10. Calculating the average fluorescence (step 11): Once a mask for each compartment is created, this mask can be used to calculate the fluorescence intensity. To this aim, the image needs to be multiplied by the mask. Then pixels outside the mask are set to zero, whereas pixels inside the mask are conserved. By summing up the values of all pixels and dividing by the number of pixels inside the mask the average intensity per pixel can be calculated.

## 4 Notes

1. The methods described herein are envisaged for the visualization of endogenous Ras-GTP levels formed in living cells. Ras is not a very abundant protein, and Ras-GTP levels will be the limiting factor for detection in many cases. We therefore recommend choosing, if possible, cells and stimuli/conditions that lead to strong accumulation of Ras-GTP. The optimal conditions for Ras activation can be explored in preliminary biochemical Ras-GTP determinations. We recommend, anyway, to perform biochemical Ras-GTP assays in parallel to the imaging experiments in order to confirm any formation of Ras-GTP visualized in living cells.
2. In preliminary experiments aimed at coming to grips with the method, the co-transfection of wild-type Ras can be useful to generate more robust probe redistribution in the cell.
3. When choosing the cell type for visualization consider that cell types which are more "cubic" in shape (e.g., MDCK, MCF-7, T-cells) will more likely give clear and sharp images of fluorescence accumulation at the PM in confocal sections, whereas flat cells, like COS, can often be ideal to observe the Golgi apparatus and other internal organelles.
4. When selecting or screening on-stage for a cell suitable for stimulation and visualization opt for cells with low fluorescence (= low reporter probe expression) rather than bright fluorescence (= high reporter probe expression). Cells with low reporter probe expression are more likely to yield a fractionally large redistribution of the probe.
5. As mentioned before, the E3–R3 reporter consisting of three wild-type RBD domains is toxic to many cell types. For example COS cells or Jurkat cells die upon expression of E3–R3, whereas PC12 cells tolerate well E3–R3 expression. In case you observe toxicity effects choose reporter probes containing the attenuated RBD versions RBD-R59A (E3–R3(A)) or RBD-(R59A/N64D) (E3–R3(A/D)).

6. For transfection of Jurkat or other suspension cells we recommend using DMRIE-C reagent (GIBCO/Life Technologies). Although electroporation or microporation will also work, we find that lipofection with DMRIE-C gives the best results in terms of responsiveness of cells and Ras-GTP visualization. For hard-to-transfect cells, transduction with viral particles can be an option, although we have not carried out that type of experiments, yet.
7. Membrane dynamics and/or cortical cytoskeletal rearrangement processes such as membrane ruffling can sometimes cause the illumination of portions of the plasma membrane which can be misinterpreted as Ras-GTP-dependent probe recruitment to those sites. One means to control for this type of artefactual signals is to co-transfect the cells with a red fluorescent, non-interacting RBD probe like C3–R1(A/D) (Subheading 2.4) (Fig. 1) [16] and to track simultaneously the redistribution of the functional (E3–R3) and nonfunctional (C3–R1(A/D)) probes in the green and red channels, respectively (Fig. 3).
8. If cell or focus drift and concomitant loss of the originally selected cell following on-stage stimulation (**steps 12–15** in Subheading 3.5) are a recurrent problem one can use dishes with a grid glass slide, which facilitates tracking and spotting the selected cell(s).
9. As observed by us and other researchers, RBD-based fluorescent reporter probes, even the large E3–R3 polypeptides, tend to accumulate in the nucleus of many cell types for yet unaccounted reasons [7, 8, 13, 16].
10. To analyze the images acquired with the confocal microscope the software supplied by the manufacturers can be used. However, despite high prices these programs often include only basic functions which cannot be customized to the need of the user. Moreover, these programs do not allow processing of a batch of files. If script- and macro-functionalities are not included in the software all parameters have to be entered for each image individually. Customization and automatization are possible by writing routines in programming environments which provide the user with libraries that contain image processing algorithms. In the examples provided with this chapter we show how images can be processed with MATLAB (The MathWorks Inc., Natick, MA, USA). The name MATLAB stands for matrix laboratory. MATLAB contains a variety of functions that allow to process matrices quickly. Since images are nothing but a two- or multidimensional matrix of pixels MATLAB is ideally suited to process images. Image processing is facilitated by a variety of toolboxes, such as the Image

Processing Toolbox, which extends the capabilities of the MATLAB environment by a collection of additional functions. Another useful toolbox is the DIP-image toolbox, which can be obtained free of charge from the website of the University of Delft (http://www.dipimg.org). The examples shown in the Appendix to this chapter are based on MATLAB and the Image Processing Toolbox.

11. This step only serves to optimize the image display. Image processing and quantification, however, must be always done with the original data sets.

## Acknowledgements

We acknowledge support by the Novartis Stiftung für Therapeutische Forschung to IR.

# Appendix

```
function SegmentCell

% ===== Initialize workspace ===============================================
% Clear workspace
close all;
% ===== Step 1: Load image =================================================
% Load fluorescence image
% In this example, the image was exported from the LSM software into the tif
% format. The channels have been assigned as follows:
% R = "red" channel of the confocal measurement = Golgi marker
% G = "green" channel of the confocal measurement = GFP
% B = "transmission" channel of the confocal measurement = transmission image.
% Load image
ImgRGB = imread('Sample.tif');
% Set blue channel (transmission image) to zero
ImgRGB(:,:,3) = 0;
% Determine pixel format
ImgSize = [size(ImgRGB,1) size(ImgRGB,2)];
% Show and export original image
figure('Name','Original Image');
imshow(ImgRGB);
imwrite(ImgRGB,'Sample-Original.jpg','jpg')

% ===== Step 2: Optimize contrast ==========================================
% Optimize contrast and brightness
ImgRGBAdj = ImgRGB;
ImgRGBAdj(:,:,1) = imadjust(ImgRGB(:,:,1))/1.5;
ImgRGBAdj(:,:,2) = imadjust(ImgRGB(:,:,2));
% Show and export the contrast enhanced image
figure('Name','Contrast enhanced image');
imshow(ImgRGBAdj);
imwrite(ImgRGBAdj,'Sample-ContrastEnhanced.jpg','jpg');

% ===== Step 3: Filter images ==============================================
% Here, to filter the images a simple median filter is used.
% Median filtering redues effetively "salt and pepper" noise.
% Here, each output pixel contains the median value in the 7-by-7 neighborhood
% around the corresponding pixel in the input image. medfilt2 pads the image
% with zeros on the edges, so the median values of the edges might appear
%distorted.
ImgRFilt = medfilt2(ImgRGB(:,:,1),[7 7]);
ImgGFilt = medfilt2(ImgRGB(:,:,2),[7 7]);

% Alternatively a gaussian filter can be used. However, median filtering
% preserves the edges better.
% To try out the gaussian filter unomment the following lines:
% FilterMatrix = fspecial('gaussian',[7 7],1);
% ImgRFilt = filter2(FilterMatrix,ImgRGB(:,:,1),'same');
% ImgGFilt = filter2(FilterMatrix,ImgRGB(:,:,2),'same');

% Show and export the contrast enhaned filtered images
figure('Name','Filtered red channel');
ImgRFiltAdj = imadjust(uint8(ImgRFilt));
imshow(ImgRFiltAdj);
imwrite(ImgRFiltAdj,'Sample-FilteredRedChannel.jpg','jpg');
figure('Name','Filtered green channel');
ImgGFiltAdj = imadjust(uint8(ImgGFilt));
imshow(ImgGFiltAdj);
imwrite(ImgGFiltAdj,'Sample-FilteredGreenChannel.jpg','jpg');

% ===== Step 4: Create mask to identify the cell of interest ==================
% In this step a mask can be created that allows the user to mark the cell
% of interest. As default the mask comprises the entire image.
MaskUser = ones(ImgSize);
% To mark a cell, uncomment the following line:
% MaskUser = roipoly(ImgRGBAdj);

% ===== Step 5: Calculate gradients ========================================
% Calculate gradient images
ImgRGrad = imgrad2(ImgRFilt,[7 7]);
ImgGGrad = imgrad2(ImgGFilt,[7 7]);
% Show contrast enhanced gradient image
figure('Name','Gradient image of the red channel');
ImgRGradAdj = imadjust(uint8(ImgRGrad));
imshow(ImgRGradAdj);
```

```
imwrite(ImgRGradAdj,'Sample-GradientsRedChannel.jpg','jpg');
figure('Name','Gradient image of the green channel');
ImgGGradAdj = imadjust(uint8(ImgGGrad));
imshow(ImgGGradAdj);
imwrite(ImgGGradAdj,'Sample-GradientsGreenChannel.jpg','jpg');

% ===== Step 6: Indentify the cell =========================================
% A binary mask based on the gradient image is created. This mask is used to
% identify the cell and the plasma membrane. In this example, the mask
% identifies pixels with a gradient above a threshold of 2.
MaskCell = ImgGGrad > 2;
% Include connected pixels with a gradient above a threshold of 1 in the mask.
SEL = strel('square',3);
MaskCell = imdilate(MaskCell,SEL);
TempMask = ImgGGrad > 1;
MaskCell = MaskCell & TempMask;
% Restrict mask to user defined area
MaskCell = MaskCell & MaskUser;
% Fill holes inside the cell
MaskCell = imfill(MaskCell,4,'holes');
% Remove small objects from the image
MaskCell = bwareaopen(MaskCell,200,8);

% ===== Step 7: Indentify the plasma membrane ===============================
% A binary mask that can be used to identify the cell perimeter is created.
MaskCellPerimeter = bwperim(MaskCell);
% Dilate perimeter
SEL = strel('disk',15);
MaskCellPerimeter = imdilate(MaskCellPerimeter,SEL);
MaskCellPerimeter = MaskCellPerimeter & MaskCell;
% Create a mask for interior parts of the cell ---
MaskInt = MaskCell & ~MaskCellPerimeter;
% Generate a mask for the whole plasma membrane
MaskPM = ImgGGrad > 3;
MaskPM = MaskPM & MaskCellPerimeter;

% ===== Step 8: Create a mask for the nucleus ===============================
% To identify the nucleus a binary mask based on the gradient image of the
% green channel is created. The mask identifies pixels with a gradient above
% a threshold of 3.
MaskNcl = ImgGGrad > 3;
% Include connected pixels with a gradient above a threshold of 1 in the mask.
SEL = strel('square',3);
MaskNcl = imdilate(MaskNcl,SEL);
TempMask = ImgGGrad > 3;
MaskNcl = MaskNcl & TempMask;
MaskNcl = imdilate(MaskNcl,SEL);
MaskNcl = MaskNcl & TempMask;
% Restrict mask to the interior of the cell
MaskNcl = MaskNcl & MaskInt;
% Fill interior gaps
MaskNcl = imfill(MaskNcl,4,'holes');
% To identify the nucleus one might think that it could be sufficient to apply a
% threshold to the intensity image of the green channel, i.e.
% MaskNcl = ImgGFilt > 47;
% This, however, would bias the result, since the intensity would be only
% determined in pixels which have been selected by their intensity.
% Restrict mask to the interior of the cell
MaskNcl = MaskCell & MaskNcl;
% Fill interior gaps
MaskNcl = imfill(MaskNcl,4,'holes');
% Remove small objects from the image
MaskNcl = bwareaopen(MaskNcl,300,8);

% ===== Step 9: Create a mask for the Golgi apparatus ========================
% To identify the Golgi apparatus it is sufficient to apply a threshold to the
% intensity image of the red channel.
MaskGolgi = ImgRFilt > 35;
% Restrict mask to the interior of the cell
MaskGolgi = MaskInt & MaskGolgi;
% Fill interior gaps
MaskGolgi = imfill(MaskGolgi,4,'holes');
% Remove small objects from the image
MaskGolgi = bwareaopen(MaskGolgi,300,8);

% ===== Step 10: Create a mask for the cytosol  ==============================
% The cytosol and other compartments should not overlap. To this aim a safety
% margin is added.
```

```
SEL = strel('disk',5);
MaskNclLarge = imdilate(MaskNcl,SEL);
MaskGolgiLarge = ImgRFilt > 30;
MaskCytosol = MaskInt & ~MaskNclLarge & ~MaskGolgiLarge;
% Avoid overlap between other compartments
MaskNcl = MaskNcl & ~MaskCellPerimeter & ~MaskGolgiLarge;
MaskGolgi = MaskGolgi & ~MaskCellPerimeter & ~MaskNclLarge;

% ===== Step 11: Integrate fluorescence ========================================
FluoP = sum(sum(immultiply(ImgRGB(:,:,2),MaskPM)))/sum(sum(MaskPM));
FluoN = sum(sum(immultiply(ImgRGB(:,:,2),MaskNcl)))/sum(sum(MaskNcl));
FluoG = sum(sum(immultiply(ImgRGB(:,:,2),MaskGolgi)))/sum(sum(MaskGolgi));
FluoC = sum(sum(immultiply(ImgRGB(:,:,2),MaskCytosol)))/sum(sum(MaskCytosol));
% Show the results
% red    = Mask for the Golgi apparatus
% green  = Mask for the cytosol
% cyan   = Mask for the nucleus
% yellow = Mask for the plasma membrane
% b&w    = Overlay of red and green fluorescence images
Fluo = ImgRGBAdj(:,:,1)/2 + ImgRGBAdj(:,:,2)/2;
OverlayR = Fluo;
OverlayR(MaskPM)      = 255;
OverlayR(MaskCytosol) = 0;
OverlayR(MaskNcl)     = 0;
OverlayR(MaskGolgi)   = 255;
OverlayG = Fluo;
OverlayG(MaskPM)      = 255;
OverlayG(MaskCytosol) = 255;
OverlayG(MaskNcl)     = 255;
OverlayG(MaskGolgi)   = 0;
OverlayB = Fluo;
OverlayB(MaskPM)      = 0;
OverlayB(MaskCytosol) = 0;
OverlayB(MaskNcl)     = 255;
ImgSeg = ImgRGBAdj;
ImgSeg(:,:,1) = OverlayR;
ImgSeg(:,:,2) = OverlayG;
ImgSeg(:,:,3) = OverlayB;
figure('Name','Segmented image');
imshow(ImgSeg);
text(10,30,'Average fluorescence intensity per pixel:',...
    'FontName','FixedWidth','FontSize',10,'Color','g');
text(10,45,['Plasma membrane (yellow area):' num2str(FluoP,'%06.2f')],...
    'FontName','FixedWidth','FontSize',10,'Color','g');
text(10,60,['Nucleus         (cyan area)   :' num2str(FluoN,'%06.2f')],...
    'FontName','FixedWidth','FontSize',10,'Color','g');
text(10,75,['Golgi           (red area)    :' num2str(FluoG,'%06.2f')],...
    'FontName','FixedWidth','FontSize',10,'Color','g');
text(10,90,['Cytosol         (green area)  :' num2str(FluoC,'%06.2f')],...
    'FontName','FixedWidth','FontSize',10,'Color','g');
saveas(gcf,'Sample-Segmented.jpg','jpg')


% ===== Subroutines ============================================================

function ImgGrad=imgrad2(Img,Shift)
% Function calculates the steepest gardient between neighbouring pixels
% Parameters transferred to this function:
% Img  : Image
% Shift: The image is shifted to all distances by this and smaller values.
% Differences between the images are calculated. The highest value is taken.
% --- Initialize constants and variables ---
[NoRow,NoCol] = size(Img);
Img      = double(Img);
ImgPad   = padarray(Img,[Shift Shift],'replicate','both');
ImgGrad  = double(zeros(NoRow,NoCol));
% --- Calculate gradient ---
for RowShiftCount = -Shift:Shift
    for ColShiftCount = -Shift:Shift
        if ~isequal(RowShiftCount,ColShiftCount)
            ImgDiff = Img-ImgPad(1+Shift+RowShiftCount:NoRow+Shift+...
                RowShiftCount,1+Shift+ColShiftCount:NoCol+Shift+ColShiftCount);
            % To calculate the true gradient the distance between the pixels
            % has to be taken into account:
            % ImgDiff = ImgDiff/sqrt(RowShiftCount^2+ColShiftCount^2);
            % However, the algorithm is more effective, if this step is omitted
            Mask = ImgDiff > ImgGrad;
            ImgGrad(Mask) = ImgDiff(Mask);
```

```
            end
        end
    end
    % --- Set border to zero ---
    ImgGrad(1:Shift+1,:)          = 0;
    ImgGrad(NoRow-Shift:NoRow,:)  = 0;
    ImgGrad(:,1:Shift+1)          = 0;
    ImgGrad(:,NoCol-Shift:NoCol)  = 0;
    % --- Convert to int8 format ---
    % ImgGrad = ImgGrad;
    % ImgGrad = uint8(ImgGrad);
```

## References


1. Karnoub AE, Weinberg RA (2008) Ras oncogenes: split personalities. Nat Rev Mol Cell Biol 9:517–531
2. Herrmann C, Martin GA, Wittinghofer A (1995) Quantitative analysis of the complex between p21ras and the Ras-binding domain of the human Raf-1 protein kinase. J Biol Chem 270:2901–2905
3. Herrmann C, Horn G, Spaargaren M et al (1996) Differential interaction of the ras family GTP-binding proteins H-Ras, Rap1A, and R-Ras with the putative effector molecules Raf kinase and Ral-guanine nucleotide exchange factor. J Biol Chem 271:6794–6800
4. Nassar N, Horn G, Herrmann C et al (1995) The 2.2 A crystal structure of the Ras-binding domain of the serine/threonine kinase c-Raf1 in complex with Rap1A and a GTP analogue. Nature 375:554–560
5. Mochizuki N, Yamashita S, Kurokawa K et al (2001) Spatio-temporal images of growth-factor-induced activation of Ras and Rap1. Nature 411:1065–1068
6. Sawano A, Takayama S, Matsuda M et al (2002) Lateral propagation of EGF signaling after local stimulation is dependent on receptor density. Develop Cell 3:245–257
7. Chiu VK, Bivona T, Hach A et al (2002) Ras signalling on the endoplasmic reticulum and the Golgi. Nat Cell Biol 4:343–350
8. Bivona TG, Perez De Castro I, Ahearn IM et al (2003) Phospholipase C gamma activates Ras on the Golgi apparatus by means of RasGRP1. Nature 424:694–698
9. Mohney RP, Das M, Bivona TG et al (2003) Intersectin activates Ras but stimulates transcription through an independent pathway involving JNK. J Biol Chem 278:47038–47045
10. Perez de Castro I, Bivona TG, Philips MR et al (2004) Ras activation in Jurkat T cells following low-grade stimulation of the T-cell receptor is specific to N-Ras and occurs only on the Golgi apparatus. Mol Cell Biol 24:3485–3496
11. Walker SA, Kupzig S, Bouyoucef D et al (2004) Identification of a Ras GTPase-activating protein regulated by receptor-mediated Ca2+ oscillations. EMBO J 23:1749–1760
12. Zhang SQ, Yang W, Kontaridis MI et al (2004) Shp2 regulates SRC family kinase activity and Ras/Erk activation by controlling Csk recruitment. Mol Cell 13:341–355
13. Augsten M, Pusch R, Biskup C et al (2006) Live-cell imaging of endogenous Ras-GTP illustrates predominant Ras activation at the plasma membrane. EMBO Rep 7:46–51
14. Genove G, Glick BS, Barth AL (2005) Brighter reporter genes from multimerized fluorescent proteins. Biotechniques 39:814, 816, 818 passim
15. Nassar N, Horn G, Herrmann C et al (1996) Ras/Rap effector specificity determined by charge reversal. Nat Struct Biol 3:723–729
16. Rubio I, Grund S, Song SP et al (2010) TCR-induced activation of Ras proceeds at the plasma membrane and requires palmitoylation of N-Ras. J Immunol 185:3536–3543
17. Leadsham JE, Miller K, Ayscough KR et al (2009) Whi2p links nutritional sensing to actin-dependent Ras-cAMP-PKA regulation and apoptosis in yeast. J Cell Sci 122:706–715
18. Broggi S, Martegani E, Colombo S (2012) Live-cell imaging of endogenous Ras-GTP shows predominant Ras activation at the plasma membrane and in the nucleus in Saccharomyces cerevisiae. Int J Biochem Cell Biol 45:384–394
19. Wu RF, Ma Z, Liu Z et al (2010) Nox4-derived H2O2 mediates endoplasmic reticulum signaling through local Ras activation. Mol Cell Biol 30:3553–3568

# Chapter 20

# Nanoclustering and Heterogeneous Membrane Diffusion of Ras Studied by FRAP and RICS Analysis

**Camilo Guzmán, Maja Šolman, and Daniel Abankwa**

## Abstract

Fluorescence Recovery After Photobleaching (FRAP) and Raster Image Correlation Spectroscopy (RICS) are two powerful techniques to study the diffusion dynamics of fluorescently labeled proteins. FRAP and RICS can be easily applied on any commercial confocal microscope. In this chapter, we describe the principles of these methods and provide the reader with a detailed guide on how to apply these methods in the study of Ras nanoclustering and diffusion in the plasma membrane of live cells.

**Key words** Diffusion, Fluorescent protein, GFP, FRAP, RICS, Nanoclustering, Ras, Galectin, Confocal microscopy

## 1 Introduction

### 1.1 Nanoclustering of Ras

The oncoproteins H-, N-, and K-Ras4B laterally segregate into distinct signaling platforms on the plasma membrane, called nanoclusters [1]. Nanoclusters are only 6–20 nm in dimension and may depend on cholesterol or the actin network [2, 3]. Therefore, they are recognized as proteo-lipid structures with properties reminiscent of the better-known "lipid-rafts". However, nanoclusters are defined as dynamic assemblies of 6–8 Ras proteins, with experimentally estimated lifetimes of 0.1 ms to 1 s [1, 4], while computational simulations suggest even shorter microsecond lifetimes [5]. Current data furthermore suggest that Ras becomes immobilized in nanoclusters, which then serve as the only sites for efficient effector recruitment [4]. Ras nanoclustering is modulated by scaffolding proteins, such as galectin-1 and -3 that stabilize Ras nanoclusters and thus increase Ras activity [6–8]. Only few other Ras nanocluster modulators are suggested or known, such as Shoc2/Sur-8 [9] and nucleophosmin [10]. Of note, nanocluster modulators may exhibit a preference for a specific Ras isoform.

Lorenza Trabalzini and Saverio Francesco Retta (eds.), *Ras Signaling: Methods and Protocols*,
Methods in Molecular Biology, vol. 1120, DOI 10.1007/978-1-62703-791-4_20, © Springer Science+Business Media, LLC 2014

Recent experimental evidence furthermore corroborates initial data, which indicated that disruption of Ras nanoclustering down-modulates its signaling output [11–13]. The extent of Ras-nanoclustering thus tunes Ras signal amplification [14], making it an indispensable and underappreciated feature of the Ras signaling architecture. We propose that pharmacological targeting of Ras nanoclusters represents a viable strategy to block oncogenic Ras signaling.

Ras nanoclustering was first described using Electron Microscopy (EM) of intact apical plasma membrane sheets followed by statistical analysis of the immunogold point patterns [2, 3]. This analysis provides high spatial resolution, which is only limited by the size of the gold particle and the antibody (i.e., <20 nm). In addition, nanoclustering has been studied using Single Fluorophore Video Tracking (SFVT) [15, 16], a method that provides good spatial and high temporal information but requires customized equipment and analysis. Due to the dense packing of the Ras proteins in the nanocluster, FRET (Förster Resonance Energy Transfer) also reports on nanoclustering [11, 17, 18]. While FRET allows for very high temporal resolution (μs) and high spatial resolution (1–10 nm), its efficiency is too low for distances above 10 nm.

In this chapter we present two methods that can be applied to study Ras nanoclustering using a conventional confocal microscope. Fluorescence Recovery After Photobleaching (FRAP) provides a simple measure of Ras diffusion and mobility. On the other hand, Raster Image Correlation Spectroscopy (RICS) provides access to Ras molar quantity and diffusion parameters that can otherwise only be obtained with more specialized equipment.

### 1.2 FRAP

FRAP is a well-established method that was initially developed to measure the two-dimensional lateral mobility of fluorescent particles [19]. In a FRAP experiment fluorescent molecules in a Region Of Interest (ROI) are specifically photobleached, i.e., their chromophore is irreversibly destroyed, by a high intensity laser beam. Bleaching is then followed by monitoring the recovery of the fluorescence signal in the ROI using the same laser but at a much lower intensity. Fluorescence recovery in the ROI takes place due to the lateral diffusion of non-bleached fluorescent molecules from the surrounding areas into the ROI. The magnitude and speed of the recovery depend on whether all molecules are mobile and on the diffusion properties of the fluorescent molecule, respectively. The magnitude of recovery is quantified by the immobile fraction (fraction of non-recovered fluorescence) and the speed by the half-time of recovery (i.e., the amount of time needed to achieve half of the final intensity). These are the parameters that we are going to determine from the FRAP data (*see* Fig. 1).

Since the development of this technique, most data acquired on membrane proteins have been interpreted by assuming free Brownian

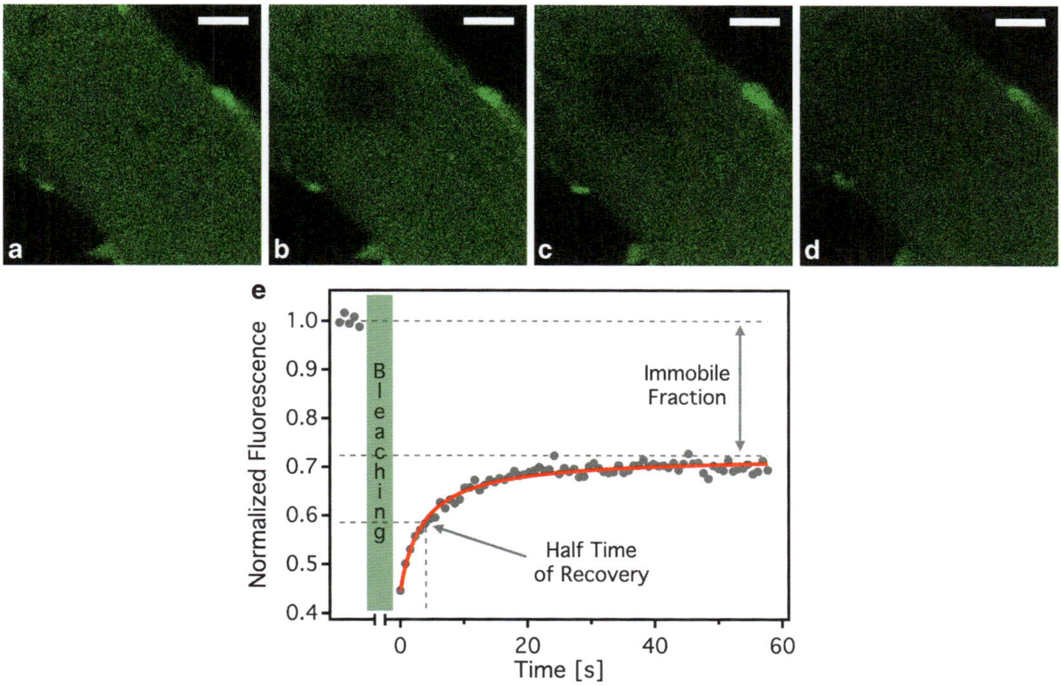


**Fig. 1** Example of a FRAP experiment performed on BHK cells expressing pmGFP-H-RasG12V. (**a–d**) Confocal images showing pre-bleaching, post-bleaching, and recovery on the region of interest (ROI). Scale bars correspond to 2 μm. Contrast has been adjusted for a better display on a printed page. (**e**) Plot of the normalized fluorescence against time. *Grey circles* correspond to experimental data points while the *red continuous line* corresponds to a fit using an anomalous diffusion model (*see* **step 14** of Subheading 3.3 for details). *Grey dotted lines* are placed to illustrate the half-time of recovery and the immobile fraction. Normalization of data is done using the reference fluorescence intensity acquired on the ROI just before bleaching (first five points of this figure)

diffusion in two dimensions. However this interpretation was shown to be incomplete in the case of cell surface receptors where receptors may diffuse freely or exhibit hindered diffusion [20]. In those cases an anomalous diffusion model, which considers the presence of diffusion traps is much more appropriate. The model has been successfully used for example to describe the diffusion of heterotrimeric G proteins [21]. In the case of Ras, diffusion is hindered due to the formation of (immobile) nanoclusters. We therefore provide details for correct data acquisition and analysis, including the necessary equations derived from the anomalous diffusion model.

### 1.3 RICS

RICS is a novel technique for extracting molecular diffusion and concentration data from confocal images [22, 23]. It allows the determination of spatial maps of concentrations, aggregation, and dynamics in living cells using confocal microscopy [22]. As compared to single point methods, such as Fluorescence Correlation Spectroscopy (FCS), the fluorescence intensity is detected successively in many focal points during scanning.

Performing a raster scan enables data acquisition from different spatial locations at different times. Because each pixel is collected at a different time, there is temporal information included in each individual image. This is referred to as the "hidden time structure" and can be used to extract temporal information from the image [24]. From each image a spatial correlation is calculated. This correlation contains a spatial component, based on the distribution of the molecules, and a temporal component, based on the diffusion of the molecules during the raster scan. By mathematical fitting and separation of these components the temporal information is obtained (*see* Fig. 2 as an example of the spatial correlation function).

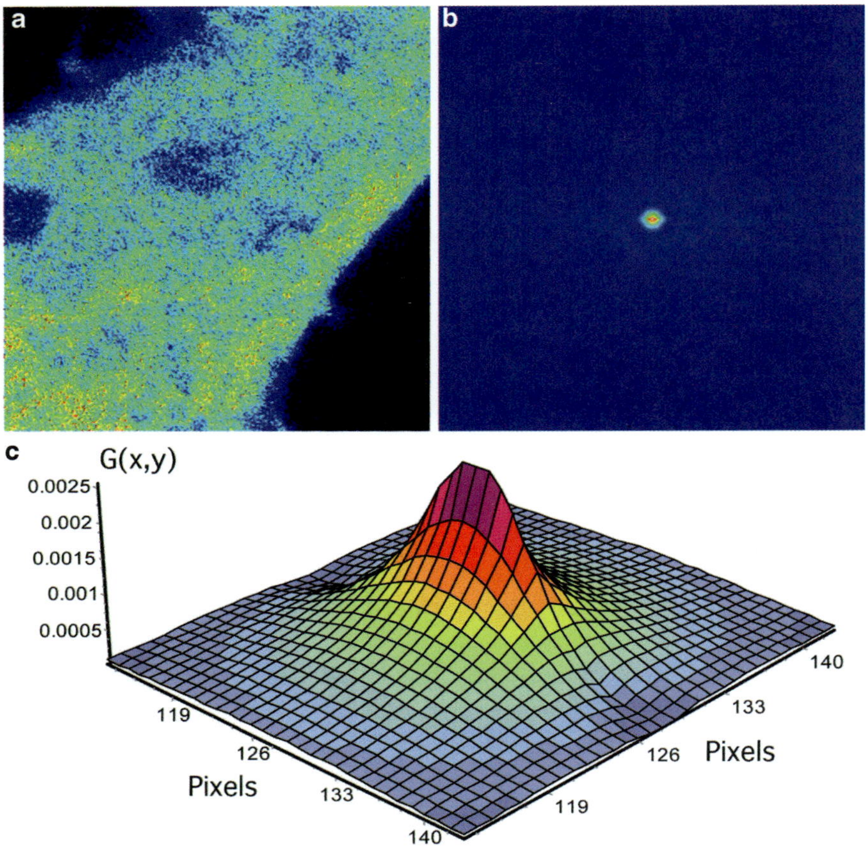


**Fig. 2** Example of a RICS experiment performed on BHK cells expressing pmRFP-galectin-1. (**a**) Single image extracted from the stack of images analyzed using the software SimFCS 2.0. Image is 256 × 256 pixels (14.08 μm × 14.08 μm) and shown in *rainbow false color*, where *more red colors* correspond to a higher intensity. (**b**) Image correlation function calculated from the acquired stack of images, with the magnitude shown in rainbow false coloring. Displacements used to calculate the correlation function are limited to only 64 pixels, and therefore the size of this correlation image is 64 × 64 pixels. (**c**) Resulting fit of the image correlation displayed as a 3-dimensional contour graph. The *height* represents the magnitude of the function, while the *horizontal axes* correspond to the horizontal and vertical correlation shifts. Fit was performed excluding the edges of the image correlation function on the central 32 × 32 pixels

In a confocal image, spatial correlations for fluorescent particles depend both on the spatial overlap and on the time interval between adjacent pixels. Random diffusion of the particle and longer time intervals between data points cause a decrease in the correlation function at shorter spatial scales but an increase of the correlation function at distant pixels. It is precisely these changes at different scales of the correlation function that contain information about molecular diffusion [22]. By correctly fitting the correlation function spatially and temporally, we can extract the diffusion constant and the concentration of the labeled molecule.

In RICS, the scanning frequency should be tuned to the diffusion properties of the molecule of interest. The scanning frequency must be high enough, so the molecules will not move too much before the laser can scan them again, but also low enough so the laser can register the movement of the molecule by detecting it in different locations. Toggling the scanning frequency to match the diffusion properties of the molecule of interest exposes the intrinsic time structure that is contained in every confocal image [22]. Typically multiple images are acquired in order to improve the signal-to-noise ratio and data statistics [24].

In the following, we will describe how we determine the nanoclustered (immobile) fraction of mGFP-tagged H-RasG12V and its dependence on the level of the nanocluster scaffolding protein galectin-1. We use FRAP to measure the immobile fraction of Ras in a BHK cell, while RICS serves to quantitate the exact concentration of mRFP-tagged galectin-1 in that specific cell.

## 2 Materials

### 2.1 Cell Culture

The plasmids and reagents described in this subsection are presented for the study of H-Ras nanoclustering and its dependence on galectin-1 expression levels. If a study of other members of the Ras family is intended, corresponding plasmids for the expression of a fluorescently tagged version of that particular Ras are required. For quantification of the overexpression by RICS, the nanocluster modulator has to be fluorescently tagged, such that it can be spectrally well separated from the fluorescence tag on the Ras protein. If the effect of other known or putative nanocluster modulators is intended to be studied, corresponding expression plasmids and knock-down reagents will be required.

1. Baby Hamster Kidney (BHK) cells (ATTC® Number CCL-10™). Other cell lines can also be used, however, in each case the endogenous levels of nanocluster modulator (galectin-1 in this case) should be considered.
2. Cell culture round plates (100 mm), sterile glass pipettes, autoclaved pipette tips, cell counting chamber (hemocytometer), and light microscope.

3. Dulbecco's Modified Eagle's Medium (DMEM) supplemented with 10 % FBS, L-glutamine, penicillin (100 U/ml), and streptomycin (100 μl/ml): Medium used for cell culture.
4. Trypsin–EDTA, Phosphate Buffer Saline (PBS): Other solutions needed to culture BHK cells.
5. JetPRIME transfection reagent (Polyplus transfection): Reagent used for plasmid transfection. If desired another transfection reagent can be used.
6. Cover glass-bottomed culture dish or chambered cover glass, with the cover glass number 1.5 (0.17–0.19 mm thick): Culture dish where cells can be both cultured and imaged. We used the Lab-Tek™ II Chambered Cover glass, 4 wells (Nunc).
7. pmGFP-H-RasG12V plasmid: Plasmid encoding for a constitutively active H-Ras, labeled with a monomeric green fluorescent protein (mGFP) [18].
8. pmRFP-galectin-1: Plasmid used for overexpression of galectin-1, labeled with the monomeric red fluorescent protein mRFP [25].
9. pcDNA3-antisense-galectin-1 [13]: Plasmid used to knockdown galectin-1.
10. Ringers buffer: 10 mM HEPES, 10 mM glucose, 2 mM $NaH_2PO_4$, 1 mM $MgCl_2 \cdot 6H_2O$, 2 mM $CaCl_2$, 5 mM KCl, 155 mM NaCl, pH 7.2, filter sterilized. Buffer used to maintain cells during confocal imaging. Similar buffers or media can be used; however, for imaging purposes translucent, phenol red-free solutions are required.

### 2.2 Confocal Imaging

In this subsection we describe the basic equipment for confocal imaging on commercially available microscopes.

1. Vibration-isolated optics table: Table used to prevent any external vibration that would reach the microscope and change the position of the laser with respect to the sample.
2. Leica TCS SP5 microscope (Leica Microsystems GmbH, Germany) equipped with live cell incubation system and FRAP experiment software ("FRAP wizard"). Different confocal microscopes can be used according to the preference of the reader (*see* **Notes 1**, **6** and **9**).
3. Oil immersion objective lens: 100× NA 1.4 Leica immersion oil objective was used. A high numerical aperture (NA) objective is required. Oil immersion, water immersion, and air objectives can be used. Objective power should be between 40× and 100× [24].
4. Immersion oil: Oil with the matching refractive index for the objective lens used here.
5. Mercury lamp: Used to illuminate the sample in bright-field mode, in order to localize cells of interest for confocal imaging.

6. Ar 100 mW laser at 488 nm: Laser line used in the FRAP experiments both to image and to bleach the mGFP labeled Ras protein.
7. HeNe 1 mW laser at 543 nm: Laser line used in RICS experiments to image the introduced mRFP tagged galectin-1, and monitor its overexpression levels.
8. In the case of RICS a photon-counting detector is required, as it has a better rejection of noise [24]. Therefore, Avalanche Photo Diode (APD) detectors, GaAsP or Hybrid-PMT (Hybrid photomultiplier tube) detectors are preferred.
9. RICS calibration sample to determine the confocal volume: Use Tetraspeck Microspheres 0.1 μm fluorescent blue/green/orange/dark red (Molecular Probes, Invitrogen cat No T7279).

### 2.3 Data Analysis

In this subsection we describe necessary software for data extraction, fitting and analysis. If desired, similar software can be used as long as the correct fitting function for FRAP and RICS experiments is included.

1. ImageJ 1.46a software (Free Software, National Institutes of Health, Bethesda, Maryland, USA [26]): Software used to extract the fluorescence intensities of a region of interest in FRAP experiments. In RICS this software is used to export the data into the appropriate format for RICS data analysis.
2. ImageJ plugin for image stack analysis named "Measure Stacks" developed by Bob Dougherty and freely available through the ImageJ Web page (*see* **Note 15**).
3. IGOR Pro 6.22a software (435 USD for academic license, WaveMetrics): Software used to fit the correct FRAP model to the extracted data. Similar mathematical software can also be used that allows defining a fitting function for data analysis with the anomalous diffusion model.
4. SimFCS 2.0 (2000 USD, Laboratory for Fluorescence Dynamics, University of California—Irvine, USA): Software created by the developers of the RICS method and used for fitting of the correlation functions that provide the desired concentration and diffusion information.

## 3 Methods

### 3.1 Cell Culture

1. Use a water bath or incubator to preheat the cell culture media, PBS, and trypsin–EDTA to 37 °C for 30 min.
2. Cell culture should be done in an aseptic environment of laminar flow hood using sterile cell culture dishes, glass pipettes, and autoclaved pipette tips.

3. Grow cells on 100 mm round cell culture plates. When they reach 80 % of confluence, remove medium from the culture plate. Carefully add 10 ml of sterile PBS to wash any residual medium from the plate. Remove PBS and add 1 ml of trypsin–EDTA to detach cells from the plate. Put the plate back into the incubator.
4. As soon as cells have detached (usually 1–5 min) add 10 ml of culture media and resuspend them by pipetting using a glass pipette.
5. Put a cover glass on the cell counting chamber and pipette an appropriate volume of cell suspension to one of the counting areas. Count the cells using the light microscope.
6. To each well of the Lab-Tek™ II Chambered Cover glass, add 50,000 cells from the cell suspension and fresh culture media to a final volume of 500 μl.
7. After 24 h, do the transfection according to the instructions of the transfection reagent used. Prepare the transfection mixtures in the polystyrene tubes. In the case of JetPrime, first mix plasmid DNA with 50 μl of JetPrime buffer. Use 0.25 μg of pmGFP-H-RasG12V plasmid. In the case of overexpression of galectin-1, use 0.25 μg of pmGFP-H-RasG12V plasmid and 0.75 μg of pmRFP-galectin-1. When down-modulating galectin-1, use 0.25 μg of pmGFP-H-RasG12V plasmid and 0.75 μg of pcDNA3-antisense-galectin-1. Mix the solution by vortexing and add 0.5 μl JetPrime transfection reagent (2 μl if galectin-1 is down-modulated or overexpressed). Vortex the mixture again and incubate for 10 min at room temperature. Add to the well drop wise, gently rock the chambered cover glass and return it to the incubator.
8. Twenty-four hours after the transfection, remove the media and add Ringers buffer to the cells.
9. Proceed with the FRAP data acquisition.

### 3.2 FRAP Data Acquisition

1. Before taking the cells to the microscope, turn on the microscope, the computer, lasers, and incubation system. Chamber temperature should be set to 37 °C. Allow at least 30 min for stabilization.
2. Select the 100× oil immersion objective.
3. Add a drop of immersion oil to the objective and place the chambered cover glass on top of it.
4. Use the bright-field lamp for focusing on the sample.
5. Select the 488 nm Ar laser and its corresponding set of filters for detection. Detection should be done in the range 500–550 nm for mGFP.
6. Activate the 488 nm laser at around 10 % of its nominal power (*see* **Notes 2** and **3**).

7. Start imaging in live mode.
8. Select a cell with a healthy morphology that shows high fluorescence intensity (high signal to noise ratio) and then zoom in onto the bottom membrane that is attached to the coverslip. Bring digital zoom up to 15× (approximate scan area 11×11 μm) (*see* **Notes 4** and **5**).
9. Enter the "FRAP wizard" or corresponding FRAP experiment software of the confocal microscope that you are using (*see* **Note 9** and **steps 12–14**).
10. Set up imaging with 512×512 pixels per frame, pixel depth 8 bits, pinhole to 1 airy unit (AU) and scanning frequency in the range 800–1,400 Hz. Take note of the time between frames that the selected scanning frequency will give. This time is crucial for correct data analysis (*see* **Notes 6** and 7).
11. Select a square region of interest (ROI) for bleaching with a length in the range 1–2.5 μm. Circular or smaller regions can also be used. However, experiments that will be compared with each other must be performed under the exact same conditions. The choice of the ROI will depend on the capabilities of the confocal microscope (*see* **Note 8**).
12. Set the "FRAP wizard" to perform at least five reference-scans before bleaching of the ROI starts. These scans measure the initial fluorescence intensity that is required to determine the immobile fraction (*see* Fig. 1 and **Note 9**).
13. Set up the "FRAP wizard" such that bleaching is performed only on the ROI using the Ar 488 nm laser at its maximum power. A sufficiently high number of bleaching frames are required so that the intensity in the ROI is reduced to 30–50 % of the intensity before bleaching. In the specific case described here, 20 frames were sufficient (*see* **Note 10**).
14. Using the "FRAP wizard" monitor the fluorescence of the whole image until recovery in the ROI has reached a plateau. For membrane proteins like Ras, a plateau was reached after 50–70 s at our settings (*see* Fig. 1 for details).
15. Using your FRAP software make sure that bleaching and recovery took place correctly. Check that there are no undesired effects such as drift of the sample, bleaching of areas outside the ROI or labeled structures such as vesicles that might have passed through the ROI during data acquisition.
16. If the recovery curve is satisfactory (*see* Fig. 1 as an example) repeat **steps 11–15** two or three more times in the same cell.
17. Save all the image sequences of your FRAP experiments. Your image format will depend on your microscope and the software it uses. In our case the format is ".lif" (Leica image file).
18. If your experiment involves overexpression of mRFP-galectin-1 (or of a similar fluorescently tagged protein) to

modulate nanoclustering, proceed with RICS data acquisition, described in Subheading 3.4 (*see* **Notes 11** and **12**).

19. Repeat FRAP data acquisition in as many cells as possible (a minimum of three or four cells per experimental condition and biological repeat should be performed).
20. Proceed with FRAP Data Analysis.

### 3.3 FRAP Data Analysis

Most FRAP data acquisition software also include a data analysis option. Unfortunately, such analysis is typically done using the classical two-dimensional free Brownian diffusion. As already mentioned in the introduction, such model is not entirely accurate for the study of Ras diffusion. In this subsection we provide you with the more accurate analysis that uses the equation for anomalous diffusion for data fitting [20].

1. Using ImageJ 1.46a software open the image frame sequence acquired in one of the FRAP experiments. Tick the boxes "Open Files Individually" and "Split Channels" (*see* **Note 13**).
2. Use the rectangular selection tool (or circular if your ROI had that shape) to delimit the photobleached ROI. We want to extract the increase of the average fluorescence from that bleached area over time after bleaching (*see* **Note 14**).
3. Go to "Plugins > Stacks > Measure Stack". This will measure the mean fluorescence intensity (in a.u.) of the ROI for each image in the image stack of that particular FRAP experiment (*see* **Note 15**).
4. Go to the results window and copy the columns 1 (frame number) and 3 (mean fluorescence).
5. Open Igor Pro 6.22a or equivalent software and create a new table to paste the two columns just mentioned. ImageJ can be closed and all the remaining steps of this analysis will be performed in Igor Pro 6.22a.
6. Plot the mean fluorescence as a function of the frame number.
7. Apply an offset to your frame number such that the frame containing the information of the first recovery point becomes your frame 0 (*see* **Note 16**).
8. Multiply your frame number by the time between frames (in seconds) that you took note of during data acquisition (*see* **step 10** of the Subheading 3.2). Now you should have the mean fluorescence as a function of time.
9. Calculate the average fluorescence of the reference points taken before bleaching (*see* Fig. 1 for details). Use this value to normalize the complete sequence of fluorescence intensities. After this your plot will be a curve of the normalized intensities ranging from 0 to 1, where 0 would represent complete bleaching and 1 the fluorescence before bleaching.

10. Go to "Analysis > Curve Fitting".
11. In the "Y data" choose the variable containing the normalized mean fluorescence and on the "X data" choose the variable that initially contained the frame number that has now become the time.
12. Define the range of your fit from time 0 until the end of the measurement (*see* **Note 17**).
13. Click on "New fit function".
14. Define a new fitting function according to the equation:

$$F(t) = \frac{F_0 + F_\infty \left( \frac{t}{t_{1/2}} \right)}{1 + \left( \frac{t}{t_{1/2}} \right)}$$

where $F(t)$ corresponds to the mean fluorescence as a function of time $t$. $F_0$ is the fluorescence at $t = 0$, meaning the time when bleaching ends and recovery starts. $F_\infty$ is the recovery achieved after a plateau is reached (by definition the fluorescence when $t \to \infty$). The parameter $t_{1/2}$ is the half-time of recovery (*see* **Notes 18** and **19** for details). Details about the deduction of this equation can be found elsewhere [20].
15. Use the newly defined fit function to analyze your data (*see* **Note 20**).
16. Taking the parameters obtained from the fit (displayed on the history of the command window), use the equation:

$$ImmFrac = \frac{1 - F_\infty}{1 - F_0}$$

to calculate the immobile fraction (*see* **Note 21**).
17. Repeat the analysis for all acquired FRAP experiments.
18. After summarizing repeat measurements for each condition, you can compare the average half-times of recovery and the immobile fractions between different biological conditions.

### 3.4 RICS Data Acquisition

As mentioned before, RICS is used here to determine the concentration of overexpressed mRFP-galectin-1. For correct data acquisition in RICS, parameters like the pixel dwell time (time the laser spends on each pixel) and the pixel size (distance the laser beam travels between pixels) are critical and must be kept within the limits given here. Please, pay special attention to **Notes 22**, **25** and **26**.

1. On the exact same cell that was used in your FRAP experiments, switch the settings of your confocal microscope to use the 543 nm laser together with the appropriate filters for the

detection of mRFP (i.e., band pass 580/75 nm). Use the Avalanche Photodiode Detector (APD) in photon counting mode (*see* **Note 22**).

2. Laser power should be adjusted according to the sensitivity of the detector, which must not become saturated. In our case, results were best when using a power around 10–20 % of the nominal power (1 mW) (*see* **Note 23**).
3. Set up acquisition with image size 256 × 256 pixels, 8 bits pixel depth, pinhole to 1 Airy Unit (AU) and digital zoom to 11× (pixel size 0.055 μm × 0.055 μm) (*see* **Notes 24** and **25**).
4. Set the scanning frequency to 80 Hz. This scanning frequency and the number of pixels set in the **step 3** of this section provide a pixel dwell time of $\tau_p$ 48.83 μs, a line scan time $\tau_l$ of 12.5 ms and a frame time of 3.2 s (*see* **Note 26**).
5. Check that the focus is still on the same membrane that you used for the FRAP experiment.
6. Set up the acquisition of 40–80 frames (*see* **Note 27**).
7. Make sure the scan is performed in raster scan mode and not in bidirectional mode.
8. Save all the image sequences of your RICS experiments. Again, the format will depend on your microscope and the software it uses (here ".lif", Leica image file).
9. Continue with the FRAP and RICS data acquisition of the next cell.
10. Once you have performed the RICS data acquisition of your last cell in that particular biological sample, remove your sample and place the reference calibration sample on the microscope (*see* **Note 28**).
11. On a cover glass, place a drop (~50 μl) of diluted ($1–2 \times 10^{10}$ particles/ml) Tetraspeck Microspheres as reference sample. The diffusion constant of the reference sample must be known, and it must be detectable with the same laser and detector settings that were used for RICS data acquisition on the living cell.
12. Set the temperature of the incubation system to 25 °C (*see* **Note 29**).
13. Acquire reference data by repeating **steps 3–7** of this section. If the mentioned Tetraspeck microspheres are used, no changes in the settings are needed. If another reference sample is used, the pixel dwell time should be changed accordingly (i.e., 8–32 μs in the case of small molecule organic fluorophores, such as FITC).
14. Save as reference data.
15. Proceed with RICS data analysis.

### 3.5 RICS Data Analysis

The analysis described in this section is based on the software SimFCS 2.0. This software was created by Enrico Gratton and colleagues, the developers of the RICS method.

1. The first step before the actual analysis is to convert the files from the format ".lif" in which they were saved (**steps 8** and **14** of the Subheading 3.4) into a format that SimFCS can recognize. This can be done using ImageJ. Open each of the individual frame series containing the RICS data and save as ".tif", which is recognized by SimFCS.
2. Open SimFCS and click on the RICS button.
3. Go to "File" and select the section "Open Multiple Images (int., tif.)"
4. Select the Tiff-converted reference calibration data saved in the **step 14** of the Subheading 3.4.
5. Go to the menu "Tools > RICS > Subtract average". This will calculate the image autocorrelation (*see* **Note 30**).
6. Go to "Fit", this will open a new window.
7. In the new window you will be asked to provide values of important acquisition parameters like the pixel size, the pixel dwell time, the line time and the frame time. *See* **steps 3** and **4** of the Subheading 3.4 for details on the values used in the configuration described here (*see* **Notes 31** and **32** for more details).
8. Enter into the box labeled as *D*l the diffusion coefficient of the reference sample (4.4 $\mu m^2/s$ at 25 °C for the Tetraspeck microspheres [27]). The "fix" box next to the value should be ticked.
9. Fit the data by clicking "Perform fit" in the fitting window. Results from this fit provide the required calibrated value of $\omega_0$ (in μm), which describes the beam waist of your laser (beam waist $\omega_0$ named W0 in SimFCS 2.0). Make sure that the "fix" box next to the value of W0 was not ticked.
10. Close all the fitting windows and open a new file by repeating **step 3** of this section. This time open the RICS experiments performed on the cells.
11. Go to the menu "Tools > RICS > Subtract moving average" (*see* **Note 30**). This will calculate the image autocorrelation function (*see* Fig. 2 for an example and details).
12. Repeat **steps 6** and 7 of this section.
13. This time enter in the box *D*l a value of 1 (starting value for the fit) and do not tick the "fix" box next to it. Doing this allows the software to fit this value according to the acquired data.
14. In the box W0 insert the value obtained from the reference sample describing the beam waist of your laser (**step 9** of this section). For W0 tick the box "fix" keeping this value constant.

15. Fit the data by clicking "Perform fit" in the fitting window. Results of G1(0) give you information about the average number of molecules in the confocal volume ($N = 1/G1(0)$) while results of *D*1 give you information about the diffusion constant of your labeled sample, in this case galectin-1.
16. To obtain the actual concentration of galectin-1 divide the average number of molecules by the confocal volume (*see* **Note 33**).
17. From the RICS analysis, you have now determined the concentration of overexpressed galectin-1, which is added on top of the endogenous expression. You can now analyze, whether there is a dependence of the immobile fraction (i.e., nanoclustered fraction) or of the recovery time on the concentration of overexpressed galectin-1 (*see* **Note 34**).

## 4 Notes

1. An incubation system on the confocal microscope is not essential. However, diffusion properties depend on the viscosity and therefore temperature of the environment. The temperature must therefore be constant for experiments that are intended to be compared with each other.
2. Laser power for observation and image acquisition should be a compromise between being as low as possible to prevent undesired bleaching and being high enough for a good signal to noise ratio.
3. Sample exposure to the laser and the halogen bright-field lamp should be minimized to limit unwanted photobleaching.
4. It is recommended to perform the experiments on the plasma membrane facing the bottom of the cell (i.e., attached to the cover glass), as this part of the plasma membrane is more even.
5. To correctly perform imaging on the focal plane of the plasma membrane, the best method is to move the position of the sample in *Z*, while keeping the system in live imaging mode. Since H-RasG12V is mainly located in the membrane, the image will appear bright, homogeneous, and flat as you pass through the membrane. Other locations will have lower intensities and not be so homogeneous.
6. In FRAP, high scanning frequencies are preferred because the time between frames is reduced and more data points are obtained during recovery. High scanning frequencies have the drawback that a higher number of scans is needed to achieve sufficient bleaching of the ROI. This may increase the size of the data files. The number of pixels per frame does not need to be high, as image content is not analyzed. We recommend $512 \times 512$ (bit depth 8 bits).

7. Time between frames should be below 1 s and knowing its exact value is necessary to perform the data analysis. It is important that all the FRAP experiments have the same time between frames.
8. The size and shape of the ROI affect the half-time of recovery of a FRAP experiment. If results from different experiments are compared, it is essential to use the same size and shape.
9. **Steps 12–14** of the FRAP method must be performed immediately after each other without any time interval in between them. Commercial confocal microscopes have usually the appropriate software module (in our case "FRAP wizard") for this. Do not intend to do this manually, as any delay between the end of bleaching and the start of monitoring the recovery will affect your results.
10. The necessary time for bleaching will depend on the amount of fluorescent molecules in the ROI, the full power of your laser, your scanning frequency and the number of pixels you are acquiring. The key parameter here is the pixel dwell time (i.e., the time the laser spends on each pixel). Insufficient bleaching or over-bleaching can lead to unrealistic results. In the former case recovery may appear faster. In the latter bleaching of molecules in pixels adjacent to the ROI renders recovery apparently slower.
11. As already mentioned, RICS is used here only to measure the expression levels of the overexpressed mRFP-galectin-1. If the experiment does not involve galectin-1, or any other nanocluster modulator, the steps of RICS data acquisition and analysis are not required. Experiments with endogenous cellular levels of galectin-1 or down-regulation of it, are not amenable to the RICS analysis.
12. RICS analysis should be performed in exactly the same cell on which the FRAP experiments are performed. Levels of mRFP-galectin-1 expression differ from cell to cell. In order to correlate the level of mRFP-galectin-1 with the amount of nanoclustering (immobilization) in a given cell, both RICS and FRAP experiments need to be performed in the same cell. Subsequent comparison of these data has to be done cell by cell.
13. Certain FRAP acquisition software like the Leica FRAP wizard, separate bleaching and imaging data into two channels. In such cases it is important to do the analysis on the channel containing the imaging data since this is the one where the recovery information is recorded. Data from the second channel (bleaching channel) can be used to properly identify the ROI during data analysis (*see* Fig. 3 for details).
14. If the boundaries of a ROI are not completely clear, information from the bleaching channel can be used to extract the coordinates of the ROI. The rectangular selection tool in

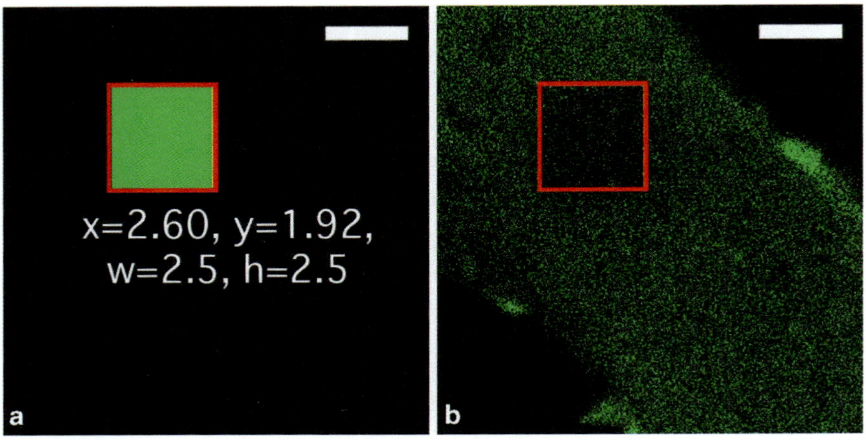


**Fig. 3** Frames from the two separate channels (bleaching channel and imaging channel) of a FRAP experiment as obtained using the "FRAP wizard" software included in the Leica TCS SP5 confocal microscope. Using ImageJ these two channels can be separated and analyzed to extract the fluorescence intensity over the region of interest (ROI) before and after bleaching. (**a**) The channel containing information of the bleaching event is used to identify the ROI. The *red box* represents the selection of the ROI with ImageJ and the numbers correspond to the coordinates (*x* and *y*) and the size of the ROI (width, *w*, and height, *h*). (**b**) After, using the coordinates obtained from the bleaching channel, we can select the ROI in the imaging channel where the recovery is analyzed

ImageJ will give you these coordinates. The same coordinates can then be applied on the imaging channel (*see* Fig. 3).

15. The "Stacks" plugin for ImageJ is not provided with the standard installation of the software but it can be downloaded from the same location as ImageJ. It is important to mention that this plugin does not keep track of the time between frames and gives all the information as a function of the frame number.
16. For the specific FRAP experiment described here, the frame number offset is done by subtracting 26 (5 initial reference frames + 20 bleaching frames + 1 because frame numbers start with 1 while time should start in 0) from the variable containing the frame number. After doing that the frame number should range from −26 to 74 and not from 1 to 100 as initially. Keep in mind that if you use a different number of frames for reference or bleaching, these values must be adjusted accordingly.
17. You can choose the correct range for the FRAP fitting either by adding the point number containing the start and end of your fit or by graphically selecting those points in the plot using the cursors provided by Igor in the "Show info" option of the "Graph" menu.
18. In FRAP, the half-time of recovery is the time at which a system reaches half of its maximum recovery. This value is strongly related to the diffusion properties of your labeled molecule. In our case, this value carries information about the diffusion constant of H-Ras.

19. The maximum recovery is the highest level of recovery that the system can reach. If complete recovery is reached the magnitude of the signal should be the same as before the bleaching event. Typically, the recovery is incomplete due to the presence of immobile particles that are bleached and stay inside the ROI. In FRAP, mobile particles inside the ROI that have been bleached diffuse out, while mobile particles from outside of the ROI replace them. The maximum recovery is observed when the FRAP curve has reached a plateau and no further significant recovery takes place (*see* Fig. 1 as an example).
20. Conventional FRAP fitting protocols included in the FRAP acquisition tool of most commercial confocal microscopes use the exponential equations derived for free Brownian diffusion in two dimensions. As already mentioned this is not a good model to describe the diffusion of membrane proteins. In terms of the immobile fraction (*see* **Note 21** for definition) the error is less than 10 %, but in terms of the characteristic diffusion time, errors can be up to 100 % or more. This is why we recommend to use the more accurate anomalous diffusion model [20].
21. The immobile fraction in a FRAP experiment corresponds to the fraction of fluorescent molecules that is not replaced by those from outside of the ROI. Since Ras molecules inside nanoclusters are immobile and free Ras is mobile, the immobile fraction is a direct indicator of the level of Ras nanoclustering. For each experimental condition, the immobile fraction provides information on the nanoclustering level of Ras.
22. We recommend the use of photon counting detectors for RICS. However, analog detection can also be used for RICS. The drawback is that analog detection introduces unwanted correlations that need to be removed [28], making the process more complicated and time consuming.
23. Other fluorescence correlation techniques like FCS use the count rate criteria to determine a good signal. In RICS this is not applied. The only criterion is that the detector may not be saturated and that the signal is visible in the image. Detectors in photon counting mode do not allow gain adjustments, thus the image only depends on the fluorescence signal.
24. The software used for data analysis in RICS (SimFCS 2.0) has a maximum number of pixels ($512 \times 512$ pixels) that it can use to calculate the correlation function. We find that our recommended $256 \times 256$ pixels are a good compromise between having sufficient pixels for the analysis, and not too many pixels so data acquisition is sufficiently fast to avoid problems of sample drift or excessive cell movement.

25. In RICS the pixel size, i.e., the distance the laser moves between collecting pixels, must be four to five times smaller than the beam waist of the laser defining the confocal volume (typically around 250 nm). For good results it is critical that consecutive pixels have a strong overlap (oversampling) and we recommend using a pixel size between 50 and 70 nm.
26. The pixel dwell time, defined as the time that the laser spends on a given pixel, must be appropriate for the diffusion coefficient of interest [24]. In the case of membrane proteins where the diffusion coefficient is between 0.1 and 10 $\mu m^2/s$, the pixel dwell time should be between 32 and 100 μs.
27. Acquiring sufficient frames in a RICS experiment allows you to have sufficient statistical information about your sample. It has been shown that a minimum of ten frames should always be acquired [23]. We recommend 50 frames per experiment.
28. RICS uses a reference sample to calibrate the focal volume of your system. This focal volume is crucial for accurate concentration measurements by RICS [24]. The focal volume is determined by the optics of the system like laser wavelength, laser power, alignment mirrors, objective lens and even the pinhole diameter. It is important that all these parameters remain the same during imaging of the cells and the calibration procedure.
29. The diffusion coefficient of most reference samples is given at 25 °C. Therefore, all calibration measurements should be conducted at that temperature. For experiments on cells, the temperature may then be reset to 37 °C.
30. SimFCS contains two useful functions to eliminate unwanted fluctuations from your data. These two functions are "Subtract average" that is used to eliminate signal from static molecules and "Subtract moving average" that is used to eliminate fluctuations caused by drift or small movements of the whole cell. Data from a cell with excessive movement (i.e., a cell changing its shape or leaving the field of view) should however not be considered for analysis.
31. The pixel size can be obtained by dividing the image size by the number of pixels (i.e., 10 μm divided by 256 pixels equal 39.0625 nm).
32. These necessary parameters are all determined by the scanning frequency, which is defined as the number of lines acquired in a second. The line time is therefore 1/frequency (i.e., 1/80 Hz = 12.5 ms), the pixel dwell time is the line time divided by the number of pixels (i.e., 12.5 ms/256 = 48.82 μs) and the frame time is the total number of lines in a frame divided by the scanning frequency (i.e., 256/80 Hz = 3.2 s).

33. The confocal volume of your system can be calculated using the calibrated value of $\omega_0$ obtained in **step 9** of the Subheading 3.5. The formula used is:

$$V_c = \pi^{3/2} \omega_0^3 k$$

where $V_c$ is the confocal volume, $\omega_0$ corresponds to the beam waist of your laser (named W0 in SimFCS 2.0), and $k$ is the eccentricity of your confocal volume ($k = z_0/\omega_0$). Typical values for $k$ are between 2 and 6 and its lower limit is determined by

$$k = \frac{2.33\,n}{NA}$$

Here $n$ represents the refractive index of your mounting medium and NA is the numerical aperture of the objective lens. For precise calibration of the eccentricity, experiments involving Fluorescence Correlation Spectroscopy (FCS) or scanning of beads are required [29].

34. It may be important to control by other methods, such as Western blotting, that the total level of nanocluster modulator (e.g., galectin-1) scales with the amount of overexpressed nanocluster modulator (here mRFP-galectin-1). In other words, verify that overexpression does not repress the expression of endogenous nanocluster modulator.

## References

1. Abankwa D, Gorfe AA, Hancock JF (2007) Ras nanoclusters: molecular structure and assembly. Semin Cell Dev Biol 18:599–607
2. Prior IA, Muncke C, Parton RG et al (2003) Direct visualization of Ras proteins in spatially distinct cell surface microdomains. J Cell Biol 160:165–170
3. Plowman SJ, Muncke C, Parton RG et al (2005) H-ras, K-ras, and inner plasma membrane raft proteins operate in nanoclusters with differential dependence on the actin cytoskeleton. Proc Natl Acad Sci U S A 102: 15500–15505
4. Hancock JF, Parton RG (2005) Ras plasma membrane signalling platforms. Biochem J 389:1–11
5. Janosi L, Li Z, Hancock JF et al (2012) Organization, dynamics, and segregation of Ras nanoclusters in membrane domains. Proc Natl Acad Sci U S A 109:8097–8102
6. Belanis L, Plowman SJ, Rotblat B et al (2008) Galectin-1 is a novel structural component and a major regulator of h-ras nanoclusters. Mol Biol Cell 19:1404–1414
7. Rotblat B, Belanis L, Liang H et al (2010) H-Ras nanocluster stability regulates the magnitude of MAPK signal output. PLoS One. doi:10.1371/journal.pone.0011991
8. Shalom-Feuerstein R, Plowman SJ, Rotblat B et al (2008) K-ras nanoclustering is subverted by overexpression of the scaffold protein galectin-3. Cancer Res 68:6608–6616
9. Matsunaga-Udagawa R, Fujita Y, Yoshiki S et al (2010) The scaffold protein Shoc2/SUR-8 accelerates the interaction of Ras and Raf. J Biol Chem 285:7818–7826
10. Inder KL, Lau C, Loo D et al (2009) Nucleophosmin and nucleolin regulate K-Ras plasma membrane interactions and MAPK signal transduction. J Biol Chem 284:28410–28419
11. Zhou Y, Cho K-J, Plowman SJ et al (2012) Nonsteroidal anti-inflammatory drugs alter the spatiotemporal organization of Ras proteins on the plasma membrane. J Biol Chem 287:16586–16595
12. Köhnke M, Schmitt S, Ariotti N et al (2012) Design and application of in vivo FRET biosensors to identify protein prenylation and nanoclustering inhibitors. Chem Biol 19:866–874
13. Paz A, Haklai R, Elad-Sfadia G et al (2001) Galectin-1 binds oncogenic H-Ras to mediate

Ras membrane anchorage and cell transformation. Oncogene 20:7486–7493
14. Tian T, Harding A, Inder K et al (2007) Plasma membrane nanoswitches generate high-fidelity Ras signal transduction. Nat Cell Biol 9:905–914
15. Lommerse PHM, Snaar-Jagalska BE, Spaink HP et al (2005) Single-molecule diffusion measurements of H-Ras at the plasma membrane of live cells reveal microdomain localization upon activation. J Cell Sci 118: 1799–1809
16. Murakoshi H, Iino R, Kobayashi T et al (2004) Single-molecule imaging analysis of Ras activation in living cells. Proc Natl Acad Sci U S A 101:7317–7322
17. Abankwa D, Vogel H (2007) A FRET map of membrane anchors suggests distinct microdomains of heterotrimeric G proteins. J Cell Sci 120:2953–2962
18. Abankwa D, Hanzal-Bayer M, Ariotti N et al (2008) A novel switch region regulates H-ras membrane orientation and signal output. EMBO J 27:727–735
19. Axelrod D, Koppel DE, Schlessinger J et al (1976) Mobility measurement by analysis of fluorescence photobleaching recovery kinetics. Biophys J 16:1055–1069
20. Feder TJ, Brust-Mascher I, Slattery JP et al (1996) Constrained diffusion or immobile fraction on cell surfaces: a new interpretation. Biophys J 70:2767–2773
21. Perez J-B, Segura JM, Abankwa D et al (2006) Monitoring the diffusion of single heterotrimeric G proteins in supported cell-membrane sheets reveals their partitioning into microdomains. J Mol Biol 363:918–930
22. Digman MA, Brown CM, Sengupta P et al (2005) Measuring fast dynamics in solutions and cells with a laser scanning microscope. Biophys J 89:1317–1327
23. Brown CM, Dalal RB, Hebert B et al (2008) Raster image correlation spectroscopy (RICS) for measuring fast protein dynamics and concentrations with a commercial laser scanning confocal microscope. J Microsc 229:78–91
24. Rossow MJ, Sasaki JM, Digman MA et al (2010) Raster image correlation spectroscopy in live cells. Nat Protoc 5:1761–1774
25. Abankwa D, Gorfe AA, Inder K et al (2010) Ras membrane orientation and nanodomain localization generate isoform diversity. Proc Natl Acad Sci U S A 107:1130–1135
26. Schneider CA, Rasband WS, Eliceiri KW (2012) NIH Image to ImageJ: 25 years of image analysis. Nat Methods 9:671–675
27. Kapusta P (2010) Absolute diffusion coefficients: compilation of reference data for FCS calibration. Application note: PicoQuant GmbH, Berlin. http://www.picoquant.com/images/uploads/page/files/7353/appnote_diffusioncoefficients.pdf
28. Gielen E, Smisdom N, vandeVen M et al (2009) Measuring diffusion of lipid-like probes in artificial and natural membranes by raster image correlation spectroscopy (RICS): use of a commercial laser-scanning microscope with analog detection. Langmuir 25:5209–5218
29. Buschmann V, Krämer B, Koberling F et al (2009) Quantitative FCS: determination of the confocal volume by FCS and bead scanning with the MicroTime 200. Application note: PicoQuant GmbH, Berlin. http://www.picoquant.com/images/uploads/page/files/7351/appnote_quantfcs.pdf

# Chapter 21

# Analyzing the Roles of Rho GTPases in Cancer Cell Migration with a Live Cell Imaging 3D-Morphology-Based Assay

**Audrey Colomba and Anne J. Ridley**

## Abstract

Rho GTPases are master regulators of cytoskeleton dynamics and therefore regulate cell motility. Rho GTPases, as well as their regulators and effectors, are often deregulated in cancers and thus contribute to tumor progression to metastasis. Cancer progression involves multiple steps, including invasion of the surrounding tissues. Several methods to investigate the invasion of tumors cells in 3D matrices in vitro have been developed. In this chapter we describe a 3D-based morphology assay that can be used for medium-throughput microscopy-based screening to identify regulators of cancer cell invasion. We use this method coupled to RNAi to investigate how Rho GTPases affect prostate cancer cell morphology in 3D Matrigel.

**Key words** Rho GTPases, siRNA knockdown, 3D-based morphology assay, Medium-throughput screen, Qualitative image analysis

## 1 Introduction

Tumor cell motility is a critical step in cancer progression. To metastasize tumor cells must move through surrounding tissues, which requires increased cell motility driven by remodelling of the cytoskeleton and interactions with the extracellular matrix (ECM). Recent studies investigating tumor cell migration and associated matrix remodelling using three-dimensional (3D) matrices have shown that the ECM strongly impacts on the invasion process and forces the cytoskeleton to adapt and generate different characteristic morphological changes [1, 2].

Indeed in 3D matrices cancer cells possess a broad spectrum of migration and invasion mechanisms, as they can move either as collective groups or as individual cells [3, 4]. The latter have two different modes of movement: mesenchymal type, characterized by an elongated cellular morphology and requiring extracellular proteolysis; and amoeboid type, in which cells have a rounded morphology and which is less dependent on proteases [5, 6].

Lorenza Trabalzini and Saverio Francesco Retta (eds.), *Ras Signaling: Methods and Protocols*,
Methods in Molecular Biology, vol. 1120, DOI 10.1007/978-1-62703-791-4_21, © Springer Science+Business Media, LLC 2014

Rho GTPases and their downstream effectors are well-known key players in metastatic progression [7, 8]. They not only regulate cell-matrix adhesion and migration via their effects on the cytoskeleton, but also stimulate secretion of matrix-degrading proteases via changes in gene transcription [9, 10]. In addition, their expression and/or their activity level are frequently altered in tumors [11, 12].

Most Rho GTPases act as molecular switches, cycling between an active GTP-bound and an inactive GDP-bound state [13]. They are activated by guanine nucleotide exchange factors (GEFs) and inactivated by GTPase activating proteins (GAPs) [14]. Some Rho GTPases are also regulated by guanine nucleotide dissociation inhibitors (GDIs), which bind to Rho GTPases in the cytosol, preventing them from interacting with their regulators/effectors [15].

In recent years several studies have investigated how some of the Rho GTPase signalling pathway components contribute to tumor cell motility in 3D matrices [16–21]. For example, Rho-ROCK signalling has been shown to promote the amoeboid mode of motility [22, 23], whereas DOCK3-Rac signalling regulates mesenchymal-type movement of melanoma cells [24, 25].

However, the Rho GTPase pathways regulating cancer cell migration in 3D environments similar to that found in vivo are not yet fully identified and are likely to vary significantly depending on the cancer type. A greater knowledge of which Rho GTPase pathways control cancer cell morphology and invasion will be beneficial in designing new multi-targeted cancer therapies.

To delineate the roles of Rho family proteins in tumor cell motility and invasion, we have investigated how Rho GTPase signalling pathways affect tumor cell morphology in a 3D matrix. In this chapter we describe a 3D Matrigel morphology-based screen assay which we used to identify components of Rho GTPase pathways involved in Met-dependent prostate cancer cell invasion using a siRNA library. However, this protocol can be used to study cancer cell morphology in any other kind of 3D matrix and in combination with functional genomics as well as chemical compound screens.

## 2 Materials

### 2.1 Cell Culture

1. PC3 prostate cancer cells (ATCC).
2. RPMI: Roswell Park Memorial Institute 1640 medium (RPMI-1640) with L-Glutamine, 25 mM Hepes, 100 μg/ml streptomycin, and 100 U/ml penicillin supplemented with 10 % heat-inactivated fetal calf serum (FCS). Store at 4 °C.
3. Trypsin–EDTA. Store at 4 °C.
4. Phosphate buffered saline (PBS) minus $CaCl_2$ and $MgCl_2$.

### 2.2 Matrigel Matrix Preparation

1. Matrigel basement membrane matrix growth factor reduced phenol red-free (BD Biosciences). Store at −20 °C and thaw on ice at 4 °C overnight. Prepare working solution on ice by diluting in chilled RPMI without FCS (*see* **Note 1**).
2. Recombinant human hepatocyte growth factor (HGF) reconstituted according to the manufacturer's instructions and stored at −40 °C in small aliquots (*see* **Note 2**).
3. 96-well flat bottom plates.
4. Multichannel pipette.

### 2.3 siRNA Reverse Transfection

1. Lipofectamine RNAiMax transfection reagent (Life Technologies) (*see* **Note 3**).
2. On target plus non-targeting siRNA #1, ON target plus siRNA targeting Met (mix of three oligos) and siGenome smart pool RNAs targeting RhoA (Dharmacon, Thermo Fisher Scientific, *see* **Note 4**).
3. Opti-MEM (modification of Eagle's minimum essential medium) with 25 mM Hepes. Store at 4 °C (*see* **Note 5**).
4. 96-well V bottom plates.

### 2.4 Automated Medium Throughput Imaging

1. Nikon Eclipse TE 2000-E with a 10× objective (Plan Fluor, dry, NA 1.2, WD 15.2) equipped with an automated stage control (Prior, *see* **Note 6**).
2. Temperature- and $CO_2$-controlled imaging chamber (custom design).
3. Digital Camera (*see* **Note 6**).
4. Metamorph software (version 7.7.3.0, Molecular Devices Corporation, *see* **Note 7**).

## 3 Methods

Cell morphology in 3D can be investigated in various experimental settings. Here we describe an example experiment illustrating the type of assays that can be performed using the following protocol. In this particular example we use the PC3 cell line, a human prostate cell line. However, the protocol can be adapted to various other mammalian cell types.

### 3.1 PC3 Cell Culture

1. Thaw PC3 cells from frozen aliquots in a water bath at 37 °C.
2. Seed into a 75 $cm^2$ flask containing 10 ml RPMI.
3. Maintain cells at 37 °C in a tissue culture incubator with humidified air supplemented with 5 % $CO_2$ and change medium after 24 h.

4. When cells reach 80 % confluency, passage them by removing the medium from the flask and washing cells twice with 10 ml PBS. Aspirate the PBS and add 1.5 ml of trypsin–EDTA for 2 min. Add 8.5 ml of RPMI to inactivate the trypsin–EDTA.
5. Reseed into culture flasks to maintain the cells or for use in experiments.
6. Split cells at between 1:2 and 1:10 and change the medium every 2–3 days.
7. To improve experimental consistency, discard cells after 1 month and defrost a fresh cell aliquot.

### 3.2 RNAi-Mediated Gene Silencing by Reverse Transfection of siRNAs

1. Count the cells using a CASY counter or similar (*see* **Note 8**). Prepare an aliquot of cells for the required number of wells, considering that cells will be seeded at 100,000 cells per well (*see* **Note 9**). Wash the cells twice with PBS and resuspend them in Opti-MEM without antibiotics and without FCS, so that the cell concentration is $1.25 \times 10^6$cells/ml.
2. In a 96-well V bottom plate, aliquot the siRNAs of interest in order to be at a final concentration of 50 nM (the final volume of the transfection is 100 μl) (Fig. 1, *see* **Note 10**).
3. In a microfuge tube dilute the Lipofectamine RNAiMax in Opti-MEM without antibiotics and without FCS. For one well, add 1 μl of Lipofectamine RNAiMax to 14 μl of medium. Prepare a mix for the required number of wells and distribute 15 μl of that mix in each well using a multichannel pipette (*see* **Note 9**).
4. Incubate the siRNAs and the mix for 20 min at room temperature so the siRNA/cationic lipids complexes can form.
5. Using a multichannel pipette, add 80 μl of the cell aliquot from **step 1** to each well. Gently mix by triturating a couple of times with the pipette.
6. Incubate for transfection at 37 °C for 4 h.
7. Spin the 96-well plate at 240 × *g* for 5 min and carefully remove 75 μl of medium with the multichannel pipette.
8. Resuspend the cells in 250 μl of RPMI and leave the plate overnight in the incubator at 37 °C.

### 3.3 Matrigel Mix Preparation

1. Coat two 96-well flat bottom plates with 40 μl/well of 7.5 mg/ml Matrigel (*see* **Note 11**). To obtain the desired concentration, dilute Matrigel in chilled RPMI without FCS (*see* **Notes 12** and **13**). Mix carefully to ensure complete homogenization of the mix and to avoid bubble formation.
2. Allow Matrigel to polymerize by incubating the plate at 37 °C for 1 h.

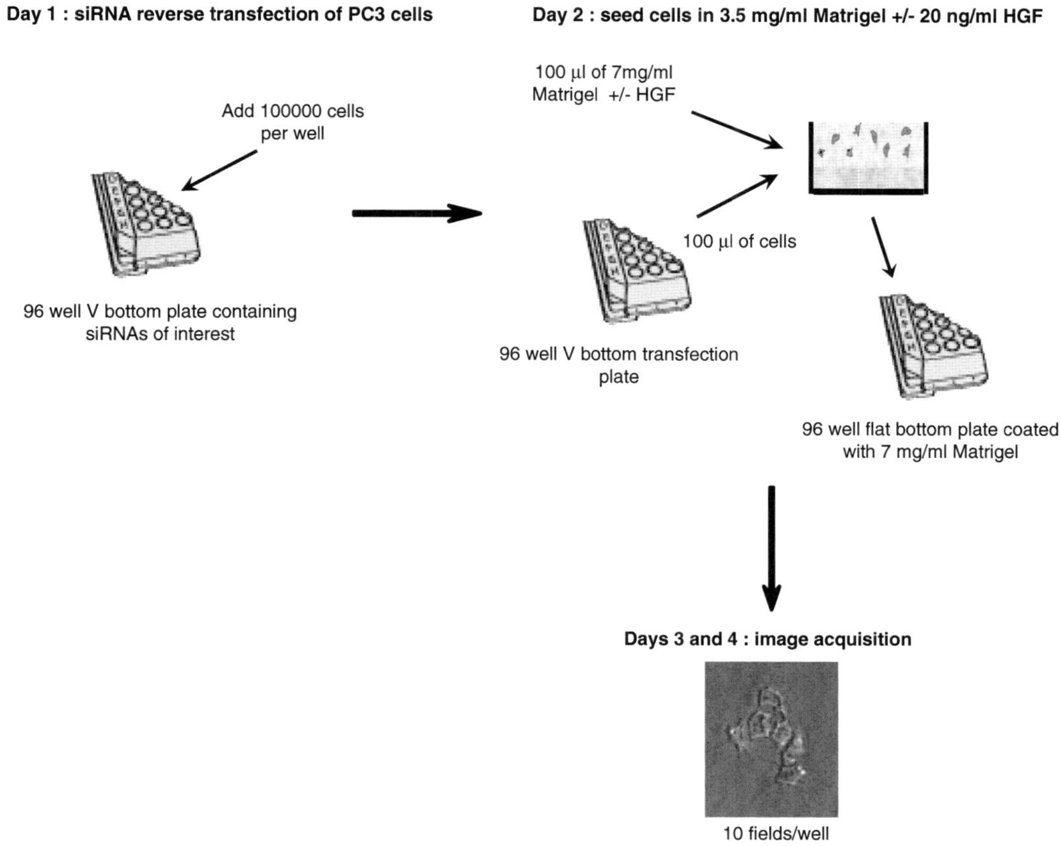


**Fig. 1** 3D morphology-based assay in Matrigel. PC3 cells were reverse-transfected with siRNA oligos in a 96-well V bottom plate. 24 h later cells were embedded in 3.5 mg/ml Matrigel with or without 20 ng/ml HGF in a 96-well flat bottom plate coated with 7 mg/ml Matrigel. 48 and 72 h after transfection, phase-contrast images were taken

3. Prepare the Matrigel mix for the required number of wells, considering that 100 μl of Matrigel mix will be used per well. Dilute the Matrigel in RPMI without FCS at a concentration of 7 mg/ml, and add 40 ng/ml HGF to half of the Matrigel mix (*see* **Note 2**). When the cells are mixed with the Matrigel mix at a 1:1 volume ratio (*see* Subheading 3.4), the final Matrigel and HGF concentrations are 3.5 mg/ml and 20 ng/ml, respectively.

### 3.4 Embedding Cells in Matrigel Mix

1. Spin the 96-well V bottom plate containing transfected cells (from Subheading 3.2) at 240 × *g* for 5 min and aspirate the medium carefully with a micropipette. Resuspend the cells in 250 μl of RPMI without FCS. Transfer 100 μl from each well to two new 96-well flat bottom plates, so there are approximately 50,000 cells per well (*see* Subheading 3.2).

2. For each well, add 100 μl of Matrigel mix without HGF to 100 μl of cells, mix gently by triturating with the pipette and transfer the 200 μl of cells mixed with the Matrigel to one of the Matrigel-coated plates (from Subheading 3.3) (*see* **Note 14**). Proceed similarly for the plate with Matrigel containing HGF.
3. Allow Matrigel to polymerize by incubating the plates at 37 °C for 2 h.
4. Add 100 μl of RPMI without FCS and with or without 20 ng/ml HGF to each well (*see* **Note 15**).
5. Keep the plate in the incubator at 37 °C for 24 h (Fig. 1).

### 3.5 Automated Image Acquisition

This automated imaging protocol is designed for application on a Nikon inverted fluorescent microscope and phase-contrast imaging with automated stage control, but can be adapted to be used with similar medium throughput imaging systems.

Morphology of the cells embedded in Matrigel is assessed 24 and 48 h after seeding into the matrix.

1. Prior to imaging, heat the climate control chamber to 37 °C.
2. Start the microscope including the camera.
3. Place the 96-well plate in the plate holder on the motorized stage and open the $CO_2$ supply (*see* **Note 16**).
4. Open Image acquisition Metamorph software (*see* **Note 7**).
5. Select the 10× objective in the appropriate tab.
6. Click on *Apps* tab and select *Multi Dimensional Acquisition*. A new window opens (Fig. 2a).
7. In the *Saving* tab, create a new folder in which images will be saved.
8. In the *Wavelengths* tab, select transmitted, gain 1, 12-bits, and set up the exposure time (*see* **Note 17**).
9. In the *Stage* tab, positions (*x,y,z* coordinates) in each well will be recorded. Move the stage in order to record the positions in the first well of the plate. Click on acquire to take a picture and optimize the exposure time to get a good image. Choose at least ten positions per well. Click on the block arrow to save each position (Fig. 2b).
10. Save the recorded positions. A .stg file is created.
11. Click on acquire.
12. Images are saved in TIFF format.

### 3.6 3D Morphology Analysis

The 3D morphology of the cells is analyzed to determine whether siRNA-mediated knockdown of proteins of interest affects the cell response to HGF stimulation. Knockdown effects on cell morphology following HGF stimulation are compared to the HGF response of the control siRNA-transfected cells. HGF induces control siRNA transfected PC3 cells to change from a rounded to

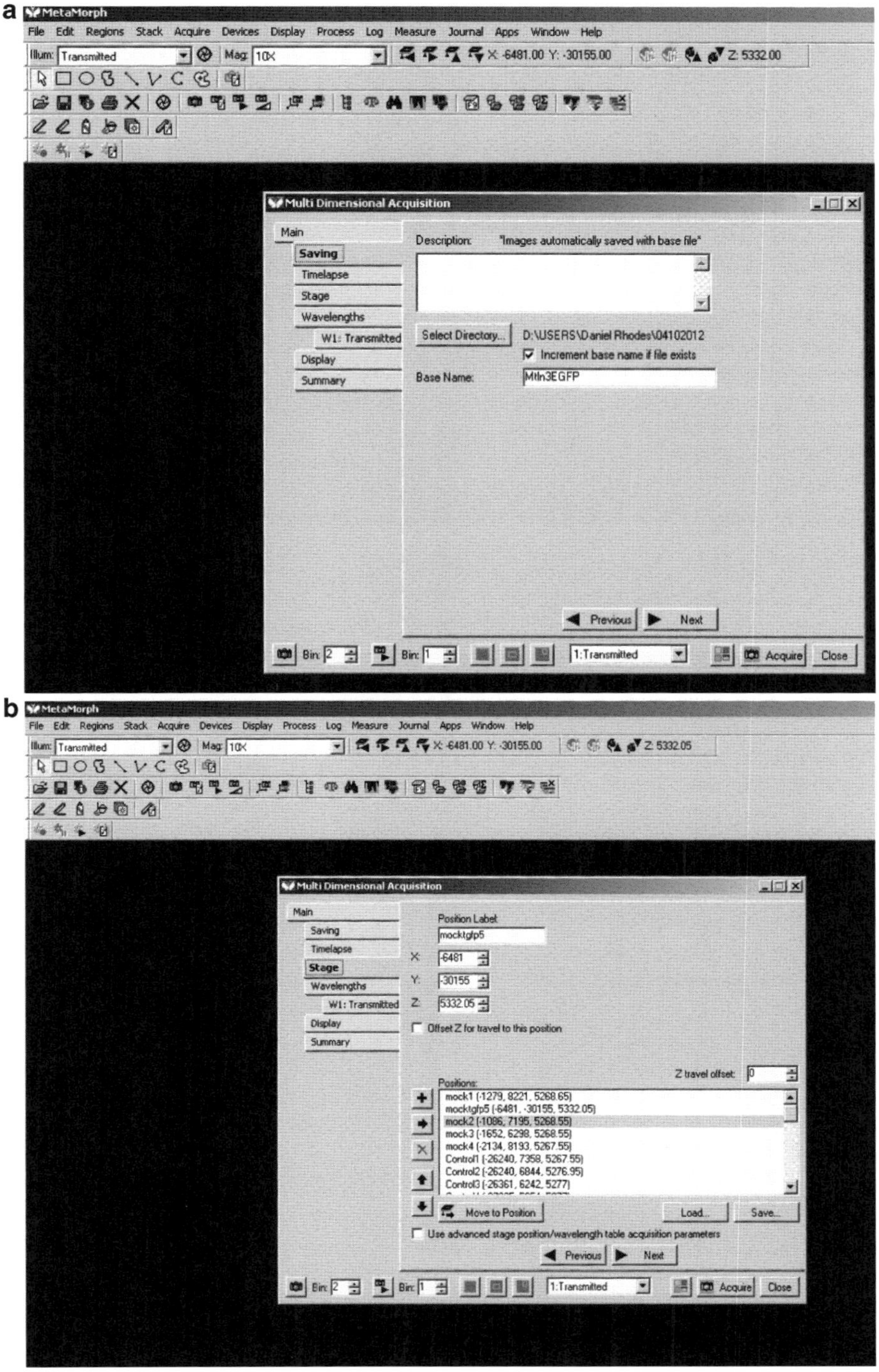


**Fig. 2** Automated image acquisition. This procedure is described using Metamorph software. (**a**) Open the Apps/Multi Dimensional Acquisition window and in the Saving panel select the folder in which images will be saved. In the Wavelengths panel select "transmitted" and set up the exposure time. (**b**) In the Stage panel define the plate by moving the stage from the first to the last well. For each well choose at least ten positions and click on the *block arrow* to save each position. Recorded positions can be deleted by clicking on the *red cross*. Click on "save" to save the stage when all the settings are satisfactory. A .stg file will be created and can be reopened for subsequent experiments to reload the recorded positions of a specific plate. Click on "acquire" to start taking pictures (Color figure online)

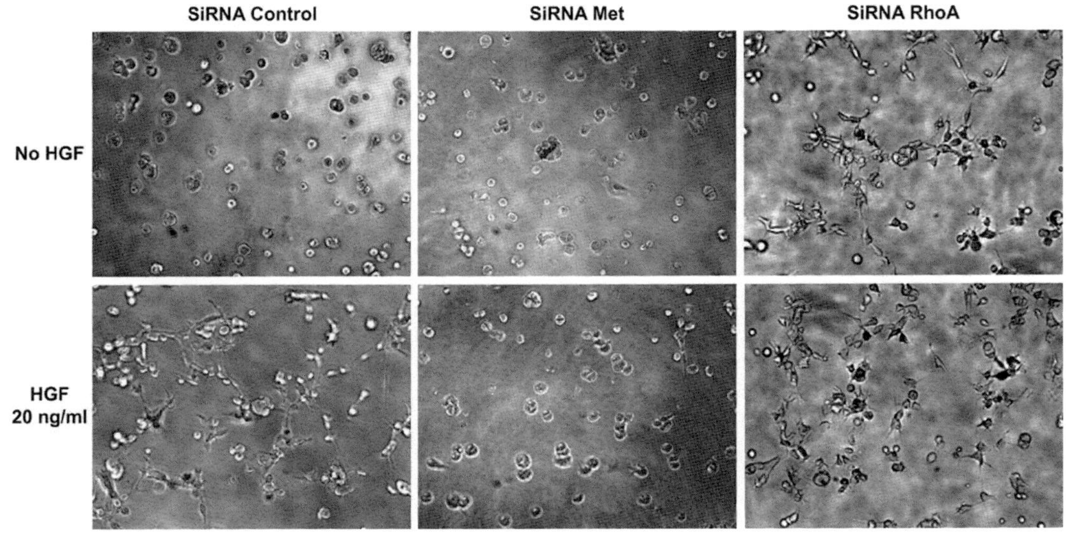


**Fig. 3** Analysis of phenotypical changes. Effects of siRNA-mediated protein depletion on the cell morphology in response to HGF were classified in three different categories. The first phenotype observed was a response to HGF stimulation, as for example for siRNA control-transfected cells. These cells displayed a round phenotype without HGF and became elongated with long protrusions in the presence of HGF. The second phenotype observed was inhibition of the response to HGF stimulation. In that case after gene silencing cells showed a rounded phenotype even in the presence of HGF. This phenotype is illustrated by the knockdown of Met, the HGF receptor, which as expected prevents the HGF-induced protrusions. A third phenotype, independent of HGF, was observed, as for example following RhoA depletion: knockdown cells displayed elongated protrusions independent of HGF stimulation, similar to control cells stimulated with HGF

elongated morphology with long protrusions invading into the Matrigel (Fig. 3, siRNA Control). This change of phenotype in response to HGF stimulation is dependent on the Met receptor, as Met knockdown prevents protrusions from forming, and the cells remain round in the presence of HGF (Fig. 3, siRNA Met).

The ten images of each well from the microscope acquisitions (*see* Subheading 3.5) are opened in Photoshop and the phenotype of the cells for each image classified as follows:

(a) Response to HGF: the cells display a phenotype similar to the control cells.

(b) No response to HGF: the cells remain round after HGF stimulation.

(c) Independent of HGF: the cells have an elongated phenotype even without HGF stimulation (Fig. 3, siRNA RhoA).

Classification in these three categories is based on a qualitative image analysis according to the phenotype of the cells in each well (*see* **Note 18**).

If siRNA smart pools are used for screening, hits need to be deconvoluted by testing each of the four single siRNA oligos in the pool in the 3D-based morphology assay to validate the phenotype observed (*see* **Note 19**).

## 4 Notes

1. As Matrigel polymerizes at room temperature, make sure chilled Matrigel is used. Cool tips, tubes, plates, and multi-channel reservoirs overnight at 4 °C.
2. In our assay we were investigating which components of Rho GTPase pathways could affect Met receptor-dependent invasion. PC3 cells were therefore stimulated with the Met ligand, HGF. Other growth factors, chemokines, or inhibitors can also be used with this assay.
3. Although Lipofectamine RNAiMax transfection reagent transfects a wide range of cell types with a high-efficiency and minimal cytotoxicity, it may not be suitable for some cell types. Other transfection reagents can be used depending on the cell type.
4. When performing siRNA experiments, it is essential to include the proper controls. In this case, non-targeting siRNA is used as a control for the specificity of knockdowns. siRNAs targeting Met receptor are chosen as a positive control, since Met knockdown should prevent the response of cells to HGF from occurring.
5. Opti-MEM I Reduced Serum Medium is ideal for use during cationic lipid transfections.
6. Data shown in this chapter were collected using a Nikon inverted microscope equipped with an automated Prior stage and a Hamamatsu camera. However, any microscope equipped with any kind of automated stage and high-resolution camera would be suitable.
7. We used Metamorph software to drive our microscope and acquire the images shown in this chapter. However, any other software that is capable of acquiring automated images from multiple recorded positions would be suitable.
8. Cells can also be counted manually using a hemocytometer.
9. Make excess to the required number of wells due to loss of volume during pipetting.
10. The siRNA concentration has to be chosen according to the knockdown efficiency of the gene of interest in the cell type used.
11. There can be variations in the concentration of Matrigel between batches. To improve experimental consistency, we recommend buying the same batch of Matrigel in bulk to carry out all the experimental repeats.
12. As Matrigel polymerizes at room temperature, always keep the matrix solution on ice while preparing it.
13. Make excess to what is required as Matrigel is a bit viscous and difficult to handle.

14. Make sure there are no bubbles after mixing. If there are, remove them with a sterile 21 G needle.
15. This protocol can be further adapted for full automation by applying robotics for automated liquid handling, e.g., cell seeding and drug exposure.
16. Prior to placing the plates on the microscope, make sure that the temperature and $CO_2$ levels in the chamber are optimal in order to minimize cell stress.
17. Choose an exposure time and take an image. If needed, make changes in the exposure time in order to get an image with satisfactory contrast and brightness.
18. The classification of cells in a phenotype category is made if at least 80 % of the cells display that phenotype in at least 70 % of the images taken for each well.
19. In assays using siRNA oligos it is essential to validate the data with several siRNA oligos targeting the genes of interest. In the method described here, some knockdowns display an intermediate phenotype. Using multiple single oligos will identify whether this is an off-target effect of a single siRNA oligo in the pool or a genuine phenotype.

## References


1. Friedl P, Wolf K (2010) Plasticity of cell migration: a multiscale tuning model. J Cell Biol 188:11–19
2. Mierke CT (2011) The biomechanical properties of 3d extracellular matrices and embedded cells regulate the invasiveness of cancer cells. Cell Biochem Biophys 61:217–236
3. Friedl P, Alexander S (2011) Cancer invasion and the microenvironment: plasticity and reciprocity. Cell 147:992–1009
4. Friedl P, Gilmour D (2009) Collective cell migration in morphogenesis, regeneration and cancer. Nat Rev Mol Cell Biol 10:445–457
5. Friedl P, Wolf K (2003) Tumour-cell invasion and migration: diversity and escape mechanisms. Nat Rev Cancer 3:362–374
6. Wolf K, Mazo I, Leung H et al (2003) Compensation mechanism in tumor cell migration: mesenchymal-amoeboid transition after blocking of pericellular proteolysis. J Cell Biol 160:267–277
7. Olson MF, Sahai E (2009) The actin cytoskeleton in cancer cell motility. Clin Exp Metastasis 26:273–287
8. Rathinam R, Berrier A, Alahari SK (2011) Role of Rho GTPases and their regulators in cancer progression. Front Biosci 16:2561–2571
9. Parri M, Chiarugi P (2010) Rac and Rho GTPases in cancer cell motility control. Cell Commun Signal 8:23
10. Price LS, Collard JG (2011) Regulation of the cytoskeleton by Rho-family GTPases: implications for tumour cell invasion. Semin Cancer Biol 11:167–173
11. Vega FM, Ridley AJ (2008) Rho GTPases in cancer cell biology. FEBS Lett 582: 2093–2101
12. Gomez del Pulgar T, Benitah SA, Valeron PF et al (2005) Rho GTPase expression in tumourigenesis: evidence for a significant link. Bioessays 27:602–613
13. Ridley AJ (2012) Historical overview of Rho GTPases. Methods Mol Biol 827:3–12
14. Bos JL, Rehmann H, Wittinghofer A (2007) GEFs and GAPs: critical elements in the control of small G proteins. Cell 129:865–877
15. Garcia-Mata R, Boulter E, Burridge K (2011) The 'invisible hand': regulation of RHO GTPases by RHOGDIs. Nat Rev Mol Cell Biol 12:493–504
16. Gadea G, Sanz-Moreno V, Self A et al (2008) DOCK10-mediated Cdc42 activation is necessary for amoeboid invasion of melanoma cells. Curr Biol 18:1456–1465
17. Yamazaki D, Kurisu S, Takenawa T (2009) Involvement of Rac and Rho signaling in cancer cell motility in 3D substrates. Oncogene 28:1570–1583
18. Vega FM, Fruhwirth G, Ng T et al (2011) RhoA and RhoC have distinct roles in

migration and invasion by acting through different targets. J Cell Biol 193:655–665
19. Mishima T, Naotsuka M, Horita Y et al (2010) LIM-kinase is critical for the mesenchymal-to-amoeboid cell morphological transition in 3D matrices. Biochem Biophys Res Commun 392:577–581
20. Hirata E, Yukinaga H, Kamioka Y et al (2012) In vivo fluorescence resonance energy transfer imaging reveals differential activation of Rho-family GTPases in glioblastoma cell invasion. J Cell Sci 125:858–868
21. Heck JN, Ponik SM, Garcia-Mendoza MG et al (2012) Microtubules regulate GEF-H1 in response to extracellular matrix stiffness. Mol Biol Cell 23:2583–2592
22. Sahai E, Marshall CJ (2003) Differing modes of tumour cell invasion have distinct requirements for Rho/ROCK signalling and extracellular proteolysis. Nat Cell Biol 5:711–719
23. Sahai E, Garcia-Medina R, Pouyssegur J et al (2007) Smurf1 regulates tumor cell plasticity and motility through degradation of RhoA leading to localized inhibition of contractility. J Cell Biol 176:35–42
24. Sanz-Moreno V, Gadea G, Ahn J et al (2008) Rac activation and inactivation control plasticity of tumor cell movement. Cell 135:510–523
25. Sanz-Moreno V, Marshall CJ (2010) The plasticity of cytoskeletal dynamics underlying neoplastic cell migration. Curr Opin Cell Biol 22:690–696

# Chapter 22

# Analysis of Rho GTPase-Induced Localization of Nanoscale Adhesions Using Fluorescence Nanoscopy

**Annica K.B. Gad, Daniel Rönnlund, Jerker Widengren, and Pontus Aspenström**

## Abstract

Rho GTPases are important regulators of the formation of focal adhesions and focal complexes, and thereby they are key regulators of cell adhesion and migration. Here, we describe a method to study the relocalization of proteins within cell–matrix adhesions at a nanoscale level, through the use of super-resolution stimulated emission depletion microscopy imaging. We furthermore describe computational tools for image processing and data analysis. Thus, the method presented in this chapter provides an unbiased approach to the quantitative evaluation of the spatial distribution of the nanoscale protein assemblies by which cells adhere to an underlying substrate.

**Key words** Rho GTPases, Focal adhesions, Actin, Super-resolution stimulated emission depletion microscopy, Nanoscopy, Image analysis

## 1 Introduction

The Rho GTPases govern the fundamental behavior of cells, such as their ability to adhere to and migrate within an extracellular environment [1]. Observations that show that defective Rho GTPase activities are linked to tumor cell invasion underline the importance of an understanding of the mechanisms by which Rho GTPases control cell–matrix adhesion [2]. Cells adhere to an underlying extracellular matrix through the use of specific protein assemblies that are known as focal adhesions. Epifluorescence and confocal microscopy have been used to identify these focal adhesions as clusters of proteins at the ends of stress fibers. However, due to the relatively low resolution of conventional microscopy, focal adhesions appear as homogenous structures. Interestingly, recent studies using cryo-electron microscopy have shown that

"Annica K.B. Gad and Daniel Rönnlund" contributed equally to this chapter.

Lorenza Trabalzini and Saverio Francesco Retta (eds.), *Ras Signaling: Methods and Protocols*,
Methods in Molecular Biology, vol. 1120, DOI 10.1007/978-1-62703-791-4_22, © Springer Science+Business Media, LLC 2014

these adhesions actually consist of numerous assembled nanoscale adhesion-related particles [3]. To clarify the adhesive properties of cells and their control of cell motility, we analyzed how the Rho GTPases regulate the spatial distribution of the nanoscale adhesion particles that constitute cell adhesions using ultrahigh resolution stimulated emission depletion (STED) microscopy followed by computational image analysis [4].

There are several methods available to date to circumvent the diffraction barrier that limits the resolution of conventional confocal microscopy. Some further examples of these techniques are structured illumination microscopy [5]; stochastic optical reconstruction microscopy [6]; photoactivated localization microscopy [7]; and super-resolution optical fluctuation imaging [8]. Although we use STED microscopy, for the type of computational analysis described herein, it should be possible to use any technique that can provide sufficiently high resolution.

In STED microscopy, the resolution is increased compared to confocal microscopy by applying an extra laser beam to remove the emission from fluorophores or fluorescent proteins outside the very center of the excitation focus. The theory behind the STED methodology can be found in many excellent articles and reviews [9–12]. As focal adhesions are positioned at the ends of stress fibers, it is advantageous to detect the focal adhesion components together with filamentous actin. This requires the simultaneous acquisition of two separately fluorescently labelled objects that have separate excitation and emission spectra. Each fluorophore therefore requires one excitation, one emission, and one STED wavelength, which results in a total of six spectrally separated optical paths that need to be handled simultaneously. It is important to avoid spectral overlap between these excitation, emission, and STED wavelengths, to prevent the detection of noise, and each laser beam has to be focused at the very same point in the sample. Thus, for such super-resolution, the alignment of the foci becomes crucial. Indeed, while a 10-nm error can be acceptable for confocal resolution (i.e., at 200–300 nm), at super-resolution, a 10-nm error will have a major impact on the resulting image and image analysis. Therefore special care has to be taken when aligning the STED instrument. For the same reason, it is critical to avoid drift in the instrument, which could arise, for instance, from vibrations or temperature changes in the laboratory. Consequently, the ideal location of a STED microscope is on top of an air-dampened table in a closed room that is equipped with an efficient temperature control system.

Super-resolution images contain far more information than their low-resolution counterparts, which thus makes the image and data analysis of super-resolution images more challenging [13]. The resolution of protein clusters, or even of individual proteins, allows the analysis of properties such as the density, size, and location of these proteins [14]. We have previously reported that focal adhesions consist of clusters of nanoscale adhesion-related particles [4]. We have also shown that different members of the Rho GTPases have specific impact on the formation of such particles [4]. This chapter describes how imaging by STED microscopy followed by computational nearest neighbor image analysis can be used to determine how Rho GTPase activities control the spatial reorganization of nanoscale adhesion-related particles.

## 2 Materials

### 2.1 Preparation of Cover Slips and Cell Seeding

1. Glass cover slips of approximately 170 μm thickness.
2. $CO_2$ incubator for tissue culture.
3. 50 % ethanol solution.
4. 60 % ethanol/40 % acetic acid solution.
5. We use the mouse fibroblast NIH3T3 cell line.
6. Six-well cell culture plates.
7. Phosphate-buffered saline (PBS) solution: 1 l contains 8 g of NaCl, 0.2 g of KCl, 1.44 g of $Na_2HPO_4 \cdot 2H_2O$, 0.2 g of $KH_2PO_4$ in distilled water.
8. 1× Trypsin/PBS solution.
9. High-glucose Dulbecco's modified Eagle's medium (DMEM) supplemented with 10 % foetal bovine serum (FBS).

### 2.2 Cell Transfection

1. JetPEI transfection reagent.
2. High-purity DNA plasmid encoding green fluorescent protein (GFP)-tagged constitutively active Rho GTPases (we use Rac1/Q61L, RhoA/Q63L, RhoD/G26V).

### 2.3 Cell Fixing and Immunostaining

1. Fixation solution: 3.7 % Paraformaldehyde in PBS. This solution needs to be freshly prepared.
2. Permeabilization solution: 0.2 % Triton X-100 in PBS.
3. Cytoskeleton fixation solution: 3.7 % paraformaldehyde/0.1 % glutaraldehyde/0.2 % Triton X-100 in PBS.
4. Bovine serum albumin (BSA): 1 % solution in PBS.
5. Mouse anti-phosphotyrosine antibody: In our experiments, we use a mouse monoclonal anti-phosphotyrosine antibody, diluted 1:100 in 0.1 % BSA. This is the same concentration we use for conventional confocal microscopy.
6. Secondary antibody: 4 μg/ml anti-mouse antibody conjugated to AlexaFluor594 and 150 mM phalloidin labelled with Atto647N in 0.1 % BSA.

7. 100 mM Tris buffer: Solubilize 2.42 g of Tris in 100 ml of water, adjust to pH 8.5 with HCl.
8. STED mounting media: Place 6 g of glycerol in a 50-ml disposable plastic tube. Add 2.4 g of Mowiol 4-88 and stir. Add 6 ml of distilled water, and leave the mixture for 2 h at room temperature, with stirring. Add 12 ml of 100 mM Tris buffer. Incubate the solution at 50 °C for 10–30 min to dissolve the Mowiol. If the Mowiol does not dissolve, place the tube in hot water until the Mowiol is dissolved. Add the anti-fading reagent, such as 25 mg/ml DABCO, and centrifuge the solution for 15 min at 4,000 ×*g*. Freeze the solution in 1 ml aliquots at −20 °C.
9. Forceps.
10. Syringes.
11. Aluminum foil.

### 2.4 Microscope Alignment

1. Super-resolution stimulated depletion (STED) microscope, assembled as described in **Note 1**.
2. Confocal microscope to detect GFP together with the ATTO647N-labelled target protein.
3. Gold beads of 80 nm diameter and fluorescent crimson beads of 20 nm diameter (*see* **Note 2**).

## 3 Methods

### 3.1 Preparation of Cover Slips and Cell Seeding

1. Prepare cover slips with an approximate thickness of 170 μm, as follows: Wash the cover slips in 50 % ethanol in a glass beaker for at least 60 min, with gentle shaking. Bring to boil twice in a microwave. Be very careful, as the solution is inflammable. Discard the ethanol solution, and wash the cover slips in distilled water four times. Store the cover slips in distilled water overnight.
2. Incubate the cover slips with the ethanol/acetic acid solution for 3 h, with gentle shaking. The ethanol/acetic acid solution can be reused several times. Wash the cover slips ten times in distilled water. Stand the cover slips in distilled water overnight.
3. Position the cover slips on clean tissue paper in a tissue culture hood, and leave them to dry. Then place the cover slips in a plastic cell culture dish, and leave the dish to stand without a lid under the UV lamp in the hood overnight.
4. Place the cover slips one per well in six-well cell culture plates, and add 2 ml of DMEM/10 % FBS culture medium to each well. Ensure that the cover slips are placed on the bottoms of the wells by pushing them to the bottom with a sterile pipette tip.

5. Trypsinize the cells necessary for the experiment, and add enough cells such that a 70 % confluent monolayer is formed within 24 h. When plating the cells, pipette them gently up and down to get an even distribution of the cells in the media. Incubate the cells for 24 h in a $CO_2$ tissue culture incubator.

### 3.2 Cell Transfection

1. Confirm that the cells are attached, that they have a normal morphology, and that they have reached the desired confluency.
2. Transfect the DNA plasmids encoding the GFP-tagged Rho GTPases. Different transfection protocols can be used, but we routinely use the JetPEI protocol with the following modifications: For each six-well plate to be transfected, dilute separately both 2.5 μg of DNA and 5 μl of JetPEI in 200 μl of NaCl; mix them together, and leave this (the transfection mixture) standing for 15 min at room temperature (*see* **Note 3**). Carefully, add the transfection mixture dropwise to the cells and mix. Continue to incubate the cells in the $CO_2$ incubator for an additional period of 24 h.

### 3.3 Cell Fixing and Immunostaining

1. Preheat the fixation solution and PBS to 37 °C.
2. Remove the cells from the incubator, discard the cell culture media, and place the plates briefly upside down on a piece of tissue paper to remove the excess fluid while making sure that the cover slips are not lost. Wash the cells with PBS, and then briefly place the plates upside down again on tissue paper. Then add the fixation solution and incubate for 15 min at 37 °C (*see* **Note 4**).
3. To permeabilize the cells, remove the fixation solution, add the permeabilization solution, and incubate the cells for 5 min at room temperature. Wash the cells once with PBS, followed by incubation in PBS for 5 min. Discard the PBS, and add 1 % BSA solution, which functions as a blocking solution. Leave the cells to stand overnight at 4 °C or for 1 h at room temperature. For specific visualization of only cytoskeletal filaments, replace the fixing and permeabilization steps above with a 15-min incubation at 37 °C in the cytoskeleton fixation solution.
4. Prepare the solution of the primary antibody. Place an 80-μl droplet onto a piece of Parafilm stretched out on a flat, solid surface. To remove the blocking solution, hold the cover slip with forceps and let an edge of the cover slip briefly touch a piece of tissue paper placed on the bench (*see* **Note 5**). Then dip the cover slip once in a well of a six-well plate filled with PBS. Let an edge of the cover slip again touch the tissue paper, and then place the inverted cover slip on the droplets (with the cell side down). Incubate for 1 h at room temperature.

5. Prepare the solution with the anti-mouse secondary antibody and Atto647N-labelled phalloidin. Pick up the cover slip with forceps, and wash it gently by briefly first letting the cover slip touch a piece of tissue paper and then dipping it in PBS. To remove the excess liquid, let an edge of the cover slip gently touch a piece of tissue paper, as described above (*see* **Note 6**). Then place the inverted cover slip on a droplet of the secondary antibody/phalloidin solution, as before. Incubate in the dark for 1 h at room temperature.
6. Wash the cover slip by briefly letting an edge of the cover slip touch a piece of tissue paper, then dipping it once in PBS, and then removing the excess PBS with tissue paper, as described above. Place the cover slip with the cells upwards in a container with PBS. Wrap the container in aluminum foil and incubate for 30 min at room temperature, with gentle shaking.
7. Place 25 μl of mounting solution on the center of an object slide. Pick up the cover slip with forceps, remove the liquid by gently touching the edge of the cover slip to a piece of tissue paper, and dip the cover slip once into distilled water. To remove as much liquid as possible, gently let both the edge and the back side of the cover slip touch the tissue paper. A little droplet of liquid will gather at the lower edge of the front side of the cover slip. Gently let this droplet touch the edge and be absorbed by the tissue paper. Place the cover slip with cells downwards on the droplet of mounting solution. To avoid bubbles, when the cover slip first touches the droplet of the mounting medium, slip a syringe under the cover slip, and gently remove the syringe while slowly placing the cover slip on the droplet. This procedure will ensure that no bubbles are trapped between the object slide and the cover slip. Leave the mounted cells on a flat surface overnight at room temperature or at 4 °C for prolonged storage (up to 1 month).

### 3.4 Microscope Alignment

1. Place the object slide with the gold bead sample in the focus of the microscope objective. The reflection from the gold beads is usually very strong, so it is advisable to start with the STED and excitation beams at low power. Then, open the beam path for the first STED beam, and turn on the white-light source. The reflections from the gold beads coming from the white-light source as well as the focus of the laser should now be visible in the CCD image (the latter reflected by the refractive index mismatch of the cover slip and the mounting medium). Move the sample, and place a gold bead in the focus of the laser (*see* **Note 7**); this should provide strong reflection of the laser light.
2. Switch off the white-light source, and close the shutter; turn on the PMT. Start scanning the bead in one dimension (*X*, *Y*,

or *Z*), open one of the beam paths, and then adjust the power so that a good signal-to-noise ratio is achieved without saturating the detector. Repeat the procedure for each of the beams.

3. Take two-dimensional images (XY, XZ, and YZ) of all of the beams to monitor their point spread functions (PSFs). Record the PSFs, and make sure that they are well focused (*see* **Note 8**). Overlap the beams by selecting one beam and then moving all of the other beams so that their respective profiles overlap the selected beam; this is performed for each beam separately by adjusting the mirror positioned closest to the objective, which only affects the beam path of the desired beam. Take two-dimensional images of the profiles, and make sure that the minima for both of the STED beams are at the very same position, together with the maxima of the excitation profiles (*see* Fig. 1a, b).
4. Turn off the PMT, and turn on the APD detectors. Remove the filters, and look at the reflection of one of the excitation beams, preferably referring to one that is close to the emission bands, to reduce chromatic errors (*see* **Note 9**). Make sure that the detectors also aim at the same point as the excitation beam by moving the detector pinholes to maximize the reflected signal of the gold bead; the signals should also be aligned for both of the detectors (*see* Fig. 1c). Put back the emission filters of the APD detectors.

### 3.5 Locating and Imaging Transfected Cells

1. Put the sample under the microscope that can detect GFP and actin staining (this step can be skipped if the super-resolution microscope in use can detect GFP). Preferentially, use a relatively low magnification objective (~20×) to provide a large field of view. This makes it easier to locate the transfected cells (*see* **Note 10**). When a transfected cell has been identified through its GFP staining, take an image of it or take notes relating to its shape and the shapes of its neighboring cells. Do this for a number of cells; we routinely select five cells from each sample.
2. Put the sample under the high-resolution microscope. Locate the transfected cells by the images made with the GFP detection microscope. Take a high-resolution image in a desirable area of the cell (*see* **Note 11a**). For our analyses of Rho GTPase-induced reorganization within focal adhesions, we analyze the peripheral areas of the cells [4]. After the image capture, turn up the excitation/STED power to bleach the acquired area.
3. Put the sample back under the GFP-detectable microscope, and locate the cells again. A bleached square area in the cells will correspond to the area of the high-resolution image, and it will be possible to confirm that the cell was transfected (Fig. 2).

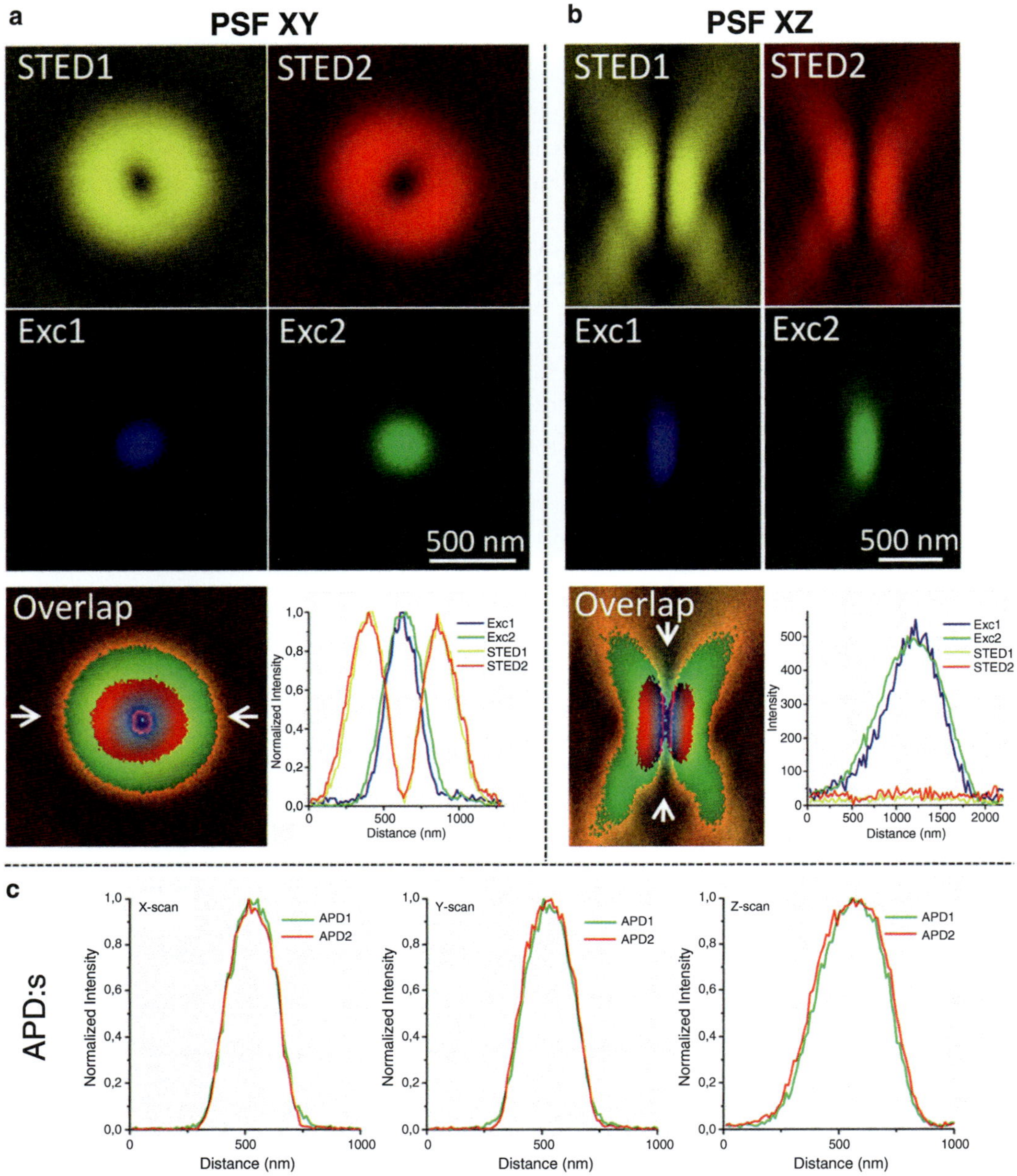


**Fig. 1** Alignment of the STED microscope. (**a**) Two-dimensional (*X–Y*) images of the focus profiles of the STED and excitation beams. The overlap of the profiles should show the minima for both of the STED beams at precisely the same position as the maxima of the excitation (*arrows*, *line profiles*). (**b**) As for (**a**), but shown as *X–Z*, the beam profiles should be straight and at the same depth position. In the overlap there should be no STED power, and there should be high excitation power in the center (*arrows*, *line profiles*). (**c**). The APD detectors should also be aligned and overlap with the same focus position as the STED and excitation beams

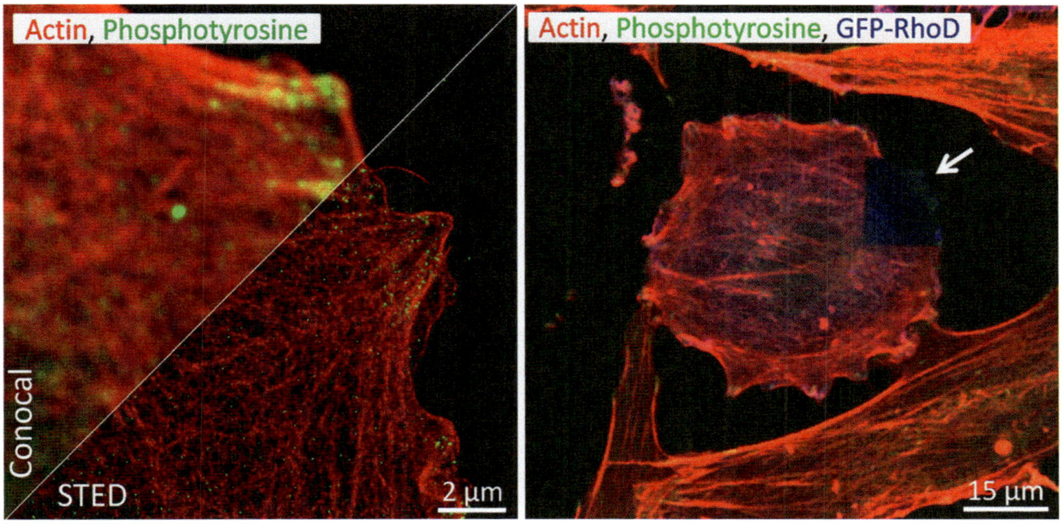


**Fig. 2** STED imaging and identification of Rho GTPase-expressing cells. Cells expressing GFP-RhoDV26 were analyzed using STED microscopy. *Left*: Representative cells showing localization of F-actin (*red*) and phosphotyrosine (*green*), with confocal and STED resolution as indicated. *Right*: To ensure that the cell analyzed by STED is transfected, the F-actin in the analyzed area is bleached so that the cell can be identified under a microscope that can detect GFP staining

### 3.6 Image Analysis

1. Export the images taken by the high-resolution microscope into a format supported by the analysis software (for example .dat or .tiff; avoid formats which compress the data, such as . JPG, since this can distort the data). Import the images into the analysis software.
2. For STED images, we recommend deconvolution of the images prior to the analysis, as this effectively reduces noise (*see* **Note 11b**). This can be performed in Matlab by the "deconvlucy" function using the Richardson–Lucy deconvolution algorithm [15] or by downloading and using the "DeconvolutionLab" plug-in for ImageJ.
3. To measure the nearest neighbor distances, use the deconvoluted phosphotyrosine image and locate the centers of the intensity peaks of the respective labelled targets. In Matlab, this can be performed by looking at each pixel in the image individually and identifying it as a peak if the pixel has a higher intensity compared to its neighboring pixel. This procedure will find all of the peaks in the image, including those arising from noise; however, the noise should have a much lower intensity compared to the labelled targets, and it can therefore be removed from the analysis by setting an intensity threshold (*see* **Note 11c**).

4. Use the locations of the peaks to find the nearest neighbor peak. This can be done by using the software to measure the distances of each peak with respect to the other peaks in the image, with the selection of the shortest distance. Store the nearest neighbor distances from the image, and perform this analysis for all of the images.
5. Combine the nearest neighbor distance data for each transfection separately.

### 3.7 Data Analysis

1. Import the distance data into the data analysis software. We routinely use Origin and Matlab. Construct a histogram of the distances to see which distances are most frequent (*see* Fig. 3a–d). In Origin, this can be performed in the menu Plot→ Statistics→ Histogram and in Matlab with the "hist" function. Construct the histograms of the distances for all of the transfections.
2. Create a cumulative curve distribution of the distance data to show what fractions of the targets are found within a certain distance. This can be performed in Matlab using the "ecdf" function (*see* **Note 11d**). Construct the curves for all of the transfections, and combine them into a single graph to visualize differences (*see* Fig. 3e).

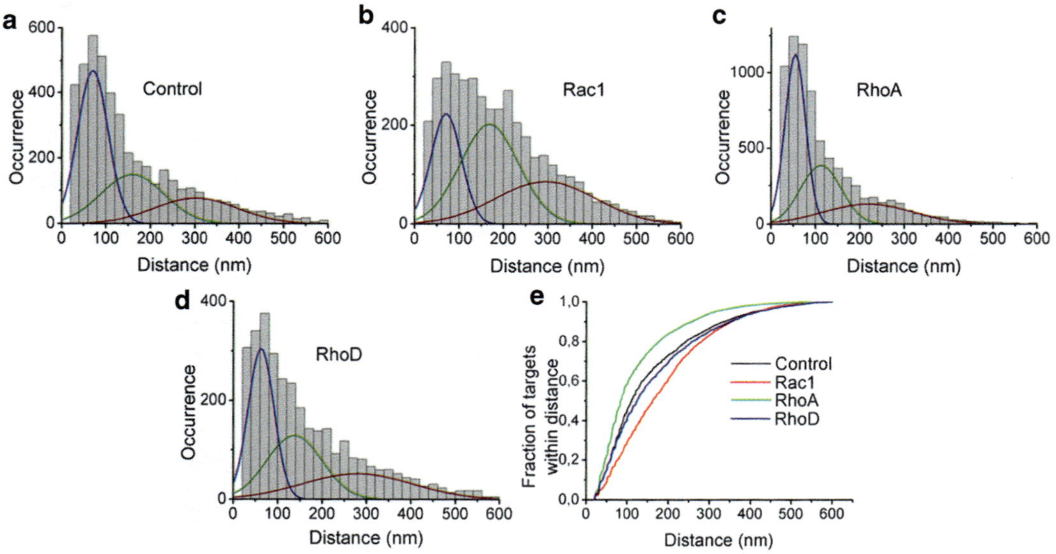


**Fig. 3** Image analysis of adhesion-related particles in cells that overexpress Rho GTPases. (**a–d**). Histograms showing the nearest neighbor distances of adhesion-related particles from images of cells expressing the GFP control and the GFP-linked active variants of Rac1, RhoA, and RhoD (as indicated). Each set of cumulative curve distance data can be fitted to three normal distributions, as shown by the curves in the histograms. (**e**) Cumulative curve distributions of the nearest neighbor distances, showing distinct separation between most of these cell types

3. Fit the cumulative curves by using cumulative normal distribution functions. This can be performed in Origin in the menu Analysis → Fitting → Nonlinear Curve Fit → Open Dialog. Select the relevant function, which in our case is a three-component normal cumulative distribution function; or if there is no premade function for the fit, formulate one using the button "Create/Edit Fitting Functions" (*see* **Note 11e**). Enter appropriate starting values for the fit through inspection of the shapes of the histograms and the cumulative curve distributions, for instance, equal distributions with one component at short distances of ~40–100 nm, one at medium distances of ~100–200 nm, and one at longer distances of ~200–400 nm, and then run the fitting. Store the results in a table (*see* Table 1).
4. Go back to the image analysis software, and use the results from the fitting to see where the adhesion particles in the different distributions are located in the image. Use the nearest neighbor distance for each adhesion particle, and give it a value that depends on which distribution it belongs to. For instance, it can be assigned a value of one if it belongs to the short-distance distribution, of two if it belongs to the medium-distance distribution, and of three if it belongs to the long-distance distribution. Show the image and compare it to the original data to see how the different distances are distributed in the cells (Fig. 4).

**Table 1**
**Analysis of distance data: Statistical analysis of the three fitted distributions to the adhesion-related particle distances**

| | Interparticle distance (nm) | | | | | | Relative proportion (%) | | |
|---|---|---|---|---|---|---|---|---|---|
| | Short (~40–100 nm) | | Medium (~100–200 nm) | | Long (~200–400 nm) | | | | |
| Condition[a] | Mean | SD | Mean | SD | Mean | SD | Short | Medium | Long |
| Control | 70 | 34 | 158 | 70 | 302 | 123 | 49 | 26 | 25 |
| Rac1 | 70 | 34 | 167 | 66 | 295 | 111 | 27 | 41 | 32 |
| RhoA | 54 | 23 | 111 | 45 | 219 | 101 | 47 | 28 | 26 |
| RhoD | 62 | 29 | 136 | 62 | 280 | 124 | 34 | 33 | 33 |

[a]Rac1, RhoA, RhoD: Cells transfected for each constitutively active Rho GTPase: Rac1/Q61L, RhoA/Q63L, RhoD/G26V, respectively

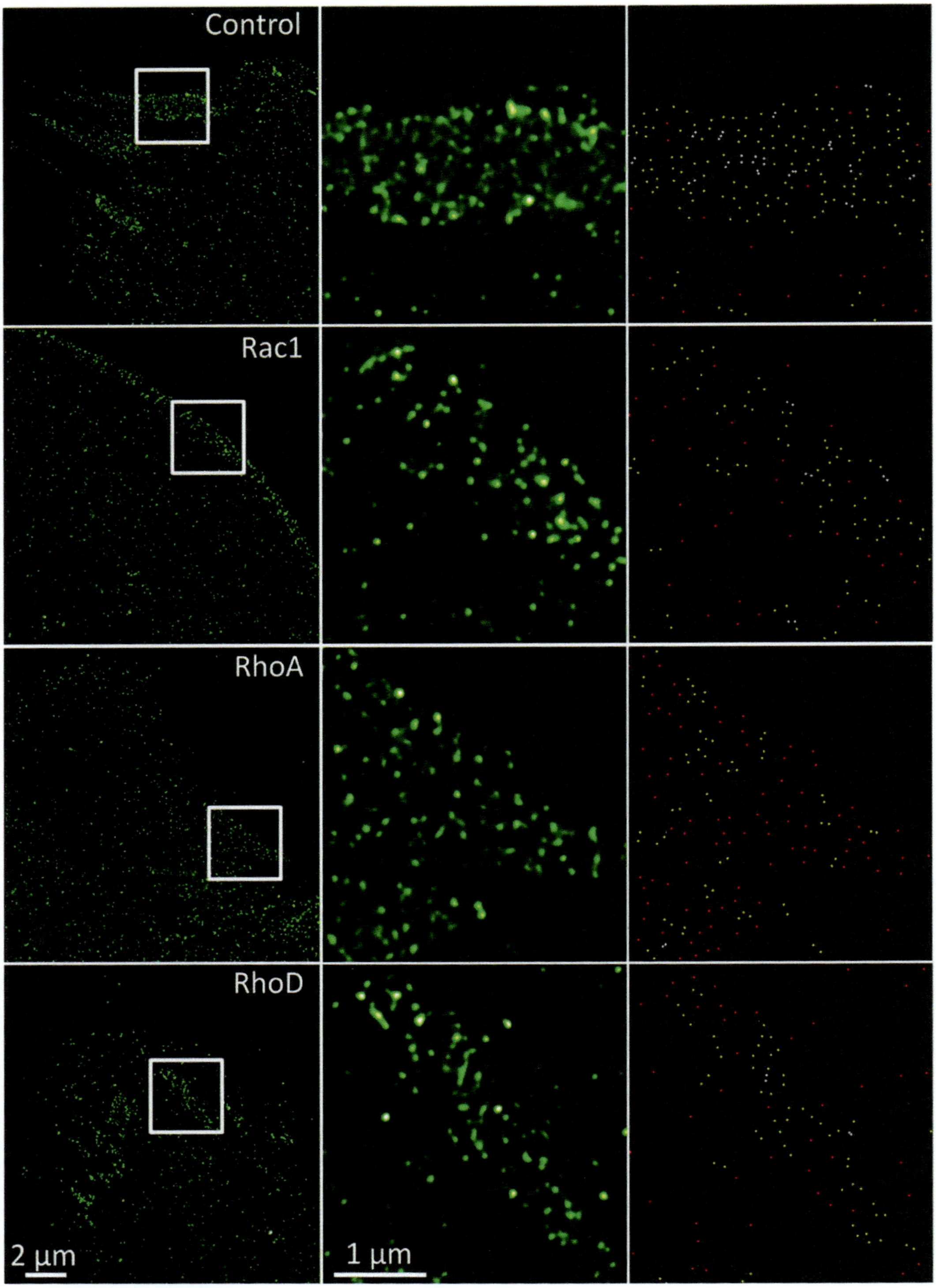


**Fig. 4** Identification of adhesion-related particles in cells that overexpress active Rho GTPases. *Left panels*: High-resolution $15 \times 15$ μm$^2$ areas of phosphotyrosine-positive particles in control cells and cells that express the different Rho GTPases (as indicated). *Middle panels*: Enlarged areas as indicated in *left panels*, showing increased phosphotyrosine signal. *Right panels*: Same enlarged areas showing the adhesion-related particles with color coding indicating to which distance distribution they belong (*white*: short, *yellow*: medium, *red*: long)

## 4 Notes

1. Super-resolution stimulated depletion microscope.
   (a) Laser: As the resolution of STED microscopy is dependent on the laser power, it is of importance to use a powerful laser. Either pulsed lasers or continuous-wave (CW) lasers can be used; however, systems based on pulsed lasers usually provide greater resolution due to the high and temporally confined pulse energies and the ability to sequentially excite and apply the STED beam. Despite this, CW lasers have the benefit of faster image acquisition time [16]. We use a pulsed Fianium SC-450-PP-HE laser (Fianium Ltd., Southampton, UK), which provides a broad spectrum from which we can select all of the desired wavelengths for our excitation and STED beams [17]. The biggest advantage of such a system is that all of the laser beams are inherently synchronized since they all originate from the same laser source. Other commonly used pulsed lasers that provide sufficiently high power are Ti:sapphire lasers.
   (b) Filters and dichroic mirrors: As STED microscopy requires additional laser beams, there is a higher demand for suitable filters and dichroic mirrors. In selecting filters and mirrors, it is important to minimize any loss in intensity of the STED beams, rather than the excitation beams. The reason for this is that while the lasers usually provide enough laser power for the excitation, as high a STED power as possible is required for the best resolution [19]. We primarily use on-stock filters and mirrors from Chroma and Semrock. However, our dichroic mirror, needed to reflect the two emission bands while transmitting both the STED and the excitation beams, had to be custom-made by Laseroptik (Garbsen, Germany).
   (c) Vortex phase plates and polarization optics: A crucial element in STED microscopy is the generation of the doughnut-shaped focus of the STED beams, with practically no intensity in the center and high intensity around the sides. The creation of such a focus can be difficult (*see* **Note 1d**), but once this is achieved, it is usually very stable. We use a vortex phase plate, which is a glass substrate with spirally increasing thickness. By matching this thickness to the wavelength of the STED beam, a phase difference of one $\pi$ can be achieved, which creates destructive interference in the center of the focus. To obtain the best destructive interference of the photons, it is also crucial to produce an evenly circular polarization of the beam. This

can be achieved by adding optics that modifies the polarization of the beam, such as $\lambda/2$ and $\lambda/4$ wave-plates.

(d) The VPP-1 vortex phase plate has several different vortices matched for different wavelengths. For our STED wavelengths at 710 ± 10 nm and 750 ± 10 nm, the closest matches on the phase plate are 678.8, 688.9, 727.7, 735.8, and 796.3 nm. For our setup, the best matches are when the vortex is slightly below the wavelength of the beam. In our case, both 678.8 and 688.9 nm work well for the 710-nm beam and 727.7 and 735.8 nm for the 750-nm beam. To obtain the best minima, good circular polarization of the beams is needed. We achieve this by inserting a $\lambda/2$ and a $\lambda/4$ wave-plate just before the objective. As the polarization is very sensitive, these wave-plates should be mounted in fine-tuning rotatable holders (e.g., PRM1, from Thorlabs). To check the polarization, it is advisable to have a detector coupled to an oscilloscope with a rotating polarizer. If there is no perfect circular polarization, the polarizer will let through light depending on its rotational position and it will be possible to detect the rotational frequency in the oscilloscope. Fine-tune the wave-plates to get as low amplitude of oscillations as possible; in the best case, there will be a flat line that indicates perfect circular polarization. There are different types of phase plates. A commercially available model was launched recently, which creates the destructive interference using polarization rather than phase shift. The advantage of using polarization is that it is more wavelength sensitive, and therefore both the excitation and the STED beams can pass through the same phase plate where only the STED beam is affected, making the setup more robust and easier to align [18].

(e) Objective and immersion oils: In confocal microscopy, the resolution is limited by the wavelength and the numerical aperture (NA) of the objective. Even if this resolution limit is surpassed by applying STED microscopy, the resolution still scales with the wavelength and NA of the objective. Thus, high NA objectives, such as immersion oil objectives, will provide the best resolutions. In our system, we use an HCX PL APO 100× objective with an NA of 1.4. To obtain the best focus of the objective, it is also important to have immersion oil with the correct refractive index. The company supplying the objective usually also supplies the oil. However, since the refractive index is dependent on many parameters, including the temperature in the laboratory, this brand of oil is not necessarily optimal. We routinely use a mixture of two types of oil in a 25/75 ratio (type B and 37; Cargille Laboratories, Cedar Gove, NJ, USA) to obtain the best focus at 23.0 °C.

(f) Scanner: Higher resolution demands smaller pixels and therefore puts higher demands on the scanner. Either beam scanning or stage scanning can be applied, and there are many high-precision scanners commercially available. We use a piezo-stage scanning unit that is coupled to a closed-loop controller that provides positional resolution down to 5 nm in three dimensions (MAX311/M and BPC203; Thorlabs AB, Sweden).

(g) Detectors: Images generated by STED typically have a higher background noise signal than their confocal counterparts. The major reason for this is that they are generated from a much smaller volume of detection, which results in less emitted light reaching the detectors. It is therefore important to have highly sensitive detectors. For the fluorescence emission, we use avalanche photodiodes. For the detection of the reflected light that we need for alignment, we use a photomultiplier tube. It is also possible to use a charge-coupled device (CCD) camera and a white-light source to record wide-field images of samples. A very simple CCD camera is sufficient, which can even be a normal webcam. We use an SPC 900 NC.

(h) Electronics: The detectors and scanning stage are coupled to a computer through two separate NIDAQ cards. The software we have used for controlling the hardware and acquiring the images is Imspector (Max-Planck-Innovation, München, Germany).

2. Gold beads and fluorescent beads for alignment: The fluorescent beads should be sonicated before they are put on a cover slip to reduce their aggregation.

3. The STED analysis allows imaging of 15 × 15 μm areas of cells. Both the imaging and the identification of the analyzed cells after the imaging are time consuming. It is therefore of importance to avoid capturing images of non-transfected cells. To reduce the risk of analyzing non-transfected cells, we modified the transfection protocol, to increase the transfection efficiency. In addition, during the DNA preparation, we routinely wash the precipitated DNA at least twice with 50 % ethanol.

4. To avoid damage to the cytoskeletal structures due to the sheer stress of solvents, it is necessary to minimize the washing steps during the staining procedure. It is also important to ensure that the fixed cells are never allowed to dry.

5. To facilitate the handling of the cover slips, we lift and manipulate them using a syringe together with the forceps.

6. To obtain reproducible results, STED analysis requires a more stringent and careful staining procedure compared to conventional confocal microscopy. Hence, it is advisable to ensure that different cover slips are treated in an identical manner during the different steps of the washing and antibody incubation.
7. Gold beads are usually not uniform in their reflective capabilities. Some will give very strong reflection, while others will give weak reflection. Some might result in a slightly tilted profile, while others provide a straight profile. These effects are caused by the differences in the sizes and shapes of the beads. As the gold bead used will be the reference for the beam profiles, it is important to select an optimal bead. Try to find a bead that gives a good reflection (too weak will cause problems with noise, and too strong will result in saturation effects) and a straight profile.
8. The PSFs of each beam should be straight (not tilted), and the sizes should be close to the diffraction limit ($\sim\lambda/2\times NA$) for the excitation and STED beams without the phase plate. A tilted PSF indicates that the beam does not go straight into the objective. This can be corrected by beam walking, which is performed using the two closest mirrors to the objective, which only affects the desired beam. First, tilt the mirror furthest away and then realign the beam with the second mirror. If the beam profile improves, continue until it is straight; if it worsens, move it in the other direction. If the PSF is much larger than the diffraction limit, the objective does not focus the laser beam correctly. It is important to check that the beam profile is large enough to fill the entire objective. If not, use a beam expander to increase the size of the beam. The beam also needs to be well collimated. Adjust the focusing lens of the laser and see if the profile is improved. The focusing of the laser is also affected by the mounting medium, the immersion oil, and the thickness of the cover slip. Always use cover slips of a thickness specified by the objective. In addition, it is necessary to experiment with different mounting mediums and immersion oil mixtures to optimize the focus of the system.
9. Due to the possibility of chromatic errors, it is important to check the detector position using fluorescence emission rather than excitation reflection, in particular if the excitation wavelength is distant from the emission wavelength. If possible, use fluorescent beads that provide a broad enough emission spectrum so that they can be detected in both of the emission channels (for our emission bands at 600–630 and 660–690 nm, we routinely use 20-nm crimson beads from Invitrogen).

10. As the field of view in our STED microscope setup only shows 15 × 15 μm areas, it is difficult to identify precisely the same cell using different microscopes. To overcome this problem, we routinely scrape off a quadrant of the cells on the cover slip prior to fixing. The borders of this wound in the cell layer are easily recognizable by both STED and confocal microscopy. Transfected cells located near, but not at, the cell borders are chosen for the analysis.
11. Data analysis.
    (a) The pixel size should not be larger than half of the resolution due to the Nyquist criterion [20]; this is 20 nm for our system. Smaller pixels will not provide additional information for the image. Instead, they will increase the acquisition time and cause more bleaching. We acquire 15 × 15 $\mu m^2$ images with a pixel dwell time of 1 ms, which gives a total image acquisition time of around 10 min. The pixel dwell time can be adjusted. A longer pixel dwell time increases the signal; however, it will also cause more bleaching and result in increased acquisition time.
    (b) A microscope image can be seen as a "blurred" version of the true image where the blurring function is due to the excitation and emission profiles. By prior knowledge of these profiles, the image can be "deblurred" by deconvolution. The deconvolution software goes through the image and enhances features that correspond to the excitation/emission profile, and at the same time it suppresses noise. When deconvoluting an image, it is therefore very important to choose the correct PSF. This can be achieved by imaging fluorescent beads, which are smaller than the resolution and sufficiently far apart to be imaged as single beads. Look at several beads, and combine the profiles to get a good average. Then fit this profile by either a Gaussian (confocal) or Lorentzian (STED) function. Deconvolution is an iterative procedure, where more iterations provide stronger deconvolution; therefore, it is important to compare deconvoluted images with the original image, so that image artifacts are avoided. In our case, we routinely use a 40-nm Lorentzian PSF and ten iterations for the STED images.
    (c) Thresholding is commonly used in image analysis to separate interesting features (foreground) from background data. The setting of the thresholding will have a large impact on the final analysis, and care should be taken to avoid bias. There are several ways to calculate thresholds automatically. For instance, it is possible to use the Otsu

method, which calculates the threshold by minimizing the intra-class variance in the foreground and background data [21]. Although this method works well for conventional microscopy, it is not fully compatible with the many fine details shown in super-resolution images. Because of this, we set the threshold manually. For this, it is important to always acquire images using exactly the same parameters (i.e., labelling, excitation intensity, pixel dwell time, and size) and to use the same threshold value for all images.

(d) Our image and data analysis are performed in Matlab (The MathWorks Inc., Natick, MA, USA) and in Origin (Originlab Corporation, Wellesley, MA, USA), although many, if not all, of the steps can be performed using other software. For image analysis, there are user-friendly open-source software available, such as ImageJ or CellProfiler, where it is possible to download a nearest neighbor plug-in/module. For data analysis, software such as Excel or Open Office can provide useful alternatives.

(e) For normal distributions, the cumulative distribution function has the form

$$Y=1/2\ \mathrm{erf}\left((x-\mu)/\sqrt{2\sigma^2}\right),$$

where erf is the error function, $\mu$ is the mean, and $\sigma$ is the deviation. For a three-parameter cumulative distribution function, the function will take the form

$$\begin{aligned} Y &= C_1/2\times\left(1+\mathrm{erf}\left((x-\mu_1)/\sqrt{2\sigma_1^2}\right)\right) \\ &\quad + C_2/2\times\left(1+\mathrm{erf}\left((x-\mu_2)/\sqrt{2\sigma_2^2}\right)\right)+C_3/2\times\left(1+erf((x-\mu_3)/\sqrt{2\sigma_3^2})\right), \end{aligned}$$

where $C_{1-3}$ are the weighting constants for each distribution. The F-test can be used to see how many parameters are required for the fit.

## Acknowledgements

A.G. has been supported through grants from the Karolinska Institutet, the Alex and Eva Wallström Foundation, the O.E. and Edla Johansson Foundation, the Magnus Bergvall Foundation, and the Swedish Society for Medical Research; P.A. through grants from the Karolinska Institutet, the Swedish Cancer Society, and the Swedish Research Council; and D.R. and J.W. through grants from the Swedish Research Council, the European Community's 7th Framework Program (Project FLUODIAMON 801237), and the Swedish Cancer Society. The authors thank Stefan W Hell, Lars Kastrup, and Andreas Schönle, MPIBPC, Göttingen, and Hans Blom, KTH, Stockholm, for important support in the buildup of the STED instrument.

## References

1. Heasman SJ, Ridley AJ (2008) Mammalian Rho GTPases: new insights into their functions from *in-vivo* studies. Nat Rev Mol Cell Biol 9:690–701
2. Parsons JT, Horwitz AR, Schwartz MA (2010) Cell adhesion: integrating cytoskeletal dynamics and cellular tension. Nat Rev Mol Cell Biol 11:633–643
3. Patla I, Volberg T, Elad N et al (2010) Dissecting the molecular architecture of integrin adhesion sites by cryo-electron tomography. Nat Cell Biol 12:909–915
4. Gad AK, Rönnlund D, Spaar A et al (2012) Rho GTPases link cellular contractile force to the density and distribution of nanoscale adhesions. FASEB J 26:2374–2382
5. Gustafsson MGL (2000) Surpassing the lateral resolution limit by a factor of two using structured illumination microscopy. J Microsc 198: 82–87
6. Rust MJ, Bates M, Zhuang XW (2006) Sub-diffraction-limit imaging by stochastic optical reconstruction microscopy (STORM). Nat Methods 3:793–795
7. Betzig E, Patterson GH, Sougrat R et al (2006) Imaging intracellular fluorescent proteins at nanometer resolution. Science 313:1642–1645
8. Dertinger T, Colyer R, Iyer G et al (2009) Fast, background-free, 3D super-resolution optical fluctuation imaging (SOFI). Proc Natl Acad Sci U S A 106:22287–22292
9. Hell SW, Wichmann J (1994) Breaking the diffraction resolution limit by stimulated-emission–stimulated-emission-depletion fluorescence microscopy. Opt Lett 19:780–782
10. Dyba M, Jakobs S, Hell SW (2003) Immunofluorescence stimulated emission depletion microscopy. Nat Biotechnol 21: 1303–1304
11. Hell SW (2007) Far-field optical nanoscopy. Science 316:1153–1158
12. Hell SW (2009) Microscopy and its focal switch. Nat Methods 6:24–32
13. Gould TJ, Hess ST, Bewersdorf J (2012) Optical nanoscopy: from acquisition to analysis. Annu Rev Biomed Eng 14:231–254
14. Rönnlund D, Yang Y, Blom H et al (2012) Fluorescence nanoscopy of platelets resolves platelet-state specific storage, release and uptake of proteins, opening for future diagnostic applications. Adv Healthc Mater 1: 707–713
15. Richardson WH (1972) Bayesian-based iterative method of image restoration. J Opt Soc Am 62:55–59
16. Moneron G, Medda R, Hein B et al (2010) Fast STED microscopy with continuous wave fiber lasers. Opt Express 18:1302–1309
17. Wildanger D, Rittweger E, Kastrup L et al (2008) STED microscopy with a supercontinuum laser source. Opt Express 16:9614–9621
18. Reuss M, Engelhardt J, Hell SW (2010) Birefringent device converts a standard scanning microscope into a STED microscope that also maps molecular orientation. Opt Express 18:1049–1058
19. Harke B, Keller J, Ullal CK et al (2008) Resolution scaling in STED microscopy. Opt Express 16:4154–4162
20. Shannon CE (1949) Communication in the presence of noise. Proc IRE 37:10–21
21. Otsu N (1979) Threshold selection method from gray-level histograms. IEEE T Syst Man Cyb 9:62–66

# Chapter 23

# Yeast as a Model for Ras Signalling

**Renata Tisi, Fiorella Belotti, and Enzo Martegani**

## Abstract

For centuries yeast species have been popular hosts for classical biotechnology processes, such as baking, brewing, and wine making, and more recently for recombinant proteins production, thanks to the advantages of unicellular organisms (i.e., ease of genetic manipulation and rapid growth) together with the ability to perform eukaryotic posttranslational modifications. Moreover, yeast cells have been used for few decades as a tool for identifying the genes and pathways involved in basic cellular processes such as the cell cycle, aging, and stress response.

In the budding yeast *S. cerevisiae* the Ras/cAMP/PKA pathway is directly involved in the regulation of metabolism, cell growth, stress resistance, and proliferation in response to the availability of nutrients and in the adaptation to glucose, controlling cytosolic cAMP levels and consequently the cAMP-dependent protein kinase (PKA) activity.

Moreover, Ras signalling has been identified in several pathogenic yeasts as a key controller for virulence, due to its involvement in yeast morphogenesis.

Nowadays, yeasts are still useful for Ras-like proteins investigation, both as model organisms and as a test tube to study variants of heterologous Ras-like proteins.

**Key words** *Saccharomyces cerevisiae*, *Schizosaccharomyces pombe*, *Candida albicans*, Cyclic AMP, Protein kinase A, Filamentous growth

## Abbreviations

| | |
|---|---|
| cAMP | Cyclic AMP |
| ER | Endoplasmic reticulum |
| GAP | GTPase activating protein |
| GDP | Guanosine diphosphate |
| GEF | Guanine nucleotide exchange factor |
| GPCR | G-Protein coupled receptor |
| GTP | Guanosine triphosphate |
| PKA | Protein kinase A |
| ROS | Reactive oxygen species |
| VPS | Vacuolar protein-sorting |

Lorenza Trabalzini and Saverio Francesco Retta (eds.), *Ras Signaling: Methods and Protocols*,
Methods in Molecular Biology, vol. 1120, DOI 10.1007/978-1-62703-791-4_23, © Springer Science+Business Media, LLC 2014

## 1 Ras Proteins Signalling Module in Budding Yeast

The *Saccharomyces cerevisiae* genome contains two *RAS* genes, *RAS1* and *RAS2* [1], encoding proteins of 309 and 322 amino acid residues, respectively, and showing significant homology to the mammalian Ras proteins in their N-terminal portions, involved in the recognition of guanine nucleotide and phosphate required for GTPase activity [2]. In the C-terminal third of the molecule, yeast Ras proteins diverge significantly from the mammalian Ras [3]. Yeast Ras proteins, like their mammalian counterparts, undergo extensive posttranslational modification—C-terminal proteolytic cleavage, farnesylation, palmitoylation, and carboxymethylation—and are deposited on the inner surface of the plasma membrane by a specialized transport mechanism [4, 5].

Ras proteins activity is finely tuned by two classes of regulatory proteins in yeast (*see* Fig. 1): Cdc25 [6] and Sdc25 [7] guanine nucleotide exchange factors (GEFs), which stimulate the GDP/

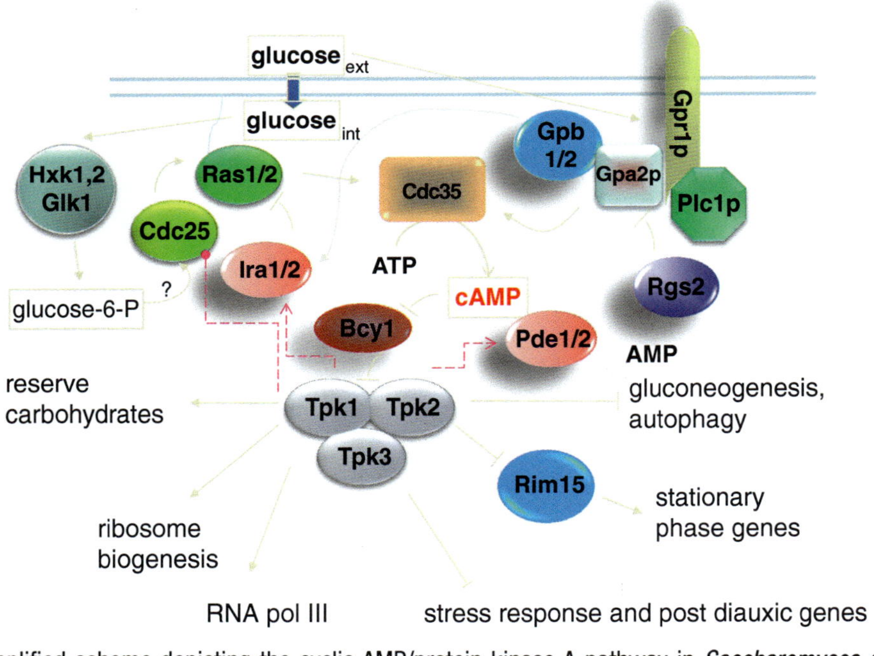


**Fig. 1** Simplified scheme depicting the cyclic AMP/protein kinase A pathway in *Saccharomyces cerevisiae*. Extracellular glucose and sucrose activate PKA through GPCR system (comprising Gpr1 receptor, Gpa2 G protein, Gpb1/2 regulatory proteins, and the Plc1 phosphoinositides-specific phospholipase), which controls adenylate cyclase (Cyr1/Cdc35) activity together with Ras signalling module, comprising Ras1/2 small G proteins and their regulators, Cdc25 RasGEF, and Ira1/2 RasGAPs. Ras signalling module is responsive to intracellular glucose but it also requires glucose phosphorylation by hexokinases (Hxk1,2) or glucokinase (Glk1). The second messenger cyclic AMP, which is degraded by Pde1/2 phosphodiesterases, releases Tpk1/2/3 catalytic subunit of protein kinase A (PKA) from Bcy1 regulatory subunit. Main effects of PKA activity on cellular processes are shown. Reported/proposed feedback loops (see the text) are indicated in purple: arrow-headed, positive controls; round-headed, negative control

GTP exchange on Ras, and Ira1 and Ira2 GTPase activating proteins (GAPs), which promote the intrinsically low Ras GTPase activity [8, 9].

The molecular mechanisms of modulation of Ras proteins activity have been conserved during evolution, as it was confirmed by the functional interchangeability of Ras proteins [10] and their regulators in mammals and budding yeast. As the only exception, the *S. cerevisiae* GAPs Iral and Ira2 [8, 11] are known, from genetic and biochemical studies [9, 12–15], to be strictly specific for the yeast Ras1 and Ras2, although mammalian RasGAP neurofibromin 1 can, in contrast, substitute for yeast GAPs.

Ras1 and Ras2 have similar functions, but their expressions differ [16]. The *RAS* genes are essential for growth; thus, *ras1 ras2* double mutants are nonviable [16, 17]. Cells with a temperature-sensitive *RAS2* mutation together with *ras1* mutation are blocked in the G1 phase of the cell cycle and accumulate as unbudded cells at nonpermissive temperatures [16].

Addition of glucose to glucose-starved cells triggers a fast increase in the GTP loading of Ras2, contemporary with the glucose-induced increase in cAMP [18, 19]. Besides, the studies on this pathway were complicated by the fact that Ras and Cdc25 proteins are not only required for cAMP synthesis upon glucose refeeding but also essential for basal adenylate cyclase activity and cell viability [17]. Therefore, suppressor mutations or multicopy suppressor genes restoring viability have to be present in cells in which *RAS* or *CDC25* genes have been deleted, possibly affecting cAMP accumulation [20] and leading to some contradictory results.

Despite extensive investigations, the mechanism by which glucose affects Ras-GTP levels remains unresolved. The intracellular levels of GTP, which quickly respond to nutrient availability, could be the metabolic signal that regulates Ras activity in response to glucose [18, 19, 21, 22]. On the other hand, the Ira proteins seem to be inhibited by the glucose signal, since an *ira1Δ ira2Δ* double deletion mutant displays a dramatic increase in Ras-GTP levels but no further increase upon glucose addition [19].

Addition of glucose to cells elicits a rapid cytoplasmic acidification, which was proposed as the second messenger activating PKA signalling [23], rather than the abundance of a specific carbon source or metabolite thereof. Intracellular acidification was reported to induce a rapid increase in GTP on the Ras proteins, likely mediated by inhibition of the Ira proteins or by another target downstream of Ira [24].

The Ira1 and Ira2 proteins are really large proteins (3,092 and 3,079 amino acids respectively) that share similar amino acids sequence and have related, even if not identical, functions. A region of approximately 360 amino acids in the middle of these gene products, called the GAP domain, is responsible for activating intrinsic GTPase activity of Ras.

Ira1 and Ira2 are both homologs of human neurofibromin 1 (NF1) protein [25], whose mutations were found to be involved in human neurofibromatosis [26]. NF1 primarily affects cells of neural crest origin and is characterized by patches of skin pigmentation (café-au-lait spots) and neurofibromas. Human NF1 is structurally closer to Ira proteins than to human rasGAPs [25, 27, 28]. Many proteins recently found to bind NF1 have a role in the cytoskeleton [29]; interestingly, Ira2 GAP activity was reported to be counteracted by tubulin binding [14].

Curiously, a screening for utilization of glycerophosphoinositol as the sole phosphate source identified Ira1 and Ira2 as required: Ira1 and Ira2 affect both the production and secretion of glycerophosphoinositol [30], but the RAS/cyclic AMP pathway does not appear to be involved in this function.

The main positive regulator of yeast Ras proteins is *CDC25* gene product. Cdc25 protein is a ~180 kDa polypeptide with a C-terminal highly conserved Ras-GEF catalytic domain (1,121–1,573 aa) [31, 32]. The C-terminal fragment (aa 1,256–1,589) is essential for normal growth and viability of *cdc25Δ* and *cdc25 ts* mutants [33, 34]. An SH3 domain (residues 65–134) is present in the N-terminal region, involved in protein–protein interactions with adenylyl cyclase, and seems to enhance its responsiveness to activation by Ras in vitro [35, 36]; a cyclin destruction box (CDB, position 149) near the SH3 domain is important for the short half-life of the protein controlled by a ubiquitin-dependent degradation process [37].

In spite of the fact that Cdc25 was the first Ras GEF identified and characterized [38, 39], it is still unclear how its activity is regulated and how the Cdc25/Ras/cAMP pathway transduces the signal originated by nutrients. Contradictory reports stated either the necessity [40–43] or the dispensability [44] of Cdc25 for Ras and adenylate cyclase activation upon glucose stimulation, and Cdc25 was finally assessed as necessary for Ras2 GTP-loading [18, 19] after glucose refeeding.

Otherwise, its N-terminal region (1–1,121 aa) was initially proposed either to be [40, 41] or not to be [42] involved in the glucose induced cAMP response, whereas afterwards a negative regulatory role has been suggested for this region [45, 46]. Nevertheless, strains with an unregulated GEF activity substituting for Cdc25 activity were viable and showed only minor defects: a complete Cdc25 protein is required for normal spore germination and it may also play a role in reentry of the cell into the cell cycle [46, 47].

Cdc25 protein is tightly bound to membranes [31, 48]; the protein is more soluble when hyper-phosphorylated within the 114–348 region and becomes less available for association with Ras [49]; however, glucose-induced hyper-phosphorylation of Cdc25 upon glucose addition has recently been reported to directly inhibit Ras GEF activity instead of affinity for Ras [50]. Cdc25 activity modulation by phosphorylation was correlated to Tpk1 PKA catalytic subunit activity [51, 52]. Hyper-phosphorylation is abolished by nutrient starvation [52] and is believed to be a part of a negative feedback loop resulting from PKA activity. Stochastic simulations show that stable oscillatory regimes for the dynamics of cAMP can be obtained through the effect of the negative feedback loops on the activity of both Ira2 and Cdc25 proteins, and when the amount of Cdc25 is within a specific range [22].

It is interesting to note that, although Ras-GRF1—one of the mammalian GEFs, originally isolated by complementation of *cdc25*$^{ts}$ mutation in yeast [53]—shows only partial homology to the yeast *CDC25* and mammalian and yeast Ras proteins have limited functional homology, both exchange factors are regulated by the PKA [54], suggesting the existence of conserved Ras-dependent signalling networks.

Recent data have shown that mammalian Ras isoforms are differentially distributed within cell surface nanoclusters and on endomembranous compartments [55]. Even if cAMP/PKA pathway has been extensively studied in budding yeast, the physiological and functional localization of Ras signalling complex components is still under investigation. Ras2 localization on internal membranes was deeply investigated: most of the Ras posttranslational modification steps occur on the cytoplasmic surface of the endoplasmic reticulum and the subsequent Ras translocation from ER to plasma membrane does not require the classical secretory pathway in yeast [4, 56]. An additional Ras effector called Endoplasmic Reticulum Ras Inhibitor 1 (Eri1) was identified [57], which functions in the biosynthesis of the glycosylphosphatidylinositol anchor, a process that occurs in the ER. Furthermore, Ras2, when overexpressed, co-fractionated with Eri1 in the ER in a GTP-dependent manner [58].

Recent works point to a possible role for mitochondria for correct Ras localization [5] and to PKA-dependent control on plasma membrane Ras localization [50]. An aberrant accumulation of activated Ras2 at the mitochondria in response to nutritional depletion was reported in cells lacking Whi2 [59, 60]. Whi2 was suggested to be involved in cell cycle regulation and to have a role in the general stress response [61, 62]. In *whi2Δ* mutant, the failure to address Ras2 to the vacuole and consequently the failure to shut down Ras signalling would lead to mitochondrial dysfunction, the accumulation of damaging ROS and cell death. Disruption of class C VPS genes, encoding for proteins playing a role in endosome and vacuole membrane fusion, also results in mitochondrial defects and an accumulation of total Ras proteins on mitochondrial membranes [5].

While Cdc25 and Ira proteins were shown to be tightly attached to membrane fraction [31, 48], adenylate cyclase is only loosely bound to the membranes and seems to rely on Ira1 for its membrane targeting [63, 64]. Recent results [65] indicate a largely intracellular localization of Ras signalling complex: in fact, only a small amount of Cyr1, Ras, Cdc25, and Ira proteins co-localized with plasma membrane in exponentially growing cells. These proteins are mainly associated to ER membranes; Ras2 and Ira proteins appear to be also largely present on mitochondria; a minor fraction of Cyr1 protein is also present in this last compartment. This is consistent with the preliminary identification of Ira1 in the mitochondrial proteome by two distinct high-throughput approaches [66, 67].

Moreover, evidence suggests that the Cdc25 protein could dimerize, and possibly be part of a large complex with other elements of the pathway, i.e., adenylate cyclase and Ira proteins [36]; in addition, Cdc25 was found to be associated to several different large protein complexes [68, 69]. As Ras2 also presents functional and structural interactions with adenylate cyclase and with the adenylate cyclase-associated protein CAP [70], adenylate cyclase itself could act as a scaffold that keeps Ras proteins available for their GEF.

More literature data suggest that upstream components of cAMP/PKA pathway interact to constitute a large multiprotein complex: after detergent extraction, adenylate cyclase behaves as a large complex on gel filtration and sucrose density centrifugation [63, 71].

Recently, a feedback regulation on Ras2 localization has been suggested, based on PKA-dependent phosphorylation: Ras2 was shown to localize mainly on the plasma membrane in glycerol-growing cells, while glucose-addition causes a rapid re-localization of large part of the protein to the cytoplasm [50]. The overexpression of the Tpk1 subunit drives both Cdc25 and Ras2 away from the plasma membrane [72], consistently with data reported in literature [49, 50]. The fraction of Cdc25 on the plasma membrane was far higher in starved cells than in exponentially growing cells, while Ras2 only partially re-localizes [72]; on the other hand, Ras2 is far more abundant than Cdc25, which could be the limitative component of the signalling module. Taken together, these results suggest that localization of these proteins could be important for signal compartmentalization. Actually, Ras signal compartmentalization was already reported to be an issue both in mammals and in yeasts [73–76].

Finally, Cdc25 is found not only associated with membranes, but also in the nucleus [77], and active Ras was also detected in the nuclear compartment in glucose-growing cells [60, 78].

The region responsible for Cdc25 nuclear localization was identified as the central uncharacterized region (residues 353–1,100), and cell stress or unbalanced PKA activity were reported to affect nuclear localization of Cdc25 likely via phosphorylation of two serine ($Ser^{825}$ and $Ser^{826}$) residues near a bipartite Nuclear

Localization Signal (NLS) in the middle of the protein [72, 77], which were previously proposed as PKA phosphorylation sites [79].

Surprisingly, Ira1, but not Ira2, nor adenylate cyclase Cyr1, was found to accumulate into nuclear fraction as Cdc25 [72], confirming previously reported differences for these apparently redundant GAPs. Ira1 is actually the yeast homolog showing the best similarity to mammalian NF1, and both show a bipartite nuclear localization sequence, which is functional in NF1 [80].

A nuclear localization of Ras proteins was reported in mammalian cells and active Ras (i.e., GTP-bound Ras) has been found in this cellular compartment in *Dictyostelium discoideum* also [81–83]. Using a fluorescent probe based on the Ras binding domain of Raf1, yeast active Ras was detected mainly in the nucleus and in the plasma membrane in glucose growing wild-type cells, while in glucose-starved cells it was localized in mitochondria, and re-localized to the plasma membrane following glucose addition, suggesting the plasma membrane as the predominant site of agonistic-induced Ras activation.

The localization of activated Ras in presence of glucose, as well as the re-localization of Cdc25 to the plasma membrane after nutrient starvation [72] supports the hypothesis that the strong and transient cAMP increase observed after glucose refeeding requires Ras signalling complex on the plasma membrane. Anyway, both Gpr1/Gpa2 and Ras complex are required to cooperate for rapid response to glucose re-addition—*see* Subheading 2. Surprisingly, in *gpa2Δ* cells, lacking the Gpr1-coupled G protein, the probe was detectable in internal membranes and mitochondria, both in growing and in starved cells, indicating that Gpa2, independently of Gpr1 and consequently of its activation state, was required for the recruitment of Ras-GTP to the plasma membrane and to the nucleus [78].

Interestingly, the mutant Ras2$^{C318S}$, which is not palmitoylated and subsequently not able to bind to the plasma membrane, supports a normal mitotic growth but fails to induce cAMP increase in response to glucose [84]. Moreover, total Ras2 protein was found to be only partially localized in the plasma membrane during growth on glucose, while it was found mainly in the plasma membrane during growth on glycerol, and shifted back to the cytoplasm after glucose addition [50]. These data are consistent with a model where Ras signalling complex during exponential growth on glucose does not show a specific plasma membrane localization, which is conversely required in transient glucose-induced signalling in derepressed cells [84], when Ras2 and Cdc25 are both available in the plasma membrane. Whilst Cdc25 sequestering to the membrane could be justified by the scarcity of this protein when compared to Ras2 level, the opposite activated Ras2 movement seems to point to a specific regulation of the pathway by compartmentalization of the signal, likely finally acting as a negative feedback loop on cAMP transient signalling.

Some other interactors of Ras signalling complex were described in literature, but their function is still unclear. First, the protein Tfs1p has been isolated as a multicopy-suppressor of a mutation in the RasGEF, Cdc25 [85]. In vivo, Tfs1 binds to Ira2 and inhibits its function towards Ras [86]. Tfs1 is a member of phosphatidylethanolamine-binding proteins family [87]. In addition, Tfs1 inhibits the vacuolar protease Carboxypeptidase Y [88]. Interestingly, point mutants abolishing Tfs1 binding to Ira2 are located in the lipid-binding pocket, suggesting that Ira2 and lipid binding are mutually exclusive [86–89]. In stationary phase, Tfs1 re-localizes from the cytoplasm to the vacuolar rim and the vacuolar lumen [87, 90], thus potentially contributing to inactivation of Ras under these conditions.

Lte1 was also identified as a multicopy suppressor for the heat shock sensitivity caused by a hyperactive Ras pathway [91]. Later findings further interconnected Lte1 and Ras activities: Lte1 binds preferentially to GTP- rather than GDP-bound Ras [92, 93]. Moreover, increased expression of Lte1 rescues the heat shock sensitivity of hyperactive Ras2 [91, 92], suggesting that Lte1 binding reduces Ras activity by preventing activation of adenylate cyclase. Lte1 is important for coupling cell cycle progression with timely activation of polarized cell growth. Nevertheless, Ras activation has been linked to Cdc42-dependent polarized growth [94, 95], so inhibition of Ras in the regulation of bud polarization cannot be excluded. A model was proposed which integrates Lte1 localization and activity [96]: interaction with Ras would be essential to localize Lte1 over the whole bud cortex. However, the physiological role of the inhibition of Ras by Lte1 is unclear at present.

## 2 Protein Kinase A Pathway in Budding Yeast

In *Saccharomyces cerevisiae*, Ras/cAMP/protein kinase A (PKA) pathway is involved in cell growth regulation in response to nutrients and mainly to glucose availability [97, 98]. Transcriptomic analysis indicate that the cAMP/PKA pathway is the main player in the transcriptional response to glucose [99]. Although the molecular link between nutrient availability/growth rate and the activity of the PKA pathway is only partially understood, several specific nutrient sensors have been identified which are responsible for rapid activation of the PKA pathway upon re-addition of the limiting nutrient [100, 101].

Glucose addition to derepressed cells induces a rapid and transient increase in cAMP level, leading to protein kinase A activation [102]. Cyclic AMP is strictly regulated by the balance between synthesis by Cyr1/Cdc35 adenylate cyclase [24, 103] and degradation by Pde1 and Pde2 phosphodiesterases [104, 105] (*see* Fig. 1).

Adenylate cyclase is a large membrane-bound protein of 2,026 amino acids. In the middle of the protein the LRR domain (amino acids 674–1,300), a repetitive domain consisting of a repeat of a 23-residue amphipathic leucine-rich motif, is the primary site of interaction with Ras, spanning a Ras associating domain (RAD) (aa 676–756), a motif of about 100 amino acids conserved among Ras-binding regions of mammalian Ral-GDS, Rin1 and afadin/AF-6 [106].

Yeast adenylate cyclase is activated by two different systems (*see* Fig. 1): the G-protein-coupled receptor system (GPCR), acting through the G-protein Gpa2 [24, 103], and the Cdc25/Ras system [17]. Cdc25 GEF activity is essential for rapid Ras activation upon glucose refeeding [19]; nonetheless, Cdc25 RasGEF is required for the regulation of cAMP level in exponentially growing cells [47], but GPCR system is equally important for this transient response [24, 46].

*GPR1* encodes a seven-transmembrane G protein-coupled receptor that physically interacts with Gpa2 [107, 108], a small GTP-binding protein homologous to the mammalian Gα subunit of the heterotrimeric G proteins [103]. Since binding of glucose to Gpr1 is characterized by a very low affinity ($K_M$ 20 mM) [109, 110], Gpa2 protein seems to be mainly involved in producing the rapid increase of cAMP levels that occurs upon addition of high glucose concentrations (100 mM) to glucose-derepressed cells [108]. Nevertheless, Cdc25 and Ras are also required for the glucose- (and fructose-)induced cAMP increase [109] and play a central role in adenylate cyclase activity regulation during yeast growth.

Glucose (or sucrose)-dependent activation of cAMP signalling, even through the GPCR system, is strictly dependent on sugar uptake and phosphorylation, the latter mediated by three isoenzymes, Hxk1, Hxk2, and Glk1 [109, 111, 112] (*see* Fig. 1). Glucose-induced Ras-GTP loading is also dependent on sugar uptake and phosphorylation, while it does not require the presence of a functional GPCR system [19].

It is now assessed that both the activating systems, GPCR and Ras complex, are required to cooperate for rapid response to glucose re-addition, while during exponential growth this is not necessary. In fact, though GPCR complex is unable to activate adenylate cyclase in the absence of active Ras, very small amounts of active Ras are sufficient for GPCR system to activate glucose-induced transient cAMP spike [47]. On the other hand, the changes in transcription following glucose addition are essentially eliminated by induction of a dominant negative *RAS2* allele while they are not affected at all by deletion of *GPR1* [113]. Finally, the GPCR system was proposed to act as an overlay on the Ras/PKA network designed to effect long-term responses to glucose availability [114].

Nonetheless, effects on active Ras localization—*see* Subheading 1—point to an uncoupling between Gpa2 and Gpr1, uncoupling which has also been previously reported using a global transcriptional analysis approach [99].

Gpa2 also interacts with the kelch repeat proteins Gpb1/Krh2 and Gpb2/Krh1; these proteins were originally thought to act as Gβ subunits [115], but their function is rather to down-regulate the PKA pathway [116–118], either by promoting the association between the regulatory (Bcy1) and catalytic (Tpk1/2/3) subunits of PKA [119, 120], or by increasing the levels of RasGAPs Ira1/2 [118]. In fact, the two Ras2 GTPase activating proteins (GAPs), Ira1 and Ira2, were identified as Gpb1/2 interacting proteins by mass spectrometry; Gpb1/2 associate with Ira1/2, resulting in inhibition of the Ras2 GTPase [118], possibly by modulating Ira1/2 proteins stability and degradation within the cell [121].

Cyclic AMP-bound Bcy1 releases the Tpk1/2/3 catalytic subunits of the kinase [122]. The three catalytic subunits of PKA are largely, but not completely, redundant [123–129]. For example, deletion of *TPK2* abolished filamentous growth, while deletion of *TPK3* caused hyperfilamentous growth [123, 130].

PKA activity is involved in posttranslational regulation of a variety of proteins, for example key enzymes for gluconeogenesis or glycolysis [97], and the Cdc25/Ras/cAMP signal transduction pathway regulates many cellular and physiological processes, such as growth, resting state [131], glycogen and trehalose content, carbohydrate and nitrogen metabolism, stress tolerance and expression of genes that are controlled by STRE-boxes in their promoters—reviewed by Thevelein [97, 132, 133]—sporulation [134–136], traffic at the *trans*-Golgi network (TGN) and endosomes [137], autophagy—a degradative process that recycles nonessential proteins and organelles during periods of nutrient starvation—[138–140], cell wall resistance to lyticase digestion [137, 141, 142], and finally regulation of Snf1 protein kinase [143–145]—which controls responses to glucose limitation and other stresses [98].

The cAMP pathway is also involved in the control of cell cycle progression at G0/G1 and/or G2 phases [146–148], G1 cyclins expression [149–153], and nutritional modulation of the critical cell size required for entry into the S phase [154–156].

Inactivation of the PKA signalling pathway causes first cycle arrest at START, a regulatory area in the G1 phase of the cell cycle, followed by entry into stationary phase (G0) [114, 157]. Stress response, nutritional availability and growth rate are intimately interconnected. Even when supplied with a rapidly fermentable carbon source, mutants with reduced PKA activity grow with small cell volume and exhibit several characteristics typical of stationary phase cells, including enhanced stress resistance, high level of storage carbohydrates and low trehalase activity, impaired filamentous growth and enhanced sporulation efficiency. Conversely, mutants with upregulated PKA pathway grow with large cell size, but poorly on non fermentable carbon source, are sensitive to various stress forms, do not arrest properly in stationary phase when deprived of nutrients, exhibit a vigorous filamentous growth, but fail to sporulate [114, 132, 141, 154].

Recently, Ras signalling has also been implicated in localization of signalling factors required for mitotic exit [92, 93]. PKA also can control mitosis exit by phosphorylating Cdc20 and preventing $APC^{Cdc20}$-mediated degradation of Pds1 and Clb2 [148, 158].

A new role for the Ras signalling pathway has been identified as a regulator of DNA damage checkpoint recovery. The DNA damage checkpoint maintains genome stability by arresting the cell cycle and promoting DNA repair. When PKA regulatory elements are hyperactivated by *IRA* deletion, after checkpoint down-regulation cells undergo a permanent mitotic arrest [159].

Growing evidence shows that hyperactivation of the Ras/cAMP pathway affects the cell mitochondrial content [160–163]. The failure to shut down Ras/cAMP/PKA pathway as cells experience starvation leads to mitochondrial dysfunction, as cells expressing the constitutively active $RAS2^{ala12\ val19}$ have been shown to exhibit altered mitochondrial membrane potential and elevated ROS levels [164, 165]. The accumulation of ROS and consequent cell death in mutant cells which show actin stabilization can be prevented by the deletion of Tpk3 PKA subunit, while actin filament stabilization leads to the hyperactivation of Ras [166, 167].

Protein kinase A activity has dramatic effects on gene expression. Glucose-dependent activation of PKA pathway promotes cell biosynthetic metabolism by inducing the transcription of genes involved in ribosome biogenesis [168–170], mainly by activation of Rap1 transcription factor [168, 169] and regulation of subcellular localization of the major regulator of ribosome biogenesis (Ribi) and ribosomal protein (RP) genes, the zinc-finger transcription factor Sfp1 [170–172]. Thus, cellular components that affect the size threshold—Ras/PKA, Sch9, and Sfp1—all converge on ribosome biogenesis, suggesting a model in which the rate of ribosome biogenesis sets the size threshold.

Moreover, Ras/cAMP/PKA signalling in yeast controls relocalization/activation of the two redundant zinc-finger transcription factors Msn2/4 [173–175], activation of the transcription factors Gis1 through Rim15 [176–179] and Hsf1 activation [178], leading to cellular protection systems activation by transcription of stress response genes, e.g., superoxide dismutases, catalase, heat shock proteins [176].

Msn2 and Msn4 bind to stress responsive elements (STRE) in the promoter of general stress-response (STRE) genes, encoding for molecular chaperones, antioxidant proteins, enzymes involved in carbohydrate metabolism and proteolysis [176, 180], in response to nutrient starvation and to several stress conditions [181–183].

PKA directly phosphorylates both the Msn2 nuclear localization signal (NLS) and its nuclear export signal (NES) [184]; as a consequence, Msn2 and Msn4 localize to the cytoplasm when PKA is active, shuttling in and out of the nucleus according to the waves of increasing and decreasing cellular cAMP levels [173, 185]. The oscillatory behavior of Msn2/4 nucleocytoplasmic shuttling is an effect of the multiple feedback loops—the best known mechanisms are inhibition of cAMP synthesis by down-regulation of adenylate cyclase and/or its regulatory proteins, e.g., Ras, RasGEF and RasGAPs, or stimulation of cAMP breakdown by phosphodiesterases [186]—exerted by PKA on a few pivotal components of the pathway, which tightly regulate the pathway. By performing stochastic simulations of a mechanistic model of the Ras/cAMP/PKA pathway, this multilevel regulation of the amplitude and frequency of oscillations in the Ras/cAMP/PKA pathway was suggested to act as a fine tuning mechanism for the downstream targets of PKA [175, 187].

Furthermore, full activation of Msn2/4 upon glucose starvation involves Yak1 kinase, a dual-specificity tyrosine-regulated kinase (DYRK) [188], acting as a negative growth regulator that antagonizes the cAMP/PKA pathway [189–191] by posttranscriptionally regulating expression of genes involved in stress response and use of alternative carbon sources. In fact, *YAK1* was identified as a gene whose deletion suppressed complete loss of function of the Ras/PKA pathway [189]. Yak1 also controls Bcy1 redistribution from nucleus to the cytoplasm upon glucose exhaustion [192]. In addition, Bcy1 apparently determines the localization of the TPK subunits associated with it [192–194]. Finally, the TOR network was suggested to control the subcellular localization of both the *TPK1* catalytic subunit and the Yak1 kinase [139].

PKA drives rapid shifts between cell growth and stress adaptation according to environmental conditions by coordinating activity of Msn2/4 and Hsf1 through Yak1 activity [188]. Even if *MSN2/4* are both deficient, PKA can still regulate several stress responsive genes by inhibiting the activity of Hsf1 and thus controlling a subset of Hsf1-dependent genes [178]. Hsf1 controls the expression of a large set of genes involved in several processes—heat-stress response, protein folding and degradation, detoxification, energy generation, carbohydrate metabolism and cell wall organization [195–200].

PKA inhibits by phosphorylation the activity of another kinase, Rim15, which is required for entry into quiescence [177]. Inactivation of *RIM15* suppresses the lethality of *tpk1 tpk2 tpk3* triple null strain, whereas overexpression of Rim15 during exponential growth causes a synthetic growth defect in mutants with reduced PKA activity [201, 202].

The Rim15-controlled expression program at the diauxic shift—due to glucose limitation—is almost completely mediated by the three transcription factors Msn2, Msn4, and Gis1

[203–205]. Gis1 binds to the so-called post-diauxic shift (PDS) elements in the promoter of stress response genes [206, 207]. Carbon source responsive cAMP/PKA pathway, nitrogen source responsive TOR pathway, and phosphate responsive Pho80/Pho85 pathway converge on Rim15, controlling its nuclear/cytoplasmic distribution [177, 204, 208].

## 3 MAPK Pathway Controlling Invasive Growth in Budding Yeast

Historically, much interest in understanding filamentous growth regulation has come from studies in fungal pathogens. Pathogens like *Candida albicans* and *Aspergillus fumigatus* represent a threat to human health [209, 210], particularly harmful to individuals whose immune system has been compromised by AIDS or by suppression resulting from chemotherapies and other drug treatments [211]. Nonetheless, fungal pathogens can also be devastating to plant communities, causing serious harvest loss.

Knowledge on the genetic pathways that regulate filamentous growth has benefited from versatility of *S. cerevisiae* as a model for the phenomenon, despite the fact that this organism cannot carry out a strictly defined dimorphic transition. In appropriate conditions, the baker's yeast *Saccharomyces cerevisiae* cells actually experience a transition from globular, yeast-like growth to a pseudohyphal, filamentous growth pattern, which is characterized by an elongated cell morphology and a polarized budding pattern [212, 213]. Whilst yeast-form cells completely separate from each other after each cell cycle, filamentous cells remain connected in chains or filaments. The situation in budding yeast differs from that of filamentous fungi, which grow as multinucleate hyphae: filamentously growing yeast cells undergo cytokinesis but remain attached to each other by virtue of adhesive contacts in the cell wall. Increased expression of *FLO11*, a cell-surface flocculin involved in cell-cell adhesion, is sufficient to induce both invasive and filamentous growth [214].

The resulting morphological changes differ slightly between haploid and diploid yeast cells: the term "invasive growth" has been applied to the filamentation phenomenon shown by haploid cells because of their ability to invade agar substrates; the term "pseudohyphal growth" is sometimes used to describe the response in diploid cells [215].

One of the triggers for filamentous growth in yeast and many other fungal species is nutrient scarcity, such as ammonium-limiting conditions and/or presence of a fermentable carbon source [212, 216], or the lack or limitation of a fermentable carbon source [217–219].

Filamentous growth is controlled by evolutionarily conserved signalling pathways. An activated version of Ras2, in which the protein is blocked in its activated (GTP-bound) state, dramatically

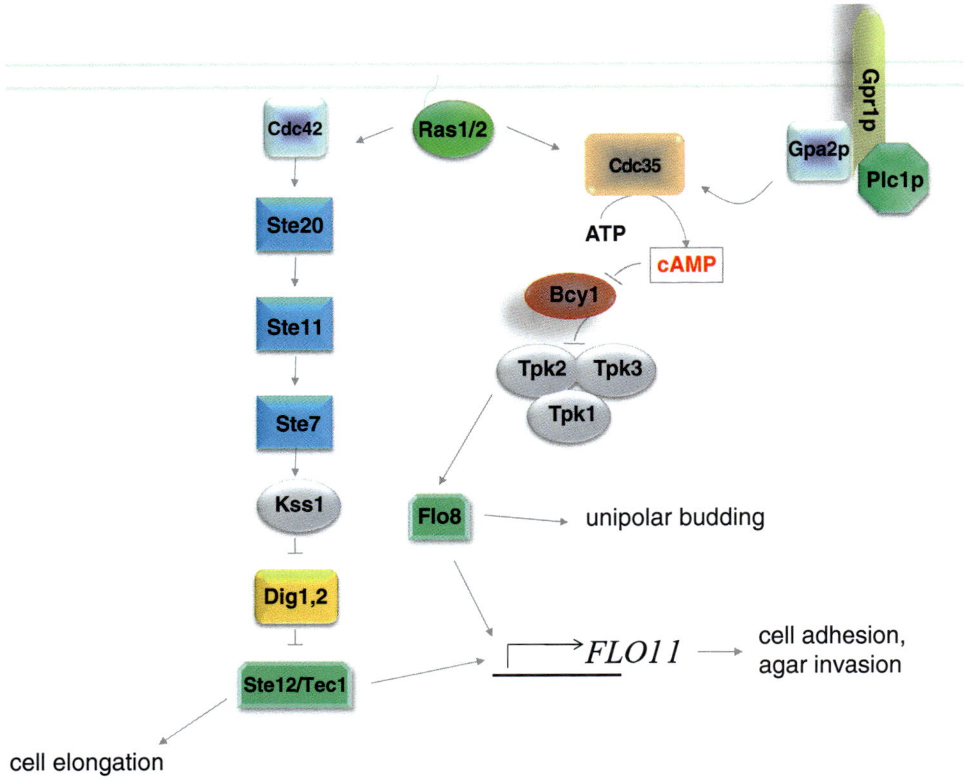


**Fig. 2** Simplified scheme describing the pseudohyphal growth regulation in *Saccharomyces cerevisiae*. PKA signalling is involved in the induction of *FLO11* transcription by controlling the activities of transcription regulators, e.g., Flo8. Nutrient limitation can activate Cdc42 Rho GTPase, which in its turn activates the Kss1 MAP kinase cascade consisting of Ste20 PAK kinase, Ste11 MAPKK kinase, and Ste7 MAPK kinase. Kss1 kinase phosphorylates Dig1/2 transcriptional repressors, allowing Ste12 and Tec1 transcriptional factors binding to *FLO11* and other filamentous growth specific genes promoters

stimulated the filamentous properties of this organism [212]. In fact Ras2, but not Ras1, can activate two pathways that act independently, the mitogen-activated protein (Cdc42/Ste20/MAPK) kinase pathway, leading to activation of filamentation-specific Kss1 MAP kinase, and the cyclic AMP-protein kinase A (cAMP–PKA) pathway [220] (*see* Fig. 2).

Bcy1, Fus3, Ksp1, Kss1, Sks1, and Tpk2 localize predominantly to the nucleus during filamentous growth. These kinases form part of an interdependent, localization-based regulatory network: deletion of each individual kinase, or loss of kinase activity, disrupts the nuclear translocation of at least two other kinases [221]. Individual PKA catalytic subunits plays distinct roles in pseudohyphal growth with Tpk2 stimulating filamentation and Tpk1 and Tpk3 inhibiting the process [123, 125, 130]. The Ras2/Tpk2 pathway and the MAPK pathway converge on the *FLO11* promoter [213, 222].

A dominant active version of Ras2, Ras2$^{Val19}$, can activate filamentous growth. This stimulation is not observed in cells containing loss-of function alleles of *CDC42*, implying that Ras2 also functions at or above Cdc42 in the filamentous growth pathway [94]. In mammalian cells, Ras2 regulates MEK-ERK signalling through the protein kinase RAF, but no homologs of Raf are present in yeast [223]. Ras2 may modulate upstream components of the pathway or may indirectly modulate MAPK activity, for example by regulating the expression of a MAPK regulatory gene. Specifically, Ras2 was found to regulate expression of the filamentation specific *MSB2* gene—the first cell-surface protein to be identified with a specific role in regulating the filamentation pathway [224]—in a Ste12-independent manner [225]. In further support of this possibility, Tpk2 was also required for *MSB2* expression [225]. Moreover, Msb2 associates with Cdc42, but not with Ras2 [224] and Ste50 associates with Cdc42, but not Ras2 [226].

Curiously, filamentous growth, particularly in haploid cells, is stimulated by glucose limitation, while PKA activity is generally associated with glucose availability. However, filamentous growth resulting from constitutive PKA activity can be further stimulated by glucose limitation, suggesting that glucose levels influence filamentation through a PKA-independent mechanism [114]. Consistently, pseudohyphal growth is also known to be controlled by the glucose-repressed Snf1 kinase, which interacts with two repressor proteins, Nrg1 and Nrg2, inhibiting filamentation and other *FLO11*-dependent processes, such as haploid invasive growth and biofilm formation [227].

Previous work has identified Gpa2-coupled Gpr1 receptor as a regulator of pseudohyphal differentiation [228, 229]. Gpr1 participates, together with Snf1-mediated signalling, in the induction of pseudohyphal growth on different fermentable carbon sources [230, 231]. Signalling through Gpr1 also triggers filamentous growth on sucrose even under ammonium-rich conditions [231].

## 4 Pathways Controlled by Ras Signalling in Other Yeasts

The fission yeast *Schizosaccharomyces pombe* has been widely used as a model for genetics of eukaryotic cells. It has only one *ras* gene, called *ras1* [232], which encodes a protein of 219 amino acid residues; thus, its size is closer to that of mammalian Ras protein than *S. cerevisiae* Ras proteins. A human *ras* cDNA sequence can complement a *ras1Δ* strain, demonstrating functional similarity between the fission yeast and human genes [233]. The *ras1* mutant cells are abnormally round and sterile, because of the involvement of Ras1 in two different downstream signalling pathways. In fact, Ras1 interacts with two different effectors controlling two distinct functions.

First, Ras1 activates the Byr2 MAP kinase kinase (MEKK) to mediate mating pheromone signalling: inactivation of this pathway blocks sexual differentiation and results in sterility. Byr2 activates another kinase called Byr1, which then activates Spk1 kinase, resulting in the regulation of pheromone signalling to induce gene expression that is crucial for sexual differentiation [234]. Thus, Ras1 directly activates a MAP kinase module, and this aspect is similar to the activation of the Raf/Mek/Erk by mammalian Ras.

The second Ras1 effector is Scd1/Ral1 [234], which is a GEF for Cdc42 [235], a member of the Rho family of Ras-like proteins. Cdc42, in turn, activates the Shk1/Orb2 protein kinase [236]. The Ras1–Scd1 pathway is further regulated by a scaffold protein, Scd2/Ral3, which positively affects protein binding between components in this pathway [234]. Inactivation of this pathway affects cell morphology which changes from elongated to round. While the Ras pathways in fibroblasts and budding yeast can regulate functioning of the actin cytoskeleton, in fission yeast the Ras1–Scd1 pathway can also affect the functioning of microtubules to affect spindle formation and chromosome segregation [237, 238].

Deleting *byr2* blocks mating pheromone signalling to cause sterility but does not affect cell morphology [239], chromosome segregation or spindle formation; conversely, *scd1Δ* cells are round and defective in chromosome segregation and spindle formation [240, 241] but still respond to mating pheromones [234]. Apart from Ras1, the two pathways seem to be completely independent. Even the RasGEFs required for activation of the two pathways seem distinguishable [242]. Like other RasGEFs from mammals (Sos and GRF) and the budding yeast (Cdc25), Ras1 exchange factor Ste6 contains a "classic" catalytic domain in its C terminus [243]. *ste6Δ* cells, like *byr2Δ* cells, are sterile but have a normal cell shape. In addition, *ste6* expression is barely detectable during vegetative growth but is induced by signals for mating, nutrient starvation, and mating pheromones [244]. These observations suggest that Ste6 is specific for Ras1 function in mating pheromone signalling. On the contrary, Efc25 acts as a Ras1 GEF specific for the Scd1 pathway [245]: while *ste6* expression is induced during mating, *efc25* expression is constitutive. Moreover, Efc25 overexpression renders cells hyper-elongated and confers sterility, which can be rescued by activated Ras1, suggesting that Efc25 can recruit Ras1 to selectively activate Scd1 at the expense of Byr2 [242].

The Ras GTPase activating protein (GAP) in *S. pombe* is encoded by *sar1/gap1* gene. Gap1 is homolog to mammalian GAP, the yeast *Saccharomyces cerevisiae* Ira proteins and the product of the human neurofibromatosis *locus*, NF1 [246].

A recent study in *S. pombe* provided the clearest demonstration that Ras signalling is compartmentalized [247]: whereas a Ras chimera (Ras1-Rit) forcedly localized to the plasma membrane selectively regulates MAP kinase pathway to mediate mating pheromone signalling, a mutant Ras ($Ras1^{C215S}$) trapped to the endomembranes activates the Cdc42 pathway, which mediates cell polarity and protein trafficking [248]. Remarkably, $Ras1^{C215S}$ restored cell morphology of *ras1* cells but did not rescue sterility, while the converse was observed with Ras1-Rit.

*Candida albicans* is the most prevalent opportunistic fungal pathogen in humans; frequently found as a harmless commensal in the gastrointestinal tract, it sometimes causes various forms of candidiasis, usually just superficial mucosal infections in the healthy individual, but may invade the body and cause life-threatening systemic disease associated with poor prognosis and high mortality in the compromised patient [249]. It is a pleiomorphic organism, characterized by its ability to grow in filamentous or yeast forms [250]. Transition to the filamentous morphology is critical for virulence [251] and depends on a multitude of extracellular factors, including carbohydrates, amino acids, salts, pH changes, temperature increases, starvation, serum and growth within a matrix [252–254].

The pathogenic forms of several yeasts have been studied extensively due to their medical impact, but only a few fungal species have been studied to search for molecular events leading up to dimorphic transition, with *Candida albicans* probably being the most studied species [255]. The molecular dissection of morphogenesis in *C. albicans* has taken advantage of the strong evolutionary conservation of fungal signal transduction pathways: many *C. albicans* signal transduction genes important for filamentous growth have been isolated using genetic screenings in *S. cerevisiae* mutants.

*C. albicans* Ras1 [256], activated by Cdc25 GEF as in *Saccharomyces cerevisiae* [257], activates both the MAPK and cAMP–PKA pathways. Amino acid sequence alignment of *S. cerevisiae* (Sc) Ras1, ScRas2, *Candida albicans* (Ca) Ras1 and CaRas2 revealed that CaRas2 is distantly related with all the others, sharing the highest identity of ~25 % to ScRas1. This level of identity is markedly lower than those scored between the other Ras proteins ranging from 50 to 60 % [258]. In spite of the poor homology with other Ras GTPases, *C. albicans* Ras2 still contains a CCIIT sequence at the C-terminus, matching the consensus lipid-conjugation motif CaaX or CCXXX commonly found at the C-terminus of Ras proteins, which is required for membrane anchorage [259]. Surprisingly, Ras2 performs functions that antagonize those of Ras1 in the control of cellular levels of cAMP, cell density in the stationary phase and responses to certain stress conditions [258].

Ras proteins regulate both a MAPK pathway—consisting of the Cst20, Ste11, Hst7, and Cek1 kinases—which regulates the activity of the transcription factor Cph1 [260–263], and the cAMP–PKA pathway—comprising Cdc35/Cyr1, Pde2, Cap1, Tpk1, Tpk2, and Bcy1—which regulates the activity of the transcription factor Efg1 [256, 264, 265]. In contrast with budding yeast, Gpa2 was initially described as part of the MAPK pathway in *C. albicans* [263], but later reports confirmed that, again, this G protein acts with Gpr1 receptor in the cAMP/PKA pathway [266], even if it is responsive to certain amino acids and not to glucose or sucrose [253].

The *C. albicans* adenylyl cyclase coordinates three developmental programs, i.e., invasive filamentous growth, phenotypic switching to a mating-competent cell type—characterized by "opaque" colonies—and biofilm formation [267]. During biofilm growth, cells exhibit changes in morphology and surround themselves with extracellular polymers that enhance protection and cohesion. Biofilms formed by a/α cells have different properties than a/a or α/α biofilms. By mutational analyses, a/α biofilms were demonstrated to be regulated by the Ras1/cAMP pathway, while a/a biofilms by the MAP kinase pathway [268, 269].

In the presence of serum, protein kinase A (PKA) induces yeast-to-hypha transition and virulence acquisition [270, 271]. Just as hypha formation is positively regulated by cAMP signalling, Ras1 and Cyr1 increase the rate of white-opaque switching in response to environmental cues, such as N-acetylglucosamine [272].

Ras1 stimulates adenylate cyclase via the Ras1 association domain or the LRR domain, which also requires Cyr1 association with Srv2 (also referred to as Cap1) [273] and actin [274]. In addition, the localization of Ras1 to different cellular membrane pools suggests that cAMP synthesis occurs at different sites within the cell [275].

Besides the several similarities with cAMP pathway in *S. cerevisiae*, clear divergence in the mechanisms that regulate the cAMP pathway has been observed in *C. albicans*. For example, while *ras1 ras2* null or *cdc35* null mutant in *S. cerevisiae* is lethal, *Caras1*Δ and *Cacdc35*Δ mutants are viable; disruption of *BCY1*, encoding the regulatory subunit of PKA, is viable in *S. cerevisiae*, but deletion of its orthologue in *C. albicans* is lethal [276]. In *S. cerevisiae* the MAP kinase pathway seems to be the most important for pseudohyphal growth induction, whereas the cAMP-protein kinase A pathway is predominant in *C. albicans* [252]. Furthermore, the antagonism between Ras1 and Ras2 suggests that in wild-type cells, a more or less constant level of cAMP is maintained thank to the balance between Ras1 and Ras2 [258].

The Ras signalling pathway has also been shown to regulate apoptosis in the pathogenic yeast *Candida albicans* [258, 277], suggesting that signals influencing cAMP production and sensation may be a common mechanism by which programmed cell death is regulated in yeast.

Another dimorphic fungus and one of the more intensively studied "nonconventional" yeast species is *Yarrowia lipolytica*. Distantly related to *Saccharomyces cerevisiae*, it can be genetically modified and can grow either in haploid or in diploid states in either yeast, pseudomycelial or mycelial forms, depending on environmental conditions; under controlled conditions, a nearly homogeneous population of yeast-like cells or hyphae can be obtained [278]. *Y. lipolytica* has important biotechnological applications, such as degradation of hydrophobic substrates, production of organic acids, and secretion of homologous and heterologous proteins [279, 280].

The dimorphic transition of *Y. lipolytica* is affected by several conditions, such as the available carbon or nitrogen source [278], blood serum [279, 281], and citrate and pH of the medium [278]. Some genes that appear to play in the dimorphic phenomenon have been identified [282–286] which have outlined the most important signalling pathways as the MAPK and PKA pathways. Differently from *S. cerevisiae* and *C. albicans*, but similar to the case of *Ustilago maydis*, a *Basidiomycota* fungus and a specific pathogen of maize [287], in *Y. lipolytica* these pathways are oppositely involved in dimorphism: while the integrity of the MAPK signalling pathway is necessary for mycelial growth, a functional PKA pathway is required for growth in the yeast-like form [288, 289].

The ascomycete *Kluyveromyces lactis* has been used in the food industry for the production of lactase (β-galactosidase) since 1950s [290]. In response to severe environmental conditions, ascomycetes can produce stress-resistant spores which survive until conditions are suitable for growth. Since sporulation implies a diploid DNA content, species with a haploid lifestyle, such as *K. lactis*, first have to induce mating in response to stress. In *K. lactis*, mating and mating-type switching are induced by the DNA-binding protein Mts1, whose expression is known to be upregulated by nutrient limitation through Msn2 transcription factor. Thus, Ras2/cAMP/Msn2 has recently been suggested to mediate the stress-induced mating and mating-type switching responses in *K. lactis* [291].

## 5 Concluding Remarks

Signalling processes in mammalian cells are characterized by a high complexity due to the large number of factors involved in the systems and their extremely sophisticated regulatory mechanisms. Compared with the same systems in mammalian cells, signalling systems in yeast cells are relatively simple and easier to reproduce in silico. However, most of the factors known in the yeast signalling networks are conserved and operate similarly in mammalian cells; thus, knowledge of the signalling systems in a simpler organism, such as the budding yeast (*S. cerevisiae*), can be useful to understand the signalling systems in mammalian cells.

Study in yeast has contributed significantly to the understanding of the Ras signalling pathway in mammalian cells: as described, the hint that NF1 was a GAP for Ras came from the study of the

Ira proteins; the study in yeast contributed to elucidate the C-terminal processing of Ras; the work on downstream effectors of Ras contributed to the identification of the Ras binding domains and reinforced the idea that one of the functions of Ras is to activate the MAP kinase cascade; finally, several works done in yeast contributed greatly to the identification and characterization of mammalian Ras GEFs (Guanine nucleotide Exchange Factors). More recently, the application of high-throughput experimental approach and computational methods, through the systems biology procedure [292], has led to realize modules that will be a useful guide for higher complexity in silico models.

## References

1. Powers S, Kataoka T, Fasano O, Goldfarb M, Strathern J, Broach J, Wigler M (1984) Genes in *S. cerevisiae* encoding proteins with domains homologous to the mammalian ras proteins. Cell 36:607–612
2. Tamanoi F, Walsh M, Kataoka T, Wigler M (1984) A product of yeast RAS2 gene is a guanine nucleotide binding protein. Proc Natl Acad Sci U S A 81:6924–6928
3. Créchet J, Cool R, Jacquet E, Lallemand J (2003) Characterization of *Saccharomyces cerevisiae* Ras1p and chimaeric constructs of Ras proteins reveals the hypervariable region and farnesylation as critical elements in the adenylyl cyclase signaling pathway. Biochemistry 42:14903–14912
4. Dong X, Mitchell D, Lobo S, Zhao L, Bartels D, Deschenes R (2003) Palmitoylation and plasma membrane localization of Ras2p by a nonclassical trafficking pathway in *Saccharomyces cerevisiae*. Mol Cell Biol 23: 6574–6584
5. Wang G, Deschenes R (2006) Plasma membrane localization of Ras requires class C Vps proteins and functional mitochondria in *Saccharomyces cerevisiae*. Mol Cell Biol 26:3243–3255
6. Jones S, Vignais M, Broach J (1991) The CDC25 protein of *Saccharomyces cerevisiae* promotes exchange of guanine nucleotides bound to ras. Mol Cell Biol 11:2641–2646
7. Damak F, Boy-Marcotte E, Le-Roscouet D, Guilbaud R, Jacquet M (1991) SDC25, a CDC25-like gene which contains a RAS-activating domain and is a dispensable gene of *Saccharomyces cerevisiae*. Mol Cell Biol 11:202–212
8. Tanaka K, Matsumoto K, Toh-E A (1989) IRA1, an inhibitory regulator of the RAS-cyclic AMP pathway in Saccharomyces cerevisiae. Mol Cell Biol 9:757–768
9. Tanaka K, Nakafuku M, Tamanoi F, Kaziro Y, Matsumoto K, Toh-e A (1990) IRA2, a second gene of *Saccharomyces cerevisiae* that encodes a protein with a domain homologous to mammalian ras GTPase-activating protein. Mol Cell Biol 10:4303–4313
10. DeFeo-Jones D, Tatchell K, Robinson LC, Sigal IS, Vass WC, Lowy DR, Scolnick EM (1985) Mammalian and yeast ras gene products: biological function in their heterologous systems. Science 228:179–184
11. Tanaka K, Nakafuku M, Satoh T, Marshall MS, Gibbs JB, Matsumoto K, Kaziro Y, Toh-e A (1990) S. cerevisiae genes IRA1 and IRA2 encode proteins that may be functionally equivalent to mammalian ras GTPase activating protein. Cell 60:803–807
12. Xu G, Lin B, Tanaka K, Dunn D, Wood D, Gesteland R, White R, Weiss R, Tamanoi F (1990) The catalytic domain of the neurofibromatosis type 1 gene product stimulates ras GTPase and complements ira mutants of *S. cerevisiae*. Cell 63:835–841
13. Tanaka K, Lin BK, Wood DR, Tamanoi F (1991) IRA2, an upstream negative regulator of RAS in yeast, is a RAS GTPase-activating protein. Proc Natl Acad Sci U S A 88: 468–472
14. Parrini M, Jacquet E, Bernardi A, Jacquet M, Parmeggiani A (1995) Properties and regulation of the catalytic domain of Ira2p, a *Saccharomyces cerevisiae* GTPase-activating protein of Ras2p. Biochemistry 34:13776–13783
15. Parrini MC, Bernardi A, Parmeggiani A (1996) Determinants of Ras proteins specifying the sensitivity to yeast Ira2p and human p120-GAP. EMBO J 15:1107–1111
16. Kataoka T, Powers S, McGill C, Fasano O, Strathern J, Broach J, Wigler M (1984) Genetic analysis of yeast RAS1 and RAS2 genes. Cell 37:437–445

17. Toda T, Uno I, Ishikawa T, Powers S, Kataoka T, Broek D, Cameron S, Broach J, Matsumoto K, Wigler M (1985) In yeast, RAS proteins are controlling elements of adenylate cyclase. Cell 40:27–36
18. Rudoni S, Colombo S, Coccetti P, Martegani E (2001) Role of guanine nucleotides in the regulation of the Ras/cAMP pathway in *Saccharomyces cerevisiae*. Biochim Biophys Acta 1538:181–189
19. Colombo S, Ronchetti D, Thevelein J, Winderickx J, Martegani E (2004) Activation state of the Ras2 protein and glucose-induced signaling in *Saccharomyces cerevisiae*. J Biol Chem 279:46715–46722
20. Nikawa J, Cameron S, Toda T, Ferguson KM, Wigler M (1987) Rigorous feedback control of cAMP levels in *Saccharomyces cerevisiae*. Genes Dev 1:931–937
21. Cazzaniga P, Pescini D, Besozzi D, Mauri G, Colombo S, Martegani E (2008) Modeling and stochastic simulation of the Ras/cAMP/PKA pathway in the yeast *Saccharomyces cerevisiae* evidences a key regulatory function for intracellular guanine nucleotides pools. J Biotechnol 133:377–385
22. Pescini D, Cazzaniga P, Besozzi D, Mauri G, Amigoni L, Colombo S, Martegani E (2012) Simulation of the Ras/cAMP/PKA pathway in budding yeast highlights the establishment of stable oscillatory states. Biotechnol Adv 30:99–107
23. Dechant R, Binda M, Lee SS, Pelet S, Winderickx J, Peter M (2010) Cytosolic pH is a second messenger for glucose and regulates the PKA pathway through V-ATPase. EMBO J 29:2515–2526
24. Colombo S, Ma P, Cauwenberg L, Winderickx J, Crauwels M, Teunissen A, Nauwelaers D, de Winde J, Gorwa M, Colavizza D, Thevelein J (1998) Involvement of distinct G-proteins, Gpa2 and Ras, in glucose- and intracellular acidification-induced cAMP signalling in the yeast *Saccharomyces cerevisiae*. EMBO J 17: 3326–3341
25. Buchberg AM, Cleveland LS, Jenkins NA, Copeland NG (1990) Sequence homology shared by neurofibromatosis type-1 gene and IRA-1 and IRA-2 negative regulators of the RAS cyclic AMP pathway. Nature 347: 291–294
26. Ballester R, Marchuk D, Boguski M, Saulino A, Letcher R, Wigler M, Collins F (1990) The NF1 locus encodes a protein functionally related to mammalian GAP and yeast IRA proteins. Cell 63:851–859
27. Nishi T, Saya H (1991) Neurofibromatosis type 1 (NF1) gene: implication in neuroectodermal differentiation and genesis of brain tumors. Cancer Metastasis Rev 10:301–310
28. Gutmann D, Collins F (1992) Recent progress toward understanding the molecular biology of von Recklinghausen neurofibromatosis. Ann Neurol 31:555–561
29. Corral T, Jiménez M, Hernández-Muñoz I, Pérez de Castro I, Pellicer A (2003) NF1 modulates the effects of Ras oncogenes: evidence of other NF1 function besides its GAP activity. J Cell Physiol 197:214–224
30. Bishop A, Surlow B, Anand P, Hofer K, Henkel M, Patton-Vogt J (2009) Neurofibromin homologs Ira1 and Ira2 affect glycerophosphoinositol production and transport in *Saccharomyces cerevisiae*. Eukaryot Cell 8:1808–1811
31. Vanoni M, Vavassori M, Frascotti G, Martegani E, Alberghina L (1990) Overexpression of the CDC25 gene, an upstream element of the RAS/adenylyl cyclase pathway in *Saccharomyces cerevisiae*, allows immunological identification and characterization of its gene product. Biochem Biophys Res Commun 172:61–69
32. Jacquet E, Parrini M, Bernardi A, Martegani E, Parmeggiani A (1994) Properties of the catalytic domain of CDC25, a *Saccharomyces cerevisiae* GDP/GTP exchange factor: comparison of its activity on full-length and C-terminal truncated RAS2 proteins. Biochem Biophys Res Commun 199:497–503
33. Lai C, Boguski M, Broek D, Powers S (1993) Influence of guanine nucleotides on complex formation between Ras and CDC25 proteins. Mol Cell Biol 13:1345–1352
34. Coccetti P, Mauri I, Alberghina L, Martegani E, Parmeggiani A (1995) The minimal active domain of the mouse ras exchange factor CDC25Mm. Biochem Biophys Res Commun 206:253–259
35. Freeman NL, Lila T, Mintzer KA, Chen Z, Pahk AJ, Ren R, Drubin DG, Field J (1996) A conserved proline-rich region of the *Saccharomyces cerevisiae* cyclase-associated protein binds SH3 domains and modulates cytoskeletal localization. Mol Cell Biol 16:548–556
36. Mintzer K, Field J (1999) The SH3 domain of the *S. cerevisiae* Cdc25p binds adenylyl cyclase and facilitates Ras regulation of cAMP signalling. Cell Signal 11:127–135
37. Kaplon T, Jacquet M (1995) The cellular content of Cdc25p, the Ras exchange factor in *Saccharomyces cerevisiae*, is regulated by destabilization through a cyclin destruction box. J Biol Chem 270:20742–20747
38. Camonis J, Kalékine M, Gondré B, Garreau H, Boy-Marcotte E, Jacquet M (1986) Characterization, cloning and sequence analysis of the CDC25 gene which controls the cyclic AMP level of *Saccharomyces cerevisiae*. EMBO J 5:375–380

39. Martegani E, Baroni M, Frascotti G, Alberghina L (1986) Molecular cloning and transcriptional analysis of the start gene CDC25 of *Saccharomyces cerevisiae*. EMBO J 5:2363–2369
40. Munder T, Küntzel H (1989) Glucose-induced cAMP signaling in Saccharomyces cerevisiae is mediated by the CDC25 protein. FEBS Lett 242:341–345
41. Schomerus C, Munder T, Küntzel H (1990) Site-directed mutagenesis of the *Saccharomyces cerevisiae* CDC25 gene: effects on mitotic growth and cAMP signalling. Mol Gen Genet 223:426–432
42. Van Aelst L, Boy-Marcotte E, Camonis J, Thevelein J, Jacquet M (1990) The C-terminal part of the CDC25 gene product plays a key role in signal transduction in the glucose-induced modulation of cAMP level in *Saccharomyces cerevisiae*. Eur J Biochem 193:675–680
43. van Aelst L, Jans A, Thevelein J (1991) Involvement of the CDC25 gene product in the signal transmission pathway of the glucose-induced RAS-mediated cAMP signal in the yeast *Saccharomyces cerevisiae*. J Gen Microbiol 137:341–349
44. Goldberg D, Segal M, Levitzki A (1994) Cdc25 is not the signal receiver for glucose induced cAMP response in *S. cerevisiae*. FEBS Lett 356:249–254
45. Chen R, Michaeli T, Van Aelst L, Ballester R (2000) A role for the noncatalytic N terminus in the function of Cdc25, a *Saccharomyces cerevisiae* Ras-guanine nucleotide exchange factor. Genetics 154:1473–1484
46. Belotti F, Tisi R, Martegani E (2006) The N-terminal region of the *Saccharomyces cerevisiae* RasGEF Cdc25 is required for nutrient-dependent cell-size regulation. Microbiology 152:1231–1242
47. Paiardi C, Belotti F, Colombo S, Tisi R, Martegani E (2007) The large N-terminal domain of Cdc25 protein of the yeast *Saccharomyces cerevisiae* is required for glucose-induced Ras2 activation. FEMS Yeast Res 7:1270–1275
48. Garreau H, Geymonat M, Renault G, Jacquet M (1996) Membrane-anchoring domains of Cdc25p, a *Saccharomyces cerevisiae* ras exchange factor. Biol Cell 86:93–102
49. Gross E, Goldberg D, Levitzki A (1992) Phosphorylation of the *S. cerevisiae* Cdc25 in response to glucose results in its dissociation from Ras. Nature 360:762–765
50. Dong J, Bai X (2011) The membrane localization of Ras2p and the association between Cdc25p and Ras2-GTP are regulated by protein kinase A (PKA) in the yeast *Saccharomyces cerevisiae*. FEBS Lett 585:1127–1134
51. Deminoff SJ, Ramachandran V, Herman PK (2009) Distal recognition sites in substrates are required for efficient phosphorylation by the cAMP-dependent protein kinase. Genetics 182:529–539
52. Jian D, Aili Z, Xiaojia B, Huansheng Z, Yun H (2010) Feedback regulation of Ras2 guanine nucleotide exchange factor (Ras2-GEF) activity of Cdc25p by Cdc25p phosphorylation in the yeast *Saccharomyces cerevisiae*. FEBS Lett 584:4745–4750
53. Martegani E, Vanoni M, Zippel R, Coccetti P, Brambilla R, Ferrari C, Sturani E, Alberghina L (1992) Cloning by functional complementation of a mouse cDNA encoding a homologue of CDC25, a *Saccharomyces cerevisiae* RAS activator. EMBO J 11:2151–2157
54. Buday L, Downward J (2008) Many faces of Ras activation. Biochim Biophys Acta 1786: 178–187
55. Rocks O, Peyker A, Bastiaens P (2006) Spatio-temporal segregation of Ras signals: one ship, three anchors, many harbors. Curr Opin Cell Biol 18:351–357
56. Zhao L, Lobo S, Dong X, Ault A, Deschenes R (2002) Erf4p and Erf2p form an endoplasmic reticulum-associated complex involved in the plasma membrane localization of yeast Ras proteins. J Biol Chem 277: 49352–49359
57. Sobering AK, Watanabe R, Romeo MJ, Yan BC, Specht CA, Orlean P, Riezman H, Levin DE (2004) Yeast Ras regulates the complex that catalyzes the first step in GPI-anchor biosynthesis at the ER. Cell 117:637–648
58. Sobering AK, Romeo MJ, Vay HA, Levin DE (2003) A novel Ras inhibitor, Eri1, engages yeast Ras at the endoplasmic reticulum. Mol Cell Biol 23:4983–4990
59. Leadsham JE, Gourlay CW (2008) Cytoskeletal induced apoptosis in yeast. Biochim Biophys Acta 1783:1406–1412
60. Leadsham JE, Miller K, Ayscough KR, Colombo S, Martegani E, Sudbery P, Gourlay CW (2009) Whi2p links nutritional sensing to actin-dependent Ras-cAMP-PKA regulation and apoptosis in yeast. J Cell Sci 122: 706–715
61. Sudbery PE, Goodey AR, Carter BL (1980) Genes which control cell proliferation in the yeast *Saccharomyces cerevisiae*. Nature 288:401–404
62. Radcliffe P, Trevethick J, Tyers M, Sudbery P (1997) Deregulation of CLN1 and CLN2 in the *Saccharomyces cerevisiae* whi2 mutant. Yeast 13:707–715
63. Mitts M, Grant D, Heideman W (1990) Adenylate cyclase in *Saccharomyces cerevisiae* is a peripheral membrane protein. Mol Cell Biol 10:3873–3883

64. Mitts M, Bradshaw-Rouse J, Heideman W (1991) Interactions between adenylate cyclase and the yeast GTPase-activating protein IRA1. Mol Cell Biol 11:4591–4598
65. Belotti F, Tisi R, Paiardi C, Rigamonti M, Groppi S, Martegani E (2012) Localization of Ras signaling complex in budding yeast. Biochim Biophys Acta 1823:1208–1216
66. Sickmann A, Reinders J, Wagner Y, Joppich C, Zahedi R, Meyer H, Schönfisch B, Perschil I, Chacinska A, Guiard B, Rehling P, Pfanner N, Meisinger C (2003) The proteome of *Saccharomyces cerevisiae* mitochondria. Proc Natl Acad Sci U S A 100:13207–13212
67. Reinders J, Zahedi R, Pfanner N, Meisinger C, Sickmann A (2006) Toward the complete yeast mitochondrial proteome: multidimensional separation techniques for mitochondrial proteomics. J Proteome Res 5:1543–1554
68. Ho Y, Gruhler A, Heilbut A, Bader GD, Moore L, Adams SL, Millar A, Taylor P, Bennett K, Boutilier K, Yang L, Wolting C, Donaldson I, Schandorff S, Shewnarane J, Vo M, Taggart J, Goudreault M, Muskat B, Alfarano C, Dewar D, Lin Z, Michalickova K, Willems AR, Sassi H, Nielsen PA, Rasmussen KJ, Andersen JR, Johansen LE, Hansen LH, Jespersen H, Podtelejnikov A, Nielsen E, Crawford J, Poulsen V, Sørensen BD, Matthiesen J, Hendrickson RC, Gleeson F, Pawson T, Moran MF, Durocher D, Mann M, Hogue CW, Figeys D, Tyers M (2002) Systematic identification of protein complexes in *Saccharomyces cerevisiae* by mass spectrometry. Nature 415:180–183
69. Archambault V, Chang EJ, Drapkin BJ, Cross FR, Chait BT, Rout MP (2004) Targeted proteomic study of the cyclin-Cdk module. Mol Cell 14:699–711
70. Shima F, Okada T, Kido M, Sen H, Tanaka Y, Tamada M, Hu C, Yamawaki-Kataoka Y, Kariya K, Kataoka T (2000) Association of yeast adenylyl cyclase with cyclase-associated protein CAP forms a second Ras-binding site which mediates its Ras-dependent activation. Mol Cell Biol 20:26–33
71. Field J, Nikawa J, Broek D, MacDonald B, Rodgers L, Wilson I, Lerner R, Wigler M (1988) Purification of a RAS-responsive adenylyl cyclase complex from *Saccharomyces cerevisiae* by use of an epitope addition method. Mol Cell Biol 8:2159–2165
72. Belotti F, Tisi R, Paiardi C, Groppi S, Martegani E (2011) PKA-dependent regulation of Cdc25 RasGEF localization in budding yeast. FEBS Lett 585:3914–3920
73. Chiu V, Bivona T, Hach A, Sajous J, Silletti J, Wiener H, Johnson RL, Cox A, Philips M (2002) Ras signalling on the endoplasmic reticulum and the Golgi. Nat Cell Biol 4:343–350
74. Bivona T, Quatela S, Bodemann B, Ahearn I, Soskis M, Mor A, Miura J, Wiener H, Wright L, Saba S, Yim D, Fein A, Pérez de Castro I, Li C, Thompson C, Cox A, Philips M (2006) PKC regulates a farnesyl-electrostatic switch on K-Ras that promotes its association with Bcl-XL on mitochondria and induces apoptosis. Mol Cell 21:481–493
75. Chang E, Philips M (2006) Spatial segregation of Ras signaling: new evidence from fission yeast. Cell Cycle 5:1936–1939
76. Omerovic J, Prior IA (2009) Compartmentalized signalling: Ras proteins and signalling nanoclusters. FEBS J 276: 1817–1825
77. Tisi R, Belotti F, Paiardi C, Brunetti F, Martegani E (2008) The budding yeast RasGEF Cdc25 reveals an unexpected nuclear localization. Biochim Biophys Acta 1783:2363–2374
78. Broggi S, Martegani E, Colombo S (2013) Live-cell imaging of endogenous Ras-GTP shows predominant Ras activation at the plasma membrane and in the nucleus in *Saccharomyces cerevisiae*. Int J Biochem Cell Biol 45:384–394
79. Gross A, Winograd S, Marbach I, Levitzki A (1999) The N-terminal half of Cdc25 is essential for processing glucose signaling in *Saccharomyces cerevisiae*. Biochemistry 38: 13252–13262
80. Li C, Cheng Y, Gutmann D, Mangoura D (2001) Differential localization of the neurofibromatosis 1 (NF1) gene product, neurofibromin, with the F-actin or microtubule cytoskeleton during differentiation of telencephalic neurons. Brain Res Dev Brain Res 130:231–248
81. Sutherland BW, Spiegelman GB, Weeks G (2001) A Ras subfamily GTPase shows cell cycle-dependent nuclear localization. EMBO Rep 2:1024–1028
82. Fuentes-Calvo I, Blázquez-Medela AM, Santos E, López-Novoa JM, Martínez-Salgado C (2010) Analysis of k-ras nuclear expression in fibroblasts and mesangial cells. PLoS One 5:e8703
83. Contente S, Yeh TJ, Friedman RM (2011) H-ras localizes to cell nuclei and varies with the cell cycle. Genes Cancer 2:166–172
84. Bhattacharya S, Chen L, Broach J, Powers S (1995) Ras membrane targeting is essential for glucose signaling but not for viability in yeast. Proc Natl Acad Sci U S A 92:2984–2988
85. Robinson L, Tatchell K (1991) TFS1: a suppressor of cdc25 mutations in *Saccharomyces cerevisiae*. Mol Gen Genet 230:241–250
86. Chautard H, Jacquet M, Schoentgen F, Bureaud N, Bénédetti H (2004) Tfs1p, a member of the PEBP family, inhibits the

Ira2p but not the Ira1p Ras GTPase-activating protein in *Saccharomyces cerevisiae*. Eukaryot Cell 3:459–470
87. Mima J, Fukada H, Nagayama M, Ueda M (2006) Specific membrane binding of the carboxypeptidase Y inhibitor I(C), a phosphatidylethanolamine-binding protein family member. FEBS J 273:5374–5383
88. Bruun A, Svendsen I, Sørensen S, Kielland-Brandt M, Winther J (1998) A high-affinity inhibitor of yeast carboxypeptidase Y is encoded by TFS1 and shows homology to a family of lipid binding proteins. Biochemistry 37:3351–3357
89. Gombault A, Godin F, Sy D, Legrand B, Chautard H, Vallée B, Vovelle F, Bénédetti H (2007) Molecular basis of the Tfs1/Ira2 interaction: a combined protein engineering and molecular modelling study. J Mol Biol 374:604–617
90. Fukada H, Mima J, Nagayama M, Kato M, Ueda M (2007) Biochemical analysis of the yeast proteinase inhibitor (IC) homolog ICh and its comparison with IC. Biosci Biotechnol Biochem 71:472–480
91. Shirayama M, Matsui Y, Tanaka K, Toh-e A (1994) Isolation of a CDC25 family gene, MSI2/LTE1, as a multicopy suppressor of ira1. Yeast 10:451–461
92. Yoshida S, Ichihashi R, Toh-e A (2003) Ras recruits mitotic exit regulator Lte1 to the bud cortex in budding yeast. J Cell Biol 161:889–897
93. Seshan A, Amon A (2005) Ras and the Rho effector Cla4 collaborate to target and anchor Lte1 at the bud cortex. Cell Cycle 4:940–946
94. Mösch HU, Roberts RL, Fink GR (1996) Ras2 signals via the Cdc42/Ste20/mitogen-activated protein kinase module to induce filamentous growth in *Saccharomyces cerevisiae*. Proc Natl Acad Sci U S A 93:5352–5356
95. Mösch HU, Kübler E, Krappmann S, Fink GR, Braus GH (1999) Crosstalk between the Ras2p-controlled mitogen-activated protein kinase and cAMP pathways during invasive growth of *Saccharomyces cerevisiae*. Mol Biol Cell 10:1325–1335
96. Geymonat M, Spanos A, Jensen S, Sedgwick SG (2010) Phosphorylation of Lte1 by Cdk prevents polarized growth during mitotic arrest in *S. cerevisiae*. J Cell Biol 191:1097–1112
97. Thevelein J, de Winde J (1999) Novel sensing mechanisms and targets for the cAMP-protein kinase A pathway in the yeast *Saccharomyces cerevisiae*. Mol Microbiol 33:904–918
98. Busti S, Coccetti P, Alberghina L, Vanoni M (2010) Glucose signaling-mediated coordination of cell growth and cell cycle in *Saccharomyces cerevisiae*. Sensors (Basel) 10:6195–6240
99. Zaman S, Lippman SI, Schneper L, Slonim N, Broach JR (2009) Glucose regulates transcription in yeast through a network of signaling pathways. Mol Syst Biol 5:245
100. Rubio-Texeira M, Van Zeebroeck G, Voordeckers K, Thevelein JM (2010) *Saccharomyces cerevisiae* plasma membrane nutrient sensors and their role in PKA signaling. FEMS Yeast Res 10:134–149
101. Kriel J, Haesendonckx S, Rubio-Texeira M, Van Zeebroeck G, Thevelein JM (2011) From transporter to transceptor: signaling from transporters provokes re-evaluation of complex trafficking and regulatory controls: endocytic internalization and intracellular trafficking of nutrient transceptors may, at least in part, be governed by their signaling function. Bioessays 33:870–879
102. Thevelein J (1991) Fermentable sugars and intracellular acidification as specific activators of the RAS-adenylate cyclase signalling pathway in yeast: the relationship to nutrient-induced cell cycle control. Mol Microbiol 5:1301–1307
103. Nakafuku M, Obara T, Kaibuchi K, Miyajima I, Miyajima A, Itoh H, Nakamura S, Arai K, Matsumoto K, Kaziro Y (1988) Isolation of a second yeast *Saccharomyces cerevisiae* gene (GPA2) coding for guanine nucleotide-binding regulatory protein: studies on its structure and possible functions. Proc Natl Acad Sci U S A 85:1374–1378
104. Ma P, Wera S, Van Dijck P, Thevelein J (1999) The PDE1-encoded low-affinity phosphodiesterase in the yeast *Saccharomyces cerevisiae* has a specific function in controlling agonist-induced cAMP signaling. Mol Biol Cell 10:91–104
105. Sass P, Field J, Nikawa J, Toda T, Wigler M (1986) Cloning and characterization of the high-affinity cAMP phosphodiesterase of *Saccharomyces cerevisiae*. Proc Natl Acad Sci U S A 83:9303–9307
106. Kido M, Shima F, Satoh T, Asato T, Kariya K, Kataoka T (2002) Critical function of the Ras-associating domain as a primary Ras-binding site for regulation of *Saccharomyces cerevisiae* adenylyl cyclase. J Biol Chem 277:3117–3123
107. Xue Y, Batlle M, Hirsch JP (1998) GPR1 encodes a putative G protein-coupled receptor that associates with the Gpa2p Galpha subunit and functions in a Ras-independent pathway. EMBO J 17:1996–2007
108. Kraakman L, Lemaire K, Ma P, Teunissen A, Donaton M, Van Dijck P, Winderickx J, de Winde J, Thevelein J (1999) A *Saccharomyces cerevisiae* G-protein coupled receptor, Gpr1, is

specifically required for glucose activation of the cAMP pathway during the transition to growth on glucose. Mol Microbiol 32:1002–1012
109. Rolland F, De Winde J, Lemaire K, Boles E, Thevelein J, Winderickx J (2000) Glucose-induced cAMP signalling in yeast requires both a G-protein coupled receptor system for extracellular glucose detection and a separable hexose kinase-dependent sensing process. Mol Microbiol 38:348–358
110. Lemaire K, Van de Velde S, Van Dijck P, Thevelein J (2004) Glucose and sucrose act as agonist and mannose as antagonist ligands of the G protein-coupled receptor Gpr1 in the yeast *Saccharomyces cerevisiae*. Mol Cell 16:293–299
111. Beullens M, Mbonyi K, Geerts L, Gladines D, Detremerie K, Jans A, Thevelein J (1988) Studies on the mechanism of the glucose-induced cAMP signal in glycolysis and glucose repression mutants of the yeast *Saccharomyces cerevisiae*. Eur J Biochem 172:227–231
112. Rolland F, Wanke V, Cauwenberg L, Ma P, Boles E, Vanoni M, de Winde J, Thevelein J, Winderickx J (2001) The role of hexose transport and phosphorylation in cAMP signalling in the yeast *Saccharomyces cerevisiae*. FEMS Yeast Res 1:33–45
113. Wang Y, Pierce M, Schneper L, Güldal CG, Zhang X, Tavazoie S, Broach JR (2004) Ras and Gpa2 mediate one branch of a redundant glucose signaling pathway in yeast. PLoS Biol 2:E128
114. Zaman S, Lippman SI, Zhao X, Broach JR (2008) How Saccharomyces responds to nutrients. Annu Rev Genet 42:27–81
115. Harashima T, Heitman J (2005) Galpha subunit Gpa2 recruits kelch repeat subunits that inhibit receptor-G protein coupling during cAMP-induced dimorphic transitions in *Saccharomyces cerevisiae*. Mol Biol Cell 16:4557–4571
116. Batlle M, Lu A, Green DA, Xue Y, Hirsch JP (2003) Krh1p and Krh2p act downstream of the Gpa2p G(alpha) subunit to negatively regulate haploid invasive growth. J Cell Sci 116:701–710
117. Lu A, Hirsch JP (2005) Cyclic AMP-independent regulation of protein kinase A substrate phosphorylation by Kelch repeat proteins. Eukaryot Cell 4:1794–1800
118. Harashima T, Anderson S, Yates J Jr, Heitman J (2006) The kelch proteins Gpb1 and Gpb2 inhibit Ras activity via association with the yeast RasGAP neurofibromin homologs Ira1 and Ira2. Mol Cell 22:819–830
119. Peeters T, Louwet W, Geladé R, Nauwelaers D, Thevelein J, Versele M (2006) Kelch-repeat proteins interacting with the Galpha protein Gpa2 bypass adenylate cyclase for direct regulation of protein kinase A in yeast. Proc Natl Acad Sci U S A 103:13034–13039
120. Budhwar R, Lu A, Hirsch JP (2010) Nutrient control of yeast PKA activity involves opposing effects on phosphorylation of the Bcy1 regulatory subunit. Mol Biol Cell 21:3749–3758
121. Phan VT, Ding VW, Li F, Chalkley RJ, Burlingame A, McCormick F (2010) The RasGAP proteins Ira2 and neurofibromin are negatively regulated by Gpb1 in yeast and ETEA in humans. Mol Cell Biol 30:2264–2279
122. Toda T, Cameron S, Sass P, Zoller M, Wigler M (1987) Three different genes in *S. cerevisiae* encode the catalytic subunits of the cAMP-dependent protein kinase. Cell 50:277–287
123. Robertson LS, Fink GR (1998) The three yeast A kinases have specific signaling functions in pseudohyphal growth. Proc Natl Acad Sci U S A 95:13783–13787
124. Robertson LS, Causton HC, Young RA, Fink GR (2000) The yeast A kinases differentially regulate iron uptake and respiratory function. Proc Natl Acad Sci U S A 97:5984–5988
125. Pan X, Heitman J (2002) Protein kinase A operates a molecular switch that governs yeast pseudohyphal differentiation. Mol Cell Biol 22:3981–3993
126. Chevtzoff C, Vallortigara J, Avéret N, Rigoulet M, Devin A (2005) The yeast cAMP protein kinase Tpk3p is involved in the regulation of mitochondrial enzymatic content during growth. Biochim Biophys Acta 1706:117–125
127. Palomino A, Herrero P, Moreno F (2006) Tpk3 and Snf1 protein kinases regulate Rgt1 association with *Saccharomyces cerevisiae* HXK2 promoter. Nucleic Acids Res 34:1427–1438
128. Voordeckers K, Kimpe M, Haesendonckx S, Louwet W, Versele M, Thevelein JM (2011) Yeast 3-phosphoinositide-dependent protein kinase-1 (PDK1) orthologs Pkh1-3 differentially regulate phosphorylation of protein kinase A (PKA) and the protein kinase B (PKB)/S6K ortholog Sch9. J Biol Chem 286:22017–22027
129. Haesendonckx S, Tudisca V, Voordeckers K, Moreno S, Thevelein JM, Portela P (2012) The activation loop of PKA catalytic isoforms is differentially phosphorylated by Pkh protein kinases in *Saccharomyces cerevisiae*. Biochem J 448:307–320
130. Pan X, Heitman J (1999) Cyclic AMP-dependent protein kinase regulates pseudohyphal differentiation in *Saccharomyces cerevisiae*. Mol Cell Biol 19:4874–4887
131. De Virgilio C (2012) The essence of yeast quiescence. FEMS Microbiol Rev 36:306–339
132. Thevelein J, Cauwenberg L, Colombo S, De Winde JH, Donation M, Dumortier F, Kraakman L, Lemaire K, Ma P, Nauwelaers

D, Rolland F, Teunissen A, Van Dijck P, Versele M, Wera S, Winderickx J (2000) Nutrient-induced signal transduction through the protein kinase A pathway and its role in the control of metabolism, stress resistance, and growth in yeast. Enzyme Microb Technol 26:819–825

133. Rolland F, Winderickx J, Thevelein J (2002) Glucose-sensing and -signalling mechanisms in yeast. FEMS Yeast Res 2:183–201
134. Jungbluth M, Mösch HU, Taxis C (2012) Acetate regulation of spore formation is under the control of the Ras/cyclic AMP/protein kinase A pathway and carbon dioxide in *Saccharomyces cerevisiae*. Eukaryot Cell 11:1021–1032
135. Honigberg SM, Purnapatre K (2003) Signal pathway integration in the switch from the mitotic cell cycle to meiosis in yeast. J Cell Sci 116:2137–2147
136. Uno I, Oshima T, Hirata A, Ishikawa T (1990) The functional domain of adenylate cyclase associated with entry into meiosis in *Saccharomyces cerevisiae*. J Bacteriol 172: 102–109
137. Aoh QL, Graves LM, Duncan MC (2011) Glucose regulates clathrin adaptors at the trans-Golgi network and endosomes. Mol Biol Cell 22:3671–3683
138. Budovskaya YV, Stephan JS, Reggiori F, Klionsky DJ, Herman PK (2004) The Ras/cAMP-dependent protein kinase signaling pathway regulates an early step of the autophagy process in *Saccharomyces cerevisiae*. J Biol Chem 279:20663–20671
139. Schmelzle T, Beck T, Martin DE, Hall MN (2004) Activation of the RAS/cyclic AMP pathway suppresses a TOR deficiency in yeast. Mol Cell Biol 24:338–351
140. Yorimitsu T, Klionsky DJ (2005) Autophagy: molecular machinery for self-eating. Cell Death Differ 12(Suppl 2):1542–1552
141. Martegani E, Vanoni M, Baroni M (1984) Macromolecular syntheses in the cell cycle mutant cdc25 of budding yeast. Eur J Biochem 144:205–210
142. Van Dijck P, Colavizza D, Smet P, Thevelein J (1995) Differential importance of trehalose in stress resistance in fermenting and nonfermenting *Saccharomyces cerevisiae* cells. Appl Environ Microbiol 61:109–115
143. Thompson-Jaeger S, François J, Gaughran JP, Tatchell K (1991) Deletion of SNF1 affects the nutrient response of yeast and resembles mutations which activate the adenylate cyclase pathway. Genetics 129:697–706
144. Hedbacker K, Townley R, Carlson M (2004) Cyclic AMP-dependent protein kinase regulates the subcellular localization of Snf1-Sip1 protein kinase. Mol Cell Biol 24:1836–1843
145. Barrett L, Orlova M, Maziarz M, Kuchin S (2012) Protein kinase A contributes to the negative control of Snf1 protein kinase in *Saccharomyces cerevisiae*. Eukaryot Cell 11: 119–128
146. Drebot M, Barnes C, Singer R, Johnston G (1990) Genetic assessment of stationary phase for cells of the yeast *Saccharomyces cerevisiae*. J Bacteriol 172:3584–3589
147. Thevelein J (1992) The RAS-adenylate cyclase pathway and cell cycle control in *Saccharomyces cerevisiae*. Antonie Van Leeuwenhoek 62:109–130
148. Anghileri P, Branduardi P, Sternieri F, Monti P, Visintin R, Bevilacqua A, Alberghina L, Martegani E, Baroni MD (1999) Chromosome separation and exit from mitosis in budding yeast: dependence on growth revealed by cAMP-mediated inhibition. Exp Cell Res 250:510–523
149. Baroni MD, Monti P, Alberghina L (1994) Repression of growth-regulated G1 cyclin expression by cyclic AMP in budding yeast. Nature 371:339–342
150. Polymenis M, Schmidt EV (1997) Coupling of cell division to cell growth by translational control of the G1 cyclin CLN3 in yeast. Genes Dev 11:2522–2531
151. Hall DD, Markwardt DD, Parviz F, Heideman W (1998) Regulation of the Cln3-Cdc28 kinase by cAMP in *Saccharomyces cerevisiae*. EMBO J 17:4370–4378
152. Tokiwa G, Tyers M, Volpe T, Futcher B (1994) Inhibition of G1 cyclin activity by the Ras/cAMP pathway in yeast. Nature 371: 342–345
153. Hubler L, Bradshaw-Rouse J, Heideman W (1993) Connections between the Ras-cyclic AMP pathway and G1 cyclin expression in the budding yeast *Saccharomyces cerevisiae*. Mol Cell Biol 13:6274–6282
154. Baroni M, Martegani E, Monti P, Alberghina L (1989) Cell size modulation by CDC25 and RAS2 genes in *Saccharomyces cerevisiae*. Mol Cell Biol 9:2715–2723
155. Baroni MD, Monti P, Marconi G, Alberghina L (1992) cAMP-mediated increase in the critical cell size required for the G1 to S transition in *Saccharomyces cerevisiae*. Exp Cell Res 201:299–306
156. Mitsuzawa H (1994) Increases in cell size at START caused by hyperactivation of the cAMP pathway in *Saccharomyces cerevisiae*. Mol Gen Genet 243:158–165
157. Santangelo GM (2006) Glucose signaling in *Saccharomyces cerevisiae*. Microbiol Mol Biol Rev 70:253–282
158. Searle JS, Schollaert KL, Wilkins BJ, Sanchez Y (2004) The DNA damage checkpoint and PKA pathways converge on APC substrates

and Cdc20 to regulate mitotic progression. Nat Cell Biol 6:138–145
159. Wood MD, Sanchez Y (2010) Deregulated Ras signaling compromises DNA damage checkpoint recovery in *S. cerevisiae*. Cell Cycle 9:3353–3363
160. Dejean L, Beauvoit B, Bunoust O, Guérin B, Rigoulet M (2002) Activation of Ras cascade increases the mitochondrial enzyme content of respiratory competent yeast. Biochem Biophys Res Commun 293:1383–1388
161. Dejean L, Beauvoit B, Alonso AP, Bunoust O, Guérin B, Rigoulet M (2002) cAMP-induced modulation of the growth yield of *Saccharomyces cerevisiae* during respiratory and respiro-fermentative metabolism. Biochim Biophys Acta 1554:159–169
162. Hlavatá L, Nyström T (2003) Ras proteins control mitochondrial biogenesis and function in *Saccharomyces cerevisiae*. Folia Microbiol (Praha) 48:725–730
163. Yoboue ED, Augier E, Galinier A, Blancard C, Pinson B, Casteilla L, Rigoulet M, Devin A (2012) cAMP-induced mitochondrial compartment biogenesis: role of glutathione redox state. J Biol Chem 287:14569–14578
164. Hlavatá L, Aguilaniu H, Pichová A, Nyström T (2003) The oncogenic RAS2(val19) mutation locks respiration, independently of PKA, in a mode prone to generate ROS. EMBO J 22:3337–3345
165. Heeren G, Jarolim S, Laun P, Rinnerthaler M, Stolze K, Perrone GG, Kohlwein SD, Nohl H, Dawes IW, Breitenbach M (2004) The role of respiration, reactive oxygen species and oxidative stress in mother cell-specific ageing of yeast strains defective in the RAS signalling pathway. FEMS Yeast Res 5:157–167
166. Gourlay CW, Ayscough KR (2006) Actin-induced hyperactivation of the Ras signaling pathway leads to apoptosis in *Saccharomyces cerevisiae*. Mol Cell Biol 26:6487–6501
167. Leadsham JE, Gourlay CW (2010) cAMP/PKA signaling balances respiratory activity with mitochondria dependent apoptosis via transcriptional regulation. BMC Cell Biol 11:92
168. Klein C, Struhl K (1994) Protein kinase A mediates growth-regulated expression of yeast ribosomal protein genes by modulating RAP1 transcriptional activity. Mol Cell Biol 14:1920–1928
169. Neuman-Silberberg FS, Bhattacharya S, Broach JR (1995) Nutrient availability and the RAS/cyclic AMP pathway both induce expression of ribosomal protein genes in *Saccharomyces cerevisiae* but by different mechanisms. Mol Cell Biol 15:3187–3196
170. Jorgensen P, Rupes I, Sharom JR, Schneper L, Broach JR, Tyers M (2004) A dynamic transcriptional network communicates growth potential to ribosome synthesis and critical cell size. Genes Dev 18:2491–2505
171. Zurita-Martinez SA, Cardenas ME (2005) Tor and cyclic AMP-protein kinase A: two parallel pathways regulating expression of genes required for cell growth. Eukaryot Cell 4:63–71
172. Marion RM, Regev A, Segal E, Barash Y, Koller D, Friedman N, O'Shea EK (2004) Sfp1 is a stress- and nutrient-sensitive regulator of ribosomal protein gene expression. Proc Natl Acad Sci U S A 101:14315–14322
173. Görner W, Durchschlag E, Martinez-Pastor MT, Estruch F, Ammerer G, Hamilton B, Ruis H, Schüller C (1998) Nuclear localization of the C2H2 zinc finger protein Msn2p is regulated by stress and protein kinase A activity. Genes Dev 12:586–597
174. Jacquet M, Renault G, Lallet S, De Mey J, Goldbeter A (2003) Oscillatory nucleocytoplasmic shuttling of the general stress response transcriptional activators Msn2 and Msn4 in *Saccharomyces cerevisiae*. J Cell Biol 161: 497–505
175. Garmendia-Torres C, Goldbeter A, Jacquet M (2007) Nucleocytoplasmic oscillations of the yeast transcription factor Msn2: evidence for periodic PKA activation. Curr Biol 17:1044–1049
176. Martínez-Pastor MT, Marchler G, Schüller C, Marchler-Bauer A, Ruis H, Estruch F (1996) The *Saccharomyces cerevisiae* zinc finger proteins Msn2p and Msn4p are required for transcriptional induction through the stress response element (STRE). EMBO J 15: 2227–2235
177. Pedruzzi I, Dubouloz F, Cameroni E, Wanke V, Roosen J, Winderickx J, De Virgilio C (2003) TOR and PKA signaling pathways converge on the protein kinase Rim15 to control entry into G0. Mol Cell 12:1607–1613
178. Ferguson SB, Anderson ES, Harshaw RB, Thate T, Craig NL, Nelson HC (2005) Protein kinase A regulates constitutive expression of small heat-shock genes in an Msn2/4p-independent and Hsf1p-dependent manner in *Saccharomyces cerevisiae*. Genetics 169:1203–1214
179. Wei M, Fabrizio P, Hu J, Ge H, Cheng C, Li L, Longo VD (2008) Life span extension by calorie restriction depends on Rim15 and transcription factors downstream of Ras/PKA, Tor, and Sch9. PLoS Genet 4:e13
180. Moskvina E, Schüller C, Maurer CT, Mager WH, Ruis H (1998) A search in the genome of *Saccharomyces cerevisiae* for genes regulated via stress response elements. Yeast 14:1041–1050
181. Boy-Marcotte E, Lagniel G, Perrot M, Bussereau F, Boudsocq A, Jacquet M, Labarre J

(1999) The heat shock response in yeast: differential regulations and contributions of the Msn2p/Msn4p and Hsf1p regulons. Mol Microbiol 33:274–283
182. Gasch AP, Spellman PT, Kao CM, Carmel-Harel O, Eisen MB, Storz G, Botstein D, Brown PO (2000) Genomic expression programs in the response of yeast cells to environmental changes. Mol Biol Cell 11:4241–4257
183. Boisnard S, Lagniel G, Garmendia-Torres C, Molin M, Boy-Marcotte E, Jacquet M, Toledano MB, Labarre J, Chédin S (2009) H2O2 activates the nuclear localization of Msn2 and Maf1 through thioredoxins in *Saccharomyces cerevisiae*. Eukaryot Cell 8: 1429–1438
184. Görner W, Durchschlag E, Wolf J, Brown E, Ammerer G, Ruis H, Schüller C (2002) Acute glucose starvation activates the nuclear localization signal of a stress-specific yeast transcription factor. EMBO J 21:135–144
185. Jacquet M, Renault G, Lallet S, De Mey J, Goldbeter A (2003) Oscillatory behavior of the nuclear localization of the transcription factors Msn2 and Msn4 in response to stress in yeast. Sci World J 3:609–612
186. Vandamme J, Castermans D, Thevelein JM (2012) Molecular mechanisms of feedback inhibition of protein kinase A on intracellular cAMP accumulation. Cell Signal 24: 1610–1618
187. Besozzi D, Cazzaniga P, Pescini D, Mauri G, Colombo S, Martegani E (2012) The role of feedback control mechanisms on the establishment of oscillatory regimes in the Ras/cAMP/PKA pathway in *S. cerevisiae*. EURASIP J Bioinform Syst Biol 2012:10
188. Lee P, Cho BR, Joo HS, Hahn JS (2008) Yeast Yak1 kinase, a bridge between PKA and stress-responsive transcription factors, Hsf1 and Msn2/Msn4. Mol Microbiol 70:882–895
189. Garrett S, Broach J (1989) Loss of Ras activity in *Saccharomyces cerevisiae* is suppressed by disruptions of a new kinase gene, YAKI, whose product may act downstream of the cAMP-dependent protein kinase. Genes Dev 3:1336–1348
190. Garrett S, Menold MM, Broach JR (1991) The *Saccharomyces cerevisiae* YAK1 gene encodes a protein kinase that is induced by arrest early in the cell cycle. Mol Cell Biol 11:4045–4052
191. Moriya H, Shimizu-Yoshida Y, Omori A, Iwashita S, Katoh M, Sakai A (2001) Yak1p, a DYRK family kinase, translocates to the nucleus and phosphorylates yeast Pop2p in response to a glucose signal. Genes Dev 15:1217–1228
192. Griffioen G, Branduardi P, Ballarini A, Anghileri P, Norbeck J, Baroni MD, Ruis H (2001) Nucleocytoplasmic distribution of budding yeast protein kinase A regulatory subunit Bcy1 requires Zds1 and is regulated by Yak1-dependent phosphorylation of its targeting domain. Mol Cell Biol 21:511–523
193. Griffioen G, Anghileri P, Imre E, Baroni MD, Ruis H (2000) Nutritional control of nucleocytoplasmic localization of cAMP-dependent protein kinase catalytic and regulatory subunits in *Saccharomyces cerevisiae*. J Biol Chem 275:1449–1456
194. Griffioen G, Thevelein J (2002) Molecular mechanisms controlling the localisation of protein kinase A. Curr Genet 41:199–207
195. Zarzov P, Boucherie H, Mann C (1997) A yeast heat shock transcription factor (Hsf1) mutant is defective in both Hsc82/Hsp82 synthesis and spindle pole body duplication. J Cell Sci 110(Pt 16):1879–1891
196. Hahn JS, Hu Z, Thiele DJ, Iyer VR (2004) Genome-wide analysis of the biology of stress responses through heat shock transcription factor. Mol Cell Biol 24:5249–5256
197. Imazu H, Sakurai H (2005) *Saccharomyces cerevisiae* heat shock transcription factor regulates cell wall remodeling in response to heat shock. Eukaryot Cell 4:1050–1056
198. Eastmond DL, Nelson HC (2006) Genome-wide analysis reveals new roles for the activation domains of the *Saccharomyces cerevisiae* heat shock transcription factor (Hsf1) during the transient heat shock response. J Biol Chem 281:32909–32921
199. Causton HC, Ren B, Koh SS, Harbison CT, Kanin E, Jennings EG, Lee TI, True HL, Lander ES, Young RA (2001) Remodeling of yeast genome expression in response to environmental changes. Mol Biol Cell 12:323–337
200. Yamamoto N, Maeda Y, Ikeda A, Sakurai H (2008) Regulation of thermotolerance by stress-induced transcription factors in *Saccharomyces cerevisiae*. Eukaryot Cell 7: 783–790
201. Reinders A, Bürckert N, Boller T, Wiemken A, De Virgilio C (1998) *Saccharomyces cerevisiae* cAMP-dependent protein kinase controls entry into stationary phase through the Rim15p protein kinase. Genes Dev 12: 2943–2955
202. Swinnen E, Wanke V, Roosen J, Smets B, Dubouloz F, Pedruzzi I, Cameroni E, De Virgilio C, Winderickx J (2006) Rim15 and the crossroads of nutrient signalling pathways in *Saccharomyces cerevisiae*. Cell Div 1:3
203. Cameroni E, Hulo N, Roosen J, Winderickx J, De Virgilio C (2004) The novel yeast PAS kinase Rim 15 orchestrates G0-associated

antioxidant defense mechanisms. Cell Cycle 3:462–468
204. Roosen J, Engelen K, Marchal K, Mathys J, Griffioen G, Cameroni E, Thevelein J, De Virgilio C, De Moor B, Winderickx J (2005) PKA and Sch9 control a molecular switch important for the proper adaptation to nutrient availability. Mol Microbiol 55:862–880
205. Orzechowski Westholm J, Tronnersjö S, Nordberg N, Olsson I, Komorowski J, Ronne H (2012) Gis1 and Rph1 regulate glycerol and acetate metabolism in glucose depleted yeast cells. PLoS One 7:e31577
206. Pedruzzi I, Bürckert N, Egger P, De Virgilio C (2000) *Saccharomyces cerevisiae* Ras/cAMP pathway controls post-diauxic shift element-dependent transcription through the zinc finger protein Gis1. EMBO J 19:2569–2579
207. Zhang N, Wu J, Oliver SG (2009) Gis1 is required for transcriptional reprogramming of carbon metabolism and the stress response during transition into stationary phase in yeast. Microbiology 155:1690–1698
208. Wanke V, Pedruzzi I, Cameroni E, Dubouloz F, De Virgilio C (2005) Regulation of G0 entry by the Pho80-Pho85 cyclin-CDK complex. EMBO J 24:4271–4278
209. Netea MG, Brown GD, Kullberg BJ, Gow NA (2008) An integrated model of the recognition of Candida albicans by the innate immune system. Nat Rev Microbiol 6:67–78
210. Gastebois A, Clavaud C, Aimanianda V, Latgé JP (2009) Aspergillus fumigatus: cell wall polysaccharides, their biosynthesis and organization. Future Microbiol 4:583–595
211. Ben-Ami R, Weinberger M, Orni-Wasserlauff R, Schwartz D, Itzhaki A, Lazarovitch T, Bash E, Aharoni Y, Moroz I, Giladi M (2008) Time to blood culture positivity as a marker for catheter-related candidemia. J Clin Microbiol 46:2222–2226
212. Gimeno CJ, Ljungdahl PO, Styles CA, Fink GR (1992) Unipolar cell divisions in the yeast *S. cerevisiae* lead to filamentous growth: regulation by starvation and RAS. Cell 68:1077–1090
213. Gancedo JM (2001) Control of pseudohyphae formation in *Saccharomyces cerevisiae*. FEMS Microbiol Rev 25:107–123
214. Palecek SP, Parikh AS, Kron SJ (2000) Genetic analysis reveals that FLO11 upregulation and cell polarization independently regulate invasive growth in *Saccharomyces cerevisiae*. Genetics 156:1005–1023
215. Cullen PJ, Sprague GF (2012) The regulation of filamentous growth in yeast. Genetics 190:23–49
216. Boeckstaens M, André B, Marini AM (2008) Distinct transport mechanisms in yeast ammonium transport/sensor proteins of the Mep/Amt/Rh family and impact on filamentation. J Biol Chem 283:21362–21370
217. Cullen PJ, Sprague GF (2000) Glucose depletion causes haploid invasive growth in yeast. Proc Natl Acad Sci U S A 97:13619–13624
218. Medintz IL, Vora GJ, Rahbar AM, Thach DC (2007) Transcript and proteomic analyses of wild-type and gpa2 mutant *Saccharomyces cerevisiae* strains suggest a role for glycolytic carbon source sensing in pseudohyphal differentiation. Mol Biosyst 3:623–634
219. Iyer RS, Das M, Bhat PJ (2008) Pseudohyphal differentiation defect due to mutations in GPCR and ammonium signaling is suppressed by low glucose concentration: a possible integrated role for carbon and nitrogen limitation. Curr Genet 54:71–81
220. Chen RE, Thorner J (2010) Systematic epistasis analysis of the contributions of protein kinase A- and mitogen-activated protein kinase-dependent signaling to nutrient limitation-evoked responses in the yeast *Saccharomyces cerevisiae*. Genetics 185:855–870
221. Bharucha N, Ma J, Dobry CJ, Lawson SK, Yang Z, Kumar A (2008) Analysis of the yeast kinome reveals a network of regulated protein localization during filamentous growth. Mol Biol Cell 19:2708–2717
222. Rupp S, Summers E, Lo HJ, Madhani H, Fink G (1999) MAP kinase and cAMP filamentation signaling pathways converge on the unusually large promoter of the yeast FLO11 gene. EMBO J 18:1257–1269
223. Zebisch A, Czernilofsky AP, Keri G, Smigelskaite J, Sill H, Troppmair J (2007) Signaling through RAS-RAF-MEK-ERK: from basics to bedside. Curr Med Chem 14:601–623
224. Cullen PJ, Sabbagh W, Graham E, Irick MM, van Olden EK, Neal C, Delrow J, Bardwell L, Sprague GF (2004) A signaling mucin at the head of the Cdc42- and MAPK-dependent filamentous growth pathway in yeast. Genes Dev 18:1695–1708
225. Chavel CA, Dionne HM, Birkaya B, Joshi J, Cullen PJ (2010) Multiple signals converge on a differentiation MAPK pathway. PLoS Genet 6:e1000883
226. Truckses DM, Bloomekatz JE, Thorner J (2006) The RA domain of Ste50 adaptor protein is required for delivery of Ste11 to the plasma membrane in the filamentous growth signaling pathway of the yeast *Saccharomyces cerevisiae*. Mol Cell Biol 26:912–928
227. Kuchin S, Vyas VK, Carlson M (2002) Snf1 protein kinase and the repressors Nrg1 and Nrg2 regulate FLO11, haploid invasive growth, and diploid pseudohyphal differentiation. Mol Cell Biol 22:3994–4000

228. Kübler E, Mösch HU, Rupp S, Lisanti MP (1997) Gpa2p, a G-protein alpha-subunit, regulates growth and pseudohyphal development in *Saccharomyces cerevisiae* via a cAMP-dependent mechanism. J Biol Chem 272: 20321–20323
229. Lorenz M, Heitman J (1997) Yeast pseudohyphal growth is regulated by GPA2, a G protein alpha homolog. EMBO J 16:7008–7018
230. Lorenz MC, Pan X, Harashima T, Cardenas ME, Xue Y, Hirsch JP, Heitman J (2000) The G protein-coupled receptor gpr1 is a nutrient sensor that regulates pseudohyphal differentiation in *Saccharomyces cerevisiae*. Genetics 154:609–622
231. Van de Velde S, Thevelein J (2008) Cyclic AMP-protein kinase A and Snf1 signaling mechanisms underlie the superior potency of sucrose for induction of filamentation in *Saccharomyces cerevisiae*. Eukaryot Cell 7: 286–293
232. Nadin-Davis SA, Yang RC, Narang SA, Nasim A (1986) The cloning and characterization of a RAS gene from *Schizosaccharomyces pombe*. J Mol Evol 23:41–51
233. Nadin-Davis SA, Nasim A, Beach D (1986) Involvement of ras in sexual differentiation but not in growth control in fission yeast. EMBO J 5:2963–2971
234. Chang EC, Barr M, Wang Y, Jung V, Xu HP, Wigler MH (1994) Cooperative interaction of *S. pombe* proteins required for mating and morphogenesis. Cell 79:131–141
235. Hirota K, Tanaka K, Ohta K, Yamamoto M (2003) Gef1p and Scd1p, the Two GDP-GTP exchange factors for Cdc42p, form a ring structure that shrinks during cytokinesis in *Schizosaccharomyces pombe*. Mol Biol Cell 14:3617–3627
236. Verde F, Wiley DJ, Nurse P (1998) Fission yeast orb6, a ser/thr protein kinase related to mammalian rho kinase and myotonic dystrophy kinase, is required for maintenance of cell polarity and coordinates cell morphogenesis with the cell cycle. Proc Natl Acad Sci U S A 95:7526–7531
237. Li Y, Chang EC (2003) *Schizosaccharomyces pombe* Ras1 effector, Scd1, interacts with Klp5 and Klp6 kinesins to mediate cytokinesis. Genetics 165:477–488
238. Yen HC, Gordon C, Chang EC (2003) *Schizosaccharomyces pombe* Int6 and Ras homologs regulate cell division and mitotic fidelity via the proteasome. Cell 112: 207–217
239. Wang Y, Xu HP, Riggs M, Rodgers L, Wigler M (1991) byr2, a *Schizosaccharomyces pombe* gene encoding a protein kinase capable of partial suppression of the ras1 mutant phenotype. Mol Cell Biol 11:3554–3563
240. Chen CR, Li YC, Chen J, Hou MC, Papadaki P, Chang EC (1999) Moe1, a conserved protein in *Schizosaccharomyces pombe*, interacts with a Ras effector, Scd1, to affect proper spindle formation. Proc Natl Acad Sci U S A 96:517–522
241. Li YC, Chen CR, Chang EC (2000) Fission yeast Ras1 effector Scd1 interacts with the spindle and affects its proper formation. Genetics 156:995–1004
242. Papadaki P, Pizon V, Onken B, Chang EC (2002) Two ras pathways in fission yeast are differentially regulated by two ras guanine nucleotide exchange factors. Mol Cell Biol 22:4598–4606
243. Hughes DA, Fukui Y, Yamamoto M (1990) Homologous activators of ras in fission and budding yeast. Nature 344:355–357
244. Hughes DA, Yabana N, Yamamoto M (1994) Transcriptional regulation of a Ras nucleotide-exchange factor gene by extracellular signals in fission yeast. J Cell Sci 107(Pt 12):3635–3642
245. Tratner I, Fourticq-Esqueöute A, Tillit J, Baldacci G (1997) Cloning and characterization of the *S. pombe* gene efc25+, a new putative guanine nucleotide exchange factor. Gene 193:203–210
246. Wang Y, Boguski M, Riggs M, Rodgers L, Wigler M (1991) sar1, a gene from *Schizosaccharomyces pombe* encoding a protein that regulates ras1. Cell Regul 2:453–465
247. Onken B, Wiener H, Philips MR, Chang EC (2006) Compartmentalized signaling of Ras in fission yeast. Proc Natl Acad Sci U S A 103:9045–9050
248. Murray JM, Johnson DI (2001) The Cdc42p GTPase and its regulators Nrf1p and Scd1p are involved in endocytic trafficking in the fission yeast *Schizosaccharomyces pombe*. J Biol Chem 276:3004–3009
249. Odds FC (1996) Epidemiological shifts in opportunistic and nosocomial Candida infections: mycological aspects. Int J Antimicrob Agents 6:141–144
250. Sudbery P, Gow N, Berman J (2004) The distinct morphogenic states of *Candida albicans*. Trends Microbiol 12:317–324
251. Lo HJ, Köhler JR, DiDomenico B, Loebenberg D, Cacciapuoti A, Fink GR (1997) Nonfilamentous *C. albicans* mutants are avirulent. Cell 90:939–949
252. Ernst JF (2000) Regulation of dimorphism in *Candida albicans*. Contrib Microbiol 5:98–111
253. Maidan M, Thevelein J, Van Dijck P (2005) Carbon source induced yeast-to-hypha transition in *Candida albicans* is dependent on the presence of amino acids and on the G-protein-coupled receptor Gpr1. Biochem Soc Trans 33:291–293

254. Berman J (2006) Morphogenesis and cell cycle progression in *Candida albicans*. Curr Opin Microbiol 9:595–601
255. Lee N, D'Souza CA, Kronstad JW (2003) Of smuts, blasts, mildews, and blights: cAMP signaling in phytopathogenic fungi. Annu Rev Phytopathol 41:399–427
256. Leberer E, Harcus D, Dignard D, Johnson L, Ushinsky S, Thomas DY, Schröppel K (2001) Ras links cellular morphogenesis to virulence by regulation of the MAP kinase and cAMP signalling pathways in the pathogenic fungus *Candida albicans*. Mol Microbiol 42:673–687
257. Goldberg D, Marbach I, Gross E, Levitzki A, Simchen G (1993) A *Candida albicans* homolog of CDC25 is functional in *Saccharomyces cerevisiae*. Eur J Biochem 213:195–204
258. Zhu Y, Fang HM, Wang YM, Zeng GS, Zheng XD, Wang Y (2009) Ras1 and Ras2 play antagonistic roles in regulating cellular cAMP level, stationary-phase entry and stress response in *Candida albicans*. Mol Microbiol 74:862–875
259. Wennerberg K, Rossman KL, Der CJ (2005) The Ras superfamily at a glance. J Cell Sci 118:843–846
260. Köhler J, Fink G (1996) *Candida albicans* strains heterozygous and homozygous for mutations in mitogen-activated protein kinase signaling components have defects in hyphal development. Proc Natl Acad Sci U S A 93:13223–13228
261. Leberer E, Harcus D, Broadbent ID, Clark KL, Dignard D, Ziegelbauer K, Schmidt A, Gow NA, Brown AJ, Thomas DY (1996) Signal transduction through homologs of the Ste20p and Ste7p protein kinases can trigger hyphal formation in the pathogenic fungus *Candida albicans*. Proc Natl Acad Sci U S A 93:13217–13222
262. Csank C, Schröppel K, Leberer E, Harcus D, Mohamed O, Meloche S, Thomas D, Whiteway M (1998) Roles of the *Candida albicans* mitogen-activated protein kinase homolog, Cek1p, in hyphal development and systemic candidiasis. Infect Immun 66:2713–2721
263. Sánchez-Martínez C, Pérez-Martín J (2002) Gpa2, a G-protein alpha subunit required for hyphal development in *Candida albicans*. Eukaryot Cell 1:865–874
264. Borges-Walmsley MI, Walmsley AR (2000) cAMP signalling in pathogenic fungi: control of dimorphic switching and pathogenicity. Trends Microbiol 8:133–141
265. Fang HM, Wang Y (2006) RA domain-mediated interaction of Cdc35 with Ras1 is essential for increasing cellular cAMP level for *Candida albicans* hyphal development. Mol Microbiol 61:484–496
266. Maidan MM, De Rop L, Serneels J, Exler S, Rupp S, Tournu H, Thevelein JM, Van Dijck P (2005) The G protein-coupled receptor Gpr1 and the Galpha protein Gpa2 act through the cAMP-protein kinase A pathway to induce morphogenesis in *Candida albicans*. Mol Biol Cell 16:1971–1986
267. Hogan DA, Muhlschlegel FA (2011) *Candida albicans* developmental regulation: adenylyl cyclase as a coincidence detector of parallel signals. Curr Opin Microbiol 14: 682–686
268. Yi S, Sahni N, Daniels KJ, Lu KL, Srikantha T, Huang G, Garnaas AM, Soll DR (2011) Alternative mating type configurations (a/α versus a/a or α/α) of *Candida albicans* result in alternative biofilms regulated by different pathways. PLoS Biol 9:e1001117
269. Giacometti R, Kronberg F, Biondi RM, Passeron S (2011) *Candida albicans* Tpk1p and Tpk2p isoforms differentially regulate pseudohyphal development, biofilm structure, cell aggregation and adhesins expression. Yeast 28:293–308
270. Hogan DA, Sundstrom P (2009) The Ras/cAMP/PKA signaling pathway and virulence in *Candida albicans*. Future Microbiol 4:1263–1270
271. Feng Q, Summers E, Guo B, Fink G (1999) Ras signaling is required for serum-induced hyphal differentiation in *Candida albicans*. J Bacteriol 181:6339–6346
272. Huang G, Yi S, Sahni N, Daniels KJ, Srikantha T, Soll DR (2010) N-acetylglucosamine induces white to opaque switching, a mating prerequisite in *Candida albicans*. PLoS Pathog 6:e1000806
273. Bahn YS, Sundstrom P (2001) CAP1, an adenylate cyclase-associated protein gene, regulates bud-hypha transitions, filamentous growth, and cyclic AMP levels and is required for virulence of *Candida albicans*. J Bacteriol 183:3211–3223
274. Zou H, Fang HM, Zhu Y, Wang Y (2010) *Candida albicans* Cyr1, Cap1 and G-actin form a sensor/effector apparatus for activating cAMP synthesis in hyphal growth. Mol Microbiol 75:579–591
275. Piispanen AE, Bonnefoi O, Carden S, Deveau A, Bassilana M, Hogan DA (2011) Roles of Ras1 membrane localization during *Candida albicans* hyphal growth and farnesol response. Eukaryot Cell 10:1473–1484
276. Cassola A, Parrot M, Silberstein S, Magee BB, Passeron S, Giasson L, Cantore ML (2004) *Candida albicans* lacking the gene encoding the regulatory subunit of protein kinase A displays a defect in hyphal formation and an altered localization of the catalytic subunit. Eukaryot Cell 3:190–199

277. Phillips AJ, Crowe JD, Ramsdale M (2006) Ras pathway signaling accelerates programmed cell death in the pathogenic fungus *Candida albicans*. Proc Natl Acad Sci U S A 103:726–731
278. Ruiz-Herrera J, Sentandreu R (2002) Different effectors of dimorphism in *Yarrowia lipolytica*. Arch Microbiol 178:477–483
279. Domínguez A, Fermiñán E, Gaillardin C (2000) *Yarrowia lipolytica*: an organism amenable to genetic manipulation as a model for analyzing dimorphism in fungi. Contrib Microbiol 5:151–172
280. Fickers P, Benetti PH, Waché Y, Marty A, Mauersberger S, Smit MS, Nicaud JM (2005) Hydrophobic substrate utilisation by the yeast *Yarrowia lipolytica*, and its potential applications. FEMS Yeast Res 5:527–543
281. Kim J, Cheon SA, Park S, Song Y, Kim JY (2000) Serum-induced hypha formation in the dimorphic yeast *Yarrowia lipolytica*. FEMS Microbiol Lett 190:9–12
282. Ramon AM, Montero M, Sentandreu R, Valentin E (1999) *Yarrowia lipolytica* cell wall architecture: interaction of Ywp1, a mycelial protein, with other wall components and the effect of its depletion. Res Microbiol 150:95–103
283. Hurtado CA, Rachubinski RA (1999) MHY1 encodes a C2H2-type zinc finger protein that promotes dimorphic transition in the yeast *Yarrowia lipolytica*. J Bacteriol 181:3051–3057
284. Hurtado CA, Beckerich JM, Gaillardin C, Rachubinski RA (2000) A rac homolog is required for induction of hyphal growth in the dimorphic yeast *Yarrowia lipolytica*. J Bacteriol 182:2376–2386
285. Hurtado CA, Rachubinski RA (2002) YlBMH1 encodes a 14-3-3 protein that promotes filamentous growth in the dimorphic yeast *Yarrowia lipolytica*. Microbiology 148:3725–3735
286. Morales-Vargas AT, Domínguez A, Ruiz-Herrera J (2012) Identification of dimorphism-involved genes of *Yarrowia lipolytica* by means of microarray analysis. Res Microbiol 163:378–387
287. Zhao X, Mehrabi R, Xu JR (2007) Mitogen-activated protein kinase pathways and fungal pathogenesis. Eukaryot Cell 6:1701–1714
288. Cervantes-Chávez JA, Ruiz-Herrera J (2006) STE11 disruption reveals the central role of a MAPK pathway in dimorphism and mating in *Yarrowia lipolytica*. FEMS Yeast Res 6: 801–815
289. Cervantes-Chávez JA, Kronberg F, Passeron S, Ruiz-Herrera J (2009) Regulatory role of the PKA pathway in dimorphism and mating in *Yarrowia lipolytica*. Fungal Genet Biol 46:390–399
290. Celik E, Calık P (2012) Production of recombinant proteins by yeast cells. Biotechnol Adv 30:1108–1118
291. Barsoum E, Rajaei N, Aström SU (2011) RAS/cyclic AMP and transcription factor Msn2 regulate mating and mating-type switching in the yeast *Kluyveromyces lactis*. Eukaryot Cell 10:1545–1552
292. Alberghina L, Mavelli G, Drovandi G, Palumbo P, Pessina S, Tripodi F, Coccetti P, Vanoni M (2012) Cell growth and cell cycle in *Saccharomyces cerevisiae*: basic regulatory design and protein–protein interaction network. Biotechnol Adv 30:52–72

# Chapter 24

# Methods to Study the Ras2 Protein Activation State and the Subcellular Localization of Ras-GTP in *Saccharomyces cerevisiae*

**Sonia Colombo and Enzo Martegani**

## Abstract

Ras proteins were highly conserved during evolution. They function as a point of convergence for different signalling pathways in eukaryotes and are involved in a wide range of cellular responses (shift from gluconeogenic to fermentative growth, breakdown of storage carbohydrates, stress resistance, growth control and determination of life span, morphogenesis and development, and others). These proteins are members of the small GTPase superfamily, which are active in the GTP-bound form and inactive in the GDP-bound form. Given the importance of studies on the Ras protein activation state to understand the detailed mechanism of Ras-mediated signal transduction, we provide here a simple, sensitive, and reliable method, based on the high affinity interaction of Ras-GTP with the Ras binding domain (RBD) of Raf1, to measure the level of Ras2-GTP on total Ras2 in *Saccharomyces cerevisiae*. Moreover, to study the localization of Ras-GTP in vivo in single *S. cerevisiae* cells, we expressed a probe consisting of a GFP fusion with a trimeric Ras Binding Domain of Raf1 (eGFP-RBD3), which was proven to be a useful live-cell biosensor for Ras-GTP in mammalian cells.

**Key words** Yeast, *Saccharomyces cerevisiae*, Nutrient, cAMP/PKA pathway, Small G proteins, Active Ras, Fluorescence microscopy

## 1 Introduction

Two different assays have been used to analyze quantitatively the guanine nucleotides bound to the Ras proteins in vivo. The first assay was developed by Gibbs et al. [1] and used by Colombo et al. [2]. After labeling the cells in vivo with [32P] orthophosphate, the Ras proteins were immunoprecipitated with antibodies against the human Ras protein (v-Hras 259). Guanine nucleotides were extracted, separated by TLC, and quantitated by phosphorimager technology. With this assay, the GTP/GDP ratio on the yeast Ras proteins could be measured only after overexpression of Ras2 protein, indicating that the sensitivity of this assay was probably not very high. A more recent assay to analyze quantitatively the guanine

Lorenza Trabalzini and Saverio Francesco Retta (eds.), *Ras Signaling: Methods and Protocols*,
Methods in Molecular Biology, vol. 1120, DOI 10.1007/978-1-62703-791-4_24, 

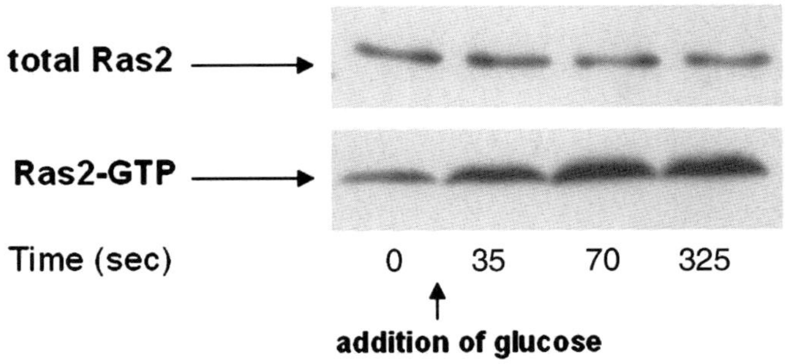


| Time (sec) | Ras2-GTP/total Ras2 (%) |
|---|---|
| **0** | **7.0** |
| **35** | **11.3** |
| **70** | **19.8** |
| **325** | **18.2** |

**Fig. 1** Total Ras2 and Ras2-GTP content, before and after addition of 100 mM glucose to starved *tpk1*$^{w1}$ *gpa2Δ* cells. Immunoblots of Ras2 in cell lysate (total Ras2) and Ras2-GTP bound to GST-RBD eluted with SDS-PAGE sample buffer

nucleotides bound to the Ras proteins in vivo was described by Taylor and Shalloway [3] and exploits the known specificity of the interaction between Ras-GTP and the Ras-binding domain (RBD) of Raf-1 to detect activated Ras. Since there is a high degree of homology between yeast and mammalian Ras proteins, the yeast Ras proteins are able to interact with the RBD of Raf-1. Consequently, we used this non-radioisotopic pull-down assay to measure the yeast Ras2-GTP loading without overexpression of this protein [4, 5]. We also set up conditions to specifically evaluate the reliability of the assay and we could clearly show that only the activated form of Ras2 was able to bind to the GST-RBD pre-bound to Glutathione Sepharose 4B, indicating that the assay was specific for Ras2-GTP [5]. Using this assay we showed that the increase in the level of cAMP after addition of glucose to glucose-starved cells is accompanied by a rapid increase in the level of Ras2-GTP/total Ras2 (an example is shown in Fig. 1). We also showed that this increase is dependent on Cdc25, glucose transport and phosphorylation and that deletion of the Ira proteins causes constitutively high GTP loading. Finally, we also provided evidence that the Ras proteins might be a target for the stringent feedback inhibition of PKA on cAMP synthesis [5]. This assay measures only the activation state of the Ras2 protein, but does not give any information about the localization of the active form of this GTPase. Spatial regulation of signalling events at the subcellular level is an important aspect of signal transduction; indeed, quite some evidence points to Ras playing distinct signalling roles depending on its subcellular

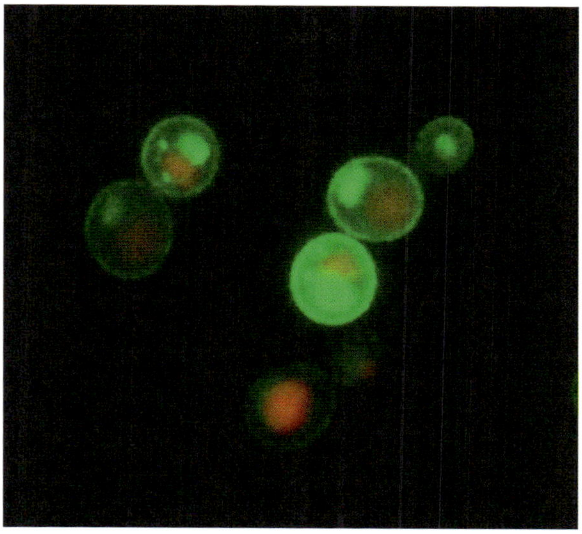

**Fig. 2** Nuclear and plasma membrane localization of active Ras in the W303-1A wild type strain. Cells were transformed with pYX212-eGFP-RBD3, grown in 2 % glucose medium at 30 °C until exponential phase and then eGFP was photographed on living cells with a Nikon fluorescence microscope. In red, the autofluorescence of the vacuole is visible

location [6]. Consequently, the scientific question we addressed was the following: where are active Ras proteins localized and when are they localized in a certain cellular compartment? Augsten et al. [7] and Rubio et al. [8] successfully performed Ras-GTP imaging in mammalian cells, using a probe consisting of a GFP fusion with a trimeric Ras Binding Domain of Raf1 (eGFP-RBD3) as a reporter for Ras activation. In order to investigate the localization of active Ras in *Saccharomyces cerevisiae*, we inserted the sequence coding for the eGFP-RBD3 fusion in multicopy plasmids for yeast and we expressed the fusion protein in wild type and different mutant strains in the cAMP/PKA pathway. Our results showed that the eGFP-RBD3 probe is a useful live-cell biosensor to report Ras activation in *S. cerevisiae*. In particular, our data showed that the probe accumulates mainly at the plasma membrane and in the nucleus in wild type cells growing on medium containing glucose (Fig. 2), while it accumulates mainly in mitochondria in glucose-starved cells and relocalizes to the plasma membrane and to the nucleus upon addition of this sugar. We also showed that Gpa2 and Hxk2 play a role in the localization of active Ras [9]. Moreover, Leadsham et al. [10] showed that cells lacking Whi2p exhibit an aberrant accumulation of activated Ras2 at the mitochondria in response to nutritional depletion. In this mutant, the failure to address Ras2 to the vacuole and consequently the failure to shut down Ras signalling would lead to mitochondrial dysfunction, the accumulation of damaging ROS and cell death.

## 2 Materials

Prepare all solutions using deionized water and analytical grade reagents. Prepare and store all reagents at room temperature (unless indicated otherwise). We do not add sodium azide to the reagents.

### 2.1 Determination of the Ras2-GTP/Total Ras2 Ratio In Vivo

#### 2.1.1 E. coli [pGEX2T-RBD] Culture Growth

1. *E.coli* strain transformed with the expression vector pGEX2T-RBD (*E. coli* [pGEX2T-RBD]), which encodes amino acids 1–149 of cRaf-1 fused to GST [3].
2. LB medium: 1 % (w/v) peptone, 0.5 % (w/v) yeast extract, 1 % (w/v) NaCl, 0.1 % (w/v) glucose.
3. Ampicillin (500×): 50 mg/ml in deionized water (store in small aliquots at −20 °C).
4. Isopropyl 1-thio-β-D-galactopyranoside (IPTG) (1000×): 100 mM in deionized water (store in small aliquots at −20 °C).

#### 2.1.2 Preparation of Cleared E. coli [pGEX2T-RBD] Lysates

1. Phosphate buffered saline (PBS; 10×): 1.37 M NaCl, 0.027 M KCl, 0.043 M $Na_2HPO_4{\cdot}2H_2O$, 0.014 M $KH_2PO_4$, pH 7.4.
2. Lysis buffer A: PBS 1×, 1 mM EDTA, 0.5 mM dithiothreitol, one tablet of Protease Inhibitor Mixture (Roche Applied Science) in 50 ml of lysis buffer (*see* **Note 1**).
3. Glass beads 425–600 μm.
4. 10 % (w/v) Triton X-100 in PBS 1×.
5. Glycerol (better 87 %, which is easier to pipet).
6. Bio-Rad protein Assay (Bio-Rad).

#### 2.1.3 Preparation of GST-RBD Pre-bound to Glutathione Sepharose 4B

1. Glutathione Sepharose 4B (GE Healthcare).
2. Phosphate buffered saline (PBS; 10×): as Subheading 2.1.2, **item 1**.
3. Phosphate buffered saline 1×, containing 1 mM EDTA (PBSE).
4. Complex phosphate buffered saline buffer (CPBS): PBS 1×, 1 % (w/v) Triton X-100, 10 % (w/v) glycerol, 1 mM EDTA, 0.5 mM dithiothreitol, 1 mM sodium vanadate, one tablet of Protease Inhibitor Mixture in 50 ml of this solution (*see* **Note 2**).
5. Lysis buffer B: 25 mM HEPES, pH 7.5, 150 mM NaCl, 1 % (w/v) Nonidet P-40, 0.25 % (w/v) sodium deoxycholate, 10 % (w/v) glycerol, 25 mM NaF, 10 mM $MgCl_2$, 1 mM EDTA, 1 mM sodium vanadate, one tablet of Protease Inhibitor Mixture in 50 ml of extraction medium (*see* **Note 3**).

#### 2.1.4 Yeast Growth and Yeast Crude Extract Preparation

1. Medium as appropriate for the cell strain under investigation.
2. Nitrocellulose filters (Whatman Protran pure nitrocellulose transfer and immobilization membrane, PerkinElmer or a similar product).

3. Sterile tooth sticks.
4. FastPrep tubes.
5. Lysis buffer B: as in Subheading 2.1.3, **item 5**.
6. Dry ice-ethanol.
7. Glass beads 425–600 μm.
8. FastPrep instrument (Savant).
9. Bio-Rad protein assay.

#### *2.1.5 Purification of Ras2-GTP*

1. Lysis buffer B: as in Subheading 2.1.3, **item 5**.
2. SDS-sample buffer, 2×: 100 mM Tris–HCl, pH 6.8, 2 % (w/v) β-mercaptoethanol, 4 % (w/v) SDS, 0.2 % (w/v) bromophenol blue, 20 % (w/v) glycerol.

#### *2.1.6 SDS-PAGE Components*

1. Resolving gel buffer: 1.5 M Tris–HCl, pH 8.8, 0.4 % SDS.
2. Stacking gel buffer: 0.5 M Tris–HCl, pH 6.8, 0.4 % SDS.
3. 30 % acrylamide–bis-acrylamide solution (29.2:0.8 acrylamide–bis-acrylamide). Store at 4 °C.
4. Ammonium persulfate: 10 % solution in deionized water (*see* **Note 4**).
5. *N*,*N*,*N*,*N*′-tetramethyl-ethylenediamine (TEMED). Store at 4 °C.
6. SDS-PAGE running buffer, 5×: 0.125 M Tris, 1.25 M glycine, 0.5 % SDS (it is not necessary to adjust the pH, which should automatically reach a value of about 8.6).

#### *2.1.7 Immunoblotting, Immunodetection, and Densitometric Analysis*

1. Nitrocellulose membrane (Whatman Protran pure nitrocellulose transfer and immobilization membrane, PerkinElmer or a similar product).
2. Western blot transfer buffer: SDS-PAGE running buffer 1×, 20 % methanol.
3. Tris buffered saline (TBS; 5×): 0.75 M NaCl, 0.05 M Tris–HCl, pH 7.4.
4. TBS 1× containing 0.05%Tween-20 (TBST).
5. Blocking solution: 5 % low fat dried milk in TBST. Store at 4 °C.
6. Anti-Ras2 polyclonal antibodies (Santa Cruz Biotechnology).
7. Donkey anti-Goat IgG-HRP (Santa Cruz Biotechnology).
8. ECL western blotting detection reagent.
9. BioMax Light Film.
10. X-Ray fixer.
11. X-Ray developer.
12. NIH-Image software (freeware http://www.nist.gov/lispix/imlab/prelim/dnld.html), or similar.

### 2.2 Localization of Active Ras

#### 2.2.1 Equipment and Instrumentation

1. Nikon Eclipse E600 microscope equipped with a 60×, 1.4 oil Plan-Apochromat objective and a standard FITC filter set (Nikon, EX 450–490, DM 505, BA 520) for GFP fluorescence and a Cy3 filter set for Rhodamine fluorescence. Images were recorded digitally using a Leica DC 350 F camera.
2. Leica TCS SP2 confocal microscope equipped with an inverted Leica DMIRE2 microscope and a PL APO 63× oil immersion objective (numerical aperture: 1.4). For GFP fluorescence, λ excitation 488 nm (Ar laser), λ emission 500–550. For Rhodamine fluorescence, λ excitation 540 nm (Ar laser), λ emission 600–650.
3. Thoma chamber.
4. Adobe Photoshop (Adobe Systems, Inc.) and ImageJ software (freeware http://imagej.en.softonic.com/).

#### 2.2.2 Cloning of eGFP-RBD3 and Yeast Transformation

1. General molecular biology reagents for restriction cloning procedures.
2. The peGFP-C2 plasmid expressing the fusion eGFP-RBD3 [7]: it may be provided by I. Rubio University of Jena, Germany.
3. The pYX expression vectors, pYX212 and pYX242 (*see* **Note 5**).
4. Restriction enzymes: NheI, SalI, and EcoRI.
5. The Klenow fragment of DNA polymerase I and the DNA ligase.
6. JETSORB Gel Extraction Kit (genomed GmbH).

#### 2.2.3 Yeast Transformation

1. YPD medium: 1 % (w/v) yeast extract, 2 % (w/v) bacto-peptone, 2 % (w/v) glucose.
2. YPDA medium: YPD with 0.005 % (w/v) adenine hemisulfate.
3. Synthetic medium (SD): 6.7 g/l YNB w/o amino acids, the proper selective drop-out CSM (Complete Synthetic Medium) as appropriate for the cell strain under investigation, 2 % (w/v) glucose, 2.5 %(w/v) agar.

   TE (10×): 0.1 M Tris–HCl, 10 mM EDTA, pH 7.5.

   Lithium acetate: prepare a 100 mM and a 1 M solution in TE 1×, sterilize by filtration.

   Polyethyleneglycol 6000 (PEG 6000): prepare a 50 % (w/v) in water, sterilize by filtration.

   Salmon Sperm DNA: 2 mg/ml Stock solution in water denatured at 100 °C for 5 min and chilled in ice.

#### 2.2.4 Mitochondria Labeling with Vital Mitochondrial Marker

1. Rhodamine B hexyl ester perchlorate: prepare a 1 mM stock solution in DMSO, aliquot, and store at −20 °C.
2. Staining solution: dilute the Rhodamine stock solution with water to make a 10 μM staining solution.

*2.2.5 Imaging of Ras-GTP*

1. Selective medium as appropriate for the cell strain under investigation.
2. Optionally: Starvation buffer: 25 mM MES, pH 6.
3. Cover glass.
4. Concanavalin A: prepare 100 μg/ml stock in deionized water, aliquot, and store at −20 °C.
5. Thoma chamber.

# 3 Methods

## 3.1 Determination of the Ras2-GTP/Total Ras2 Ratio In Vivo

### 3.1.1 *E. coli* [pGEX2T-RBD] Culture Growth (1 l)

1. Inoculate 40 ml of LB broth containing ampicillin (100 μg/ml) and grow the culture at 37 °C overnight with vigorous shaking.
2. Inoculate 1 l of LB broth containing ampicillin (100 μg/ml) 1:25 with the noninduced overnight culture and grow at 37 °C with vigorous shaking until an $OD_{600}$ of 0.5 is reached (about 2–3 h).
3. Induce expression by adding IPTG to a final concentration of 0.1 mM.
4. Incubate the culture for an additional 4–5 h, at 37 °C with vigorous shaking.
5. Harvest the cells by centrifugation at 4,000 ×*g* for 20 min at 4 °C.
6. Store cell pellet at −80 °C or proceed immediately to protocol for preparation of cleared *E. coli* lysates (next **step**).

### 3.1.2 Preparation of *E. coli* [pGEX2T-RBD] Lysates Under Native Conditions

1. Thaw the cell pellet and resuspend in 40 ml ice-cold Lysis buffer A.
2. Separate the cell suspension into two 50 ml tubes.
3. Put the two 50 ml tubes at −80 °C for 15 min and defrost the cell suspensions in $H_2O$ at room temperature.
4. Separate the cell suspension into four 50 ml tubes (about 10 ml cell suspension each tube) and add 4 ml glass beads to each tube.
5. Break the cells using a vortex: five cycles, 1 min each, with a 1 min cooling period between each cycle.
6. Add TritonX-100 to a final concentration of 1 % (use 10 % TritonX-100 in PBS 1×).
7. Incubate at 4 °C for 30 min, mixing gently on a rotary shaker.
8. Centrifuge lysate at 10,000 ×*g* for 20 min at 4 °C to pellet the cellular debris. Save supernatant.
9. Add glycerol to a final concentration of 10 % (*see* **Note 6**).

10. Store in aliquots at −80 °C or proceed immediately to protocol for preparation of GST-RBD fusion protein pre-bound to Glutathione Sepharose 4B (next **step**).

*3.1.3 Preparation of GST-RBD Pre-bound to Glutathione Sepharose 4B (Ten Samples)*

1. Sediment 300 μl of Glutathione Sepharose 4B bed volume by centrifugation at 500 × *g* for 5 min at 4 °C and discard the supernatant (*see* **Note** 7).
2. Wash the matrix with ten bed volumes of ice-cold PBSE.
3. Centrifuge the suspension at 500 × *g* for 5 min at 4 °C to sediment the matrix. Discard the wash.
4. Repeat the wash twice for a total of three washes.
5. Wash the matrix with ten bed volumes of ice-cold CPBS.
6. Repeat the wash once for a total of two washes.
7. Completely remove the wash (*see* **Note 8**).
8. Add 600 μl of cleared *E. coli* [pGEX2T-RBD] lysate each 20 μl of bed volume (each sample) and mix gently on a rotary shaker at 4 °C for 60 min (*see* **Note 9**).
9. Centrifuge the suspension at 500 × *g* for 5 min at 4 °C to sediment the matrix. Remove the supernatant.
10. Wash the pellet with ten bed volumes of ice-cold CPBS.
11. Centrifuge the suspension at 500 × *g* for 5 min at 4 °C to sediment the matrix. Discard the wash.
12. Repeat the wash once for a total of two washes.
13. Wash the pellet with ten bed volumes of ice-cold Lysis buffer B.
14. Centrifuge the suspension at 500 × *g* for 5 min at 4 °C to sediment the matrix. Discard the wash.
15. Repeat the wash twice for a total of three washes and completely remove the wash (*see* **Note 8**).
16. Resuspend the GST-RBD fusion protein pre-bound to Glutathione Sepharose 4B in ice-cold Lysis buffer B to a final concentration of 20 %.
17. Transfer 100 μl of the 20 % GST-RBD pre-bound to Glutathione Sepharose 4B slurry into each 1.5 ml tube (*see* **Note 10**).
18. Store the 20 % GST-RBD pre-bound to Glutathione Sepharose 4B slurry aliquots in ice until use (until the purification of Ras2-GTP step, Subheading 3.1.5).

*3.1.4 Harvesting of the Yeast Cells and Crude Extract Preparation*

1. Harvest about $3 \times 10^8$ yeast cells per treatment (each sample), either by centrifugation (3,000 × *g* for 5 min at 4 °C) or by fast filtration on nitrocellulose filters in a refrigerated environment (4 °C, cold room) (*see* **Note 11**).

2. If cells are harvested by centrifugation, transfer them to a 1.5 ml FastPrep tube and add 300 μl ice-cold Lysis buffer B. If cells are harvested by filtration, using a sterile toothpick, transfer them to a 1.5 ml FastPrep tube containing 300 μl ice-cold Lysis buffer B.
3. Immediately freeze the cells in dry ice-ethanol (cell pellet can be stored at −80 °C until use).
4. Add 300 μl glass beads to each tube without defrosting the cells.
5. Disrupt the cells, using a FastPrep instrument, set on speed 6: three cycles, 20 s each, with a 1 min cooling period between each cycle.
6. Centrifuge lysate at 13,000 × *g* for 1 min at 4 °C to pellet the cellular debris. Transfer the supernatant to a clean tube.
7. Quickly wash the glass beads with 600 μl of ice-cold Lysis buffer B and add it to the supernatant of the previous step.
8. Centrifuge at 13,000 × *g* for 10 min at 4 °C and transfer the supernatant to a clean tube.
9. Centrifuge at 13,000 × *g* for 20 min at 4 °C and transfer the supernatant to a clean tube.
10. Measure the protein concentration of the cleared yeast crude extract preparation.
11. Transfer a volume of yeast crude extract containing 240 μg of total proteins to a clean tube and add ice-cold Lysis buffer B to a final volume of 600 μl (consequently, the protein concentration of the sample will be 0.4 μg/μl).
12. Store in ice until use (until next step, the purification of Ras2-GTP step).

#### 3.1.5 Purification of Ras2-GTP (see **Note 11**)

1. Add 500 μl of yeast crude extract preparation (Subheading 3.1.4, **step 11**) to 100 μl of the 20 % GST-RBD fusion protein pre-bound to Glutathione Sepharose 4B slurry (Subheading 3.1.3, **step 18**) and incubate at 4 °C for 1 h, mixing gently on a rotary shaker.

   Take 40 μl of yeast crude extract preparation (Subheading 3.1.4, **step 11**), add 40 μl of 2× SDS-sample buffer and boil for 10 min. This is the "total Ras2 protein sample." Freeze until use (until the SDS-PAGE will be performed).
2. Centrifuge at 500 × *g* for 5 min at 4 °C to sediment the matrix. Discard the supernatant.
3. Wash the matrix with 1 ml of ice-cold Lysis buffer B. Be careful not to lose the matrix.
4. Centrifuge at 500 × *g* for 5 min at 4 °C to sediment the matrix. Discard the wash.

5. Repeat the wash twice for a total of three washes.
6. Completely remove the wash (*see* **Note 8**).
7. Add 80 μl of 1.25× SDS-sample buffer and boil for 10 min. This is the "Ras2-GTP protein sample." Freeze until use (until the SDS-PAGE will be performed).

#### 3.1.6 SDS-PAGE and Western Blotting

1. Perform a 10 % SDS-PAGE. Load 7.5 μl of "total Ras2 protein sample" (corresponding to 1.5 μg of total proteins) (Subheading 3.1.5, **step 1**) and 12 μl of "Ras2-GTP protein sample" (Subheading 3.1.5, **step 7**) (*see* **Note 12**).
2. Blot for 1 h and a half at 100 V.

#### 3.1.7 Immunodetection and Densitometric Analysis

1. Block the membrane with blocking solution for 1 h at room temperature with gentle shaking.
2. Add Anti-Ras2 polyclonal antibodies diluted 1:200 with blocking solution to the membrane and incubate overnight at 4 °C with gentle shaking (*see* **Note 13**).
3. Briefly rinse the membrane once with TBST and then wash three times for 10 min with fresh changes of TBST at room temperature with gentle shaking.
4. Add Donkey anti-Goat IgG-HRP diluted at 1:5,000 with blocking solution to the membrane and incubate 1 h at room temperature with gentle shaking.
5. Briefly rinse the membrane once with TBST and then wash three times for 10 min with fresh changes of TBST at room temperature with gentle shaking.
6. Incubate the membrane in the ECL western blotting detection reagent, exactly as described by the manufacturer.
7. Place a sheet of autoradiography film on top of the membrane, close the cassette and expose for a proper time (*see* **Note 14**).
8. Quantify the bands on the film by densitometric analysis (NIH-Image software or similar) and determine the ratios of Ras2-GTP/total Ras2 taking in account the amounts of proteins loaded. A representative example is shown in Fig. 1.

### 3.2 Localization of Active Ras

#### 3.2.1 Cloning of eGFP-RBD3 and Yeast Transformation

1. Digest the peGFP-C2 plasmid with *Nhe*I to recover the fragment encoding the fusion eGFP-RBD3.
2. Fill in the 3′ recessing ends with the Klenow fragment of DNA polymerase I.
3. Purify the blunted DNA with the JETSORB Gel Extraction Kit (Genomed GmbH) and digest again with *Sal*I.
4. Digest both the pYX212 and pYX242 with *Eco*RI.
5. Fill in the 3′ recessing ends with the Klenow fragment of DNA polymerase I.

6. Purify the DNA with the JETSORB Gel Extraction Kit (Genomed GmbH) and digest with *Sal*I.
7. Ligate the fragment containing the fusion eGFP-RBD3 (**step 3**) with the linearized vectors (**step 6**): pYX212-eGFP-RBD3 and pYX242-eGFP-RBD3 are obtained.

#### *3.2.2 Yeast Transformation*

Transform yeast cells with a modified lithium acetate method [11] (*see* **Note 15**):

1. Inoculate cells in 5 ml of YPDA medium and grow overnight at 30 °C. Dilute in 25 ml of fresh YPDA at $OD_{600nm}$ of about 0.4 and regrow at 30 °C till $OD_{600nm}$ of about 1.5.
2. Harvest cells by centrifugation at 2,000 ×*g* for 10 min at 4 °C and wash by suspending the cells in 12.5 ml of sterile water and subsequent centrifugation as above.
3. Suspend the cell pellet in 0.5 ml of 100 mM lithium acetate, transfer the suspension in a microfuge tube and centrifuge at 1,000 ×*g* for 15 s.
4. Suspend the cells in 0.25 ml of 100 mM lithium acetate and divide in 50 μl aliquots in microfuge tubes. Centrifuge at 1,000 ×*g* for 15 s and then add to each tube the following solutions in the given order: 240 μl of 50 % PEG 6000, 36 μl of 1 M lithium acetate, 25 μl of 2 mg/ml denatured salmon sperm DNA, 50 μl of sterile water containing 0.1 μg of plasmid DNA.
5. Vortex briefly and incubate at 30 °C for 30 min, then at 42 °C for 20 min.
6. Spin down for 15 s, wash the cells in 1 ml of sterile water, centrifuge and suspend the cell in 0.1 ml of sterile water and then plate on selective solid medium (SD medium, 2 % agar, without uracil or leucine).
7. Incubate the plates at 30 °C. Colonies of transformed yeast will grow in 2–3 days.

#### *3.2.3 Labeling of Mitochondria with Vital Dye (Optional)*

1. Add the staining solution to the cells at a final concentration of 100 nM.
2. Incubate in the dark for about 5–10 min before imaging.

#### *3.2.4 Imaging of Ras-GTP Either by Fluorescence Microscope or by Confocal Microscope*

1. Prepare in advance concanavalin A-coated cover glass as follows:
   (a) Cover the glass with 100 μg/ml concanavalin A solution and incubate 1 h at room temperature.
   (b) Remove concanavalin A solution (we recover and reuse the solution 2–3 times) and wash with deionized water. Glasses can be stored in deionized water in a sealed container for 1 day at 4 °C.
2. Grow cells at 30 °C in the appropriate medium until exponential phase.

3. Optional: if required, label organelles with vital dyes (*see* Subheading 3.2.3).
4. Seed about 50 μl of cells suspension on concanavalin A-coated cover glass for 10 min.
5. Wash the cover glass four times using about 1 ml of the proper medium each time and put on top of a Thoma chamber (*see* **Note 16**).
6. Pipet, between the cover glass and the Thoma Chamber, 5 μl of the proper medium on each side.
7. Focus the cell layer using transmission light and select a field for imaging Ras-GTP (optionally, acquire an image).
8. Switch to GFP fluorescence setting and acquire an image. If mitochondria have been labeled, switch to red fluorescence setting and acquire an image (*see* **Note 17**).
9. Process the images using either Adobe Photoshop (pictures acquired by fluorescence microscope) or ImageJ software (pictures acquired by confocal microscope). A representative example of a picture taken using a fluorescence microscope is shown in Fig. 2.

#### *3.2.5 Glucose Induction Imaging of Ras-GTP Using a Fluorescence Microscope*

1. Prepare in advance concanavalin A-coated cover glass as described above (*see* Subheading 3.2.4).
2. Grow cells at 30 °C in the appropriate medium until exponential phase.
3. Collect cells by centrifugation (max 3,000 × *g*, 3 min), resuspend in 25 mM MES buffer, pH 6 ($5 \times 10^7$ cells/ml, final concentration) and incubate for at least 1 h at 30 °C.
4. Optional: if required, label organelles with vital dyes (*see* above Subheading 3.2.3).
5. Seed about 50 μl of cells suspension on concanavalin A-coated cover glass for 10 min.
6. Wash the cover glass four times with 25 mM MES buffer (about 1 ml each time) and put on top of a Thoma chamber (*see* **Note 16**).
7. Pipet, between the cover glass and the Thoma Chamber, 5 μl of 25 mM MES buffer on each side.
8. Focus the cell layer using transmission light, select a field for imaging Ras-GTP and acquire an image.
9. Switch to GFP fluorescence setting, acquire an image and switch back to transmission light. This is the control image before stimulation.
10. Add 20 μl of 3 % glucose dissolved in starvation buffer by pipetting between the cover glass and the Thoma Chamber. Be careful not to touch the Thoma chamber. Immediately,

switch to green fluorescence setting and acquire images. To avoid bleaching, acquire the fluorescence images for 1 s every minute for at least 10 min and keep the shutter off in the meantime (*see* **Note 18**).

11. Precess the time series of images using Adobe Photoshop.

## 4 Notes

1. We suggest to prepare Lyses buffer A fresh each time, starting from stocks solutions prepared in advance. Stock solutions: PBS 10× (*see* Subheading 2.1.2.), 0.5 M EDTA in deionized water, 0.5 M dithiothreitol in deionized water, one tablet of Protease Inhibitor Mixture in 2 ml of deionized water (25×). With the exception of PBS 10×, store the other stock solutions in small aliquots at −20 °C.
2. We suggest to prepare CPBS fresh each time, starting from stocks solutions prepared in advance, with the exception of sodium vanadate that it is best to prepare fresh each time. Stock solutions: PBS 10× (*see* Subheading 2.1.2.), 10 % (w/v) Triton X-100 in PBS 1×, glycerol (better 87 %, which is easier to pipet), 0.5 M EDTA in deionized water, 0.5 M dithiothreitol in deionized water, 100 mM sodium vanadate in deionized water, one tablet of Protease Inhibitor Mixture in 2 ml of deionized water (25×). Cut end of a blue tip to aspirate Triton X-100 and glycerol. Store 0.5 M EDTA, 0.5 M dithiothreitol, and Protease Inhibitor Mixture 25× in small aliquots at −20 °C.
3. We suggest to prepare Lysis buffer B fresh each time, starting from stocks solutions prepared in advance, with the exception of sodium vanadate and Nonidet P-40 that it is best to prepare fresh each time. Stock solutions: 200 mM HEPES, 5 M NaCl in deionized water, 10 % Nonidet P-40 in deionized water, 5 % sodium deoxycholate in deionized water, glycerol (better 87 %, which is easier to pipet), 0.5 M NaF, 75 mM $MgCl_2$, 0.5 M EDTA in deionized water, 100 mM sodium vanadate in deionized water, one tablet of Protease Inhibitor Mixture in 2 ml of water (25×). Cut end of a blue tip to aspirate Nonidet P-40 and glycerol. Store 0.5 M EDTA and Protease Inhibitor Mixture 25× in small aliquots at −20 °C.
4. We suggest to prepare this fresh each time.
5. Vectors pYX212 and pYX242 are 2μ vectors; they contain a TPI promoter and only differ for the yeast selectable marker, which are respectively *URA3* and *LEU2*.
6. Measure the protein concentration using the Bio-Rad protein Assay to make sure that it is about 3 μg/μl. If the protein concentration is less, *see* **Note 8**.

7. Use 20 μl of bed volume per sample. Start with at $N+5$ bed volumes, where $N$ is the number of samples to be processed. The reason for this is that, due to the repeated washing, part of the resin will inevitably get lost.
8. We suggest to use a proper Hamilton syringe, but be careful not to aspirate the matrix with the needle.
9. Use $N+2$ bed volumes, where $N$ is the number of samples to be processed. The reason for this is that, due to the repeated washing, part of the resin will inevitably get lost. 600 μl of cleared *E. coli* [pGEX2T-RBD] lysate should contain approximately 1.8 mg of total proteins. If the protein concentration of cleared *E. coli* [pGEX2T-RBD] lysate is less than 3 μg/μl, increase the volume proportionally.
10. Each tube corresponds to a treatment (sample). It is very important to verify, before continuing to the next step, that the amount of GST-RBD pre-bound to Glutathione Sepharose 4B is exactly the same, 20 μl bed volume, in each tube.
11. The fast filtration is particularly recommended when performing kinetics to study the Ras-GTP levels after addition of 100 mM glucose.
12. The amounts of "total Ras2 protein sample" and "Ras2-GTP proteins sample" to be loaded on the SDS-PAGE are indicative and may be varied depending either on the cell strain under investigation or the experimental conditions used.
13. Since both active Ras1 and Ras2 proteins bind to the GST-RBD pre-bound to Glutathione Sepharose 4B, theoretically this protocol gives the possibility to purify both Ras1-GTP and Ras2-GTP. Nevertheless, at the moment there are not specific yeast anti-Ras1 antibodies available.
14. Normally, if the amount loaded is in the range suggested at point 1, Subheading 3.1.6., an exposition of 1–5 min should be enough.
15. Importantly, with the only exception of the *gpa2Δ* mutant, the eGFP-RBD3 expression did not modify the growth rate in the wild type and in the mutant strains analyzed, indicating that this probe does not alter endogenous Ras signalling [9].
16. Drain the excess of medium from the cover glass before putting it on top of the Thoma chamber.
17. To test the specificity of the probe, we expressed eGFP in wild type cells and eGFP-RBD3 in GG104 *cdc25Δ* cells (in this strain almost all Ras proteins should be present in the inactive form). In both cases we observed a diffuse staining, indicating that the nuclear and plasma membrane staining, shown for the wild type strain growing exponentially on glucose medium and for other strains, occurs as a result of activated forms of Ras [9]. As a control, we used strain GG104 *cdc25Δ* since, for

unknown reason, the probe was poorly expressed in a Ras1,2 deletion mutant. Interestingly, in *S. cerevisiae* the nuclear accumulation is specific. This is in contrast with what happens in mammalian cells [7, 8, 12, 13], where RBD-based fluorescent reporter probes tend to accumulate aspecifically in the nucleus of many cell types.

18. We suggest to add glucose under transmission light to be able to check immediately for possible cell or focus drift as a consequence of these manipulations.

## Acknowledgement

We thank A. Wittinghofer (Max-Planck Institute, Dortmund, Germany) for providing the expression vector for the production of GST-RBD and Ignacio Rubio (Friedrich Schiller-University, Jena, Germany) for his kind gift of the peGFP-C2 construct expressing the fusion eGFP–RBD3. This work was supported by FAR grant to E.M. and S.C., UniCellSys and by founding of Program SysBioNet, Italian Roadmap Research Infrastructure 2012.

### References

1. Gibbs JB, Schaber MD, Marshall MS et al (1987) Identification of guanine nucleotides bound to ras-encoded proteins in growing yeast cells. J Biol Chem 262:10426–10429
2. Colombo S, Ma P, Cauwenberg L et al (1998) Involvement of distint G-proteins, Gpa2 and Ras, in glucose- and intracellular acidification-induced cAMP signalling in the yeast *Saccharomyces cerevisiae*. EMBO J 17:3326–3341
3. Taylor S, Shalloway D (1996) Cell cycle-dependent activation of Ras. Curr Biol 6: 1621–1627
4. Rudoni S, Colombo S, Coccetti P et al (2001) Role of guanine nucleotides in the regulation of the Ras/cAMP pathway in *Saccharomyces cerevisiae*. Biochim Biophys Acta 1538:181–189
5. Colombo S, Ronchetti D, Thevelein JM et al (2004) Activation state of the Ras2 protein and glucose-induced signalling in *Saccharomyces cerevisiae*. J Biol Chem 279:46715–46722
6. Casar B, Arozarena I, Sanz-Moreno V et al (2008) Ras subcellular localization defines ERK1/2 substrate specificity through distinct utilization of scaffold proteins. Mol Cell Biol 29:1338–1353
7. Augsten M, Pusch R, Biskup C et al (2006) Live-cell imaging of endogenous Ras-GTP illustartes predominnt Ras activation at the plasma membrane. EMBO Rep 7:46–51
8. Rubio I, Grund S, Song SP et al (2010) TCR-induced activation of Ras proceeds at the plasma membrane and requires palmitoylation of N-Ras. J Immunol 185:3536–3543
9. Broggi S, Martegani E, Colombo S (2013) Live-cell imaging of endogenous Ras-GTP shows predominant Ras activation at the plasma membrane and in the nucleus in *Saccharomyces cerevisiae*. Int J Biochem Cell Biol 45:384–394
10. Leadsham JE, Miller K, Ayscough KR et al (2008) Whi2p links nutritional sensing to actin-dependent Ras-cAMP-PKA regulation and apoptosis in yeast. J Cell Sci 122:706–715
11. Gietz D, St Jean A, Woods RA, Schiestl RH (1992) Improved method for high efficiency transformation of intact yeast cells. Nucleic Acids Res 20:1425
12. Chiu VK, Bivona T, Hach A et al (2002) Ras signalling on the endoplasmic reticulum and the Golgi. Nat Cell Biol 4:343–350
13. Bivona TG, Perez De Castro I, Ahearn IM et al (2003) Phospholipase C-gamma activates Ras on the Golgi apparatus by means of RasGRP1. Nature 424:694–698

# Chapter 25

# Ras Proteins Signaling in the Early Metazoan *Dictyostelium discoideum*

**Enrico Bracco and Barbara Pergolizzi**

## Abstract

Since the discovery of Ras, Ras-mediated transforming activity has been the major investigative area of interest. Soon thereafter it has emerged that Ras family members regulate different biological processes, other than cell growth, like development and fine-tune the balance between cell death and survival. The lower metazoan *Dictyostelium discoideum* is a powerful and genetically accessible model organism that has been used to elucidate the roles played by different Ras members in some biological processes, such as cell motility and development. In the following chapter we describe some very basic techniques aiming to identify novel Ras signaling components, throughout insertional mutagenesis screening, and to investigate their role(s) in development and chemotaxis processes.

**Key words** *Dictyostelium discoideum*, Saturation mutagenesis, Development, Chemotaxis, Adenylate-cyclase, guanylate-cyclase, Second messenger (cAMP, cGMP)

## 1 Introduction

Ras family GTPases, like heterotrimeric G-proteins, are simple molecular switches cycling between active (GTP-bound) and inactive state (GDP-bound) by means of their intrinsic hydrolytic activity which removes γ-phosphate from GTP. Their state is tightly regulated by two classes of proteins: (a) GTPase-Activating Proteins (GAPs) which inactivate Ras members by increasing the rate of GTP hydrolysis; and (b) Guanine-nucleotide Exchange Factors (GEFs) which catalyze the exchange of GDP for GTP, thus returning Ras proteins to their active state. Ras proteins have been the subject of intensive research because of their critical role in human oncogenesis, but they represent also crucial players in growth factor-elicited differentiation of pheochromocytoma PC12 cell line. Indeed, PC12 differentiation process can be mimicked by oncogenic proteins and inhibited by dominant negative Ras proteins, respectively. These developmental effects are paralleled by genetic evidence from *Drosophila* and *Caenorhabditis*, implying

Lorenza Trabalzini and Saverio Francesco Retta (eds.), *Ras Signaling: Methods and Protocols*,
Methods in Molecular Biology, vol. 1120, DOI 10.1007/978-1-62703-791-4_25, © Springer Science+Business Media, LLC 2014

phylogenetically and functionally conserved Ras-MAPK pathways in cell fate specification [1, 2]. Actually, Ras family members and their regulators are evolutionarily conserved throughout all eukaryotes from single cell organisms, such as yeasts, to mammalian.

*D. discoideum* is a social amoeba, and most likely, it is among the most ancient metazoan living organisms whose multicellular stage is triggered by nutrient starvation. *Dictyostelium* cells grow vegetatively as single-celled amoebae, but when food sources (i.e., bacteria, yeasts, or axenic medium) become scarce, a developmental program initiates. Upon starvation single-celled amoebae start spontaneously to synthesize, secrete, and sense an extracellular messenger, cyclic AMP (cAMP), which acts as chemoattractant. This event leads cells to move and aggregate towards an aggregation center by means of a chemotaxis process and to form hemispherical multicellular mound aggregates, of approximately $10^5$ cells, tipped at their top. Mound aggregates might undergo morphogenetic changes which lead to a slug-shaped organism capable of migrating, guided either by light (phototaxis) or by temperature (thermotaxis), in search of food. In case of prolonged food deprivation the "slugs" start a process by which the amoebae differentiate in at least two kinds of cells: (a) spores, (b) stalk cells, the latter ones localized into the anterior part of the "slug." Eventually, differentiated cells give rise to mature fruiting bodies in which a sorocarp, containing a mass of spores is supported by a stalk, rising several millimeters upwards, formed by vacuolized death cells. On the whole the developmental program requires approximately 24 h. When food source becomes available again eventually, the spores germinate and the life cycle starts for a second time as separated free-living amoebae [3, 4] (*see* Fig. 1).

Conversely to other metazoan model organisms (i.e., worm, fruit fly, or mouse) the genome of *D. discoideum* is haploid, composed of six chromosomes, with a size of approximately 34 Mb. Over the years these very simple genetic features allowed to set up simple and feasible genetic tools for (1) targeted gene disruption and (2) functional genetic screening by insertional mutagenesis. The whole *D. discoideum* genome has been fully sequenced by the year 2005, and currently other social amoebae genomes sequences (i.e., *Polysphondylium pallidum*, *D. purpureum*, *D. fasciculatum*) are available [5–7]. From the complete annotated genome sequence, a number of approximately 13,000 genes encoding for proteins, of at least 50 amino acids in length, has been predicted. The comparative analysis between *Dictyostelium* and human genome revealed that many mammalian encoding genes are present also in the simple social amoeba. A *Dictyostelium* genome survey reveals that, despite its relatively small size, it harbors an unusual large amount of GTPases encoding genes with some overlapping functions [8]. It contains approximately 120 genes encoding small GTPases encompassing all five subfamilies. Ras members are 11 and the

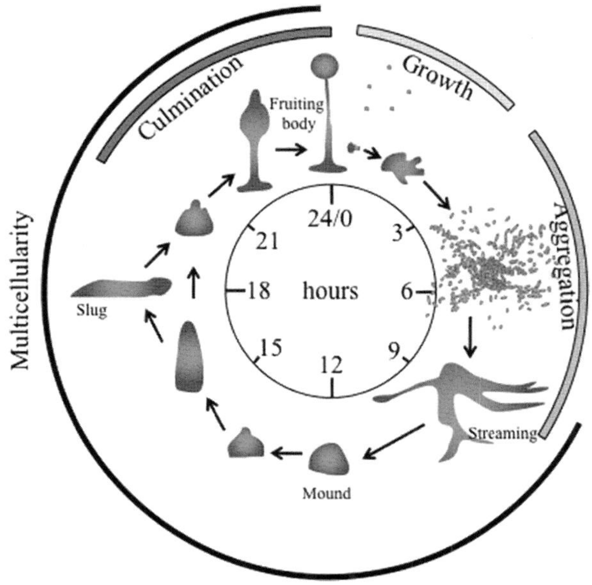


**Fig. 1** *Dictyostelium discoideum* life cycle. *Dictyostelium* lives as a single celled amoeba when food sources are available (growth phase), but upon starvation single amoebae start to aggregate towards an aggregation center by means of an orchestrated chemotaxis process (aggregation phase). Aggregation represents the transition from unicellularity to multicellularity, eventually leading to the formation of a multicellular organism, known as mound, in which the precursor of the mature spore and stalk cells differentiate and sort forming a tipped aggregate or tipped mound. As development proceeds the tipped mound changes shape until it culminates in the development of a mature fruiting body, thus completing the morphogenesis process. The whole process, from starvation of vegetative cells to the formation of a mature fruiting body, takes approximately 24 h (modified from ref. 25)

number of their regulators, represented by Ras-GEFs and Ras GAPs containing proteins, is around 35.

Aggregation of *Dictyostelium* cells requires chemotaxis, which implies orchestrated cytoskeletal rearrangement, towards a cAMP source and the relay of the cAMP signal, by the activation of adenylyl cyclase (ACA) [9, 10], which in turns activates Protein Kinase-A (PKA) and leads to remarkable gene expression changes through the modulation of a plethora of transcription factors. Both processes, cytoskeletal rearrangements and cAMP synthesis, have been broadly investigated and involve Ras signaling. Indeed, single gene disruption, of either Ras family members or their specific regulators, leads to strains with an aggregation-deficient phenotype and thus often incapable to complete the developmental program [11–15]. Ras family members are also implied in regulating cellular events either at late multicellular stage, contributing to the differentiation processes [16] or during the unicellular stage regulating cell growth throughout the control of karyokinesis [17].

*Dictyostelium* therefore provides a useful experimental model organism for the dissection of Ras signaling networks, not only because it is genetically tractable, due to its haploid genome, but also because of its fairly large family of Ras proteins and its relatively simple life cycle, allowing to study both cell growth and development. Information gleaned from its study should be relevant to organisms such as animals which require more complex signaling networks. Here, we illustrate some very basic techniques aiming to (1) identify novel Ras signal transduction components by means of Restriction Enzyme Mediated Insertional (REMI) mutagenesis, as a powerful genetic tool to generate mutants; and (2) phenotypically characterize Ras-dependent processes such as chemotaxis and cAMP/cGMP synthesis.

### 1.1 REMI Mutagenesis

Restriction Enzyme-Mediated Integration (REMI) Mutagenesis, originally set up in *S. cerevisiae* [18], allows to generate mutants by means of site-specific integration of linear DNA into a host genome using restriction enzymes to stimulate the integration process.

After REMI mutagenesis a phenotypical visual screening selection of mutants is performed. For a screening aiming to identify novel Ras pathway/s components affecting cell chemotaxis and/or development it is advisable focus the attention on those mutants presenting aberrant phenotypes (developmental abnormalities or impaired aggregation).

REMI is a powerful technique widely used in different organisms, ranging from yeast to other fungi, to perform functional saturation mutagenesis by making use of a forward genetic approach. Basically, the major advantage of this technique is due to the very simple recovery of the genomic region surrounding the site of insertion and thus the identification of the gene responsible for the phenotype. Genomic flanking regions recovery is made possible because to the shuttling properties of the plasmid used for mutagenesis (*see* Fig. 2).

Since its introduction in 1992 [19], REMI procedure has remained almost unchanged, though few modifications have improved its utility. Dominant selectable markers, such as the Blasticidin cassette, allow efficient selection of *Dictyostelium* transformants when integrated at a single-copy level and work very well in REMI transformation.

### 1.2 Chemotaxis Assay

Upon starvation *Dictyostelium* cells elongate and polarize, moving towards an aggregation center by means of cAMP synthesis and release. In vitro the function of the aggregation center can be mimicked using a micropipette, coupled to a micromanipulator, filled with cAMP which will diffuse working as chemoattractant source. The whole process can be recorded to study cells' behavior (speed, directionality, cell polarization, etc.).

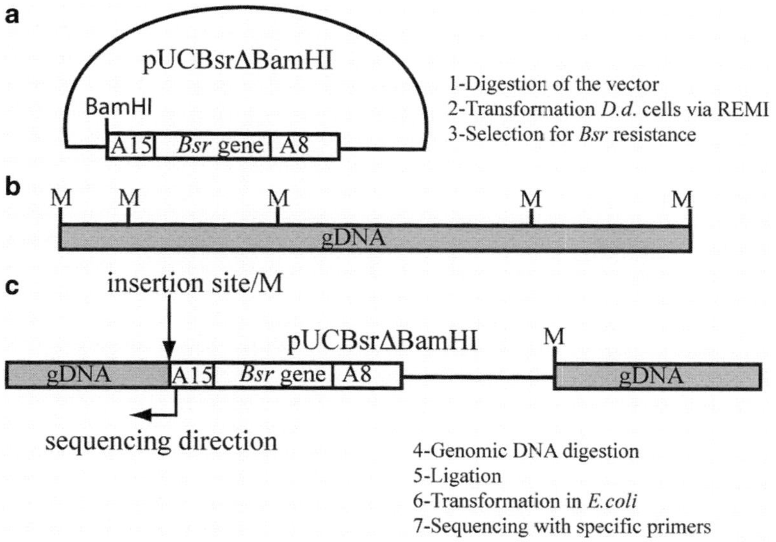


**Fig. 2** Strategy for tagging and cloning genes involved in Ras signaling by Restriction Enzyme Mediated Integration. *Bam*HI linearized-plasmid (**a**) integration into genomic DNA (**b**) occurs within *Mbo*I (M), or other isoschizomers, restriction sites. The integrated plasmid (**c**) can then be rescued from the genome by digesting the mutant's genomic DNA with a restriction enzyme not cutting within the vector and leaving some genomic DNA flanking the insertion site attached to the integrated plasmid. Afterwards, the fragments generated are circularized by ligation and transformed into *E. coli*, amplified and sequenced using specific primers

### 1.3 AC/GC Stimulation by GTPγS Binding

Upon ligand binding to their cognate receptors, G-proteins are activated because they become loaded with GTP and thus capable to transduce signals to the downstream effectors. In *Dictyostelium* cAMP represents the ligand for several G-protein coupled receptors (cARs) which activates downstream effectors, such as Adenylyl-Cyclase (AC) and Guanylyl-Cyclase (GC), throughout G-protein activation. "In vitro" assays by the use of a nonhydrolyzable GTP analogue—GTPγS—[9, 13, 20, 21] allow to mimic the effect of cAMP-stimulated binding of GTP to G-proteins. Basically, GTPγS AC and GC stimulation assays are very similar.

## 2 Materials

### 2.1 Culture, Genetic Manipulation, and Preparation of Dictyostelium discoideum Cells for Development and Chemotaxis

1. Cell culture Axenic Medium (used for routine axenic growth of *Dictyostelium* cells): 14.3 g/L proteose peptone, 18 g/L maltose, 7.15 g/L yeast extract, 0.62 g/L $Na_2HPO_4 \times 2H_2O$, 0.49 g/L $KH_2PO_4$, in distilled water. Adjust pH with 1 M $K_2PO_4$ to pH 6.7. Autoclave for 20 min to sterilize [22]. Add tetracycline from the stock solution (*see* **item 4a**, Subheading 2.2) to a final concentration of 15 μg/mL just prior use to prevent contaminations.

2. Agar phosphate plates (used for *Dictyostelium* multicellular development on solid substrata). Dissolve 20 % Bacto-Agar in Soerensen phosphate buffer (*see* **item 1**, Subheading 2.2). Sterilize by autoclaving and store at 4 °C.
3. NB agar plates (used for growth of *Dictyostelium* on bacterial lawns and for cells cloning): 20 g/L Bacto-Agar, 1 g/L Bacto-Peptone, 1 g/L glucose in 0.017 M Soerensen phosphate buffer.
4. NB Bacteria Nutrient Broth (used for growth of *Dictyostelium* food bacteria). Make just like NB agar leaving out agar.
5. Saturated *E. coli* bacterial culture: inoculate 30 mL of NB broth in a 100-mL flask with *Escherichia coli* strain *B2* 2 days ahead to use. Incubate overnight (ON) at 23 °C under shaking condition. The culture if kept under sterile condition should be usable for up to 7 days. About 0.2 mL of this culture plated on a NB agar plate and incubated at 23 °C for 24–48 h give rise to a thick lawn of bacteria suitable for *Dictyostelium* cells' growth.

### 2.2 Buffers, Solutions, and Antibiotics Stocks for Dictyostelium Cells Growth, Development, Transfection, and Mutants Selection

1. 0.017 M Soerensen phosphate buffer: 0.015 M $KH_2PO_4$, 0.002 M $Na_2HPO_4$, pH 6.1 ± 0.05. Autoclave and sterilize. Store at 4 °C. Stable for up to 1 year (*see* **Note 1**).
2. H-50 buffer: 0.356 M HEPES, 0.05 M KCl, 0.010 M NaCl, 0.001 M $MgSO_4$, 0.005 M $NaHCO_3$, 0.0013 M $NaH_2PO_4$. Adjust pH to 7.0 with HCl or NaOH as appropriate. Autoclave and store at −20 °C.
3. cAMP: 1 mM cAMP solution in 0.017 M Soerensen phosphate buffer. Adjust pH to 6.5 with NaOH and filter-sterilize. Store in aliquots at −20 °C. Stable for 1 year.
4. Antibiotics stock solutions are usually prepared at the following concentrations and stored at −20 °C for several months:
   (a) 7.5 mg/mL Tetracycline. Working concentration is 15 μg/mL.
   (b) 10 mg/mL Blasticidin S hydrochloride. Working concentration is 10 μg/mL.
5. Adenylyl and Guanylate Cyclases (AC and GC) assay buffers:
   (a) 2× AC lysis buffer: 20 mM Tris–HCl, pH 8.0, 4 mM $MgCl_2$. Store at 4 °C.
   (b) 2× GC lysis buffer: 40 mM Hepes–NaOH, 6 mM $MgSO_4$, 2 mM EGTA, pH 7.0.
   (c) GTPγS 1 mM. Store at −20 °C.
   (d) 2× AC assay mixture (use for adenylyl cyclase): 20 mM DTT, 1 mM ATP, 2 mM $MgCl_2$, 10 mM Tris, pH 8.0, 0.4 mM IBMX (phosphodiesterase specific inhibitor).

   (e) 2× GC assay mixture (use for guanylyl cyclase): 10 mM DTT, 0.6 mM GTP in deionized water.

   (f) Perchloric acid (PCA) 3.5 % v/v. Store at room temperature.

6. Tris/EDTA–TE buffer: 10 mM Tris–HCl, pH 7.0, 1 mM EDTA.
7. 0.1 M EDTA, adjust to pH 8.0 with 10 N NaOH.

### 2.3 Equipment

1. Gene Pulser Electroporation device: here we refer to the conditions used with a Gene Pulser Electroporator from Bio-Rad Laboratories, but generally speaking conditions that are proved to be useful for REMI mutagenesis are as follows: (1) field strength 2.5–10 kV/cm (e.g., 1 kV for a 0.4-cm gap cuvettes); (2) 20–100 Ω resistance (most models do not allow adjustment of resistance); (3) charge capacity between 3 and 50 μF (*see* **Note 2**).
2. 35-mm diameter glass-bottom dish (12–27 mm glass).
3. Pump with variable speed drive, equipped with a tube of 1 mm diameter, for accurately dispensing drops of approximately 0.02 ml every 6 min (*see* **Note 3**).
4. Micromanipulator and micropipettes. For experiments' reproducibility, commercially available micropipettes, with a diameter of tip's opening 0.5 ± 0.2 μm, are used (Eppendorf femtotips) (*see* Fig. 3).

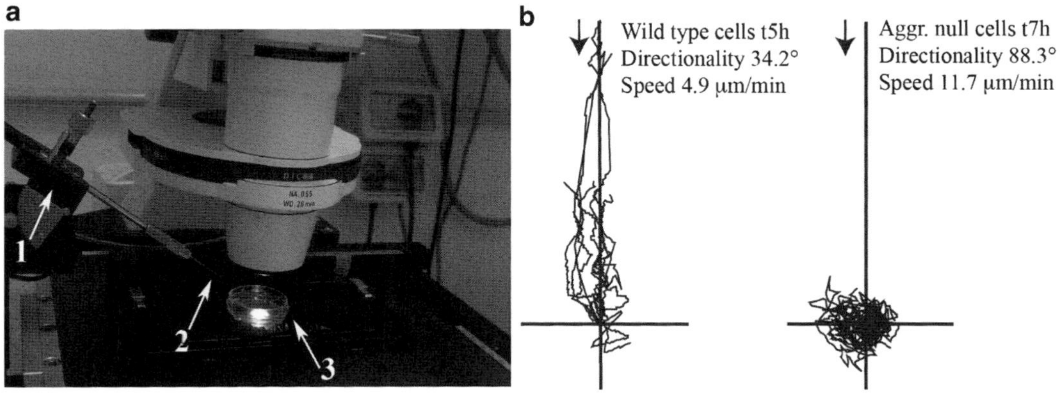


**Fig. 3** Microscope and micromanipulator working station for chemotaxis assay. (**a**) Microscope and micromanipulator with micropipette working station: glass-based dish containing chemotacting competent cells [3] is placed under an inverted microscope. By means of a micromanipulator [1] the cAMP filled micropipette can be precisely positioned and moved to stimulate different fields within the glass-based dish. (**b**) Cells chemotax toward the cAMP gradient produced by diffusion of the cAMP from the capillary: Wild type (*left panel*) and aggregation null mutant (*right panel*) are starved under shaking for the indicated times (5 and 7 h, respectively) and seeded on glass-based dish. Cells' tracks for each strain are displayed. Orientation is determined by measuring the angle between two subsequent positions of each cell in relation to the position of the cAMP-delivering capillary. *Arrows* indicate the source of the cAMP gradient

5. Inverted microscope equipped with DIC or phase-contrast optics, 40–100× oil-immersion objective lens.
6. Charge-coupled device (CCD) camera and computer with image acquisition software for acquiring images (*see* **Note 4**).
7. Stereomicroscope (for visual colony screening) with a magnification range of 6.5–25×.
8. Connect Swinnex 25 mm filter holder for cells lysis and polycarbonate 25 mm diameter membrane filter with a 3 μm pore size (Osmonics Inc.).

## 3 Methods

### 3.1 Cell Culture and Development

Harvest *D. discoideum* spores from fruiting bodies and make them to germinate either in axenic medium in a 10 cm petri dish or on a bacteria lawn (*see* **items 3** and **5** in Subheading 2.1) where single clones can be isolated if the density of cells per plate is sufficiently low. Once the spores germinate, grow cells at 23 °C either in shaken suspension at a speed of 150 rpm, or stationary on cell culture dishes with axenic medium. Culture strains carrying Blasticidin resistance expression cassette with 10 μg/mL of antibiotic. Eventually, cell culture can be scaled up to 100 ml, or even a larger volume, under shaking condition in a sterile flask (*see* **Note 5**).

### 3.2 REMI Mutagenesis

#### 3.2.1 REMI Mutagenesis with Blasticidin Selection

The protocol reported below is based on the technique developed by Adachi and coworkers making use of antibiotic (Blasticidin) selection; usually this protocol allows a frequency of transformants around $10^{-4}/10^{-5}$. Currently, numerous Blasticidin-based *E. coli*/*Dictyostelium* shuttle plasmids are available. However, all of them are based on the original blasticidin-resistance cassette [23].

1. Prepare linearized REMI plasmid (pUC-BsrΔBam) [23] by digesting with the appropriate restriction enzyme (*Bam*HI) followed by phenol–chloroform purification and ethanol precipitation. Resuspend the linearized plasmid in TE buffer at a concentration of approximately 1 mg/mL.
2. Grow cells as described in Subheading 3.1 (*see* **Note 5**).
3. Harvest exponentially growing cells by centrifugation at 500 × *g*, 4 °C for 4–5 min.
4. Wash the cells twice with a half volume of ice-cold H-50 buffer. Pellet the cells by centrifugation as described in the above **step 4** (*see* **Note 6**) and then resuspend the cells in H-50 buffer at a concentration of $1 \times 10^7$ cells/mL.
5. Transfer 800 μL of the cell suspension to an ice-cold sterile glass tube and add 25 U of *Mbo*I, or other commercial isoschizomers such as *Nde*II, *Sau*3AI, or *Dpn*II, and 8 μg of linearized plasmid.

Mix very carefully and transfer the whole suspension into a 0.4 cm gap electroporation cuvette. Alternatively to glass tubes low-binding DNA tubes can be used.

6. Electroporate using the following conditions: 1 kV and 3 μF. Wait for 5 s and then electroporate the suspension for second time using the same conditions. Time constant should range between 0.5 and 1.0 ms. For electroporation of *Dictyostelium* cells, there is no need for a resistor.
7. Incubate the cuvette on ice for 5–10 min.
8. Transfer the cells into a 10 cm petri cell culture dish by adding few hundreds of microliters of axenic medium, without antibiotics, in the cuvette and by pipetting up and down to mix the cell suspension.
9. Incubate ON transformed cells at 23 °C and the day after add selective Blasticidin antibiotic.

#### 3.2.2 Selection and Harvesting of Dictyostelium Mutant Transformants

1. After antibiotic selection, collect the transformant colonies by simply pipetting a stream of axenic medium, or Soerensen phosphate buffer, over the surface of each plate.
2. As transformation efficiency varies from experiment to experiment, the primary transformants suspension should be diluted before plating. So determine cell density for each plate by cell counting and then dilute the cells into NB medium to a concentration of approximately $5–8 \times 10^2$ cell/mL.
3. Mix 0.1 mL of the cell suspension previously prepared with 0.2 mL of saturated *E. coli* bacterial culture and spread onto NB agar plates. Incubate the plates at 23 °C. Colonies will appear as translucent pinpoint plaques within 2–3 days, forming larger plaques over the next few days.

   Visual screening, carried out with the help of stereomicroscope, will facilitate the identification of colonies showing aberrant phenotypes (i.e., aggregation deficient, late or incomplete developer, etc.).
4. With a sterile toothpick harvest 1–2 mm diameter ball from the edge of each colony-plaque and transfer to axenic culture medium with Tetracycline, to prevent bacterial contamination, and Blasticidin to keep transformants under selection. Cells are initially grown in small volume (1 mL if seeded in 24-well plate or 2 mL if grown in tube as shaking suspension) and later on each individual culture representing a clone is expanded into a larger volume.
5. The cells can now be stored frozen with 10 % DMSO in axenic medium at −80 °C or in liquid nitrogen (two tubes each clone).

#### 3.2.3 Rescue of the Genomic DNA Flanking Sequences at the Mutagenesis Insertion Site

Plasmid insertion site is usually identified by molecular cloning, though other alternative strategies are possible (i.e., inverted PCR). For successful achievement at this stage it is very important to avoid contamination of plasmid DNA during the different steps that are summarized below. For this reason it is highly recommended that no reusable glassware be used.

The procedure we report here refers to the original by Kuspa and Loomis 1992 [19].

1. Isolate genomic DNA from the mutated strain according to Reymond [24].
2. Digest the isolated genomic DNA with restriction enzyme recognizing restriction sites different from those present within the plasmid used for insertional mutagenesis (*see* **Note 7**). Purify the digestion products by phenol–chloroform and ethanol precipitation.
3. Ligate about 1 μg of digested DNA in a final volume of 50 μL ligation reaction at 14–16 °C ON.
4. Transform *E. coli* bacteria by electroporation using 1/10 of the ligation reaction (*see* **Note 8**).
5. Screen the cloned plasmid for the appropriate structure. Rescued plasmids are larger than the REMI plasmid. Sequencing the DNA across the plasmid–genomic DNA borders provides an unambiguous location of the insertion site.

### 3.3 Phenotypical Characterization: Chemotaxis Assay

1. Prepare aggregation competent cells as follows: harvest cells by centrifugation at 500 × *g* for 3 min at 4 °C, wash twice with ice-cold Soerensen phosphate buffer, re-pellet, and resuspend at a final concentration of $1 \times 10^7$ cells/mL. Afterwards, incubate cell suspension on a rotatory shaker at 150 rpm for 1 h and then pulse or not with 20 nM (final concentration) of cAMP every 6 min for the next 4 h (*see* **Note 3**).
2. After 5 h starvation settle cells in a 35-mm diameter glass-based dish at a density of approximately $1 \times 10^5$ cell/cm$^2$ and let them adhere for 10 min.
3. Place the glass-based dish on an inverted microscope and use the micromanipulator to position a micropipette containing 1 mM cAMP to create a chemoattractant gradient to stimulate the cells.
4. Capture time-lapse images (40× or 100× objective lens) of the cells.
5. Analyze the recorded time-lapse movies using the appropriate software (*see* **Note 4**). Analyze cell tracks for at least 20 cells in each field and calculate cell speed by measuring the distance over time. Determine orientation by measuring the angle between two subsequent positions of each cell in relation to the position of the cAMP-delivering capillary (*see* Fig. 3).

### 3.4 Biochemical Characterization

#### 3.4.1 "In Vitro" Adenylyl Cyclases Activity Assay (With or Without GTPγS-Stimulation)

1. Prepare aggregation competent cells as described in Subheading 3.3, **step 1**.
2. Carry out cells' "basalation" as follows:
   (a) Add 5 mM caffeine.
   (b) Incubate under shaking condition at 200 rpm for 30 min.
3. Wash out caffeine with ice-cold Soerensen phosphate buffer (500 × *g* at 4 °C, repeat twice), resuspend cells pellet at a density of $1 \times 10^8$ cells/mL in Soerensen phosphate, and keep on ice until starting the assay.
4. Label 1.5-mL tubes and add to each tube 40 μl of 2× AC assay mixture. Remember to include in the assay cell lysates from growth phase cells as control for native enzymatic activity (*see* **Note 10**).
5. Add 1 volume of 2× AC lysis buffer to the cells collected at **step 3**.
6. Prepare 2.5 mL syringe with connect Swinnex 25 mm filter holder mounted with polycarbonate nucleopore membrane filter.

   Lyse the cells, in the absence or presence of 30 μM GTPγS (*see* **Note 11**), by pressing them through the nucleopore polycarbonate filter device, and incubate the cell lysate on ice for 5 min.
7. Add 40 μl of cell lysates to the tubes, previously prepared and labelled as specified at the **step 4**, containing 40 μl of 2× AC assay mixture, mix, and incubate the reaction for 5 min at 20 °C.
8. Add 40 μL of 0.1 M EDTA, pH 8.0 to terminate the reaction and boil samples for 2 min.
9. Use isotope-dilution assay for the determination of cAMP level. Commercial kits are available.

Usually, we make use of Biotrak cAMP assay kit (GE Healthcare) which provides all the reagents. The reagents are dissolved as described by the manufacturer, but in the assay all volumes are divided by 2. A representative result is shown in Fig. 4.

#### 3.4.2 "In Vitro" Guanylyl Cyclase Activity Assay (with or Without GTPγS-Stimulation)

Perform the guanylate cyclase assay as the adenylate cyclase assay, with the few following exceptions and not preincubating cell lysates on ice (*see* **Note 9**).

1. Lyse cells in 2× GC lysis buffer.
2. Use the 2× GC assay mixture.
3. Carry out enzymatic reaction incubating the samples 1 min at 20 °C.
4. Measure GC stimulation by adding 100 μM GTPγS (final concentration).

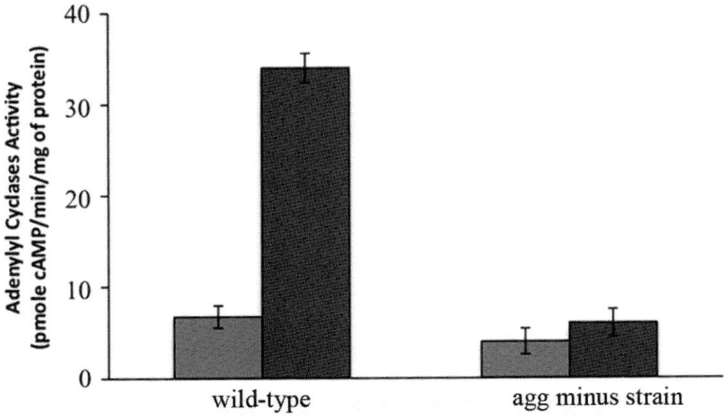


**Fig. 4** GTPγS stimulated adenylyl cyclase activity. Starving cells are treated for 5 h with 20 nM cAMP pulses every 6 min. Cell lysates from wild-type (*left panel*) and aggregation-deficient strain (*right panel*) are assayed in the absence (*grey bars*) or in the presence (*black bars*) of 30 μM GTPγS and cAMP amount measured with radioimmunoassay. Addition of GTPγS to aggregation null strain lysates fails to stimulate enzymatic activity, whereas a threefold activation is observed in wild-type cell lysate

5. Stop reactions by adding 1 volume of 3.5 % v/v perchloric acid.
6. Please strictly refer to the Subheading 3.4.1 for reaction volumes, cell amounts, and other protocol details.

## 4 Notes

1. A 50-fold concentrate stock can be prepared by dissolving 99.86 g of $KH_2PO_4$ and 14.2 g of $Na_2HPO_4$ in a total volume of 1 L of deionized and distilled water. The stock can be stored at 4 °C. After dilution at 1× concentration check pH.
2. There are many types of electroporator devices and each of them differs in the controls of setting the voltage and capacitance, so these parameters should be optimized differently according to the device.
3. Drop size depends on the volume of starving cells. The concentration of the cAMP solution must be calculated according to the volume of pulsed cells in the shaking flask. To deliver the cAMP pulses, thin tube is taped to the mouth of the flask and is connected through a peristaltic pump to a reservoir of cAMP previously diluted in Soerensen phosphate buffer.
4. Currently, several software packages are commercially available (i.e., Openlab, Improvision; IPLab Spectrum, Scanalytics) for either Apple or other PCs. Time-lapse capture image of cells every 3–6 s, depending on cell motility speed, for 30 min.

5. The cells are usually grown to mid-log phase (1–4 × $10^6$ cells/mL). Some mutant strains (i.e., myosin null mutant) grow only under adhering conditions. To prevent contamination by yeasts, bacteria, or fungi, it is recommended to work in a sterile hood and supplement media with an antibiotic (i.e., tetracycline, penicillin, streptomycin, etc.). Optimal growth of *Dictyostelium* cells is observed at a temperature range of 21–23 °C and it is significantly inhibited above 25 °C and below 18–20 °C. When culturing *Dictyostelium* cells, it is important that they remain in an exponential growth phase (less than 4 × $10^6$ cells/mL) rather than allowing them to reach stationary phase. Cultures of higher densities begin to secrete factors that trigger entry into development. On the other hand, it is not recommended to dilute the cells to less than 1 × $10^4$ cells/mL, since cultures of low cellular density lags for a period that can be difficult to estimate. Like for all cell type culture, it is a good practice to maintain a regular schedule for splitting the cultures to ensure a healthy cell population. Determine the cells density by counting using hemocytometer and calculate how much the cultures need to be diluted in order to have the necessary number of cells the next day(s).
6. Electroporation is a shock for cells, so, always, they must be handled very gently, carefully resuspending the pellet with a pipette rather than using a vortex.
7. When choosing restriction enzymes for genomic DNA digestion please consider that *Dictyostelium* genome is very rich in A and T (roughly 70–75 % of the genome is composed of A and T), and so avoid, if possible, restriction enzymes recognizing sequences rich in G and C nucleotides (i.e., *Sac*II, *Xma*III, *Nar*I, etc.) because they are represented very rarely within the genome and thus give rise to circularized fragments with a very large size which possibly recombine in bacteria.
8. The volume of DNA should be no more than 1/10th of the total bacteria transformation volume. However, if the volume of DNA is too small, the mixing might not be effective. Remember to transform a sample lacking DNA (negative control) as well as an empty vector (positive control). Due to its very A and T DNA contents some bacteria strains (*rec⁻* strains are strongly recommended) support better growth of *Dictyostelium* containing plasmids. Usually the number of colonies depends on the size of the rescued plasmid.
9. To robustly activate AC by GTPγS a short incubation time on ice (5 min) of cell lysates is required. On the contrary, in the case of GTPγS GC stimulation the short incubation on ice is not necessary because the enzyme is promptly activated.
10. Adenylyl Cyclase A activity sharply increases during early development reaching a peak around 4–8 h after starvation; by

contrast at vegetative stage (named time 0 h or *t*0) the activity is negligible and thus considered native.

11. The maximal activation occurs when GTPγS is present in the lysate directly after cell lysis and the lysate is preincubated with GTPγS for 5 min prior to assay.

## References


1. Han M, Sternberg PW (1990) *let-60*, A gene that specifies cell fates during *C. elegans* vulval induction, encodes a ras protein. Cell 63: 921–31
2. Bourne HR, Sanders DA, McCormick F (1990) The GTPase superfamily: a conserved switch for diverse cell functions. Nature 348:125–32
3. Raper K (1984) The dictyostelids. Princeton University Press, Princeton, NJ
4. Kessin RH (2001) *Dictyostelium*—evolution, cell biology, and development of multicellularity. Cambridge Univerisity Press, Cambridge
5. Eichinger L, Pachebat JA, Glöckner G et al (2005) The genome of the social amoeba *Dictyostelium discoideum*. Nature 435:43–57
6. Sucgang R, Kuo A, Tian X et al (2011) Comparative genomics of the social amoebae *Dictyostelium discoideum* and *Dictyostelium purpureum*. Genome Biol 12:R20
7. www.dictybase.org
8. Khosla M, Spiegelman GB, Insall RH et al (2000) Functional overlap of the *Dictyostelium* RasG RasD and RasB proteins. J Cell Sci 113:1427–1434
9. Pergolizzi B, Peracino B, Silverman J et al (2002) Temperature-sensitive inhibition of development in *Dictyostelium* due to a point mutation in the *piaA* gene. Dev Biol 251:18–26
10. Galardi-Castilla M, Pergolizzi B, Bloomfield G et al (2008) SrfB, a member of the serum response factor family of transcription factors, regulates starvation response and early development in *Dictyostelium*. Dev Biol 316:260–74
11. Charest P, Firtel RA (2007) Big roles for small GTPases in the control of directed cell movement. Biochem J 401:377–390
12. Insall RH, Borleis J, Devreotes PN (1996) The aimless RasGEF is required for processing of chemotactic signals through G-protein-coupled receptors in *Dictyostelium*. Curr Biol 6:719–729
13. Arigoni M, Bracco E, Lusche DF et al (2005) A novel *Dictyostelium* RasGEF required for chemotaxis and development. BMC Cell Biol 6:43
14. Lee S, Parent CA, Insall R et al (1999) A novel Ras-interacting protein required for chemotaxis and cyclic adenosine monophosphate signal relay in *Dictyostelium*. Mol Biol Cell 10: 2829–2845
15. Bolourani P, Spiegelman GB, Weeks G (2006) Delineation of the roles played by RasG and RasC in cAMP-dependent signal transduction during the early development of *Dictyostelium discoideum*. Mol Biol Cell 17:4543–4550
16. Jaffer ZH, Khosla M, Spiegelman GB et al (2001) Expression of activated Ras during *Dictyostelium* development alters cell localization and changes cell fate. Development 128:907–916
17. Sutherland BW, Spiegelman GB, Weeks G (2001) A Ras subfamily GTPase shows cell cycle-dependent nuclear localization. EMBO Rep 2:1024–1028
18. Schiestl RH, Petes TD (1991) Integration of DNA fragments by illegittime recombination in *S. cereviasiae*. Proc Natl Acad Sci 88: 7585–7589
19. Kuspa A, Loomis WF (1992) Tagging developmental genes in *Dictyostelium* by restriction enzyme-mediated integration of plasmid DNA. Proc Natl Acad Sci 89:8803–8807
20. Lilly PJ, Devreotes PN (1994) Identification of CRAC, a cytosolic regulator required for guanine nucleotide stimulation of adenylyl cyclase in *Dictyostelium*. J Biol Chem 269:14123–14129
21. Janssens PM, De Jong CC, Vink AA et al (1989) Regulatoty properties of magnesium-dependent guanylate cyclase in *Dictyostelium discoideum* membranes. J Biol Chem 264:4329–4335
22. Watts DJ, Ashworth JM (1970) Growth of myxameobae of the cellular slime mould *Dictyostelium discoideum* in axenic culture. Biochem J 119:171–4
23. Adachi H, Hasebe T, Yoshinaga K et al (1994) Isolation of *Dictyostelium discoideum* cytokinesis mutants by restriction enzyme-mediated integration of the blasticidin S resistance marker. Biochem Biophys Res Commun 205:1808–14
24. Reymond CD (1987) A rapid method for the preparation of multiple samples of eucaryotic DNA. Nucl Acid Res 15:8118
25. Chisholm RL, Firtel RA (2004) Insights into morphogenesis from a simple developmental system. Nat Rev Mol Cell Biol 5:531–541

ERRATA

# Biophysical and Proteomic Characterization Strategies for Cysteine Modifications in Ras GTPases

G. Aaron Hobbs, Harsha P. Gunawardena, and Sharon L. Campbell

Lorenza Trabalzini and Saverio Francesco Retta (eds.), *Ras Signaling: Methods and Protocols*,
Methods in Molecular Biology, vol. 1120, DOI 10.1007/978-1-62703-791-4_6, 

**DOI 10.1007/978-1-62703-791-4_26**

The publisher regrets that the online and print versions of this book were published with the name and affiliation of one of the authors of Chapter 6 omitted both on the chapter opening page and in the front matter, and should read as follows:

Xian Chen
Department of Biochemistry and Biophysics, University of North Carolina, Chapel Hill, North Carolina 27599, USA b Program in Molecular Biology and Biotechnology, University of North Carolina, Chapel Hill, North Carolina 27599, USA

The following grant information was also omitted from the Acknowledgements text in this chapter:

NCI CPTAC 1U24CA160035 to XC

The online version of the original chapter can be found at http://dx.doi.org/10.1007/978-1-62703-791-4_6

# INDEX



Lorenza Trabalzini and Saverio Francesco Retta (eds.), *Ras Signaling: Methods and Protocols*, Methods in Molecular Biology, vol. 1120, DOI 10.1007/978-1-62703-791-4, © Springer Science+Business Media, LLC 2014

22518362R00243

Printed in Great Britain
by Amazon